Soil, Plants and Microbes

Soil, Plants and Microbes

Edited by Henry Wang

SYRAWOOD
PUBLISHING HOUSE
New York

Published by Syrawood Publishing House,
750 Third Avenue, 9th Floor,
New York, NY 10017, USA
www.syrawoodpublishinghouse.com

Soil, Plants and Microbes
Edited by Henry Wang

International Standard Book Number: 978-1-68286-506-4 (Hardback)

Cataloging-in-Publication Data

Soil, plants and microbes / edited by Henry Wang.
p. cm.
Includes bibliographical references and index.
ISBN 978-1-68286-506-4
1. Soils. 2. Plants. 3. Microorganisms. I. Wang, Henry.
S591 .S65 2018
631.4--dc23

Printed in China.

TABLE OF CONTENTS

PREFACE

The main aim of this book is to educate learners and enhance their research focus by presenting diverse topics covering this vast field. This is an advanced book which compiles significant studies by distinguished experts in the area of analysis. This book addresses successive solutions to the challenges arising in the area of application, along with it; the book provides scope for future developments.

Soil is necessary for the existence of life on Earth. It is a natural element that is necessary for the growth of plants. The interaction between soil, plants and microbes is known as soil ecology. Plants and microbes are both essential for the nutrient recycling in soil ecosystems. For all readers who are interested in the study of soil ecology, the case studies included in this book will serve as an excellent guide to develop a comprehensive understanding. This book, with its detailed analyses and data, will prove immensely beneficial to professionals and students involved in this area at various levels.

It was a great honour to edit this book, though there were challenges, as it involved a lot of communication and networking between me and the editorial team. However, the end result was this all-inclusive book covering diverse themes in the field.

Finally, it is important to acknowledge the efforts of the contributors for their excellent chapters, through which a wide variety of issues have been addressed. I would also like to thank my colleagues for their valuable feedback during the making of this book.

Editor

1

Effects of Added Organic Matter and Water on Soil Carbon Sequestration in an Arid Region

Liming Lai[1,⁋], Yufei Li[2,⁋], Yuan Tian[3], Lianhe Jiang[4], Xuechun Zhao[1], Linhai Zhu[1], Xi Chen[3], Yong Gao[5], Shaoming Wang[6], Yuanrun Zheng[1*], Glyn M. Rimmington[7]

1 Key Laboratory of Resource Plants, Beijing Botanical Garden, West China Subalpine Botanical Garden, Institute of Botany, Chinese Academy of Sciences, Xiangshan, Beijing, China, **2** Institute of RS and GIS, Peking University, Beijing, China, **3** Xinjiang Institute of Ecology and Geography, Chinese Academy of Sciences, Urumqi, Xinjiang, China, **4** State Key Laboratory of Vegetation and Environmental Change, Institute of Botany, Chinese Academy of Sciences, Beijing, China, **5** Inner Mongolia Agricultural University, Hohhot, China, **6** Key Laboratory of Oasis Ecological Agriculture of Xinjiang Bingtuan, Shihezi, Xinjiang, China, **7** Global Learning Office, College of Liberal Arts and Sciences, Wichita State University, Wichita, Kansas, United States of America

Abstract

It is generally predicted that global warming will stimulate primary production and lead to more carbon (C) inputs to soil. However, many studies have found that soil C does not necessarily increase with increased plant litter input. Precipitation has increased in arid central Asia, and is predicted to increase more, so we tested the effects of adding fresh organic matter (FOM) and water on soil C sequestration in an arid region in northwest China. The results suggested that added FOM quickly decomposed and had minor effects on the soil organic carbon (SOC) pool to a depth of 30 cm. Both FOM and water addition had significant effects on the soil microbial biomass. The soil microbial biomass increased with added FOM, reached a maximum, and then declined as the FOM decomposed. The FOM had a more significant stimulating effect on microbial biomass with water addition. Under the soil moisture ranges used in this experiment (21.0%–29.7%), FOM input was more important than water addition in the soil C mineralization process. We concluded that short-term FOM input into the belowground soil and water addition do not affect the SOC pool in shrubland in an arid region.

Editor: Hauke Smidt, Wageningen University, The Netherlands

Funding: This work was supported by China National Key Basic Research Program (2009CB825103, 2012CB956204). The funders had no role in study design, data collection and analysis, decision to publish, or preparation of the manuscript.

Competing Interests: The authors have declared that no competing interests exist.

* E-mail: zhengyr@ibcas.ac.cn

⁋ These authors contributed equally to this work.

Introduction

Soils contain more carbon (C) than the atmosphere and biomass [1]. Approximately 75–100 Pg C yr^{-1} is released from this large C pool [2], and it has been estimated that this has increased by 0.1 Pg C yr^{-1} in the last twenty years [3]. Given the large amounts of C in, and released by, soil, the soil's response to environmental change will play a key role in determining future concentrations of atmospheric CO_2 [4]. It is worth examining the possibility of transferring CO_2 from the atmosphere to terrestrial C storage as a possible way of restricting increases in atmospheric CO_2 concentrations and the resulting global warming. Generally, elevated CO_2 is assumed to stimulate primary production and increase C input to soil [5], and more soil C sequestration is anticipated. However, as warming can also accelerate soil C decomposition, the effects of climate change on the soil C pool remain uncertain [6]. Therefore, accurately estimating soil C dynamics is critical for evaluating the potential effects of global climate change on the terrestrial biosphere [7].

An imbalance between the input of plant C and its decomposition rate results in changes in stored soil C [8]. A soil will give net C sequestration from increased C input if the organic C decomposition rate does not increase [9]. However, many studies have concluded that soil C does not necessarily increase when the organic C input from vegetation increases, [5,10,11,12], and have indicated that increasing the C input has a stimulating effect on C decomposition. Fresh organic C input can supply energy and nutrients for the soil microbes, and can, therefore, accelerate soil organic C (SOC) mineralization[13], which has been called a "priming effect" [14]. This process is mainly associated with stimulation of the microbes by the easily decomposable organic matter, and causes an increase in CO_2 efflux [15]. Fontaine et al. also reported that fresh C input may accelerate soil C decomposition and cause the soil C store to decrease [16]. Sayer et al. found that litter inputs can significantly increase soil C release in a tropical forest, but that doubling litter inputs caused a smaller increase in soil C release than adding less than 30% litter [12]. The relationship between organic C input and soil C sequestration is, therefore, quite uncertain.

Besides C input, soil moisture is also an important factor affecting ecosystem C exchange and soil organic matter decomposition [8,17]. Water availability can regulate plant growth and net ecosystem productivity [18], and, therefore, affects C input to the soil. Water limitations in arid and semiarid ecosystems will be exacerbated by global warming and lower soil water availability [19]. Microbial activity requires a certain soil water content range, and is limited by soil continually drying out [20,21]. Contact between the microbes and the available substrate and the physiological performance of microbes are limited at low soil water contents [22]. Therefore, droughts could reduce the decrease in soil C by inhibiting microbial decomposition. After long-term observations, it has been found that precipitation has significantly increased in eastern America, northern Europe, and central Asia, but has decreased in many other regions [23]. Because precipitation is highly variable, spatially and temporally, specific environmental factors in the area where a study is conducted must be considered.

Although many studies have been conducted on the effects of climate change on the soil C cycle, arid regions have received relatively little attention [24]. Arid regions occupy approximately 20% of the global terrestrial surface, and they are expanding because of climate change, fire, and land-use change [25]. Arid ecosystems have been predicted to be amongst the most responsive ecosystem types to global climate change [26]. The amount of soil C stock in arid ecosystems is huge [27], and has been estimated to include 241 Pg of organic C and a similar or larger amount of inorganic C [1]. The significant potential C sink capacity in arid ecosystem soils using restorative management could make arid regions an important factor in global climate change [27].

The Xinjiang Uighur Autonomous Region (XUAR) in northwest China covers over one-sixth of China's land area and includes the majority of the country's arid areas [28]. This region is characterized by widely distributed saline/alkaline soils and low precipitation [29,30]. High alkalinity, high salinity [31], and low soil water content [20,21] can inhibit the activities of microbes, and can therefore decrease soil organic matter decomposition. Soils may receive more plant C input because of increased primary production under elevated CO_2 and increased precipitation. Therefore, we wondered whether these soils could sequester more C because of global warming. If soils sequester more C, large areas of XUAR will become a huge C sink, and could partly offset global warming. Precipitation in the arid regions of central Asia is likely to increase significantly in the next 50 years [23,32]. However, the relationship between C sequestration and increased precipitation in arid regions is not clear. Identifying the response of soil C to increased plant C inputs and precipitation would help us to gain a better understanding of C cycling in arid regions under global climate change. We applied plant litter inputs and water addition to soil to simulate the main effects of climate change on soil C cycling in an arid region of northwest China. Specifically, we aimed to address the following questions: (1) what are the effects on the soil C pool of increased plant litter input, increased precipitation, and their interaction? and (2) how do microbes change seasonally under increased plant litter inputs and water addition?

Materials and Methods

Ethics Statement

All necessary permits were obtained for the field studies described. The study sites are managed by the Fukang Station of Desert Ecology, Xinjiang Institute of Ecology and Geography, Chinese Academy of Sciences.

Study site description

The study was conducted at the Chinese Academy of Sciences' Fukang Station of Desert Ecology (87° 56′ E, 44° 17′ N, elevation 461 m), which is located at the southern edge of the Junggar basin, in the hinterland of northwest China. The region has an arid climate with mean annual rainfall of 160 mm and a mean annual temperature of 6.6°C [29]. Soil is clay-loam textured with heavy salinity/alkalinity.

The field site is shrubland, and canopy coverage is below 40%. It is dominated by *Reaumuria soongorica*, which is the dominant native vegetation species in the XUAR. The main species present include *Haloxylon ammodendron*, *Nitraria sibirica*, *Salsola collina*, *Ceratocarpus arenarius*, and *Suaeda glauca*. The soil physical and chemical properties are shown in Table 1.

Experimental design

The experiment used a nested design with fresh added organic matter (FOM) and water (P) as the two controlled parameters. In early June 2011, 27 randomly distributed study plots (3 m×3 m) were established in a *R. soongorica* community with uniformly distributed vegetation. Adjacent plots were at least 2 m apart, to mitigate buffering effects.

Spring and autumn are windy in the study region, so it was impossible to add FOM to the soil surface without it being disturbed by the wind. The main root of *R. soongorica* is 0–30 cm deep, and approximately 76% of feeder roots are located above 20 cm [29]. Because the SOC in deep soil layers (> 20 cm) is relatively stable and can only receive extremely small amounts of plant C [16], we chose 20–30 cm as the FOM input layers.

FOM was added to the 20–30 cm deep soil to increase the SOC by 0% (F0, no FOM input), 5% (F1, 33.64 g/m^2), and 10% (F2, 67.28 g/m^2). Water addition treatments were initiated at the start of the experiment, and provided different amounts of water each week, based on the long-term mean weekly precipitation records at the nearby weather station at the Fukang research station. At each water addition plot, water was added with the sprinkling can manually. The three water addition treatments were equivalent to 0% (P0, no additional irrigation), 50% (P1, 2.43 mm), and 100% (P2, 4.87 mm) increased precipitation. A randomized-block design was used for the plots, with the three FOM input treatments nested within the three irrigation treatments, and with three replications.

The FOM was plant litter collected from an adjacent *R. soongorica* community site at the end of the growing season 2010. After collection, impurities such as lumps of soil were removed and the plant material was dried at 65°C in an oven for 48 h, and then the material was stored at 4°C until use. Before use in the field experiments, the plant materials were ground and passed through a 2 mm mesh sieve. The C, N content and the C: N of the FOM were 44.42±3.31% (mean±SE, n = 5), 3.18±0.08% (n = 5) and 13.94±0.97 (n = 5), respectively.

To prepare the plots, we first collected the dead plant tissues and litter from the soil surface (a very small amount), and these were replaced after the initial plot treatments had been completed. Next, we carefully removed the top 0–20 cm soil in layers, taking care to keep the soils in their original shape and structure, and ensuring that the plants and roots were not badly affected. The FOM was then added. Finally, the top soil was replaced in its original position, and gaps between the soil blocks were filled with soil from the 0–10 cm layers collected from an adjacent area. The F0 plots were treated the same as the added FOM treatment plots. Adding the FOM to the upper soil layers was difficult without damaging the root systems [11]. The canopy radius of *R. soongorica* near the study site was 0.26±0.05 m, and the root biomass out of this range was about 25% of total root biomass. For avoiding drastically damage the root systems, soil under the plant canopy was not removed.

To evaluate the effect of soil disturbance on treatments, we chose extra three natural plots adjacent experimental site, and soil respiration was measured on natural plots and treatment plots with an LI-8100 automated soil CO_2 flux system from 8:00 to 20:00 in two-hour rounds at the first sampling time in late June and July

Table 1. Descriptions of soil physical/chemical properties in the study sites.

Soil depth (cm)	Soil organic carbon (SOC, g/kg)	Soil bulk density (g/cm^3)	pH	Soil EC (dS/m)
0–10	4.66±0.61	1.29±0.05	9.36±0.12	2.11±0.19
10–20	3.26±0.33	1.29±0.07	9.34±0.17	1.66±0.20
20–30	2.15±0.24	1.39±0.03	9.43±0.34	0.87±0.09

Values represent mean±SE (n = 5).

(with high soil moisture). The results showed that no significant difference was found between the soil respiration data on the treatment plots and natural plots (P>0.05). The mean soil moisture for the natural and F0P0 plots were 21.54±0.10% (v/v) and 21.06±0.11% (v/v) in June, and 24.01±0.25% (v/v) and 23.86±0.21% (v/v) in July, respectively. As the soil respiration significantly correlated with microbial and root biomass [8], the impacts by the treatments on soil could be limited.

Root biomass measurements

Root biomass was determined in soil cores. Because the shrub roots were highly spatially variable, root samples were collected from one large square core (25 cm×25 cm, and 60 cm deep) from each plot (27 plots) in October 2011, and separated into 0–20 cm, 20–40 cm, and 40–60 cm layers. The roots were carefully separated from these samples, retaining only apparently living material based on the color, texture, and shape of the roots. Fine roots were classified by their diameter (<2 mm). All roots were dried at 70°C to a constant mass and weighed. We excluded roots deeper than 60 cm because very few roots reach that depth.

Soil temperature and moisture measurements

The soil temperature (°C) and soil water content (v/v, %) at 5, 15 and 25 cm depths of each single irrigation treatments (F0P0, F0P1 and F0P2) were monitored using sensor (ECH$_2$O EC-TM, Decagon Devices, Inc. Pullman, WA USA). Outputs of each probe was adjusted with soil collected near the plot where the probes would be equipped, using the method described by Starr and Paltineanu [33], and there were no significant difference for outputs of used probes in same soil in this experiment. Means of data at three depths were taken as soil temperature and moisture of 0–30 cm depth.

Analysis of soil physical and chemical properties

The first soil sampling was carried out in late June (20 days after experiment setup) for reestablishment of the soil microbial community. Soil samples were collected from 0–30 cm deep. Three samples were taken from each of the three replicate plots and mixed to make three samples for analysis for each plot treatment. Sampling was performed monthly from June to October. Each sample was passed through a 2 mm sieve immediately and separated into two parts, one of which was stored at 4°C for microbial biomass C (MBC) analysis and the other naturally dried at room temperature. MBC was determined using the fumigation-extraction method. The pH (using a 1 5 soil: water ratio) and electrical conductivity (EC; using a 1 5 soil: water ratio) were determined with a Eutech PC700 pH/EC meter (Thermo Fisher Scientific Inc., Waltham, Massachusetts, USA). SOC was measured using the method described by Nelson and Sommers [34].

Data analysis

Linear regression analysis, ANOVA, and homogeneity of variance tests were performed using SPSS 13.0 software [35]. Multiple comparisons of means for different treatments were analyzed using Tukey's test.

Results

Temperature and soil moisture

During the study period (June–October), water addition treatments had no significant effects (P>0.05) on soil temperature, although we did observe slightly lower temperatures in the plots with added water. However, water addition significantly increased soil moisture (P<0.001, Figure 1), by an average of 1.1% and 4.5% (v/v) for the P1 and P2 water addition regimes, respectively. When analyzed monthly, soil moistures in the P1 and P2 plots were significantly higher than the P0 plots for the first 3 months (June–August), then the soil moisture in the P0 and P1 plots were similar for the last two months (September and October). The amount of rain at day 21 was 28 mm, and increased the soil moisture over 5% (Figure 1). Under the P0, P1, and P2 water addition treatments, the monthly mean soil temperature ranged from 16.31±0.55°C (mean±SE) to 27.08±0.19°C, 15.93±0.48°C to 26.61±0.18°C, and 15.51±0.54°C to 26.65±0.21°C, respectively, and the soil moisture values ranged from 21.00±0.05% to 21.06±0.11%, from 20.77±0.05% to 22.70±0.06%, and 24.10±0.06% to 25.77±0.07% (v/v), respectively.

SOC

The input of FOM, water addition, and their interactions had no significant effects on the SOC. SOC content declined significantly with depth, and approximately 76% of the SOC in the soil from 0–30 cm deep was found above 20 cm. However, the interactions between FOM, water addition, and soil depth had no significant effects on the SOC. The SOCs in the F2 treatment soils were slightly higher than in soils from the same layers in the other treatment plots, but the differences were not significant (Figure 2).

Microbial biomass C

The FOM inputs, water addition, and their interactions had significant effects on the soil MBC (P<0.001, Table 2). The soil MBC content was at its highest at the start of the experiment, in June, and then declined with time. The soil MBC content was 4.6%–68% lower in the same treatment plots at the end of the experiment (October) than that at the beginning (Figure 3).

The FOM-induced increase in MBC ranged from 3.7% to 140.2%, 3.6% to 51.4%, 4.7% to 58.8%, 1.1% to 38.4%, and 3.0% to 7.8% in June, July, August, September, and October, respectively. FOM had no significant effect on the MBC in the treatments without water addition. The increase in MBC caused by water addition ranged from −7.9% to 169.7%, −15.4% to

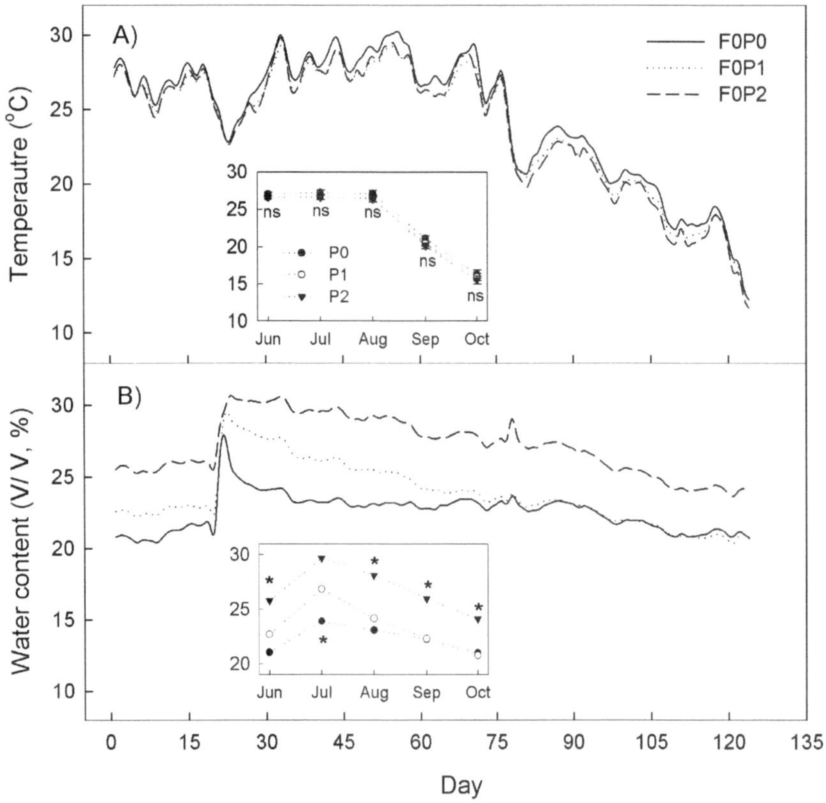

Figure 1. Dynamics of Temperature (A) and soil moisture (B) under different water addition regimes without FOM treatments during the study period. Inserts represent the monthly mean values. The day 0 represents the first sampling time. FOM: fresh organic matter; F0: no FOM input; P0, P1, P2: no added water, 50% and 100% increase in precipitation, respectively. * means significant ($P<0.001$), and ns mean not significant ($P>0.05$) among the three water addition levels in the same month.

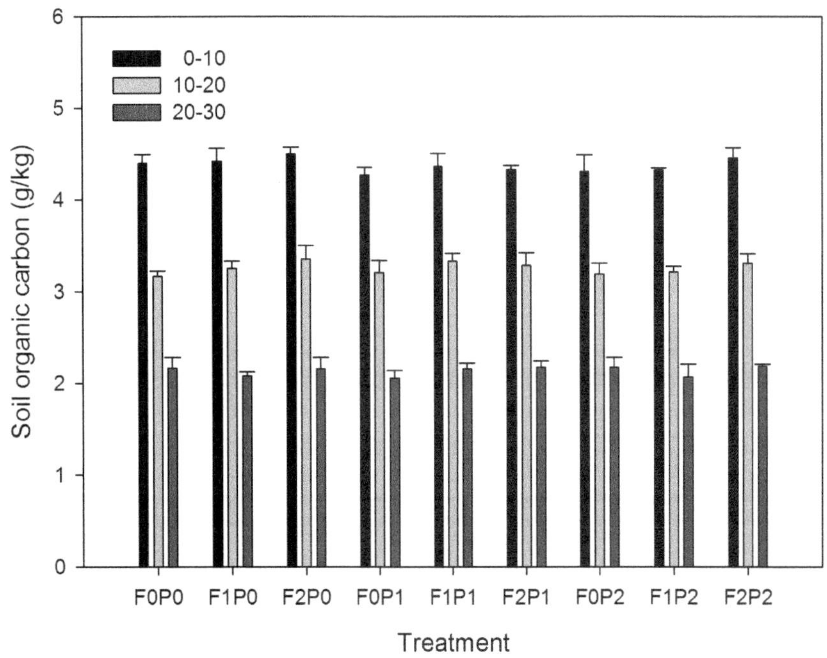

Figure 2. Final levels of soil organic carbon (mean±SE) in response to added precipitation and fine particulate matter after five months of manipulation. F1 and F2: 5% and 10% increase in SOC. Other abbreviations are same as Figure 1.

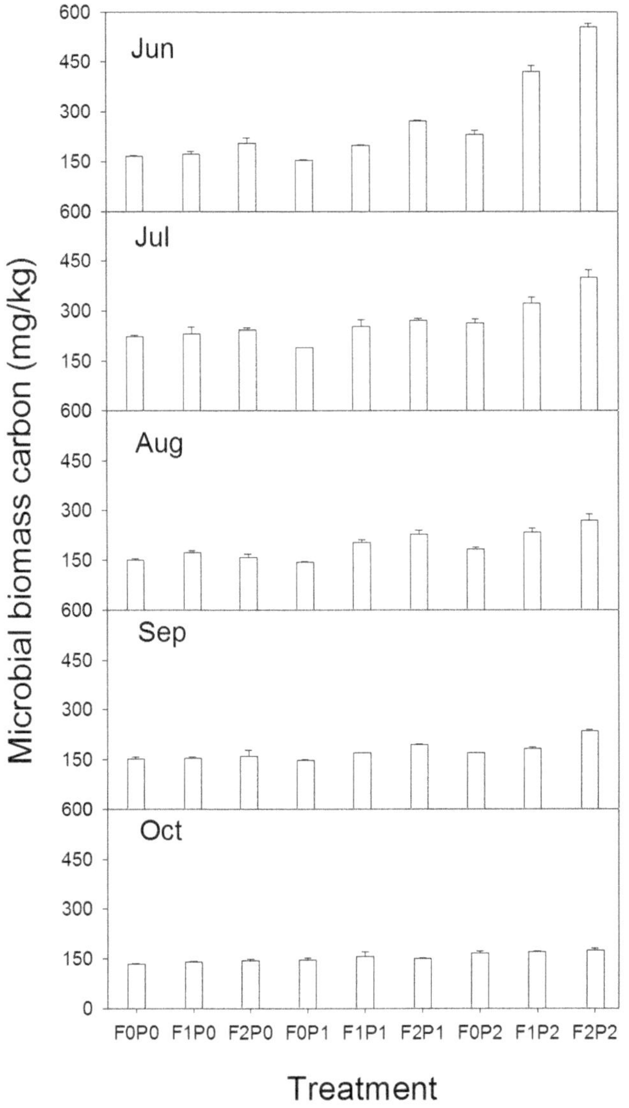

Figure 3. Seasonal dynamics (mean±SE) of microbial biomass carbon under different treatments. Other abbreviations are same as Figure 1.

64.5%, −5.3% to 70.1%, −3.8% to 47.5%, and 5.0% to 24.2% in June, July, August, September, and October, respectively (Figure 4).

The Pearson correlations were not significant between the MBC and monthly mean soil moisture for all treatments (P>0.05).

Fine root biomass

FOM input and water addition had significant effects on the fine root biomass separately (P<0.001), but their interactions did not (P>0.05, Table 2). Fine root biomass was greatest in the upper 0–20 cm of the soil and decreased significantly with soil depth.

Fine root biomass between 0 and 60 cm deep ranged from 133.3±5.6 g/m^2 in the F0P0 treatment to 170.8±3.4 g/m^2 in the F2P2 treatment. There were few significant differences in fine root biomass between the treatments with no FOM inputs or no water addition (Figure 5). Under the P1 and P2 water addition regimes, increased FOM inputs caused significant increase of fine root biomass (Figure 6A). Similar results were found for water addition (Figure 6B). Water addition increased the root biomass much more than FOM input did (Figure 6).

Discussion

Effects of FOM input on soil C

Soil organic matter plays a key role in determining soil quality and preserving plant nutrients [7]. In the natural environment organic matter inputs to the soil are supplied by plants. Although the response of the soil organic pool to climate change is quite uncertain [16], many laboratory-based incubation and field experiments have provided evidence for changes in SOC storage with increased soil organic matter. After supplying fresh C into soil, Fontaine et al. observed a faster soil C decomposition rate and a decrease in soil C content [36]. Xiao et al. reported that SOC changed slightly with C inputs in a temperate grassland [11]. Other authors have stated that soil C does not necessarily increase with increased organic C input [10,12]. Similarly, in this experiment, no remarkable changes in the SOC in the same soil layers were found from the different FOM input regimes. During the experimental period nearly all FOM input to the soil decomposed, regardless of the water content, and did not affect the SOC in other layers that didn't have direct FOM input. Trumbore and Czimczik stated that because it is physically isolated from microbes, soil organic matter can persist at a steady level in soil, whereas FOM input to soil is more easily decomposed and can be the major source of C loss from soils [7]. This is likely to be a reason for the steady SOC levels at the end of our experiment. We also have to consider that in the natural environment plant litter only drops onto the soil surface, and cannot be fully accessed by the microbial decomposers. This aboveground C input seems not to be an effective way of promoting soil C sequestration, at least in the short-term [9].

The SOC stock in soils results from the balance between the C inputs and outputs [6]. Plants supply the organic matter inputs, which are then transformed by the soil fauna and microorganisms [13]. As decomposers of SOC, microorganisms have received much interest because of their important role in the dynamics of the soil C pool [7,20,27]. MBC increased remarkably with increased FOM compared with the other treatments at the start of our experiment (Figure 4A). This may indicate that the soil system is resource limited, which is the case in most soil systems [37]. After the initial increase, a large decrease in MBC was found in most treatments, and no significant differences were found between treatments with the same water addition regimes in the last month of the study (Figure 4A). Seasonal MBC decreases have been observed by other authors, and this can partly be attributed to changes in environmental factors or plant physiology [37,38].

Fine roots are usually defined as those with diameters<2 mm. Although the amounts of these make up a small proportion of the total root biomass, they can make a major contribution to C and nutrient input to the soil [39]. Nutrient-rich FOM input can affect plant growth because it can lead to an increase in nutrient availability in the soil [40]. There have been a few reports showing different fine root responses to organic C input or measures that affect organic C input (e.g., ground litter addition or removal) in several ecosystems [11,41,42], and significant effects of FOM input on the fine root mass was found in this experiment.

Effects of water addition on soil C

Soil water availability can regulate root growth and the performance of microbes, and can indirectly affect the below-ground soil C cycle [18,21]. Consistent with previous studies [19,43], this study showed that water addition plays an important

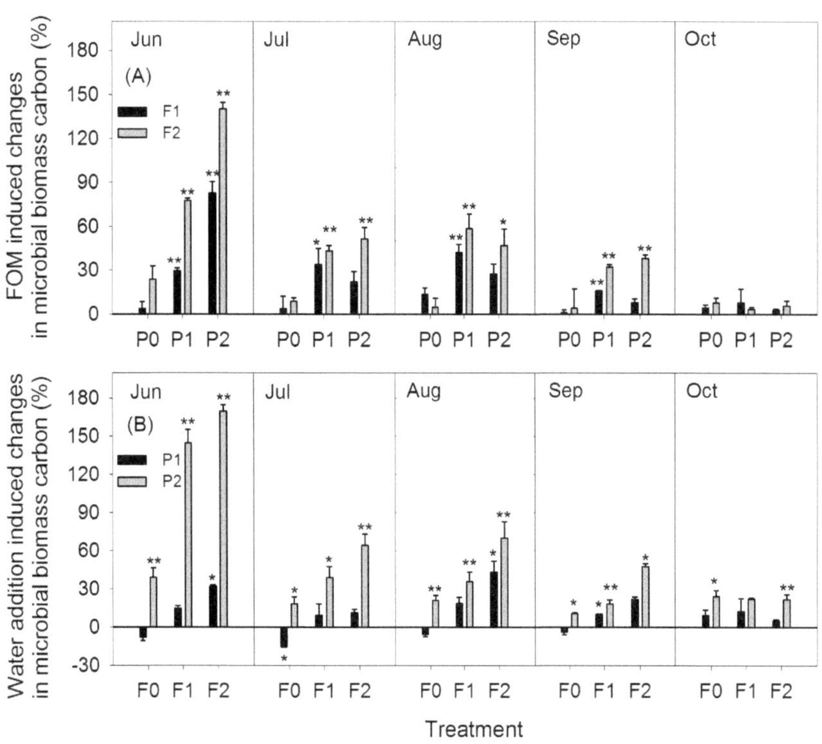

Figure 4. FOM (A) and water addition (B) induced changes in microbial biomass carbon (mean±SE). * and ** represent significant difference compared with controls at $P<0.05$ and $P<0.01$. Other abbreviations are same as Figure 1.

role in belowground C stock in arid land with low precipitation. Although water addition had little effect on the SOC, the treatments did significantly affect the MBC and fine root growth.

Seasonal variability of the soil microbial biomass is important for soil nutrient release and mineralization [37], and it closely correlates with precipitation [44]. Some studies have reported that microbial biomass is highest in dry periods and lowest in rainy periods, and have attributed the reduced microbial biomass and increased turnover of microbes to enhanced microbivory and nutrient competition from plants in wet periods [44,45]. In contrast to those results, we found significant positive responses of MBC to water addition both with and without FOM input treatments in the first four months (Figure 4B). This can be explained because, in arid regions, soil solute availability is reduced by the low water content, and this is likely to reduce substrate availability to the microbes [37]. The soils in the study area are highly saline and alkaline, and this also has strong negative effects on microbial biomass [31]. The osmotic stress caused by the salinity and alkalinity of the soils would have been relieved by water addition. The microbial biomass, therefore, increased in the irrigated plots.

Table 2. Results (*F*-Values) of ANOVA with fresh organic matter (FOM), water addition and their interactions on soil organic carbon (SOC) (g/kg), microbial biomass carbon (MBC) (mg/kg) and fine root biomass (g/m^2, above 60 cm soil depth).

	SOC		MBC		Fine root biomass	
Source	df	*F*	df	*F*	df	*F*
FOM (F)	2	1.80^{ns}	2	176.20**	2	9.68 **
Water addition (P)	2	0.27^{ns}	2	342.73**	2	21.29**
F×P	4	0.46^{ns}	4	35.37**	4	1.20 ^{ns}

ns, * and ** represents $P>0.05$, $P<0.05$ and $P<0.001$, respectively.

Continuous absorption of water is essential for the growth and survival of most plants because of their daily water consumption. In arid ecosystems, soil moisture is generally considered to be the primary factor limiting root growth, especially for desert shrubs, which are often exposed to long periods of drought [46]. In this study, water irrigation showed a significant stimulating effect on the fine root biomass. This was similar to studies in other water limited ecosystems [46,47]. In the treatments without FOM inputs, more water addition caused an approximately 10% increase in fine root biomass (Figure 6B). These results indicate that significantly increased precipitation in the arid regions of central Asia would stimulate plant root growth and indirectly affect soil C sequestration [32].

Interactive effects of FOM input and water addition

Given that both FOM input and water addition showed positive effects on the belowground C cycle, their interactive effects were assumed to be larger still. FOM input caused larger changes in MBC in the treatments with water addition than in those without water addition (Figure 4A). Similar trends were also found for changes in the MBC caused by water addition (Figure 4B). Soil in the arid climate region only receives a small amount of plant litter input because vegetation is sparse and water is limited [27,48]. Increases in soil microbes are limited by the substrate and the lower soil solute availability [37]. The interactive effects of the two treatments, therefore, caused larger changes in MBC than the single treatment did.

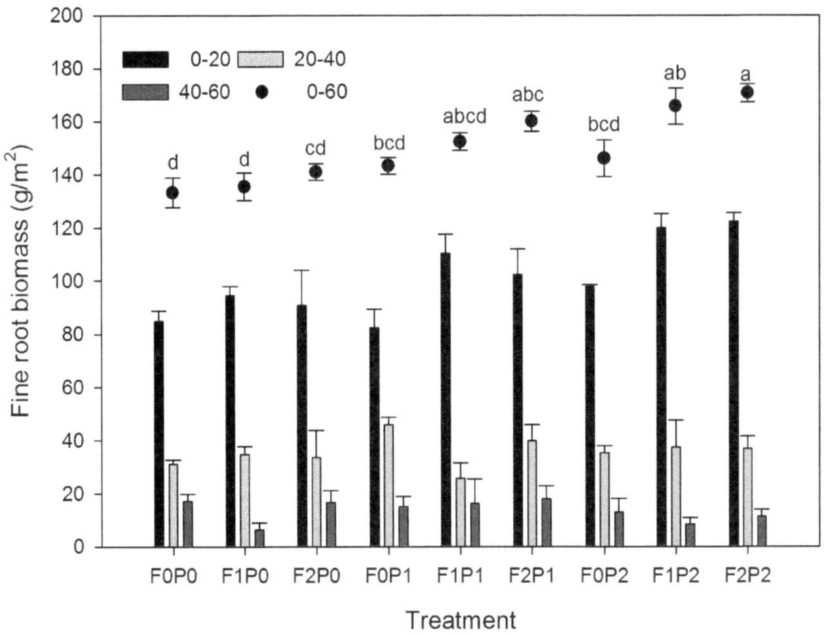

Figure 5. Effects of FOM and water addition on fine root biomass (mean±SE). Dots with different lowercase letters are significantly different from each other of fine root biomass in 0–60 cm soil layers in the different treatments at $p<0.05$ (Tukey Test). Other abbreviations are same as Figure 1.

The root biomass is closely correlated with soil moisture and nutrient availability [41,42]. Wilcox et al. stated that soil moisture is the primary factor influencing root growth in arid ecosystems, and other factors, such as nutrients and soil temperature, are secondary limiting factors [46]. After comparing the single factor effects on fine root biomass, water addition played the most important role in the root growth process in our study (Figure 6). After the soil moisture was increased by adding water, FOM inputs also caused significant increases in fine root biomass. As mentioned above, FOM decomposition through microbial activity was inhibited at low soil moisture contents, which led to weaker effects from FOM input on fine root growth, in contrast to the water addition effects.

Possible mechanism: function of microbes

The microbial biomass plays a key role in the processes that lead to priming effects [15]. Because of comparatively rapid changes in microbial biomass and detectable differences in the amount of biomass present, it can be used as a sensitive indicator of soil SOC variations [44]. Therefore, measuring microbial biomass is an effective way of understanding soil C pool dynamics associated with climate change.

Fountain et al. found that adding ^{13}C labeled organic matter significantly increased microbial biomass, and that the increase mainly came from the formation of labeled biomass [36]. After the supplied organic C was exhausted, many specialized microbes died or became dormant because they were unable to use SOC [36]. In our study, the differences in microbial biomass caused by adding FOM were smaller in August to October than those in June and July. These changes might be due to the exhaustion of FOM, leading to a decrease in the population of microbes that used this resource.

At the end of the experiment, we found that FOM input did not enhance soil C accumulation. A similar result was reported by

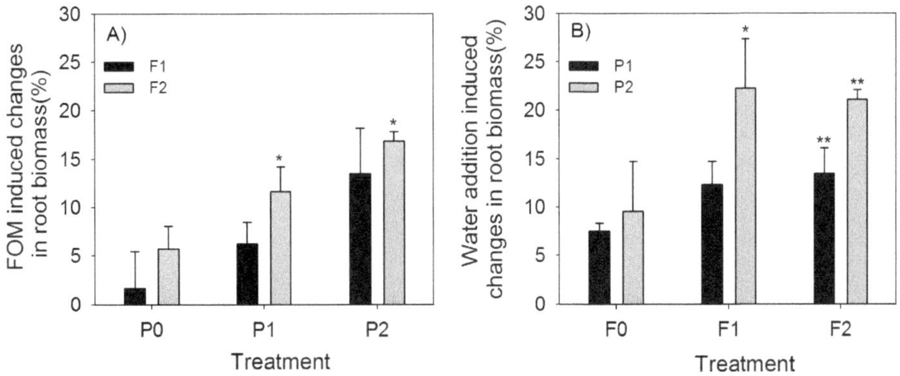

Figure 6. FOM (A) and water addition (B) induced changes in fine root biomass (mean±SE). * and ** represent significant different compared with controls at $P<0.05$ and $P<0.01$. Other abbreviations are same as Figure 1.

Xiao et al. in a semi-arid grassland [11]. This might be caused by the FOM inducing only a weak priming effect that did not influence SOC mineralization, and by most of increase in microbial populations induced by the FOM input being only FOM decomposing microbes that cannot survive on SOC [13,16]. Further, soil organic matter (SOM) is predominantly stabilized in the microaggregates, which can protect SOM to avoid being easily mineralized by microbes [49]. This could be another reason for the stabilization of the SOC.

Conclusions

We found minor changes in the SOC pool caused by adding FOM and water. Adding FOM had a stronger stimulating effect on the soil C mineralization process than did adding only water. However, adding water had an important indirect effect on the decomposition of organic matter, fine root growth, and microbial biomass dynamics. Research is required on the nature of the plant litter decomposition process in the field. These results increase our understanding of how soil C dynamics may change with global climate change.

Acknowledgments

We acknowledge the Fukang Station of Desert Ecology, Xinjiang Institute of Ecology and Geography, Chinese Academy of Sciences for providing the study site and technical advice. We appreciated Prof. Smidt and the anonymous reviewer for insightful comments and great efforts to improve this manuscript, and we thank Yan Zhao, Dr Yugang Wang, Xuecan Liu and Yuanli Li for their help with the fieldwork and laboratory analysis.

Author Contributions

Conceived and designed the experiments: LJ XC YG YZ. Performed the experiments: LL YL YT XZ LZ. Analyzed the data: YL SW YZ GR. Wrote the paper: LL YZ.

References

1. Eswaran H, Reich PF, Kimble JM, Beinroth FH, Padamnabhan E, et al. (2000) Global carbon stocks. In: Global Climate Change and Pedogenic Carbonates (eds Lal R, Eswaran H, Stewart BA), pp. 15–26. CRC Press, Boca Raton, FL.
2. IPCC: Climate Change 2001: The Science of Climate Change. Contribution of Working Group I to the Second Assessment Report of the Intergovernmental Panel on Climate Change, Cambridge Univ. Press, Cambridge.
3. Bond-Lamberty B, Thomson AM (2010) Temperature-associated increases in the global soil respiration record. Nature 464: 579–582.
4. Cox PM, Bettts RA, Jones CD, Spall SA, Totterdell IJ (2000) Acceleration of global warming due to carbon-cycle feedbacks in a coupled climate model. Nature 408: 184–187.
5. Gill RA, Polley HW, Johnson HB, Anderson LJ, Maherali H, et al. (2002) Nonlinear grassland responses to past and future atmospheric CO_2. Nature 417: 279–282.
6. Davidson EA, Janssens IA (2006) Temperature sensitivity of soil carbon decomposition and feedbacks to climate change. Nature 440: 165–173.
7. Trumbore SE, Czimczik CI (2008) An Uncertain Future for Soil Carbon. Science 321: 1455–1456.
8. Yuste JC, Baldocchi DD, Gershenson A, Goldstein A, Misson L, et al. (2007) Microbial soil respiration and its dependency on carbon inputs, soil temperature and moisture. Glob Change Biol 13: 2018–2035.
9. Garten Jr CT (2009) A disconnect between O horizon and mineral soil carbon – Implications for soil C sequestration. Acta Oecol 35: 218–226.
10. Jones CD, Cox PM, Essery RLH, Roberts DL, Woodage MJ (2003) Strong carbon cycle feedbacks in a climate model with interactive CO_2 and sulphate aerosols. Geophys Res Lett 30: 321–324.
11. Xiao C, Janssens IA, Liu P, Zhou Z, Sun OJ (2007) Irrigation and enhanced soil carbon input effects on below-ground carbon cycling in semiarid temperate grasslands. New Phytol 174: 835–846.
12. Sayer EJ, Heard MS, Grant HK, Marthewsand TR, Tanner EVJ (2011) Soil carbon release enhanced by increased tropical forest litterfall. Nature Climate Change 1: 304–307.
13. Fontaine S, Mariotti A, Abbadie L (2003) The priming effect of organic matter: a question of microbial competition? Soil Biol Biochem 35: 837–843.
14. Bingemann CW, Varner JE, Martin WP (1953) The effect of the addition of organic materials on the decomposition of an organic soil. Soil Sci Soc Amer J 17: 34–38.
15. Kuzyakov Y, Friedel JK, Stahr K (2000) Review of mechanisms and quantification of priming effects. Soil Biol Biochem 32: 1485–1498.
16. Fontaine S, Barot S, Barré P, Bdioui N, Mary B, et al. (2007) Stability of organic carbon in deep soil layers controlled by fresh carbon supply. Nature 450: 277–280.
17. Xia J, Wan S (2012) The effects of warming-shifted plant phenology on ecosystem carbon exchange are regulated by precipitation in a semi-arid grassland. PLoS ONE 7(2): e32088.
18. Weltzin JF, Loik ME, Schwinning S, Williams DG, Fay PA, et al. (2003) Assessing the response of terrestrial ecosystems to potential changes in precipitation. Bioscience 53: 941–952.
19. Niu S, Wu M, Han Y, Xia J, Li L, et al. (2008) Water-mediated responses of ecosystem carbon fluxes to climatic change in a temperate steppe. New Phytol 177: 209–219.
20. Reichstein M, Tenhunen JD, Roupsard O, Ourcival JM, Rambal S, et al. (2002) Ecosystem respiration in two Mediterranean evergreen Holm Oak forests, drought effects and decomposition dynamics. Funct Ecol 16: 27–39.
21. Yuste JC, Janssens IA, Carrara A, Meiresonne L, Ceulemans R (2003) Interactive effects of temperature and precipitation on soil respiration in a temperate maritime pine forest. Tree Physiol 23: 1263–1270.
22. Robertson GP, Klingensmith KM, Klug MJ, Paul EA, Crum JR, et al. (1997) Soil resources, microbial activity, and primary production across an agricultural ecosystem. Ecol Appl 7: 158–170.
23. Solomon S, Qin D, Manning M, Chen Z, Marquis M, et al. (2007) Technical summary. In: Climate Change 2007: The Physical Science Basis. Contribution of Working Group I to the Fourth Assessment Report of the Intergovernmental Panel on Climate Change (eds Solomon S, Qin D, Manning M, Chen Z, Marquis M, et al.), pp. 20–91. Cambridge University Press, Cambridge, UK/New York, NY, USA.
24. Sponseller RA (2007) Precipitation pulses and soil CO_2 flux in a Sonoran Desert ecosystem. Glob Change Biol 13: 426–436.
25. Smith SD, Huxman TE, Zitzer SF, Charlet TN, Housman DC, et al. (2000) Elevated CO_2 increases productivity and invasive species success in an arid ecosystem. Nature 408: 79–82.
26. Melillo JM, McGuire AD, Kicklighter W, Moore B, Vorosmarty CJ, et al. (1993) Global climate change and terrestrial net primary production. Nature 363: 234–240.
27. Lal R (2009) Sequestering carbon in soils of arid ecosystems. Land Degrad Develop 20: 441–454.
28. Zheng YR, Xie ZX, Jiang LH, Shimizu H, Drake S (2006) Changes in Holdridge Life Zone diversity in the Xinjiang Uygur Autonomous Region (XUAR) of China over the past forty years. J Arid Environ 66: 113–126.
29. Xu H, Li Y (2006) Water-use strategy of three central Asian desert shrubs and their responses to rain pulse events. Plant Soil 285: 5–17.
30. Lai L, Zhao X, Jiang L, Wang Y, Luo L, et al. (2012) Soil Respiration in Different Agricultural and Natural Ecosystems in an Arid Region. PLoS ONE 7(10): e48011.
31. Rietz DN, Haynes RJ (2003) Effects of irrigation-induced salinity and sodicity on soil microbial activity. Soil Biol Biochem 35: 845–854.
32. Qian W, Zhu Y (2001) Climate change in China from 1880 to 1998 and its impact on the environmental condition. Climatic Change 50: 419–444.
33. Starr JL, Paltineanu IC (2002) Methods for measurement of soil water content: capacitance devices. In: Methods of Soil Analysis: Part 4 Physical Methods (eds Dane JH, Topp GC), pp. 463–474. Soil Science Society of America, Inc., Madison, Wis, USA.
34. Nelson DW, Sommers LE (1982) Total carbon, organic carbon, and organic matter. In: Page AL, Miller RH, Keeney DR, eds. Methods of soil analysis. Madison, WI, USA: American Society of Agronomy and Soil Science Society of America, pp.101–129.
35. SPSS (2000) SPSS for windows, version 13.0. SPSS, Chicago.
36. Fontaine S, Bardoux G, Abbadie Luc, Mariotti A (2004) Carbon input to soil may decrease soil carbon content. Ecol Lett 7: 314–320.
37. Wardle DA (1992) A comparative assessment of factors which influence microbial biomass carbon and nitrogen levels in soil. Biol Rev 67: 321–358.
38. Myers RT, Zak DR, White DC, Peacock A (2001) Landscape-level patterns of microbial community composition and substrate use in upland forest ecosystems. Soil Sci Soc Amer J 65: 359–367.
39. Thomas K, Antonio F, Manuel M, Carlos R, Eliel S (1995) Fine-root dynamics, soil moisture and soil carbon content in a *Eucalyptus globulus* plantation under different irrigation and fertilisation regimes. Forest Ecol Manage 74: 1–12.
40. Shemesh H, Arbiv A, Gersani M, Ovadia O, Novoplansky A (2010) The Effects of Nutrient Dynamics on Root Patch Choice. PLoS ONE 5(5): e10824.
41. Sayer EJ, Tanner EVJ, Cheesman AW (2006) Increased litterfall changes fine root distribution in a moist tropical forest. Plant Soil 281: 5–13.
42. Lima TTS, Miranda LS, Vasconcelos SS (2010) Effects of water and nutrient availability on fine root growth in eastern Amazonian forest regrowth, Brazil. New Phytol 187: 622–630.

43. Domingo F, Serrano-Ortiz P, Were A, Villagarcía L, García M et al. (2011) Carbon and water exchange in semiarid ecosystems in SE Spain. J Arid Environ 75: 1271–1281.
44. Sharma P, Rai SC, Sharma P, Sharma E (2003) Effects of land-use change on soil microbial C, N and P in a Himalayan watershed. Pedobiologia 48: 83–92.
45. Wagai R, Kristofor RB, Grower ST, Norman JM, Bundy LG (1998) Land use and environmental factors influencing soil surface CO_2 flux and microbial biomass in natural and managed ecosystems in Southern Wisconsin. Soil Biol Biochem 30: 1501–1509.
46. Wilcox CS, Ferguson JW, Fernandez GCJ, Nowak RS (2004) Fine root growth dynamics of four Mojave Desert shrubs as related to soil moisture and microsite. J Arid Environ 56: 129–148.
47. Schenk HJ, Jackson RB (2002) Rooting depths, lateral root spreads and below-ground/above-ground allometries of plants in water-limited ecosystems. J Ecol 90: 480–494.
48. Raich JW, Schlesinger WH (1992) The global carbon dioxide flux in soil respiration and its relationship to vegetation and climate. Tellus 44B: 81–99.
49. Six J, Bossuyt H, Degryze S, Denef K (2004) A history of research on the link between (micro) aggregates, soil biota, and soil organic matter dynamics. Soil Till Res 79: 7–31.

2

Comparison of Soil Quality Index Using Three Methods

Atanu Mukherjee*, Rattan Lal

Carbon Management and Sequestration Center, School of Environment and Natural Resources, The Ohio State University, Columbus, Ohio, United States of America

Abstract

Assessment of management-induced changes in soil quality is important to sustaining high crop yield. A large diversity of cultivated soils necessitate identification development of an appropriate soil quality index (SQI) based on relative soil properties and crop yield. Whereas numerous attempts have been made to estimate SQI for major soils across the World, there is no standard method established and thus, a strong need exists for developing a user-friendly and credible SQI through comparison of various available methods. Therefore, the objective of this article is to compare three widely used methods to estimate SQI using the data collected from 72 soil samples from three on-farm study sites in Ohio. Additionally, challenge lies in establishing a correlation between crop yield versus SQI calculated either depth wise or in combination of soil layers as standard methodology is not yet available and was not given much attention to date. Predominant soils of the study included one organic (Mc), and two mineral (CrB, Ko) soils. Three methods used to estimate SQI were: (i) simple additive SQI (SQI-1), (ii) weighted additive SQI (SQI-2), and (iii) statistically modeled SQI (SQI-3) based on principal component analysis (PCA). The SQI varied between treatments and soil types and ranged between 0–0.9 (1 being the maximum SQI). In general, SQIs did not significantly differ at depths under any method suggesting that soil quality did not significantly differ for different depths at the studied sites. Additionally, data indicate that SQI-3 was most strongly correlated with crop yield, the correlation coefficient ranged between 0.74–0.78. All three SQIs were significantly correlated ($r = 0.92$–0.97) to each other and with crop yield ($r = 0.65$–0.79). Separate analyses by crop variety revealed that correlation was low indicating that some key aspects of soil quality related to crop response are important requirements for estimating SQI.

Editor: Upendra M. Sainju, Agricultural Research Service, United States of America

Funding: This research was part of a collaborative project supported by the USDA-NIFA, award number: 2011-68002-30190, "Cropping Systems Coordinated Agricultural Project (CAP): Climate Change, Mitigation, and Adaptation in Corn-based Cropping Systems" (sustainablecorn.org). The funders had no role in study design, data collection and analysis, decision to publish, or preparation of the manuscript.

Competing Interests: The authors have declared that no competing interests exist.

* Email: mukherjee.70@osu.edu

Introduction

A wide range of agricultural soils represents diversely managed arable lands while the main goal to improve soil quality, crop yield, and reduce the ecological foot print. Soil quality is defined as the soil's capacity to function within natural or managed ecosystem boundaries and to sustain plant productivity while reducing soil degradation [1–4]. As soil quality is a complex functional concept and cannot be measured directly in the field or laboratory [5] but can only be inferred from soil characteristics [6], a range of soil parameters or indicators has been identified to estimate soil quality. However, soil quality is often related to the management goal and practices as well to soil characteristics. Thus, a mathematical or statistical framework was put forward in early 1990s to estimate soil quality index (SQI) [1,3,4]. The SQI was assessed so that the management goals are not only focused on productivity per se, which may result in soil degradation [7], but also on environmental issues. Thus, an appropriate SQI may have three component goals: environmental quality, agronomic sustainability, and socio-economic viability [8].

Estimation of SQI is a complex process and difficult task [9], especially when linked with several functional goals. Yet a considerable progress has been made towards estimating SQI across a number of soil types and management practices [8–14]. Most studies indexed soil quality employing only one method, with a few exceptions [8,15,16]. As computation of SQI is difficult, a strong need exists for developing a user-friendly and credible SQI through comparison of various available methods. Thus, the objective of this study was to compare SQIs computed by three methods which are conceptually different from each other. The study is based on the hypothesis that SQIs computed from three methods have similar relationship with crop yields. Data are scarce on validation of SQI against crop yield as most studies focused on the environmental aspects of the soil as end point variable and that SQI of various soil layers has not been computed [13,16–18]. Nevertheless, there is a challenge in validating SQI against crop yield as SQI computed only from surface soils (0–10 and/or 0–20 cm) may not evaluate realistic relationship between soil quality and crop yield because root system can extend to deeper layers [19]. Thus, the other objective of this study was to compute SQI from multiple depths of the soil to examine its relationship with crop yield.

Materials and Methods

Soil sampling and analyses

Soil samples from the field (on-farm) were collected from Logan county, Ohio that included an organic (Martisco Variant silt loam: Mc, organic parent material, >26% organic matter, 40°25′12.4″N, 83°40′55.9″W), and a mineral soil (Crosby silt loam: CrB: *Fine, mixed, active, mesic Aeric Epiaqualfs*, sand: 23%, silt: 35%, clay: 42%, 40°24′52.4″N, 83°39′21.9″W), and from Franklin county, Ohio that included a Kokomo silty clay loam (Ko: *Fine, mixed, superactive, mesic Typic Argiaquolls*, sand: 9%, silt: 62%, clay: 29%, 40°00′41.2″N, 83°12′23.1″W) in May, 2013. These locations were farmer-owned field sites and before collecting the soils from specific sites permissions of the farmers were obtained from the land-owners or farmers. Initial permission was granted through The Ohio State University's extension managers and thereafter communication was established directly by the primary author. Additionally, no endangered or protected species were involved in the current study sites, and thus no such permission was required from any other regulatory agencies. The properties of the soils and other management details were presented elsewhere [20,21] and briefly presented in the discussion section. Soil samples were obtained from 0–10, 10–20, 20–40, and 40–60 cm depths under no-tillage (NT) and conventional tillage (CT) practices for Mc and CrB soil series and under NT for one year cover crop (CC) and no-cover crop (NCC) for Ko series. The organic soil (Mc) has been under practice of CT corn (*Zea mays*) for 10 years which was compared with the adjacent grassland soil (termed as NT). The mineral soil (CrB) was under NT and recent introduction (one year) of CT corn practices. Both soils received fertilizer (N:P:K as 5.5:26:30) with the rates of 336 (30, 142 and 164 kg ha^{-1} of N, P, and K, respectively) and 280 (25, 118 and 137 kg ha^{-1} of N, P, and K, respectively) kg ha^{-1} for corn, for Mc and CrB soils, respectively. The other site (Ko) was under NT corn and soybean (*Glycin max L.*) annual rotation. A mixture of pea (*Pisum sativum L.*) and turnip (*Brassica rapa L.*) cover crops with the seeding rate of 140 kg ha^{-1} was seeded prior to growing soybean in 2013. Roundup-ready soybean was seeded on 20 April, 2013 and harvested on 30 September, 2013. A row spacing of 38 cm was used during planting for corn and soybean. Soil core (inside diameter: 5.3 cm, height: 5.9 cm) samples were collected manually using soil sampler before planting (7 May, 2013) from four depths (0–10, 10–20, 20–40, and 40–60) for determination of bulk density (BD) by dividing dry mass of soil in the core by core volume [22] and hydrologic properties. Undisturbed soil cores were used to determine the water retention at field capacity (0.033 MPa), while sieved samples (<2-mm size) were used to determine the permanent wilting point (1.5 MPa). The potential available water capacity (AWC) of the soil was calculated as the difference in volumetric water content at 0.033 and 1.5 MPa moisture potentials. The water content of a soil layer was calculated by multiplying thickness of the layer with the volumetric water capacity. A minimum of three field penetration resistance (PR) measurements for soil's mechanical strength were made for each depth using a CP40II cone penetrometer [23]. Bulk samples were obtained to measure the aggregate size distributions and fraction of water stable aggregates (WSA) using the wet sieving method [24]. Five sieves of diameter sizes 4.75, 2, 1, 0.5, and 0.25 mm openings were placed into a Yoder apparatus. Air-dried soils (5–8 mm, 51 g) were slowly wetted by capillarity action by adjusting the water level in the container so that the base of the top sieve just touched the water. The sieve combination was oscillated mechanically in the water at 60 oscillations per minute for 30 min. Aggregates retained in the sieves were transferred to glass beakers and the weight of each of five fractions was measured after drying at 60°C overnight. The data were used to compute WSA, mean weight diameter (MWD), and geometric mean diameter (GMD) [25]. Soil water retention at matrix potentials of 0.033 and 1.5 MPa were measured using a pressure plate apparatus [26]. The pH and electrical conductivity (EC) of soil were determined in 1:2 soil:water ratio slurry using a Thermo-scientific Orion Star Series pH/Conductivity Meter. Concentrations of soil organic C and total N were determined using an Elemental analyzer (Vario Max, Elementar Americas, Inc., Germany) by the dry combustion method (900°C) after grinding subsamples to 0.25 mm. Total C and N stocks were calculated by multiplying the respective elemental concentrations by BD and the thickness of the soil layer [27].

Soil quality index (SQI) calculations

All the SQI methods involved a set of 72 soil samples and a number of soil quality indicators as parameters. The 13 parameters used for developing SQIs were pH, EC, BD, WSA, GMD, MWD, PR, SOC concentration, N concentration, C-Stock, N-Stock, AWC, and soil water content. As some of these parameters were synthesized and redundant, only nine parameters were chosen omitting MWD, C and N concentrations, and water content from the SQI-1 and SQI-2 calculations to avoid redundancy. However, SQI-3 was not a primarily additive approach and redundancy of the parameters was eliminated through the processes of elimination as prescribed before. Thus, all 13 parameters were initially included in SQI-3 model and only four non-redundant soil parameters with maximum variations in the dataset were finally retained in the model as described in the following section. The average and standard deviation values of these parameters grouped by soil types are presented in Table 1. Under the proposed framework an ideal soil would have SQI value of 1 for the highest quality soil and 0 for the severely degraded soil [1–4].

Simple additive SQI (SQI-1)

Simple additive SQI was estimated following the method outlined by Amacher et al. [28]. In this method, soil parameters were given threshold values based primarily on the literature review and expert opinion of the authors. The threshold levels, interpretations, and associated unitless soil index score values are listed in Table 2. The individual index values were then summed up to obtain a total SQI:

$$\sum SQI = \sum \textit{Individual soil parameter index values} \quad (1)$$

The scaled SQI (SQI-1) of individual soil, was computed by Eq. 2:

$$SQI-1 = (\sum SQI - SQI_{Min})/(SQI_{Max} - SQI_{Min}), \quad (2)$$

whereas, SQI_{Min} = Minimum value of SQI, and SQI_{Max} = Maximum value of SQI from the total dataset.

Weighted additive SQI (SQI-2)

In this approach, each soil parameter was first assigned unitless score ranging from o to 1 by employing linear scoring functions [8]. Non-linear scoring functions were avoided because of their lower capacity of predicting the end point variable or crop yield [8]. Soil parameters were divided into groups based on three mathematical algorithm functions: (a) 'more is better' (e.g., WSA, GMD, C-Stock, N-Stock, and AWC) (b) 'less is better' (e.g., BD, PR), and (c) 'optimum' (e.g., pH and EC). 'Optimum' properties are those which have positive influence up to a certain level beyond which the influence could be considered detrimental [13]. For 'more is better' parameters, each observation was divided by the highest observed value of the entire dataset so that the highest observed value would have a score of 1; for 'less is better'

Table 1. Descriptive statistics of all the soil indicators collected from four depths under three soil types used to estimate SQI.

Ko

	0–10 cm		10–20 cm		20–40 cm		40–60 cm	
	Mean	SD	Mean	SD	Mean	SD	Mean	SD
pH	6.0	0.5	6.2	0.6	6.4	0.6	6.4	0.6
EC (µs cm^{-1})	190.6	151.1	165.6	84.2	151.7	71.8	175.9	92.2
BD (Mg m^{-3})	1.4	0.2	1.5	0.1	1.6	0.1	1.6	0.1
WSA %	73.1	14.8	79.5	12.0	73.3	22.5	84.7	5.6
GMD (mm)	1.4	0.4	1.6	0.4	1.4	0.3	1.5	0.3
MWD (mm)	2.6	1.7	3.1	1.8	2.3	1.7	3.4	1.1
PR (Mpa)	2.2	0.3	2.4	0.2	2.3	0.2	2.2	0.4
N (%)	0.2	0.0	0.2	0.0	0.2	0.0	0.2	0.0
SOC (%)	2.4	0.2	2.1	0.2	2.0	0.2	2.0	0.3
N-Stock (Mg/ha)	3.5	0.5	3.0	0.3	6.4	0.6	6.4	1.1
C-Stock (Mg/ha)	33.7	3.7	30.6	2.8	62.5	6.8	63.7	7.2
AWC (%)	38.5	4.1	36.7	2.1	35.8	6.4	39.9	5.8
Water content (cm)	3.8	0.4	3.7	0.2	7.2	1.3	7.9	1.0

CrB

	0–10 cm		10–20 cm		20–40 cm		40–60 cm	
	Mean	SD	Mean	SD	Mean	SD	Mean	SD
pH	7.4	0.2	7.4	0.2	7.4	0.2	7.4	0.1
EC (µs cm^{-1})	250.3	44.6	217.8	26.8	212.3	31.8	223.5	45.0
BD (Mg m^{-3})	1.5	0.2	1.5	0.1	1.5	0.3	1.5	0.3
WSA %	82.4	4.4	76.4	9.7	76.5	9.5	76.5	17.7
GMD (mm)	1.2	0.1	1.1	0.2	1.1	0.2	1.2	0.3
MWD (mm)	2.0	0.5	1.7	0.8	1.5	0.7	2.0	1.1
PR (Mpa)	2.4	0.2	2.5	0.5	2.6	0.2	2.9	0.7
N (%)	0.2	0.0	0.2	0.0	0.2	0.0	0.2	0.0
SOC (%)	2.4	0.3	2.1	0.4	1.8	0.4	1.8	0.5
N-Stock (Mg/ha)	3.2	0.2	3.1	0.3	5.2	1.6	5.1	2.2
C-Stock (Mg/ha)	34.3	3.2	32.2	4.1	53.8	16.8	54.6	24.4
AWC (%)	17.8	9.0	10.1	6.8	10.7	8.2	9.0	3.2
Water content (cm)	1.8	0.9	1.0	0.7	2.1	1.6	1.8	0.6

Mc

Table 1. Cont.

Ko								
	0–10 cm		10–20 cm		20–40 cm		40–60 cm	
	Mean	SD	Mean	SD	Mean	SD	Mean	SD
	0–10 cm		10–20 cm		20–40 cm		40–60 cm	
	Mean	SD	Mean	SD	Mean	SD	Mean	SD
pH	7.4	0.1	7.4	0.1	7.4	0.1	7.5	0.1
EC (µs cm^{-1})	616.3	40.1	570.0	92.8	633.2	126.4	599.2	129.2
BD (Mg m^{-3})	0.6	0.0	0.7	0.0	0.6	0.1	0.6	0.1
WSA %	91.2	0.5	91.0	2.5	89.7	2.5	82.5	8.7
GMD (mm)	2.1	0.0	2.2	0.0	2.2	0.0	2.2	0.1
MWD (mm)	5.5	0.1	5.5	0.2	5.7	0.4	5.0	0.5
PR (Mpa)	1.3	0.1	1.5	0.3	1.6	0.3	1.9	0.1
N (%)	1.0	0.1	0.9	0.1	0.9	0.1	1.0	0.1
SOC (%)	15.1	1.1	15.0	0.8	14.9	0.9	15.9	0.6
N-Stock (Mg/ha)	6.1	0.8	6.3	0.7	11.2	1.7	12.6	1.9
C-Stock (Mg/ha)	95.7	11.7	102.8	8.8	186.6	36.2	196.6	28.0
AWC (%)	248.3	151.3	84.8	42.4	48.2	13.8	36.1	22.1
Water content (cm)	24.8	15.1	8.5	4.2	9.6	2.8	7.2	4.4

Abbreviations: SD: standard deviation, EC: electrical conductivity, BD: bulk density, WSA: water stable aggregates, GMD: geometrical mean diameter, MWD: mean weight diameter, PR: penetration resistance, N: nitrogen, SOC: soil organic carbon, AWC: available water capacity.

Table 2. Soil indicators, threshold values, interpretations and scores.

Indicators	Range	Interpretation	Score	Reference
pH	5.5–7.2	Slightly acidic to neutral: Optimum for plant growth	2	[28]
	>7.2<8.0	Slightly to moderately alkaline: Preferred by some plants, possible P and some metal deficiencies	1	
EC (us/cm)	<200	Low salt level	0	[42]
	200–500	Optimum salt level for plants	1	
	>500	High salt level, adverse effect likely	0	
BD (Mg/m^3)	<1.0	High organic soil, supports plant roots	2	[28,45–47]
	1.0–1.5	Adverse effects unlikely	1	
	>1.5	Adverse effects likely	0	
WSA (%)	<50	Infiltration and soil erosion problems likely	0	[30,42]
	50–70	Moderate constraints	1	
	70–90	Good soil	2	
	>90	Excellent soil	3	
GMD (mm)	<1.0	Infiltration and soil erosion problems likely	0	[30,42]
	1–2	Moderate limitations	1	
	>2	No limitation	2	
MWD (mm)	<1.0	Infiltration and soil erosion problems likely	0	[30,42]
	1–2	Moderate limitations	1	
	2–5	Slight limitations	2	
	>5.0	No limitation	3	
PR (Mpa)	1–2	Adverse effect on plant root unlikely	2	[42,48]
	2–3	Moderate adverse effect on plant root	1	
	>3.0	Severe adverse effect on plant roots	0	
N (%)	0.2–0.3	Moderate limitation	1	[41,42]
	>0.3	Slight to no limitation	2	
SOC (%)	2–3	Moderate limitation	1	[41,42]
	>3.0	Slight to no limitation	2	
N-Stock (Mg/ha)	<5.0	N deficient	1	Authors' opinion
	5–10	Moderate to optimum N level	2	
	>10.0	N-rich soil	3	
C-Stock (Mg/ha)	<50.0	C deficient	1	Authors' opinion
	50–100	Moderate to optimum C level	2	
	>100	C-rich soil	3	
AWC (%)	<20	Water-stress to plants	0	[31,42]
	20–50	Moderate water availability	1	
	>50	Good water capacity for plants	2	
Water content(cm)	<5.0	Water-stress to plants	0	Authors' opinion
	5–10	Moderate water availability	1	
	>10	Good water capacity for plants	2	

Abbreviations are same as Table 1.

parameters, the lowest observed value in the entire dataset was divided by each observation so that the lowest observed value received a score of 1; and 'optimum' parameters were scored up to a threshold value as 'more is better', and thereafter above the threshold values were scored as 'less is better' [8,29]. For example, pHs up to 5.5–7.2 were scored as 'more is better', and pHs >7.2 were scored as 'less is better'.

After normalizing soil parameters, the scores were integrated into a single index value for each soil using a weighted additive approach initially suggested by Karlen and Stott [4], but modified later by Fernandes et al. [13]. The following weighted additive function was used for development of SQI-2 (Eq. 3):

$$SQI-2 = [(\text{Weight1}) * \text{RDC}] + [(\text{Weight2}) * \text{WSC}] + [(\text{Weight3}) * \text{NSC}], \quad (3)$$

where, RDC (root development capacity) is the rating for the soil's ability to allow plant root development, WSC (water storage capacity) is the rating for the soil's ability to store water, NSC

Table 3. Model of SQI-2.

Soil function	Weight	Soil Indicators	Sub-weight	Scaled score	B×C	∑ B×C	D×A	%	SQI
	A		B	C		D			
RDC	0.4	BD	0.35	0.41	0.14	0.46	0.18	52.94	0.34
		PR	0.35	0.52	0.18				
		WSA	0.20	0.54	0.11				
		GMD	0.10	0.34	0.03				
WSC	0.2	AWC	1.00	0.06	0.06	0.06	0.01	2.94	
NSC	0.4	pH	0.30	0.79	0.24	0.38	0.15	44.12	
		EC	0.30	0.25	0.08				
		C-stock	0.20	0.11	0.02				
		N-stock	0.20	0.18	0.04				

Abbreviations: RDC: root development capacity, WSC: water storage capacity, NSC: nutrient storage capacity; all other abbreviations are same as Table 1.

(nutrient supply capacity) is the rating for the soil's ability to supply nutrients. Weight 1, 2 and 3 are the respective numerical weights for each soil function (RDC, WSC, and NSC). The numerical weights were assigned to each soil function according to their importance in fulfilling the management goal(s) of maintaining soil quality. While the summation of all these numerical weights (Weight 1, 2 and 3) are supposed to be 1 and distributed evenly as 0.33, 0.33, and 0.34 [4], however, a modified approach was taken for the current study following Fernandes et al. [13], where lower weightage was given to the functional attribute which had lower number of representative indicators in the model. Similar to Fernandes et al. [13] weight values were arbitrarily chosen for RDC, WSC, and NSC as 0.4, 0.2, and 0.4 in the current study. The WSC received lower weight value than others because of low number of representative indicators (only AWC) [13] among all included parameters.

The soil parameters or indicators selected for (i) RDC were BD, PR, WSA, and GMD, (ii) WSC was AWC and (iii) NSC were pH, EC, C-Stock, and N-Stock. Within this network, the sub-weight values were given to each indicator based on their importance under the particular soil functional property, field versus laboratory measurements and scope of redundancy. The sub-weight values of different soil indicators or parameters were summed up to 1 under each soil functional property [13]. The rank of the subweight values were as follows: field measurement indicators (BD, PR) > laboratory measurement indicators (WSA, AWC, pH, EC) >> synthesized parameters (GMD, C-Stock, and N-Stock). Field indicators were given the maximum weightage as these were the most representative of the soil's natural conditions and synthesized parameters received the lowest weightage to avoid data redundancy in the model. Thereafter, scaled SQI (SQI-2) of individual soil was computed by Eq. 2. An example of this model of development of SQI is shown in Table 3.

Statistically modeled SQI (SQI-3)

A statistics-based model was used to estimate SQI using principal component analysis (PCA) [8,12,14,18,30,31]. The PCA-model is used to create a minimum data set (MDS) to reduce the indicator load in the model and avoid data redundancy [8]. The main difference between the first two versus the PCA method is that the first two rely mainly on subjective expert opinion and literature review, while the PCA method is more objective of using a number of statistical tools (multiple correlation, factor and cluster analyses) which could avoid any biasness and data redundancy by choosing an MDS using mathematical formulae [8,32]. A relative advantage and disadvantage of each SQI method is explained in the later section. The preliminary function of PCA is to reduce the dimensionality of the entire data set consisting of a large number of interrelated variables, while retaining as much as possible of the variations present in the data set. This is achieved by transformation to a new set of variables, the principal components (PCs), which are uncorrelated, and ordered so that the first few retain most of the variation present in all of the original variables [33,34]. In other words, the PCA method was chosen as a data reduction tool to select the most appropriate indicator(s) to represent and estimate SQI [18].

All the original observations (untransformed data) of each soil were included in the PCA model using SPSS, version 21.0 [35]. The PCs with high eigenvalues represented the maximum variation in the dataset [8,12,14,18,30,31]. While most studies have assumed to examine PCs with eigenvalue >1.0 following Kaiser [36], the present study had the third and fourth components with eigenvalues of 0.98 and 0.89 and variances >5% (Table 4, Fig. 1). These were examined with first two PCs (eigenvalues 8.23, and 1.67) as

Table 4. Results of principal component analyses (PCA).

	PC-1	PC-2	PC-3	PC-4
Eigenvalues	8.23	1.67	0.98	0.89
% Variance	63.34	12.87	7.56	6.82
Cumulative variance	63.3	76.2	83.8	90.6
Eigen vectors or factor loading				
pH	0.423	0.351	−0.378	**0.711**
EC	**0.925**	0.113	−0.153	0.147
BD	**−0.881**	−0.034	0.212	0.002
WSA	0.590	−0.020	**0.669**	0.343
GMD	**0.927**	0.024	0.276	−0.024
MWD	**0.904**	0.001	0.342	0.046
PR	−0.770	0.268	−0.131	0.092
SOC	**0.975**	0.073	−0.109	−0.025
N	**0.972**	0.050	−0.103	−0.053
C-Stock	0.831	0.412	−0.107	−0.268
N-Stock	0.722	0.440	−0.067	−0.385
AWC	0.555	**−0.782**	−0.189	0.057
Water content	0.624	−0.693	−0.199	−0.066

Abbreviations are same as Table 1; PC: principal component; bold values under each component are highly weighted and underlined bold values are selected in minimum data set.

prescribed for the cases of fewer than three components with eigenvalue >1.0 [8,37]. Under a given PC, each variable had corresponding eigenvector weight value or factor loading (Table 2). Only the 'highly weighted' variables were retained to include in the MDS (Table 2). The 'highly weighted' variables were defined as the highest weighted variable under a certain PC and absolute factor loading value within 10% of the highest values under the same PC [37]. Thus, the bold-face values (Table 4: EC, BD, GMD, MWD, SOC, and N for PC-1, AWC for PC-2, WSA for PC-3, and pH for PC-4) were considered highly weighted eigenvectors and therefore were initially selected in the MDS. However, when more than one variable was retained under a particular PC, multivariate correlation matrix (Table 5) was used to determine the correlation coefficients between the parameters [8,12]. If the parameters were significantly correlated ($r>0.60$, $p<0.05$), then the one with the highest loading factor was retained in the MDS and all others were eliminated from the MDS to avoid redundancy. Following this procedure, except SOC, all other eigenvectors from PC-1 were eliminated from MDS due to high and significant correlation between each other (Table 5). The non-correlated parameters under a particular PC were considered important and retained in the MDS [10,12]. All the bold-faced and underlined soil parameters in Table 4 were selected in the final MDS.

After selection of parameters for the MDS, all selected observations were transformed using linear scoring functions (less is better, more is better and optimum) as described in SQI-2 method. Once the selected observations were transformed in numerical scores (ranged 0–1), a weighted additive approach was used to integrate them into indices for each soil [8,12]. Each PC explained a certain amount of variation in the dataset (Table 4), which was divided by the maximum total variation of the all PCs selected for the MDS to get a certain weightage value under a particular PC [8]. For example, the % variance (63.3) was divided by total cumulative variance (90.6) to obtain the weight value of 0.7 for PC-1 (Table 4). Thereafter, the weighted additive SQI was computed using Eq. 4:

$$SQI-3\,(PCA)=\sum Weight * Individual\,oil\,parameter\,score \quad (4)$$

Validation of SQI

The SQIs estimated from three different methods were validated against soybean (*Glycin max*) and corn (*Zea mays*) yield data of that particular year of collection of soil and crops by computing correlation coefficients [8]. The SQI values were also compared through pearson correlation to understand the effectiveness of each other.

Statistical analyses

All values are presented as means ± standard deviations of three field or laboratory measurements. Significant differences between treatments were analyzed using Tukey's test in PROC GLM in SAS version 9.2 [38]. Treatment differences were deemed significant at $p<0.05$. The PCA was performed in SPSS version 21 [35]. Descriptive statistics and linear regressions were computed in Microsoft Excel [39] and all the figures were obtained using Sigmaplot Version 12.0 [40].

Results and Discussions

Most data of three farm sites have been presented and the detailed discussions of soil type, crop and management effects on soil properties are discussed elsewhere [20,21]. The mean and

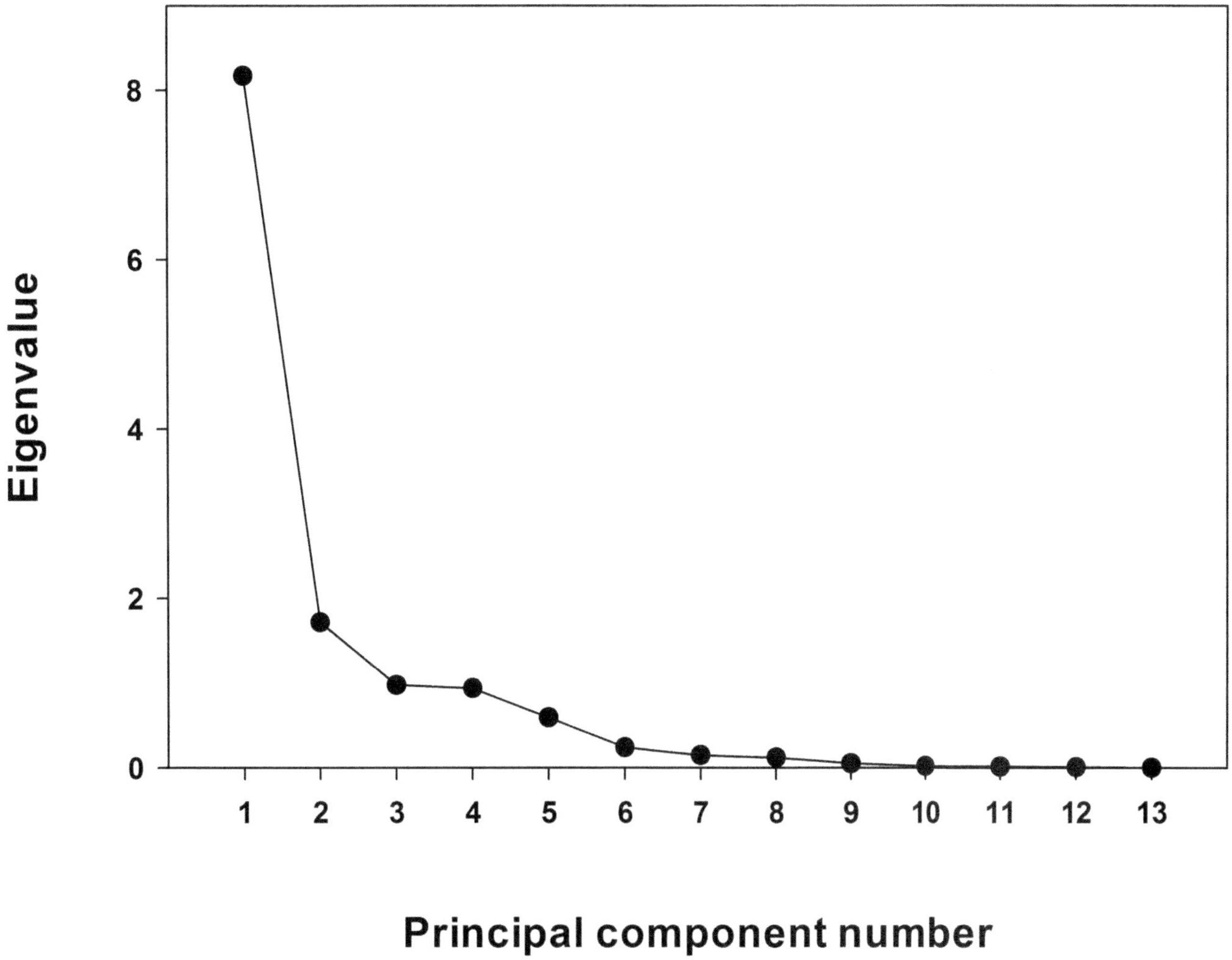

Figure 1. Scree plot of principal component analyses.

standard deviation values of the soil parameters by depth under three soil types are listed in Table 1. Most properties of mineral soils (Ko and CrB) were significantly different compared to those of the organic soil (Mc) and except for AWC and water content two mineral soils' characteristics were similar across soil profile up to 60 cm depth (Table 1). The water characteristics of Ko soil were significantly higher than those of CrB. On the other hand, organic soil (Mc) had much improved (lower BD and PR, and higher WSA, SOC, N concentrations and stock, AWC and water content) soil properties compared to those of the mineral counterparts in all four different depths (Ko and CrB). Impacts of all these parameters on overall soil quality in four soil depths calculated by three indexing methods are discussed in the following sections.

Under the framework of SQI-1, nine soil quality indicators were integrated numerically after scoring them primarily using the information from literature review. However, scoring data on some physicochemical indicators (e.g., C and N stocks) is scarce in the literature and thus, authors' opinion was used in those cases. Further, different studies used different scoring on the same indicator based on soil type, and management goals. For an example, while Amacher and Perry [28] had eight different range classes and scores for pH, Feiza et al. [41] had only four classes and scores available for pH. Thus, scoring on indicators in the current study (Table 2) had importance on experts' opinion although the knowledge was primarily based on available literature and authors' experience [28,42]. The high variability in the observations (Table 1) was reflected in the SQI-1 values. There was no statistical difference ($p>0.05$) across treatments (NCC versus CC, and CT versus NT) and depth under specific soil type (Table 6). Overall performance of indexing methods was determined from averaging SQI values obtained from calculations for each depth and presented in Fig. 2. The, SQI-1 was significantly higher in the following order: Mc>Ko>CrB (Fig. 2), indicating that soil quality was influenced more by soil type than management and sampling depth under the SQI-1 approach.

Overall percentage of various soil functional influences in SQI-2 is presented in Fig. 3 and depthwise percentage of influence in SQI-2 is presented in Fig. S1. Generally, mineral soils (Ko and CrB) did not have any depthwise significant changes for WSC but 0–10 cm layer of organic (Mc) soil had significantly higher influence of WSC in SQI-2 compared to lower depths. On the other hand, while both Ko and Mc had significantly higher influence of RDC in SQI-2 in the upper layers than the lower depths, the opposite trend was found for the NSC (Fig. S1). In the

Table 5. Pearson correlation coefficients (r) for all soil indicators.

Correlation (r) matrix	pH	EC	BD	WSA	GMD	MWD	PR	N	SOC	N-Stock	C-Stock	AWC
EC	0.547											
BD	−0.394	−0.88										
WSA	0.248	0.483	−0.338									
GMD	0.279	0.8	−0.762	0.673								
MWD	0.286	0.782	−0.725	0.734	0.975							
PR	−0.129	−0.655	0.684	−0.456	−0.71	−0.687						
N	0.405	0.922	−0.908	0.47	0.877	0.835	−0.737					
SOC	0.438	0.93	−0.914	0.473	0.881	0.842	−0.731	0.995				
N-Stock	0.271	0.632	−0.551	0.32	0.642	0.595	−0.396	0.709	0.707			
C-Stock	0.382	0.764	−0.702	0.367	0.742	0.695	−0.489	0.822	0.833	0.965		
AWC	0.091	0.447	−0.449	0.275	0.439	0.441	−0.525	0.504	0.489	0.106	0.181	
Water content (cm)	0.092	0.487	−0.485	0.282	0.501	0.496	−0.557	0.558	0.546	0.28	0.333	0.967
Significance level												
EC	<0.001											
BD	<0.001	<0.001										
WSA	0.018	<0.001	0.002									
GMD	0.009	<0.001	<0.001	<0.001								
MWD	0.007	<0.001	<0.001	<0.001	<0.001							
PR	0.141	<0.001	<0.001	<0.001	<0.001	<0.001						
N	<0.001	<0.001	<0.001	<0.001	<0.001	<0.001	<0.001					
SOC	<0.001	<0.001	<0.001	<0.001	<0.001	<0.001	<0.001	<0.001				
N-Stock	0.011	<0.001	<0.001	0.004	<0.001	<0.001	<0.001	<0.001	<0.001			
C-Stock	<0.001	<0.001	<0.001	0.001	<0.001	<0.001	<0.001	<0.001	<0.001	<0.001		
AWC	0.224	<0.001	<0.001	0.011	<0.001	<0.001	<0.001	<0.001	<0.001	0.195	0.07	
Water content (cm)	0.22	<0.001	<0.001	0.009	<0.001	<0.001	<0.001	<0.001	<0.001	0.01	0.003	<0.001

Abbreviations are same as Table 1.

Table 6. Treatment effects on soil quality index (SQI) under three soil types (Ko: Kokomo, CrB: Crosby silty loam, Mc: Muck) and four soil depths.

Treatments									
Ko		NCC	CC	NCC	CC	NCC	CC	NCC	CC
		0–10 cm		10–20 cm		20–40 cm		40–60 cm	
	SQI-1	0.36a	0.36a	0.40a	0.40a	0.45a	0.48a	0.48a	0.50a
	SQI-2	0.40a	0.37a	0.39a	0.36a	0.40a	0.36a	0.43a	0.38a
	SQI-3	0.21a	0.23a	0.20a	0.22a	0.19a	0.20a	0.21a	0.21a
CrB		CT	NT	CT	NT	CT	NT	CT	NT
		0–10 cm		10–20 cm		20–40 cm		40–60 cm	
	SQI-1	0.36a	0.38a	0.36a	0.17a	0.45a	0.29a	0.38a	0.26a
	SQI-2	0.42a	0.40b	0.40a	0.37a	0.41a	0.40a	0.41a	0.40a
	SQI-3	0.26a	0.24a	0.24a	0.22a	0.22a	0.22a	0.22a	0.21a
Mc		CT	NT	CT	NT	CT	NT	CT	NT
		0–10 cm		10–20 cm		20–40 cm		40–60 cm	
	SQI-1	0.88a	0.88a	0.90a	0.90a	0.90a	0.90a	0.90a	0.81a
	SQI-2	0.70a	0.72a	0.65a	0.61a	0.65a	0.66a	0.67a	0.65a
	SQI-3	0.85a	0.85a	0.82a	0.78a	0.78a	0.77a	0.82a	0.81a

Abbreviations: NCC: no cover crop, CC: cover crop, CT: conventional tillage, NT: no tillage.
Different letters under specific row and depth indicate significant differences at $p<0.05$ level, underline value is significant at $p<0.1$ level.

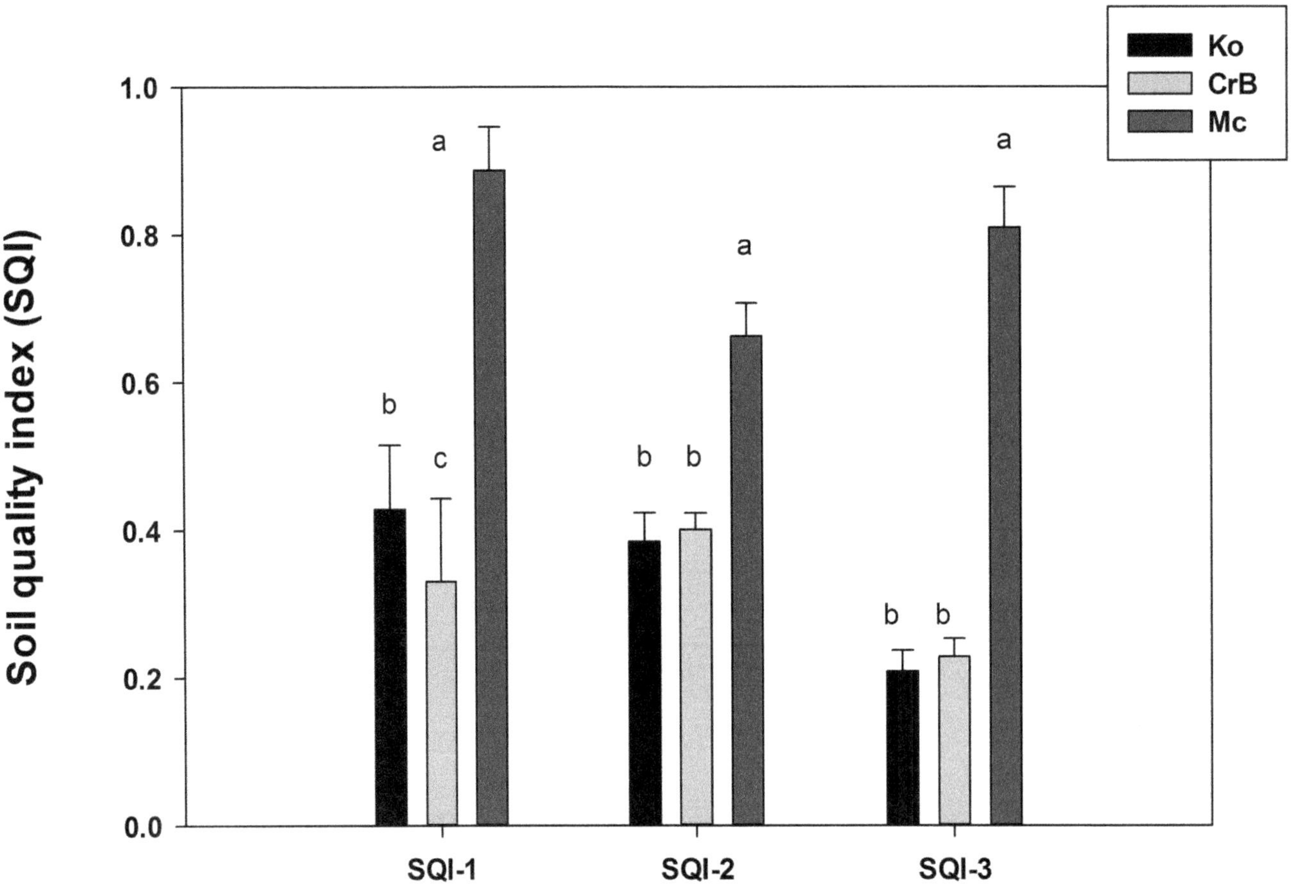

Figure 2. Overall soil quality index (SQI) values under different soil types. Different letters indicate significant differences at p<0.05 level for particular indexing method.

general composition of SQI-2, the main contributors were RDC, which ranged from 39–57% and NSC ranged from 35–59%. However, the WSC was relatively small and ranged merely from 0.1–12% (Fig. 3). The relative influence of root development functions in SQI-2 has been significantly higher in Ko than CrB, and Mc by 8 and 9%, respectively and the same for nutrient storage functions in SQI-2 has been significantly higher in CrB than Ko and Mc by 13 and 8%, respectively (Fig. 3). On the other hand, Mc soil had significantly higher water storage functional contribution to SQI-2 by 53 and 500%, respectively, compared to that of mineral (Ko and CrB) soils (Figs. 3). Thus, while the mineral soil quality was more sensitive to the functional attributes dedicated to root development and nutrient storage, muck soil quality was more influenced by soil attributes related to water storage under the scheme of SQI-2. Further, there was no significant effect on management practices on %RDC and %NSC to SQI-2 in different soil/management/crop combinations (Fig. 4). Note that CC/NCC management was used on field cultivated to soybean but CT/NT was used to soil under corn. However, contribution of WSC to SQI-2 for the case of CrB soil under NT management was significantly higher (178%) than that of CT practice (Fig. 4).

Correlation coefficients of SQIs to each other and depthwise SQIs and dry grain and straw are given in Table 7. Crop yield is obviously related to soil quality but soil quality in different layers may not be same and can influence crop yield accordingly. Data on SQI variation for different soil layers are scarce and thus no comparison to the data presented in the current study (Table 6) was possible. However, while numerical values of SQI were almost unchangeable for different depths at Mc site, SQI generally variable with depth in Ko and CrB sites (Table 6). One goal of this study was to validate SQI against crop yield as only a few studies have done so with some exceptions [8,14]. Challenge lies in establishing a correlation between crop yield versus SQI calculated either depthwise or in combination of soil layers as standard methodology is not yet available and was not given much attention to date. For example, one can have only one crop yield value for multiple soil layers and thus to date crop yield has been correlated only against SQI calculated from surface layer. Low (ranged from 0.11–0.32) and high (ranged from 0.83–0.96) correlation coefficients of SQI and crop yield across a number of soil/crop combinations were reported [8,31]. The reasons attributed to low correlation between yield and SQI were: (i) the assumptions that the indicators considered good for a soil quality may not always lead to desired outcomes and flexibility in scoring functions needed [8], (ii) indicators which may not directly relate to crop performance were given more weightage by the method employed, or (iii) some key soil indicators were not included in the MDS due to the design of the study [9,43]. Nevertheless, in order to evaluate the index performance in the current study, SQIs were correlated against end point variable (i.e., yield) at variable depths (Table 7). In the present study, depth of soil sampling had

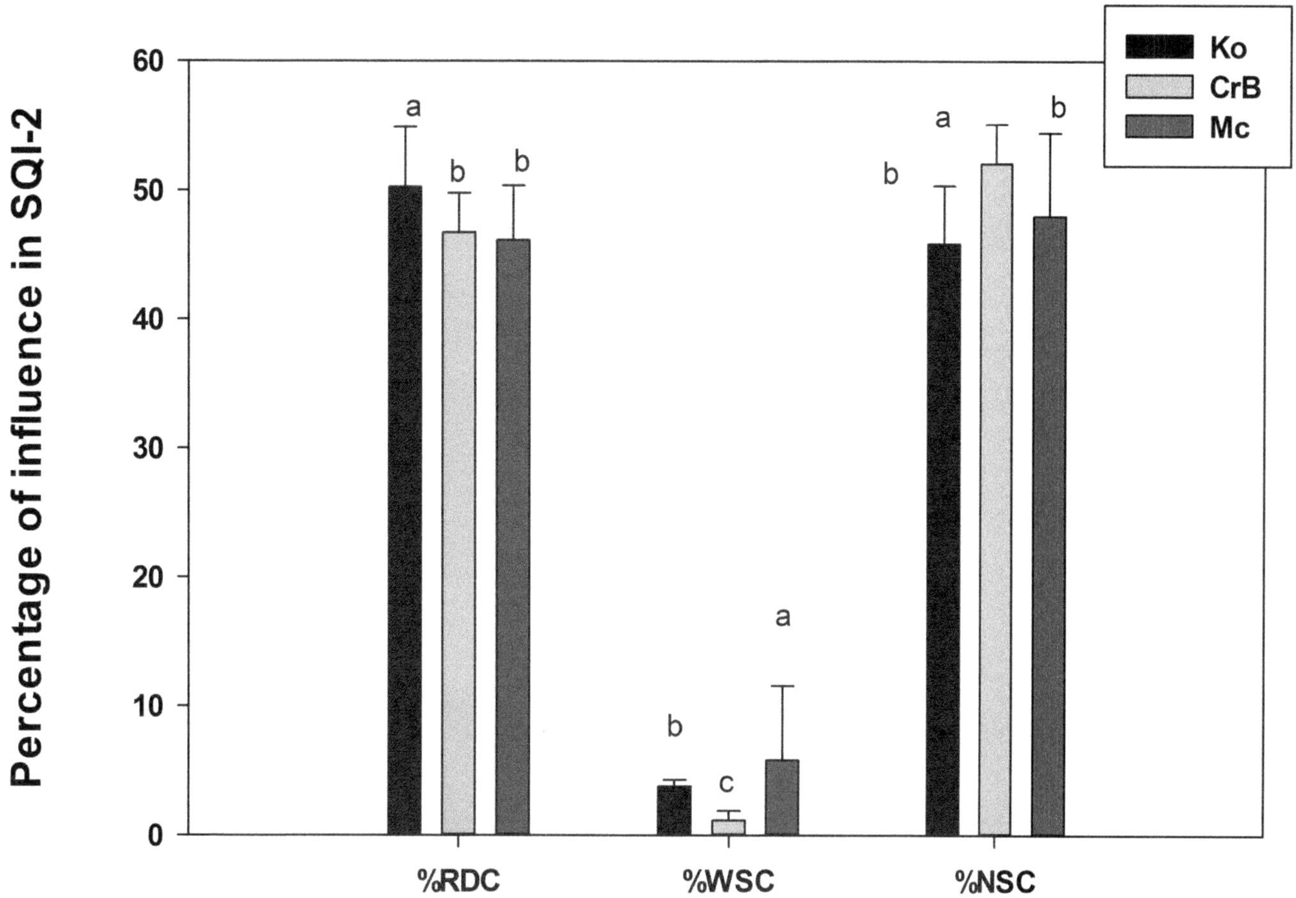

Figure 3. Percentage contribution of each soil function in SQI-2 under different soils. Different letters indicate significant differences at $p<0.05$ level for particular soil function. Abbreviations: RDC: root development capacity, WSC: water storage capacity, NSC: nutrient storage capacity.

no significant effect in SQI values (Table 6) and thus the individual SQI values were integrated by combining 0–20 cm and 0–60 cm profiles by averaging 0–10, and 10–20 cm for obtaining values for 0–20 cm depth and 0–10, 10–20, 20–40 and 40–60 cm for obtaining values for 0–60 cm depths and correlated the combined values with crop yield (Table 7). In the present study, grain and biomass yields were all significantly ($p<0.05$, $r=0.65$–0.79) correlated under any SQI methods employed indicating that either of these methods was successful in predicting crop yield (Table 7). However, correlation coefficients in Table 7 are derived from the complete datasets including both corn and soybean yield but the correlation was considerably low ($r<0.20$ for soybean grain and $r<0.60$ for corn grain; data not shown), when dataset was further divided into corn and soybean. As discussed above, low correlation of SQI and crop yield was observed before and one of the listed reasons was some key soil indicators were not included in the dataset as per the experimental design and objective [8]. Crop yield is invariably related to soil fertility status and thus SQI may not always lead to high correlation with yield if some of the related characterizations are not observed as identified before [9,43] which is supported by the data of the current study (Table 7). Thus, the low correlation coefficient values may be because soil fertility (ion exchange capacities, macro and micro nutrients) and microbial (microbial biomass, microbial C, and N, and soil respiration) aspects were not monitored in the current studied sites, especially in the case of soybean grain yield. Nevertheless, relatively higher correlation coefficient was observed in the 0–10 cm soil layer than the deeper soil profile (Table 7) probably due to higher plant root density in the top layer of the soil [44].

Additionally, SQI computed from different methods were also highly correlated with each other (Table 7) suggesting that (i) a relatively easy and user-friendly SQI (SQI-1) can be computed to evaluate and compare soil quality, which is similarly useful to other approaches, (ii) giving certain appropriate weightage on scores (SQI-2) could be similarly useful in predicting particular soil quality and (ii) a minimum group of carefully selected soil indicators (SQI-3) may be used to evaluate soil quality across various soil types and management practices. As all three SQI methods were significantly correlated to each other, it is difficult to conclude which one is the best approach, however, it highly depends on factors such as design of the study, choice of soil parameters included in the model to compute SQI and end point variable (environmental aspects versus crop yield). The advantage of using SQI-1 is that the soil quality could be assessed after measuring any number (low to high) of soil parameters and this procedure is relatively easier compared to other methods as the scoring requires literature review and expert opinions only. The disadvantage of SQI-1 is that it is subjective and relies mainly on researcher's point of view. On the other hand, advantage of SQI-2 is that it includes weightage based on the design of the study, system or the dataset to offset the subjectivity of the approach

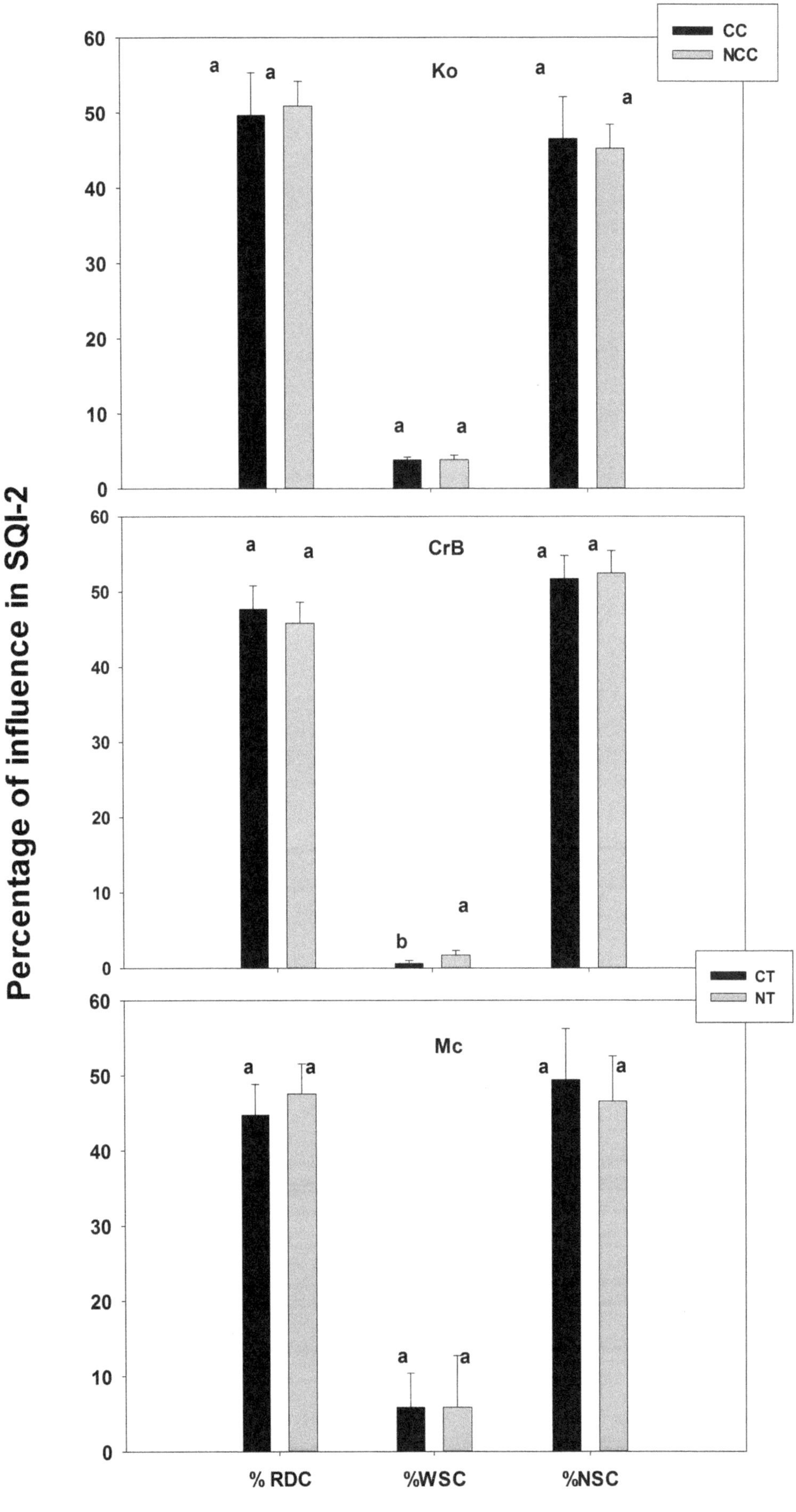
Percentage of influence in SQI-2
CC
NCC
Ko
CrB
CT
NT
Mc
% RDC
%WSC
%NSC

Figure 4. Percentage contribution of each soil function in SQI-2 under various management practices in different soil. Different letters indicate significant differences at p<0.05 level for particular soil function. Abbreviations: RDC: root development capacity, WSC: water storage capacity, NSC: nutrient storage capacity; NCC: no cover crop, CC: cover crop, CT: conventional tillage, NT: no tillage.

present in SQI-2. However, the disadvantage of SQI-2 is that it requires multiple numbers of soil parameters under different soil functional systems which may be expensive and time consuming in practical case. The SQI-3 is advantageous in the aspect of its ability to predict soil quality based on a reduced dataset with low number of soil parameters. Additionally, it is mostly objective approach as the statistical procedure would select a low number of soil parameters needed to calculate SQI based on the variances present in the whole dataset. So, in a long-term aspect within a particular soil/crop system, SQI-3 can be used effectively once it evaluated the most influential soil parameters required to assess soil quality of a particular soil/crop/management scenario. However, current data suggest that relative higher SQI values were obtained for Ko and Mc by SQI-1 than SQI-2 or SQI-3 and for CrB by SQI-2 than SQI-1 or SQI-3 (Fig. 2). Although SQI-1 and SQI-2 had higher values than SQI-3 in different scenarios, however, based on the current experimental design, and selected soil parameters SQI-3 successfully predicted crop yield relatively higher [Table 7 and also when grouped by crops (data not shown)] than SQI-1 and 2. Depth had little influence in any SQIs in the present study and additional chemical and biological parameters are also needed to strengthen the validity of SQI. Thus, SQI-3 appears to be the best method among the three under long-term scenario, especially due to its objective approach, relative higher correlations with crop yield and lower number of indicator selection which is more cost and time-effective over time than other approaches.

Conclusions

The data presented support the following conclusions: The SQI calculated by three different methods indicated that studied muck soil has significantly higher soil quality than that of mineral soils under on-farm conditions. The SQI was affected more by management and soil type than by depths in the studied on-farm sites. The SQI computed using three established methods were all significantly correlated to each other indicating that relatively easy and user-friendly SQI (SQI-1) can be similarly useful to evaluate soil quality, appropriate weightage on scores can predict soil's quality (SQI-2) with high performance but requires a number of soil parameters under different soil functional components and carefully chosen MDS with small numbers of soil variables (SQI-3) may adequately evaluate its quality. All three SQIs were highly and significantly correlated with crop yield. In general, however, analyses by crop type (corn and soybean) revealed that correlation was low (r<0.20 or 0.60 for soybean or corn grain yield, respectively; data not shown) suggesting choice of crop-specific and key soil parameters used in computing SQI. Under the current experimental on-farm conditions SQI-1 (for Ko and Mc) and SQI-2 (for CrB) values were higher than SQI-3, however, SQI-3 can be regarded as the best and easiest model given its relatively higher success to predict crop yield and objectivity approach with lower number of indicator selection ability which should be regarded as a relatively less expensive procedure over time compared to SQI-1 and 2. In addition, in order to effectively predict particular crop yield one must include soil fertility and microbial parameters in the model of SQI.

Table 7. Pearson correlation coefficients (r) of soil quality index (SQI) versus crop yield and correlations between different SQI values which were averaged up to a certain depth; all numbers are significant at $p<0.05$ level.

SQI-1	**Dry Grain**	**Dry Straw**
0–60 cm	0.65	0.67
0–20 cm	0.71	0.71
0–10 cm	0.73	0.71
SQI-2	**Dry Grain**	**Dry Straw**
0–60 cm	0.75	0.73
0–20 cm	0.75	0.72
0–10 cm	0.79	0.75
SQI-3	**Dry Grain**	**Dry Straw**
0–60 cm	0.76	0.74
0–20 cm	0.76	0.74
0–10 cm	0.78	0.76
	SQI-1	**SQI-2**
SQI-2	0.92	
SQI-3	0.93	0.97

Supporting Information

Figure S1 Percentage contribution of each soil function in SQI-2 under different soils in four soil layers; different letters indicate significant differences at p<0.05 level for particular soil function. Abbreviations: RDC: root development capacity, WSC: water storage capacity, NSC: nutrient storage capacity.

Acknowledgments

Authors would like to thank Dr. Shiguo Jiang for his help with PCA and Mr. Basant Rimal for his assistance in the laboratory. Authors would also like to thank three anonymous reviewers and an academic editor for their constructive suggestions to improve the manuscript.

Author Contributions

Conceived and designed the experiments: AM RL. Performed the experiments: AM. Analyzed the data: AM. Contributed reagents/materials/analysis tools: AM. Contributed to the writing of the manuscript: AM RL.

References

1. Doran J, Coleman D, Bezdicek D, Stewart B (1994) A framework for evaluating physical and chemical indicators of soil quality.
2. Doran JW, Parkin TB (1994) Defining and assessing soil quality. SSSA special publication 35: 3–3.
3. Karlen DL, Mausbach MJ, Doran JW, Cline RG, Harris RF, et al. (1997) Soil Quality: A Concept, Definition, and Framework for Evaluation (A Guest Editorial). Soil Sci Soc Am J 61: 4–10.
4. Karlen DL, Stott DE (1994) A framework for evaluating physical and chemical indicators of soil quality. In: Doran JW, Coleman DC, Bezdicek DF, Stewart BA, editors. Defining soil quality for a sustainable environment. Madison, WI: Soil Science Society of America. pp. 53–72.
5. Stocking MA (2003) Tropical Soils and Food Security: The Next 50 Years. Science 302: 1356–1359.
6. Diack M, Stott D (2001) Development of a soil quality index for the Chalmers Silty Clay Loam from the Midwest USA. Purdue University: USDA-ARS National Soil Erosion Research Laboratory. pp. 550–555.
7. Larson WE, Pierce FJ (1991) Conservation and enhancement of soil quality. Evaluation of Sustainable Land Management in the Developing World. International Board for Soil Research and Management, Bangkok, Thailand.
8. Andrews S, Karlen D, Mitchell J (2002) A comparison of soil quality indexing methods for vegetable production systems in Northern California. Agriculture, Ecosystems & Environment 90: 25–45.
9. Bhardwaj AK, Jasrotia P, Hamilton SK, Robertson GP (2011) Ecological management of intensively cropped agro-ecosystems improves soil quality with sustained productivity. Agriculture Ecosystems & Environment 140: 419–429.
10. Andrews SS, Carroll CR (2001) Designing a soil quality assessment tool for sustainable agroecosystem management. Ecological Applications 11: 1573–1585.
11. Andrews SS, Karlen DL, Cambardella CA (2004) The soil management assessment framework: A quantitative soil quality evaluation method. Soil Science Society of America Journal 68: 1945–1962.
12. Andrews SS, Mitchell JP, Mancinelli R, Karlen DL, Hartz TK, et al. (2002) On-Farm Assessment of Soil Quality in California's Central Valley. Agron J 94: 12–23.
13. Fernandes JC, Gamero CA, Rodrigues JGL, Mirás-Avalos JM (2011) Determination of the quality index of a Paleudult under sunflower culture and different management systems. Soil and Tillage Research 112: 167–174.
14. Mandal UK, Ramachandran K, Sharma K, Satyam B, Venkanna K, et al. (2011) Assessing Soil Quality in a Semiarid Tropical Watershed Using a Geographic Information System. Soil Science Society of America Journal 75: 1144–1160.
15. Zobeck TM, Halvorson AD, Wienhold B, Acosta-Martinez V, Karlen DL (2008) Comparison of two soil quality indexes to evaluate cropping systems in northern Colorado. Journal of Soil and Water Conservation 63: 329–338.
16. Zornoza R, Mataix-Solera J, Guerrero C, Arcenegui V, Mataix-Beneyto J, et al. (2008) Validating the effectiveness and sensitivity of two soil quality indices based on natural forest soils under Mediterranean conditions. Soil Biology and Biochemistry 40: 2079–2087.
17. Jokela W, Posner J, Hedtcke J, Balser T, Read H (2011) Midwest Cropping System Effects on Soil Properties and on a Soil Quality Index. Agronomy Journal 103: 1552–1562.
18. Navas M, Benito M, Rodriguez I, Masaguer A (2011) Effect of five forage legume covers on soil quality at the Eastern plains of Venezuela. Applied Soil Ecology 49: 242–249.
19. Brady NC, Weil RR (1984) The nature and properties of soils New York: Macmillan.
20. Mukherjee A, Lal R (2014) Long and short-term tillage effects on quality of organic and mineral soils under on-farm conditions in Ohio, Submitted. Soil and Tillage Research.
21. Mukherjee A, Lal R (2014) Short-term effects of cover cropping on quality of a Typic Argiaquolls in Central Ohio, Submitted.
22. Grossman RB, Reinsch TG (2002) Bulk density and linear extensibility. In: Dane JH, Topp GC, editors. Methods of soil analysis, Part 4: Physical methods. Madison, WI: SSSA. pp. 201–254.
23. Herrick JE, Jones TL (2002) A dynamic cone penetrometer for measuring soil penetration resistance. Soil Science Society of America Journal 66: 1320–1324.
24. Yoder RE (1936) A direct method of aggregate analysis of soils and a study of the physical nature of erosion losses. J Am Soc Agron 28: 337–351.
25. Youker RE, McGuinness JL (1957) A short method of obtaining mean weight-diameter values of aggregate analyses of soils. Soil Science 83: 291–294.
26. Dane JH, Hopmans JH (2002) Water retention and storage. In: Dane JH, Topp GC, editors. Methods of soil analysis Part 4, SSSA Book Ser 5. Madison, WI.: SSSA. pp. 671–717.
27. Chhabra A, Palria S, Dadhwal VK (2003) Soil organic carbon pool in Indian forests. Forest Ecology and Management 173: 187–199.
28. Amacher MC, Perry CH (2007) Soil vital signs: A new Soil Quality Index (SQI) for assessing forest soil health.
29. Liebig MA, Varvel G, Doran J (2001) A simple performance-based index for assessing multiple agroecosystem functions. Agronomy Journal 93: 313–318.
30. Li Q, Xu MX, Liu GB, Zhao YG, Tuo DF (2013) Cumulative effects of a 17-year chemical fertilization on the soil quality of cropping system in the Loess Hilly Region, China. Journal of Plant Nutrition and Soil Science 176: 249–259.
31. Masto RE, Chhonkar PK, Singh D, Patra AK (2008) Alternative soil quality indices for evaluating the effect of intensive cropping, fertilisation and manuring for 31 years in the semi-arid soils of India. Environmental monitoring and assessment 136: 419–435.
32. Doran JW, Parkin TB (1996) Quantitative indicators of soil quality: a minimum data set. SSSA special publication 49: 25–38.
33. Dunteman GH (1989) Principal components analysis: Sage.
34. Jolliffe I (2005) Principal component analysis: Wiley Online Library.
35. SPSS (2014) IBM SPSS, Version 21.0, Chicago, USA.
36. Kaiser HF (1960) The application of electronic computers to factor analysis. Educational and Psychological Measurement 20: 141–151.
37. Wander MM, Bollero GA (1999) Soil quality assessment of tillage impacts in Illinois. Soil Science Society of America Journal 63: 961–971.
38. SAS (2012) SAS Institute Inc.; SAS version 9.2.
39. Microsoft (2007) Microsoft Excel. Redmond, Washington.
40. Sigmaplot (2012) Scientific Software Solutions Internationals; Sigmaplot version 12.
41. Feiza V, Feiziene D, Kadziene G, Lazauskas S, Deveikyte I, et al. (2011) Soil state in the 11th year of three tillage systems application on a cambisol. Journal of Food, Agriculture & Environment 9: 1088–1095.
42. Lal R (1994) Methods and guidelines for assessing sustainable use of soil and water resources in the tropics; Washington D.C.: USDA/SMSS Technical Monograph 21.
43. Zornoza R, Mataix-Solera J, Guerrero C, Arcenegui V, García-Orenes F, et al. (2007) Evaluation of soil quality using multiple lineal regression based on physical, chemical and biochemical properties. Science of The Total Environment 378: 233–237.
44. Maurya PR, Lal R (1980) Effects of No-Tillage and Ploughing on Roots of Maize and Leguminous Crops. Experimental Agriculture 16: 185–193.
45. Brzezinska M, Sokolowska Z, Alekseeva T, Alekseev A, Hajnos M, et al. (2011) Some characteristics of organic soils irigated with municipal wastewater. Land Degradation & Development 22: 586–595.
46. Sokoowska Z (2011) Effect of phosphates on dissolved organic matter release from peat-muck soils.
47. Walczak R, Rovdan E, Witkowska-Walczak B (2002) Water retention characteristics of peat and sand mixtures. International Agrophysics 16: 161–166.
48. Carter MR (2006) Quality critical limits and standardization. In: Lal R, editor. Encyclopedia of Soil Science. New York: Marcel Dekker. pp. 1412–1415.

3

Early Response of Soil Properties and Function to Riparian Rainforest Restoration

Rose Gageler[1], Mark Bonner[1], Gunnar Kirchhof[1], Mark Amos[2], Nicole Robinson[1], Susanne Schmidt[1]*, Luke P. Shoo[3]

1 School of Agriculture and Food Sciences, The University of Queensland, St Lucia, Australia, **2** Lake Baroon Catchment Care Group, Maleny, Australia, **3** School of Biological Sciences, The University of Queensland, St Lucia, Australia

Abstract

Reforestation of riparian zones is increasingly practiced in many regions for purposes of biodiversity conservation, bank stabilisation, and improvement in water quality. This is in spite of the actual benefits of reforestation for recovering underlying soil properties and function remaining poorly understood. Here we compare remnant riparian rainforest, pasture and reforestation plantings aged 2–20 years in an Australian subtropical catchment on ferrosols to determine the extent to which reforestation restores key soil properties. Of the nine soil attributes measured (total nitrogen, nitrate and ammonium concentrations, net nitrification and ammonification rates, organic carbon, bulk density, fine root biomass and water infiltration rates), only infiltration rates were significantly lower in pasture than remnant riparian rainforest. Within reforestation plantings, bulk density decreased up to 1.4-fold and infiltration rates increased up to 60-fold with time post-reforestation. Our results suggest that the main outcome of belowground processes of early reforestation is the recovery of the soils' physical structure, with potential beneficial ecosystem services including reduced runoff, erosion and associated sediment and nutrient loads in waterways. We also demonstrate differential impacts of two commonly planted tree species on a subset of soil properties suggesting that preferential planting of select species could accelerate progress on specific restoration objectives.

Editor: Kurt O. Reinhart, USDA-ARS, United States of America

Funding: LPS was supported by funding from the Australian Government's National Environmental Research Program (Tropical Ecosystems and Environmental Decisions Hubs). The funders had no role in study design, data collection and analysis, decision to publish, or preparation of the manuscript.

Competing Interests: The authors have declared that no competing interests exist.

* Email: Susanne.Schmidt@uq.edu.au

Introduction

Globally, riparian zones are widely recognized for their critical role in water regulation [1–5], and conservation of biodiversity [6–9]. Clearing of vegetation from stream banks can cause increased soil erosion and decreased water quality through the loss of filtration services, in addition to the loss of critical species habitat [10]. In direct response, degraded stream-banks are increasingly being restored to reinstate ecological functions [11]. Riparian zones are also regarded as foci of restoration opportunity as they are often preferentially renounced from agricultural production over agronomically more valuable areas [9].

Soil physical, chemical and functional properties change with succession from pasture to secondary forest including reforestation plantings in the tropics, subtropics and temperate regions [12–16]. Soil properties such as bulk density, soil water content, fine root biomass, porosity, stocks of nitrogen, carbon and phosphorus, nitrogen cycling, availability of macro- and micronutrients, as well as toxins (metals, salts), have been studied in context of land use change [17–19]. However, most studies focus on mid- to long-term effects of reforestation with less consideration of the short-term effects (<10–20 years) of reforestation plantings, and associated reinstatement of soil functions that might motivate landholders to become actively engaged in restoration. Another knowledge gap is the relative contribution of tree species to outcomes of riparian restoration plantings, and such information could inform decisions on species selection and reforestation goals.

In Australia, riparian reforestation is a major focus of habitat restoration in the tropics and subtropics; over 70% of rainforest restoration projects have targeted banks of rivers and streams [20]. However, much evaluation of the success of reforestation plantings has focused on aboveground attributes [21–23], with less regard to belowground processes including soil function (but see Paul [14] for the recovery of soils via different rainforest restoration pathways). Understanding belowground processes is important to quantify the capacity of reforestation to recover ecosystem function.

Soil properties and functions would intuitively be expected to change over time following reforestation towards a state convergent on intact rainforest ecosystems. Studying riparian soil under remnant rainforest, pasture, and within 20 years post reforestation, we addressed three questions (i) how do soil attributes differ between pasture and remnant riparian rainforest, (ii) how do soils change with time following reforestation, and (iii) do tree species affect soils differentially following reforestation?

The study was motivated by the need for restoration of forest cover in water catchments due to the improvements in water quality that can be achieved [24]. The region targeted for our study is located in the Australian subtropics and has experienced large-scale clearing of rainforest since the mid 19th century. The region is an important water catchment that supplies drinking water to more than 100000 people in the region [25], and numerous land owners and stakeholders are committed to ongoing restoration of riparian zones. Our research aimed to generate knowledge of soil properties to direct future reforestation activities.

Materials and Methods

2.1. Study site

This study focused on riparian areas in the Maleny region in southeastern Queensland, Australia (26°45′S, 152°48′E). The climate is classified as subtropical with an annual rainfall of 1900 mm and annual daily mean temperature maximum and minimum of 23.2°C and 14.2°C, respectively. The soils in the region are classified as permeable red clay-loam ferrosols [26,27] or Nitisol in the US soil taxonomy. Prior to European settlement in the mid-1800s and subsequent land clearing for timber production and agriculture (especially dairy production), the region was dominated by cool-subtropical rainforest classified as complex notophyll vine forest [28,29].

2.2. Sampling design

Field measurements and soil sampling were carried out between May and June 2013. The sampling design was based around riparian reforestation projects established in the region by the Lake Baroon Catchment Care Group (LBCCG). Reforestation sites were grouped into five spatial clusters across the Maleny Plateau (range of elevation 196 to 441 m a.s.l.) to capture broad variation in landform and land use history. A subset of one to three reforestation sites ranging in age between 2 and 20 years were then selected from each cluster for a total of 10 sites (Table 1). Sites were also established in remnant riparian rainforest where available (2 of 5 clusters) and samples were collected from pasture (adjacent to plantings or remnant forest) at all sites in each cluster.

At each reforestation or remnant site, five subplots were established. Each subplot was deliberately located under a different tree species. The five tree species targeted here were derived from a pool of species commonly used in local rainforest plantings, which in turn were selected to include early to late successional species, nitrogen fixing and/or waterlogging tolerant species. The tree species chosen to investigate species-specific effects on soil properties included *Acacia melanoxylon* R.Br. (Fabaceae, N_2 fixer, early succession species), *Ficus coronata* Spin (Moraceae, nitrophile, waterlogging tolerant, early succession species), *Flindersia schottiana* F. Muell. (Rutaceae, mid-late succession), *Homalanthus nutans* (G.Forst.) Guill. (Euphorbiaceae, early succession species) and *Podocarpus elatus* R.Br. ex Endl (Podocarpaceae, mid-late succession, gymnosperm). In the event that one of the chosen tree species was absent from a site, a closely related species was substituted or, if not available, subplots were located under a random tree within the site. Subplots were positioned under the canopy of each selected tree (mean distance from stem 89.5 cm, SD 45.2 cm).

At each reforestation or remnant forest site, we performed the following procedures: (1) collected a single 8 cm diameter core to a depth of 30 cm (hereafter "textural core") and divided it into 0–15 cm and 15–30 cm fractions for textural analysis; (2) performed an infiltration test; and, (3) collected two samples of the surface soil using a 12.5 cm diameter, 8.5 cm tall, 1043 cm^3 volume bulk density ring (hereafter "bulk density ring") from each of five subplots for analysis of physical and chemical properties of the soil. In addition, a single texture core and two to four bulk density ring subplots were collected from adjacent pasture and infiltration measurements performed. For the purposes of analysis, samples and infiltration measures from pasture were calculated as means for each cluster.

Overall, we sampled 10 reforestation sites, 2 remnant riparian rainforest sites (Table 1) and 5 pasture clusters comprising 24 cores for soil texture analysis, 150 cores for soil chemical and physical analyses and 24 infiltration tests. All sites were private land holdings and owners approval was obtained by Mark Amos (Lake Baroon Catchment Care Group, Maleny, info@lbccg.org.au). Upon returning from the field, all soil samples were stored at 4 °C in sealed plastic bags prior to respective analyses.

Table 1. GPS coordinates and ages of reforestation and remnant sites.

Site type	Age (years)	Site location
Reforestation	3	26°46′02 S, 152°49′38 E
Reforestation	3	26°46′17 S, 152°49′51 E
Reforestation	4	26°45′06 S, 152°50′46 E
Reforestation	6	26°42′01 S, 152°53′58 E
Reforestation	7	26°46′10 S, 152°49′57 E
Reforestation	8	26°44′05 S, 152°50′54 E
Reforestation	11	26°44′46 S, 152°51′13 E
Reforestation	11	26°46′12 S, 152°49′39 E
Reforestation	12	26°46′08 S, 152°49′43 E
Reforestation	20	26°42′01 S, 152°53′58 E
Remnant	-	26°45′07 S, 152°51′02 E
Remnant	-	26°44′50 S, 152°51′11 E

All sites were located on private land in the Maleny region. Pasture sites were located immediately adjacent to each forested site. Ages are based on the year that they were planted relative to 2013.

2.3. Texture analysis

Changes in soil texture mostly occur due to weathering over very long time periods and as demonstrated by Maloney [13]. Unless there is erosion or deposition of soil material, texture is unlikely to vary over time since reforestation. For this reason, the purpose of conducting a soil texture analysis at each site was not to determine changes due to reforestation, but to characterize and evaluate background variation in soil types among sites that might warrant differential treatments of sites in the analysis. Soil samples were not dispersed prior to analysis and as such, the calculated percentages represent the relative size of secondary particles (aggregates) present rather than primary particle size *sensu stricto*. This was considered a more relevant measure of soil function than particle size determined using standard methods, as it gives a relative measure of soil aggregation and structure. Furthermore, the ferrosols in this study are recognized as being subplastic, with clay-sized particles being primarily composed of sesquioxides that do not readily break down into primary particles [30]. These analyses of soil aggregation were conducted on air-dried soil from 0–15 and 15–30 cm depth cores according to the hydrometer method for soil texture analysis [31,32]. For each sample, 40 g of air-dried soil was ground and mixed with 100 ml 50 g l^{-1} Sodium hexametaphosphate solution and filled to 500 ml with deionized water. Samples were shaken overnight and then emptied into a 1 l cylinder for analysis. Cylinders were filled to 1 l with deionized water, mixed with a plunger and left to settle. Temperature and specific gravity readings were taken 40 seconds and 7 hours after plunging [32]. Using temperature and specific gravity readings, the fraction of sand, silt and clay sized particles in each soil sample was then calculated as a percentage.

2.4. Soil physical properties: bulk density and infiltration rates

Soil bulk density was determined from the second bulk density ring in each subplot. Fresh soil was oven dried to a constant weight at 105 °C. Soil bulk density is expressed as grams of dry soil per volume. To quantify fine root biomass, soil from one bulk density ring per subplot was passed through a sieve (8 mm mesh) and roots were manually removed. Roots were then oven dried at 65 °C to a constant weight, and results expressed as grams of fine roots per kilogram of dry soil.

Infiltration rates were measured using a ponded disk permeameter [33]. Soils were assumed to be at field capacity as the region had received in excess of 30 mm in the week preceding the first set of measurements and over 100 mm in the week prior to the second sampling visit. Infiltration tests were performed at a random, relatively flat area representative of the site. Tests in the adjacent pasture were performed at a random point beyond the farthest extent of the neighbouring forest canopy. Upon setting up of the permeameter, grass and other ground covers were clipped to just above the soil surface and the soil pre-moistened with rainwater to reduce the time needed to reach steady state infiltration. The permeameter was filled with rainwater collected from water tanks on local landholder properties to ensure that water quality was similar to that occurring under natural condition. Once the disk permeameter was in place, measurements of water level changes were recorded until the time interval between measurements was stable. As soils were close to field capacity at the start of measurements, steady state was reached quickly. Infiltration rate was calculated as the average volume change per second and then converted into millimeters per hour.

2.5. Soil chemical properties: nitrogen and organic carbon

Total concentrations of soil nitrogen (TN) and organic carbon (SOC) were determined using a TruSpec CHN analyzer (LECO Australia Pty. Ltd., Castle Hill, NSW 2154, Australia), using dried, ground and sieved soil (2 mm sieve). Nitrate (NO_3^-) and ammonium (NH_4^+) were extracted with potassium chloride (KCl) and quantified via colourimetric assays. To remove NO_3^- and NH_4^+ from soil exchange sites, 15 g of fresh sieved soil was mixed with 1 M KCl in a 15 ml falcon tube and shaken on an orbital shaker for 1 hour. Samples were centrifuged at 4000 rpm for 4 minutes and 1 ml of the supernatant was frozen until analysis via colourimetric assays according to standard procedures [34,35]. Net nitrification and ammonification rates were calculated from the difference in concentration between fresh soil samples and those incubated for 20 days at 28°C. Briefly, 15 g of soil were weighed into 50 ml falcon tubes and incubated for 20 days. Soils were kept at field moisture by watering daily with deionized water. Seeds that germinated during the incubation period were removed. Concentrations of NO_3^- and NH_4^+ in the incubated soils following KCl extraction were quantified using colourimetric assays as described above. Net nitrification and ammonification rates are expressed as mg nitrate or ammonium generated per kg dry soil per day.

2.6. Statistical analysis

Statistical analyses were performed in R version 3.0.1 [36]. To avoid non-independence among subsamples, we derived the aggregated mean of soil attributes for each site (reforestation or remnant rainforest) or cluster (pasture) and used these values as inputs in all analyses, excepting those determining species-specific effects which used values from a single subplot within each site. Pairwise correlation tests (Spearman's) were performed to examine associations between tested soil attributes. Soil attributes under both pasture and remnant riparian rainforest were directly compared using a one-way analysis of variance (ANOVA). Developmental trajectories of soil attributes within reforested sites (site level and separately for specific tree species within sites) were examined using linear regressions between either age or log-transformed age and individual soil attributes. The age term was log-transformed to account for potential non-linear developmental trajectory of reforestation plantings; often showing a rapid change immediately post planting followed by diminishing returns as the system reaches natural equilibrium. For the purposes of quantifying response trajectories with age, we assume that pasture represents 0 years post reforestation.

Results

Of the 56 subplots within sites, 47 (84%) were associated with one of the five target trees. *Homalanthus nutans* was the species most commonly absent, being present in 50% of reforested sites. Soil aggregation showed only minor variation among sites and land uses so was not considered further in subsequent analyses (sand sized aggregates: mean = 60%, range = 46–76%, clay sized aggregates: mean = 30%, range = 14–46%).

3.1. Correlations between variables

Infiltration rates were inversely correlated with bulk density ($P = 0.005$) but were not correlated with soil particle size (percentage sand, silt and clay) at either 0–15 cm or 15–30 cm depth ($P > 0.5$, not shown). Bulk density was correlated with both SOC and TN ($P = 0.02$ and 0.01 respectively) (Table 2). Pairwise

Table 2. Pairwise correlations (Spearman's) between soil attributes across pasture, remnant rainforest and reforestation sites (n = 5, 2, 10 respectively).

Variable	Variable	Spearman's correlation coefficient	P value
Bulk density	SOC	−0.554	0.023
Bulk density	TN	−0.615	0.010
Infiltration	Bulk density	−0.654	0.005
TN	SOC	0.784	<0.001
Ammonification rate	Nitrification rate	−0.664	0.005
Ammonification rate	Fine root biomass	0.662	0.005
Nitrate	Ammonification rate	−0.529	0.031
Nitrate	TN	0.495	0.045
Nitrate	SOC	0.630	0.008

Only significant pairwise correlations (P<0.05) are shown.

correlations revealed strong associations between soil nitrogen-related variables.

3.2. Differences between pasture and remnant rainforest

There was no apparent difference between pasture and remnant riparian rainforest in the tested soil attributes except infiltration rates (Table 3). Infiltration rates were significantly greater in the remnant rainforest (1421±995 mm h^{-1}) than in pasture sites (220±151 mm h^{-1}) ($F = 9.526$, $DF = 5$, $P = 0.027$, Table 3).

3.3. Recovery of soil properties under reforestation plantings

Only bulk density and infiltration rates showed a relationship with age within 20 years post reforestation (Table 4). In each case, models with a log-transformed age term explained more of the variation than models with an un-transformed age term. Bulk density of the soils ranged from 0.312 to 0.991 g cm^{-3}, and decreased significantly with time post reforestation ($F = 6.312$, $DF = 13$, $P = 0.026$, Figure 1). Conversely, infiltration rates significantly increased ($F = 10.56$, $DF = 13$, $P = 0.006$) within 20 years post reforestation (Figure 1). Infiltration rates in reforestation sites ranged from 149 mm h^{-1} at a 7-year-old reforestation site to over 1800 mm h^{-1} at an 11-year-old reforestation site. Infiltration rates of pasture sites varied from 3 mm h^{-1} to over 650 mm h^{-1}, although this variation was slightly less when data was aggregated by cluster.

3.4. Species specific effects on the recovery of soils under reforestation plantings

Species-specific effects were apparent for two of five tree species when tested soil attributes were modeled against log-transformed age (Figure 2). Ammonium concentrations in soil associated with *Flindersia schottiana* decreased with time post reforestation ($F = 4.97$, $DF = 8$, $P = 0.056$). Fine root biomass in soil associated with *Ficus coronata* decreased through time within 20 years post reforestation ($F = 4.921$, $DF = 8$, $P = 0.057$).

Discussion

Replanting riparian areas can be driven by a variety of objectives including increasing landscape connectivity for wildlife movements and biogeochemical functions such as trapping nutrients, sediment, pesticides, bank stabilisation, improved water quality and flood mitigation [10]. Here, we studied the trajectory of riparian reforestation to identify key changes in soil attributes through the transition from pasture to riparian tree plantings of increasing age and remnant rainforest. We showed that the major

Table 3. ANOVA results from a comparison of soil attributes between remnant riparian rainforest (n = 2) and pasture (n = 5).

Variable	ANOVA P value	Mean pasture ±1 SD	Mean remnant ±1 SD	Contrast remnant (R) vs. pasture (P)
Fine root biomass (g kg^{-1})	0.09	6.64±4.05	13.00±0.38	NS
Bulk density (g cm^{-3})	0.36	0.69±0.08	0.61±0.12	NS
SOC (%)	0.60	6.15±1.83	7.31±4.15	NS
TN (%)	0.85	0.71±0.14	0.74±0.26	NS
Nitrate (mg kg^{-1})	0.26	11.8±7.24	22.7±17.9	NS
Nitrification rate (mg kg^{-1} soil d^{-1})	0.65	0.67±0.75	0.39±0.03	NS
Ammonium (mg kg^{-1})	0.46	4.36±4.27	7.85±8.10	NS
Ammonification rate (mg kg^{-1} soil d^{-1})	0.73	0.65±0.98	0.65±0.98	NS
Infiltration rate (mm h^{-1})	**0.03**	**220±151**	**1421±995**	**R>P**

Bold text indicates significance at 0.05 level, NS not significant if P>0.05, SD standard deviation.

Table 4. Relationship between tested soil attributes and the age of the reforestation sites (n = 15).

Variable	Age				Log Age			
	Slope	R^2 value	F statistic	P value	Slope	R^2 value	F statistic	P value
Fine root biomass (g kg^{-1} soil)	0.02	0.001	0.02	0.90	1.14	0.03	0.34	0.57
Bulk density (g cm^{-3})	−0.009	0.28	5.16	**0.04**	−0.12	0.32	6.31	**0.03**
SOC (%)	−0.03	0.01	0.15	0.70	−0.03	<0.001	0.002	0.97
TN (%)	−0.001	0.004	0.05	0.83	0.005	<0.001	0.004	0.95
Nitrate (mg kg^{-1})	0.09	0.002	0.03	0.86	3.58	0.03	0.38	0.55
Nitrification rate (mg kg^{-1} d^{-1})	0.03	0.13	1.93	0.19	0.30	0.07	0.95	0.35
Ammonium (mg kg^{-1})	−0.09	0.04	0.51	0.49	−1.72	0.11	1.55	0.24
Ammonification rate (mg kg^{-1} d^{-1})	−0.03	0.11	1.54	0.24	−0.47	0.15	2.25	0.16
Infiltration rate (mm h^{-1})	54.9	0.39	9.77	**0.008**	678	0.45	10.6	**0.006**

Pasture is assumed to be 0 years post reforestation. Results from linear models with both age and log-transformed age are shown. P values in bold are significant at the 0.05 level.

belowground outcomes within 3 to 20 years since planting are physical and associated functional attributes of the soil.

4.1. Differences in soil properties between pasture and remnant rainforest

Infiltration rate was the only soil variable studied here that differed significantly between pasture and remnant riparian rainforest. Pasture infiltration rates were ≈15% of those of riparian rainforest, consistent with the finding of an infiltration rate in pasture 20% of that in remnant rainforest by Peñuela and Drew [15]. We detected no statistically significant differences in other soil properties between pasture and remnant rainforest soils. Limited effects of land use on selected chemical properties are consistent with previous findings in Australian on ferrosol soils [14]. However, it was expected that attributes such as bulk density and fine roots would differ between land use types [14]. It is possible that these attributes differ at a scale inscrutable beyond background variability between sites, possibly necessitating longitudinal study of permanent plots following reforestation. Limited availability of remnant forest within sampling clusters, a consequence of extensive clearing of riparian rainforest in the region, was also a contributing factor in the comparatively low statistical power for detecting changes in measured attributes. Nevertheless, the data also indicate that there is little degradation of soil with respect to N and C stock in the studied systems.

4.2. Recovery of soils under reforestation plantings

Bulk density and infiltration rates both changed significantly with time under the reforestation plantings. Bulk density showed a negative trend with age whereas infiltration increased with time in the reforestation plantings. The observed trend for decreasing bulk density is consistent with previous studies globally [9,14,37,38]. The higher bulk density and lower infiltration rates of the pastures may be partially attributed to compaction by livestock and/or machinery [39]. Lower bulk densities are often associated with soil of forests and reforestation plantings because greater presence of roots and soil biota and higher concentrations of organic matter enhance soil porosity. In remnant forest and reforestation plantings, the high turnover of roots results in more open soil pores as roots grow and subsequently die [40]. Additionally, pastures generally have more shallow and non-woody roots while forests have a broader range of fine to large woody roots including with larger, woody roots that leave conduits in the soil post death [41–43].

Higher densities and diversities of soil organisms in forests and reforestation plantings [44,45] contribute to the aeration of soil and reduce bulk density. High soil organic matter is also associated with low bulk density [18], a trend seen in this study as bulk density was negatively correlated with SOC. This was expected as SOC is indicative of litter dynamics, turnover and decomposition rates and is responsible for providing low-density mass to the soil and improving soil structure [15,18,37,46]. The bulk density values in our study (0.65 g cm^{-3}) are low in comparison to average soils (>1.2 g cm^{-3}). Paul [14] also reported low bulk densities (0.6–0.8 g cm^{-3}) for similar ferrosols in reforestation plantings and remnant rainforest, confirming the low bulk density of this soil type. We attribute these low bulk densities to the friable nature of the ferrosols in the study region, high porosity and the high organic matter content in the topsoil. We also propose that the low bulk density contributes to the trend of increasing infiltration rates with time following reforestation as the two variables were inversely correlated.

Similar to prior studies, we found a significant increase in infiltration rates with reforestation age, with infiltration rates of the

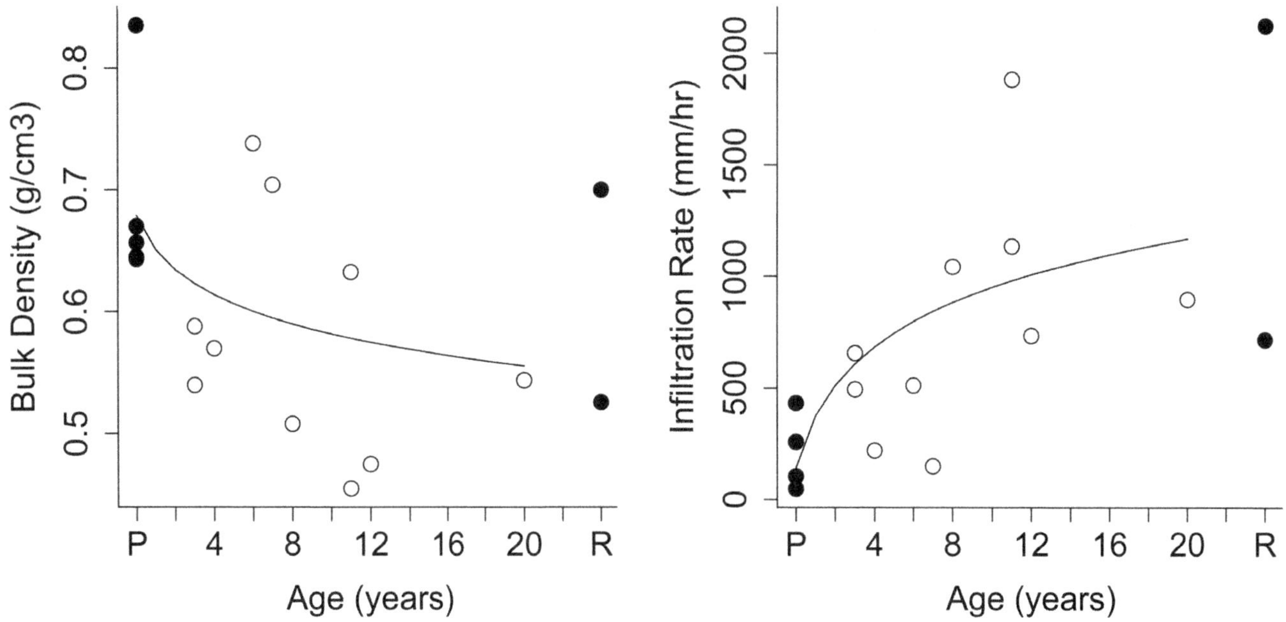

Figure 1. Differences in bulk density (left panel) and infiltration rates (right panel) among pasture (P), reforestation sites aged 2 to 20 years, and remnant rainforest (R) (n = 5, 10 and 2 respectively). Lines show fitted values of the modeled relationship between soil properties and log-transformed age of reforestation. Open circles show reforestation plantings, filled circles indicate pasture clusters and remnant riparian rainforest.

pastures being always lower than those of adjacent reforestation sites. We note a strong positive pattern with age within the first 5–10 years post reforestation, indicating that even young tree plantings are able to promote increased infiltration rates [37,47]. The trajectory in infiltration rates from the plantings (up to 20 years old) indicate that increases of 182 mm h^{-1} can be expected for each year in the first decade since planting, with most gains seen within the first 10 years post-reforestation. This trend is consistent with findings of Bharati [48] of short term changes in infiltration rates under riparian buffers when compared to adjacent cultivated fields and pastures, noting a 5-fold increase in infiltration rates under riparian buffers after 6 growing seasons. Our finding that infiltration rates are strongly enhanced soon after reforestation has meaningful consequences for decision making on land use, water quality and conservation with reduced run-off, slowed surface erosion and reduced non-point source pollution as beneficial outcomes [49].

Here, two of the tested nine soil attributes varied between pasture and reforestation sites. One possible explanation is the recentness of the plantings. Reforestation sites were relatively recent compared to studies that focused on plantings of much greater age (up to 60 years) [13,38,43] and this may have contributed to the uniformity between sites in our study. However, the common lack of differences between the studied pasture and remnant riparian rainforest sites suggests that inherent high soil fertility may be a larger contributor to soil properties than land cover, as has been suggested previously [43,50].

4.3. Species specific effects on the recovery of soils under reforestation plantings

In previous studies, species-specific effects were not used to help elucidate changes in soil properties post-reforestation [14,51], although it is known that species-specific effects can modify soil properties [52]. We detected two species-specific effects in the five tree species and eight variables examined. *Ficus coronata* and *Flindersia schottiana* showed negative relationships with time since reforestation for fine root biomass and soil ammonium levels respectively. We hypothesized that as reforested sites become more established, fine root biomass increases to access nutrients in the surface soil layer of rainforests [41,43,53]. The studied *F. coronata* specimens displayed the opposite trend to that expected, showing a negative fine root relationship with time post-reforestation. This decreasing trend exhibited by *F. coronata* may be explained by a need for greater structural support through large woody roots in which they invest carbon at the expense of fine surface roots, or the ability to access nutrients from the deeper soil, however greater sampling at depth would be needed to confirm these hypotheses. Nevertheless, if verified, an ability to shift root production into in deeper layers could be a useful trait providing physical impediment in the subsoil, particularly where greater soil stability is a desired goal of reforestation.

Another species-specific effect was observed with *F. schottiana* showing a decreasing relationship with ammonium concentrations and time post-reforestation. This result was not entirely unusual as ammonium is readily converted to nitrate in some soils and nitrate may be preferentially used by plants [54]. However, we did not detect higher nitrate concentration in soil under *F. schottiana*, which would be expected if nitrification is simulated. This may be due to high N uptake and demand of *F. schottiana*, which was not investigated here. However, this finding conforms broadly to the results of prior studies [14,38,55], which saw that ammonium levels were highest in pasture and lowest in forest.

It is noteworthy that the species-specific effects detected were not the same as those that were seen at a site level. When means at the site level were considered, only infiltration and bulk density were significant, however at a species level, both *F. schottiana* and *F. coronata* showed significant variations through time for ammonium levels and fine root biomass respectively. This suggests that while individual species may be able to preferentially alter their immediate surroundings, their influence is diminished at a

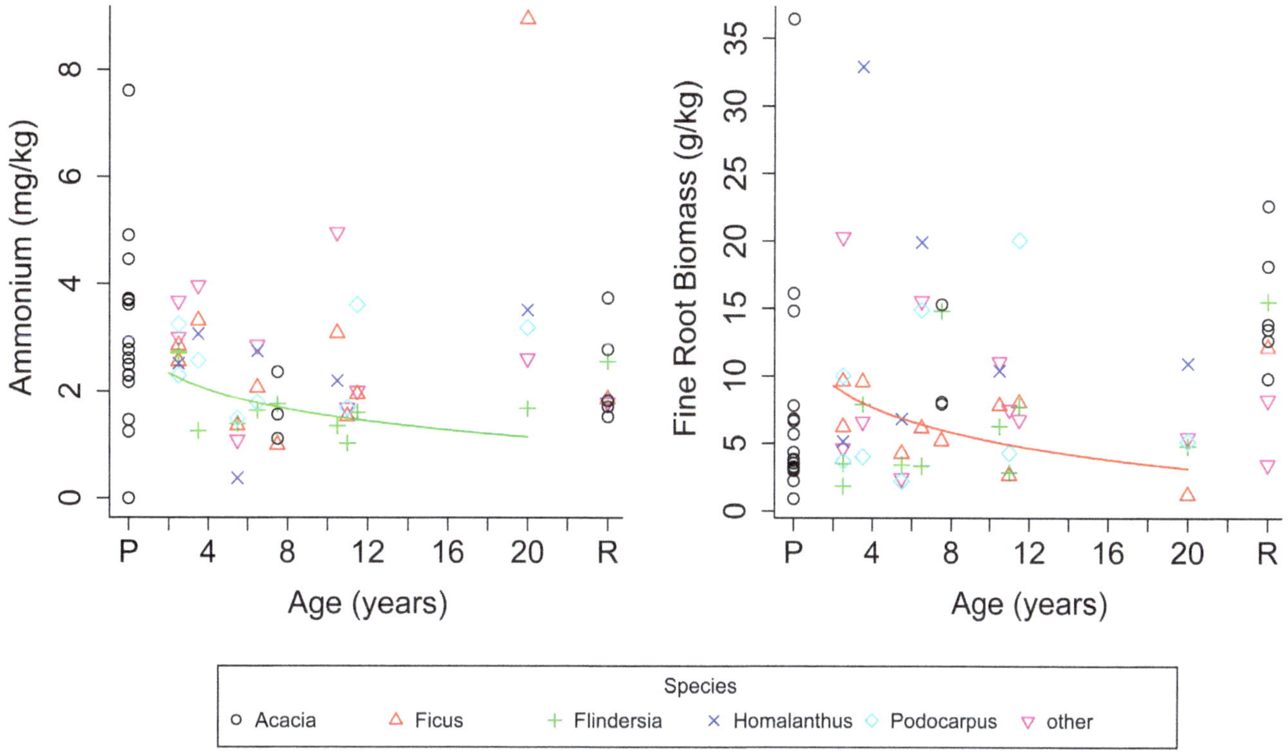

Figure 2. Differences in ammonium (left panel) and fine root biomass (right panel) over time under five tree species in pasture (P), reforestation sites aged 2 to 20 years, and remnant rainforest (R). Genus names refer to the selected tree species under which the soil samples were taken. "Other" refers to soil samples taken from non-specified tree species or pastoral sites. Lines show fitted values of the modeled relationship between individual soil properties and log-transformed age for species with significant models - *Flindersia schottiana* (a) and *Ficus coronata* (b).

site level. A possible explanation is that the diverse mix of native tree species, representing a variety of physiotypes, and the individual species-specific effects on soil are less influential than the effects of the overall plantings on soil properties. Had the reforestation planting been of a simpler species mix or the relative proportion of select species been greater, soil traits may have reflected more strongly particular species-specific effects. This indicates that particular reforestation outcomes may be facilitated by explicit selection of certain tree species. Further studies have to determine the efficacy and benefits of a less species-rich reforestation species mixture.

Conclusion

Reduced bulk density and enhanced infiltration rates were the key immediate outcomes resulting from riparian reforestation plantings. It is noteworthy that the most change occurred within just 5 to 10 years post-planting suggesting that this form of active restoration can reinstate important soil functions within a comparatively short time. Given the linkages between increased infiltration and ecosystem services such as nutrient buffering, filtration of runoff and sediments, we conclude that the reforestation is, by extension, achieving the objective of improving water quality. This finding has consequences for land managers and conservationists interested in lowering runoff, non-point source erosion and limiting loss of topsoil, all of which are major concerns and motivators for reforestation. Further, soil infiltration rates are intricately linked with several soil properties influenced by slight changes in soil structure well before changes are perceivable in other soil properties [56]. This signifies that the reforestation plantings are on a positive trajectory to restoring other soil properties and ecosystem function with infiltration rates as front-runner in these changes.

Acknowledgments

We thank the landholders that participated in the study for generously allowing access to their land. We also thank the Lake Baroon Catchment Care Group (LBCCG) for liaising with landholders, being so knowledgeable and motivated, and accommodating to our needs. We acknowledge the advice and thank Jessica Vogt and Richard Brackin on sample processing.

Author Contributions

Conceived and designed the experiments: RG GK MA SS LPS. Performed the experiments: RG MB MA SS LPS. Analyzed the data: RG GK LPS. Contributed reagents/materials/analysis tools: RG MA GK SS LPS NR. Contributed to the writing of the manuscript: RG MB GK MA SS LPS.

References

1. Connor S, Nelson PN, Armour JD, Hénault C (2012) Hydrology of a forested riparian zone in an agricultural landscape of the humid tropics. Agr Ecosyst Environ. 180: 111–122.
2. Meynendonckx J, Heuvelmans G, Muys B, Feyen J (2006) Effects of watershed and riparian zone characteristics on nutrient concentrations in the River Scheldt Basin. Hydrol Earth Syst Sci Discuss 3: 653–679.
3. Orzetti LL, Jones RC, Murphy RF (2010) Stream condition in Piedmont streams with restored riparian buffers in the Chesapeake Bay watershed. J Am Water Resour As 46: 473–485.
4. Russo TA, Fisher AT, Roche JW (2012) Improving riparian wetland conditions based on infiltration and drainage behavior during and after controlled flooding. J Hydrol 432: 98–111.
5. Vidon P (2012) Towards a better understanding of riparian zone water table response to precipitation: surface water infiltration, hillslope contribution or pressure wave processes? Hydrol Processes 26: 3207–3215.
6. Charron I, Lalonde O, Roy AG, Boyer C, Turgeon S (2008) Changes in riparian habitats along five major tributaries of the saint Lawrence river, Québec, Canada: 1964–1997. River Res Appl 24: 617–631.
7. Griscom HP, Griscom BW, Ashton MS (2009) Forest Regeneration from Pasture in the Dry Tropics of Panama: Effects of Cattle, Exotic Grass, and Forested Riparia. Restor Ecol 17: 117–126.
8. Scott SJ, McLaren G, Jones G, Harris S (2010) The impact of riparian habitat quality on the foraging and activity of pipistrelle bats (Pipistrellus spp.). J Zool 280: 371–378.
9. Rodrigues Nogueira Jr L, de Moraes Goncalves JL, Lex Engel V, Parrotta J (2011) Soil dynamics and carbon stocks 10 years after reforestation of degraded land using Atlantic Forest tree species. Forest Syst 20: 536–545.
10. Naiman RJ, Décamps H, McClain ME (2005) Riparia: ecology, conservation and management of streamside communities. New York: Academic Press.
11. Sweeney BW, Bott TL, Jackson JK, Kaplan LA, Newbold JD, et al. (2004) Riparian deforestation, stream narrowing, and loss of stream ecosystem services. Proc Nat Acad Sci USA 101: 14132–14137.
12. Guo LB, Gifford M (2002) Soil carbon stocks and land use change: a meta analysis. Glob Change Biol 8: 345–360.
13. Maloney KO, Garten CT, Ashwood TL (2008) Changes in soil properties following 55 years of secondary forest succession at Fort Benning, Georgia, USA. Restor Ecol 16: 503–510.
14. Paul M, Catterall CP, Pollard PC, Kanowski J (2010) Recovery of soil properties and functions in different rainforest restoration pathways. Forest Ecol Manage 259: 2083–2092.
15. Peñuela MC, Drew AP (2004) A model to assess restoration of abandoned pasture in Costa Rica based on soil hydrologic features and forest structure. Restor Ecol 12: 516–524.
16. Zhu B, Li Z, Li P, Liu G, Xue S (2010) Soil erodibility, microbial biomass, and physical-chemical property changes during long-term natural vegetation restoration: a case study in the Loess Plateau, China. Ecol Res 25: 531–541.
17. Bronick CJ, Lal R (2005) Soil structure and management: a review. Geoderma 124: 3–22.
18. Murty D, Kirschbaum MUF, McMurtrie RE, McGilvray H (2002) Does conversion of forest to agricultural land change soil carbon and nitrogen? a review of the literature. Glob Change Biol 8: 105–123.
19. Schoenholtz SH, Van Miegroet H, Burger JA (2000) A review of chemical and physical properties as indicators of forest soil quality: challenges and opportunities. Forest Ecol Manag 138: 335–356.
20. Catterall C, Harrison DA (2006) Rainforest Restoration Activities in Australia's Tropics and Subtropics. Cairns, Australia: Cooperative Research Centre for Tropical Rainforest Ecology and Managemen, Rainforest CRC. 0864437692 0864437692.
21. Catterall CP, Kanowski J, Wardell-Johnson GW, Proctor H, Reis T, et al. (2004) Quantifying the biodiversity values of reforestation: perspectives, design issues and outcomes in Australian rainforest landscapes. In: Conservation of Australia's Forest Fauna (ed Lunney, D), Vol. 2. Royal Zoological Society of New South Wales, Sydney 2: 359–393.
22. Kanowski J, Catterall C, Wardell-Johnson G, Proctor H, Reis T (2003) Development of forest structure on cleared rainforest land in eastern Australia under different styles of reforestation. Forest Ecol Manage 183: 265–280.
23. Kanowski J, Catterall CP (2010) Carbon stocks in above-ground biomass of monoculture plantations, mixed species plantations and environmental restoration plantings in north-east Australia. Ecol Manage Restor 11: 119–126.
24. Wickham JD, Wade TG, Riitters KH (2011) An environmental assessment of United States drinking water watersheds. Landscape Ecol 26: 605–616.
25. Australian Bureau of Statistics (2011) Census.
26. Ellison L, Coaldrake JE (1954) Soil mantle movement in relation to forest clearing in Southeastern Queensland. Ecology 35: 380–388.
27. Willmott WF (2007) Rocks and landscapes of the Sunshine Coast. Brisbane: Geological Society of Australia, Queensland Division.
28. Nakamura A, Catterall CP, House APN, Kitching RL, Burwell CJ (2007) The use of ants and other soil and litter arthropods as bio-indicators of the inpacts of rainforest clearing and subsequent land use. J Insect Conserv 11: 177–186.
29. Smith I (2001) Juggling-balancing Tourism and Biodiversity in a Small, Valuable Reserve - Mary Cairncross Scenic Reserve. In: Boyes B, editor. *Biodiversity Conservation "From Vision to Reality"*. Gatton, Queensland: Lockyer Watershed Management Association (LWMA) Inc. - Lockyer Landcare Group. 81–84.
30. McIntyre D (1976) Subplasticity in Australian soils. I. Description, occurrence, and some properties. Aust J Soil Res 14: 227–236.
31. DeForest JL (2011) Soil texture - Hydrometer method. Lab protocols for testing soils Ohio University, Athens, Ohio: The Department of Environmental and Plant Biology.
32. Robertson GP, Coleman DC, Bledsoe CS, Sollins P (1999) Standard soil methods for long-term ecological research. New York: Oxford University Press.
33. Perroux KM, White I (1988) Designs for Disc Permeameters. Soil Sci Soc Am J 52: 1205–1215.
34. Kandeler E, Gerber H (1988) Short-term assay of soil urease activity using colorimetric determination of ammonium. Biol Fert Soils 6: 68–72.
35. Miranda KM, Espey MG, Wink DA (2001) A rapid, simple spectrophotometric method for simultaneous detection of nitrate and nitrite. Nitric Oxide 5: 62–71.
36. R Core Team (2013) R: A language and environment for statistical computing. Vienna, Austria: R Foundation for Statistical Computing.
37. Neris J, Jiménez C, Fuentes J, Morillas G, Tejedor M (2012) Vegetation and land-use effects on soil properties and water infiltration of Andisols in Tenerife (Canary Islands, Spain). Catena 98: 55–62.
38. Reiners WA, Bouwman AF, Parsons WFJ, Keller M (1994) Tropical rain forest conversion to pasture: changes in vegetation and soil properties. Ecol Appl 4: 363–377.
39. Bell LW, Kirkegaard JA, Swan A, Hunt JR, Huth NI, et al. (2011) Impacts of soil damage by grazing livestock on crop productivity. Soil Tillage Res 113: 19–29.
40. Beven K, Germann P (1982) Macropores and water flow in soils. Water Resour Res 18: 1311–1325.
41. Finér L, Ohashi M, Moguchi K, Hirano Y (2011) Factors causing variation in fine root biomass in forest ecosytems. Forest Ecol Manage 261: 265–277.
42. Jiménez EM, Moreno FH, Peñuela MC, Patiño S, Lloyd J (2009) Fine root dynamics for forests on contrasting soils in the Colombian Amazon. Biogeoscience 6: 2809–2827.
43. Powers JS, Perez-Aviles D (2012) Edaphic factors are a more important control on surface fine roots than stand age in secondary tropical dry forests. Biotropica 45: 1–9.
44. Colloff MJ, Pullen KR, Cunningham SA (2010) Restoration of an ecosystem function to revegetation communities: The role of invertebrate mscropores in enhancing soil water infiltration. Restor Ecol 18: 65–72.
45. Nakamura A, Proctor H, Catterall CP (2003) Using soil and litter arthropods to assess the state of rainforest restoration. Ecol Manage Restor 4: S20–S28.
46. Laub BG, McDonough OT, Needelman BA, Palmer MA (2013) Comparison of designed channel restoration and riparian buffer restoration effects on riparian soils. Restor Ecol: 1–9.
47. Wood HB (1977) Hydrologic differences between selected forested and agricultural soils in Hawaii. Soil Sci Soc Am J 41: 132–136.
48. Bharati L, Lee K-H, Isenhart TM, Schultz RC (2002) Soil-water infiltration under crops, pasture, and established riparian buffer in Midwestern USA. Agrofor Syst 56: 249–257.
49. Schultz R, Isenhart T, Simpkins W, Colletti J (2004) Riparian forest buffers in agroecosystems–lessons learned from the Bear Creek Watershed, central Iowa, USA. Agrofor Syst 61: 35–50.
50. Lamb D (1980) Soil nitrogen mineralisation in a secondary rainforest succession. Oecologia 47: 257–263.
51. Paul M, Catterall CP, Pollard PC, Kanowski J (2010) Does soil variation between rainforest, pasture and different reforestation pathways affect the early growth of rainforest pioneer species? Forest Ecol Manage 260: 370–377.
52. Bezemer TM, Lawson CS, Hedlund K, Edwards AR, Brook AJ, et al. (2006) Plant species and functional group effects on abiotic and microbial soil properties and plant–soil feedback responses in two grasslands. J Ecol 94: 893–904.
53. Claus A, George E (2005) Effect of stand age on fine-root biomass and biomass distribution in three European forest chronosequences. Can J Forest Res 35: 1617–1625.
54. Piccolo MC, Neill C, Cerri CC (1994) Net nitrogen mineralization and net nitrification along a topical forest-to-pasture chronosequence. Plant Soil 162: 61–70.
55. Rasiah V, Florentine SK, Williams BL, Westbrooke ME (2004) The impact of deforestation and pasture abandonment on soil properties in the wet tropics of Australia. Geoderma 120: 35–45.
56. Radke JK, Berry EC (1993) Infiltration as a tool for detecting soil changes due to cropping, tillage, and grazing livestock. Am J Alternative Agr 8: 164–174.

4

Relationships between Nutrient-Related Plant Traits and Combinations of Soil N and P Fertility Measures

Yuki Fujita[1]*, Peter M. van Bodegom[2], Jan-Philip M. Witte[1,2]

1 Team Ecohydrology, KWR Watercycle Research Institute, Nieuwegein, The Netherlands, **2** Department of Systems Ecology, VU University Amsterdam, Amsterdam, The Netherlands

Abstract

Soil fertility and nutrient-related plant functional traits are in general only moderately related, hindering the progress in trait-based prediction models of vegetation patterns. Although the relationships may have been obscured by suboptimal choices in how soil fertility is expressed, there has never been a systematic investigation into the suitability of fertility measures. This study, therefore, examined the effect of different soil fertility measures on the strength of fertility–trait relationships in 134 natural plant communities. In particular, for eight plot-mean traits we examined (1) whether different elements (N or P) have contrasting or shared influences, (2) which timescale of fertility measures (e.g. mineralization rates for one or five years) has better predictive power, and (3) if integrated fertility measures explain trait variation better than individual fertility measures. Soil N and P had large mutual effects on leaf nutrient concentrations, whereas they had element-specific effects on traits related to species composition (e.g. Grime's CSR strategy). The timescale of fertility measures only had a minor impact on fertility–trait relationships. Two integrated fertility measures (one reflecting overall fertility, another relative availability of soil N and P) were related significantly to most plant traits, but were not better in explaining trait variation than individual fertility measures. Using all fertility measures together, between-site variations of plant traits were explained only moderately for some traits (e.g. 33% for leaf N concentrations) but largely for others (e.g. 66% for whole-canopy P concentration). The moderate relationships were probably due to complex regulation mechanisms of fertility on traits, rather than to a wrong choice of fertility measures. We identified both mutual (i.e. shared) and divergent (i.e. element-specific and stoichiometric) effects of soil N and P on traits, implying the importance of explicitly considering the roles of different elements to properly interpret fertility–trait relationships.

Editor: Ben Bond-Lamberty, DOE Pacific Northwest National Laboratory, United States of America

Funding: Knowledge for Climate (http://www.klimaatonderzoeknederland.nl/), Dutch Water Utility sector (B222044). The funders had no role in study design, data collection and analysis, decision to publish, or preparation of the manuscript.

Competing Interests: The authors have declared that no competing interests exist.

* E-mail: Yuki.Fujita@kwrwater.nl

Introduction

Soil nutrients, such as nitrogen (N) and phosphorus (P), constitute basic requirements for plants to support their life. In the short term, soil nutrients induce plastic responses of individual plants (e.g. as related to recruitment, growth, and reproduction). In the longer run, soil nutrients affect the species composition of a plant community, because each species has evolved to adapt to certain environments and therefore has contrasting requirements for nutrients. How and why plant community composition changes from fertile to unfertile soil has been a central concern of ecologists for many decades. A general trend derived from qualitative studies is that species with a rapid growth strategy dominate in fertile soils, whereas species with a conservative strategy dominate in infertile soils (for a review see [1]). In quantitative trait-based studies on a global scale, between-site variations of leaf traits were well explained by soil fertility if combined with climate factors (R^2 between 59 and 78% for specific leaf area and leaf N and P concentrations) [2]. Within a climatic region, however, the effects of soil fertility on leaf traits were moderate (e.g. 22–23% of between-site variance in an integrated leaf economy measure were explained by nutrient indicator value [3]; 31% and 50% of between-site variance in leaf N and P concentrations were explained by soil total P if combined with growth form [4]). Accordingly, the relatively weak relationship between soil fertility and plant traits remains an uncertain link in trait-based species distribution models [5].

Generally weak linkages between environmental factors and traits could be due to stochastic processes being more dominant than environmental filtering effects during the assembly of these plant communities [6,7]. However, these relatively weak relationships could also be the results of using suboptimal (or inappropriate) measure of soil fertility [3]. Soil fertility, or availability of nutrients for plants, is typically expressed as concentrations of dissolved N or P in soil, N mineralization rates for a certain period, soil total N or P, or soil C:N or C:P ratios. Although these soil fertility measures are generally correlated [2,8], it is likely that the choice of a fertility measure influences the strength of the correlation between soil fertility and plant traits (e.g. [9]). Nevertheless, whether the most relevant fertility measure was used to examine the relation with a specific plant trait has never been explicitly examined. For example, the two nutrient elements, N and P, both have pivotal yet different roles in how plants function. Previous studies showed that soil N and P have different degrees of contributions to each trait: specific leaf area was better predicted by soil N supply, whereas leaf N and P concentrations showed stronger relationships with soil P [2,4]. This implies that element-specific mechanisms or stoichiometric effects (i.e. effects of relative availability of N and P) may play important roles in regulating these traits.

Furthermore, any measure of soil fertility reflects the nutrient status over a certain timescale. This timescale is implicitly (e.g. total N) or explicitly (e.g. N mineralization over 5 years) reflected in the way the measure is expressed. Measures from chemical soil bulk analysis at a certain moment, such as pool size and extractable amount of nutrients, provide only partial information about the nutrient availability for plants over a long time span [10]. Nutrient mineralization rates from soil could be a more relevant measure of nutrient availability for living plants as they reflect a major flow within soil nutrient cycles [11]. However, they fluctuate considerably over time and are controlled by abiotic factors such as temperature and soil moisture [12]. Thus, the timescale associated with soil fertility measures could influence the relationship between soil fertility and plant traits.

Moreover, since plants use soil nutrients in a multidimensional manner (e.g. different ratios of elements, at different timings and periods, etc.), certain combinations of soil fertility measures may better reflect plant trait variability. Hence, an aggregated measure derived from multiple fertility measures may better explain trait variability than a single fertility measure.

This study aims to investigate whether the relationships between soil fertility and nutrient-related plant traits can be improved by analyzing the mutual and divergent impacts of different soil fertility measures. To this aim, we tested (1) whether nutrient elements in soil (N or P) have specific or shared impacts on plot-mean plant traits (i.e. averaged or aggregated values of traits of all plants in each plot), (2) whether the timescale of a soil fertility measure is important in explaining plot-mean plant traits, and (3) whether use of an integrated soil fertility measure improves the fertility vs. trait relationships. Subsequently, we infer from our findings which fertility measure or combinations of fertility measures are the best suited to describe among-plots trait variations. Throughout the paper, we considered vascular plants only, since non-vascular species (i.e. mosses and lichens) do not directly take up nutrients from soils and are thus less relevant in the context of soil fertility–trait relationships.

Materials and Methods

Ethic Statement

We obtained permissions for soil and vegetation sampling in Hoge Veluwe national park (permission from Stichting het nationale park de Hoge Veluwe), Zuid-Kennemerland national park (permission from Provinciaal Waterleidingbedrijf Noord-Holland), and nature reserves owned by Staatsbosbeheer.

Site selection

We selected 36 sites in the Hoge Veluwe and Zuid-Kennemerland national parks and nature reserves owned by Staatsbosbeheer (dataset 1). Additionally, we used 47 sites of Olde Venterink et al. [11] (dataset 2) and 51 sites of Ordonez et al. [3] (dataset 3); these studies used almost the same methods to measure soil and plant properties (see Table 1 for a methodological overview). In total, we thus acquired information for 134 sites and 372 plant species in natural ecosystems in the Netherlands and Belgium, consisting of 104 grasslands, 17 shrub lands (including heath), and 13 forests. These sites cover the range of ecosystems typical in this region except for those influenced by brackish water. Vascular plant species were recorded in a plot size of 4 m^2 for herbaceous, 4, 9, or 25 m^2 for shrub (depending on the size of the woody species), and 100 m^2 for forest stands. Soil cores of 10 cm depth (datasets 1 and 3) or 15 cm depth (dataset 2) were taken within or next to the plot. Large roots were removed from the soil cores. For datasets 1 and 3, several soil cores were mixed to make a composite soil to eliminate the effects of local soil heterogeneity. In the peak growing season (July or August), above-ground standing biomass of vascular plants (datasets 1 and 2) or leaves of dominant vascular species (dataset 3) were harvested.

Soil fertility measures

Dissolved N ($N–NO_3 + N–NH_4$, mg N kg^{-1} soil) and dissolved P ($P–PO_4$, mg P kg^{-1} soil) were measured to indicate the short-term availability of N and P, respectively. Two different extraction methods were used to measure dissolved P: ammonium lactate-acetic acid (ALA) extraction and the Olsen extraction. Therefore, we converted the values estimated with the ALA extraction to be comparable to those with Olsen extraction, using an empirically derived relationship [13] (Olsen-P = 2.35+0.45 ·ALA-P). The potential influence of using two different extraction methods is tested in Fig. S1 in Appendix S1. Olsen extraction is meant for neutral to alkaline soils; however, this method was also used for some of our acid ($pH<5$) soils. The potential interference of these acid soils on our results is tested in Fig. S2 in Appendix S2.

As a longer-term indication of N availability, we measured in-situ N mineralization rates in the mid growing season ('summer Nmin') for 6 weeks (datasets 1 and 3) or 8 weeks (dataset 2) (see [14] for details about the method). The difference in dissolved N between the beginning and the end of the incubation period was considered as mineralized N from organic N. Mineralized N was expressed as rates per week (mg N kg^{-1} soil $week^{-1}$). Because denitrification in wet soil cores could have caused a considerable loss of N from the incubation tubes [14], we corrected for the N loss by adding modelled denitrification rates simulated by DAYCENT [15] to the measured N mineralization rates.

Additionally, to estimate nutrient availability for much longer terms, we used a modified version of a SOM model, CENTURY [14]. The CENTURY model simulates decomposition of soil organic C and associated flows of organic N and P. Soil total C, N, and P, soil texture, temperature, and moisture of the top soil were used as model input values. Soil temperature and moisture were simulated with a hydrological model SWAP [16]. The daily groundwater level of each site, required for the SWAP simulation, was estimated by temporal inter- and extrapolation of the observed groundwater levels in nearby wells (mostly within 30 m from the plot) using MENYANTHES software [17]. With the CENTURY model, cumulative net mineralization rates of N and P were estimated for the year of the sampling (from the first of January [mid-winter] until the end of August [end of summer], as plant traits were measured by then) ('annual Nmin', mg N kg^{-1} soil 243-day^{-1} and 'annual Pmin', mg P kg^{-1} soil 243-day^{-1}) and for the preceding five years (from September five years before the sampling year to August of the sampling year) ('5yr Nmin', mg N kg^{-1} soil 5-$year^{-1}$ and '5yr Pmin', mg P kg^{-1} soil 5-$year^{-1}$). Transformation processes of mineralized N and P (e.g., nitrification, denitrification, adsorption and precipitation of inorganic P) were not considered.

Finally, as very rough measures of soil fertility in the long time span, we used soil total N and P (% of total soil mass), soil N:C ratio, and soil P:C ratio.

Plant traits

N and P concentration in leaves (LNC and LPC, respectively; mg g^{-1}) were determined for dataset 3 only (Table 1). Nutrient concentrations were measured in each site for dominant species,

Table 1. Overview of methodology of soil and plant trait measurements for three datasets used in this study.

		Dataset 1	Dataset 2 [11]	Dataset 3 [3]
N. of sites		**36**	**47**	**51**
N. of combined soil cores per site		**3**	**1**	**5**
Soil fertility measurements				
Soil C	%	CNS analyzer*1*2	0.5 · Loss on ignition at 550°C	CNS analyzer*1
Soil N	%	CNS analyzer*1	Kjeldahl digestion	CNS analyzer*1
Soil P	%	HNO_3 + HCl digestion	Kjeldahl digestion	HNO_3 + HCl digestion
Dissolved N ($N-NO_3$ + $N-NH_4$)	mg N kg^{-1} dry soil	1 M KCl extraction	1 M KCl extraction	1 M KCl extraction
Dissolved P ($P-PO_4$)	mg P kg^{-1} dry soil	Olsen extraction (0.5 M $NaHCO_3$)	ALA extraction (0.1 M NH_4OH + 0.1 M lactic acid + 0.4 M acetic acid)	Olsen extraction (0.5 M $NaHCO_3$)
Summer N mineralization	mg N kg^{-1} dry soil $week^{-1}$	6 weeks in-situ incubation in May–July (d 15 cm x ø 4 cm)	8 weeks in-situ incubation in July–August (d 10 cm x ø 4.8 cm)	6 weeks in-situ incubation in June–August (d 15 cm x ø 6 cm)
Annual N and P mineralization	mg N (or mg P) kg^{-1} dry soil 243-day^{-1}	Simulated with CENTURY	Simulated with CENTURY	Simulated with CENTURY
5-year N and P mineralization	mg N (or mg P) kg^{-1} dry soil 5-year^{-1}	Simulated with CENTURY	Simulated with CENTURY	Simulated with CENTURY
Soil texture*3	clay/silt/sand in fraction	Laser particle sizer	Estimated*4	Estimated*4
Plant trait measurements				
LNC	mg g^{-1}	-	-	CNS analyzer*1
LPC	mg g^{-1}	-	-	HNO_3 + HCl digestion
WNC	mg g^{-1}	CNS analyzer*1	Kjeldahl digestion	-
WPC	mg g^{-1}	HNO_3 + HCl digestion	Kjeldahl digestion	-
IV_{nut}	-	Derived from species composition		
CSR	-	Derived from species composition		

*1CNS analyzer (Carlo Erba NA 1500, Rodana).
*2C in $CaCO_3$, determined with thermogravimetric analysis (TGA-601, Leco Corporation), was subtracted.
*3Used as model input values for CENTURY simulation.
*4Estimated based on top layer characteristics of the soil physical unit, derived from 1:50,000 soil map.

and weighted averages (weighted by species' relative cover) were used as the plot-mean LNC and LPC. The dominant species were sampled until the cover sampled exceeded 50% of the total vascular plant cover. Note that the 50% of coverage being sampled is rather low compared to the recommended threshold of 80% [18], and therefore the LNC and LPC values in our study may slightly deviate from the true plot-mean values.

The entire above-ground plant biomass (i.e. all vascular plants of the community together) was sampled to determine whole-canopy N and P concentrations in each plot (WNC and WPC, respectively; mg g^{-1}) for datasets 1 and 2 (Table 1). Since WNC and WPC reflect the aggregated characteristics of the community, they are considered as plot-mean trait values. One plot in dataset 1, which did not have nutrient concentration data, was excluded from all analyses concerning WNC and WPC. Woody species were included in WNC and WPC when they were young seedlings less than one year old, so that WNC and WPC reflect the annual uptake of the nutrients. LNC, LPC, WNC, and WPC were log-transformed prior to analyses to correct for the right-skewed distributions.

Combinations of multiple plant traits are constrained by physiological trade-offs [19]; thus, integrative traits (e.g. nutrient use efficiency) or strategy types help to express trait variability in fewer dimensions. Here we used two types of integrative plant traits: one based on species occurrences in different habitats (i.e. indicator values for nutrients, IV_{nut} [20]), and the other based on life history traits (Grime's CSR strategy [21]).

IV_{nut} is comparable to the Ellenberg indicator value for nutrients, but is tailored for Dutch flora and has a continuous scale ranging from 1 (prevailing at nutrient poor sites) to 3 (prevailing at nutrient rich sites) [20]. Plot-mean IV_{nut} values were computed as arithmetic means of IV_{nut} for each site, rather than weighted means of IV_{nut} based on species abundance, as the former was shown to be sufficient for this trait with ordinal-scale values [22]. 26 species (out of total 372 species recorded in our study), for which IV_{nut} value was not available, were excluded from the calculation of plot-mean IV_{nut} values. The trait coverage (i.e. percentage of species with IV_{nut} values within each plot) ranged from 76% to 100% (median 94%).

The CSR scheme represents the adaptive strategy of plant species along gradients of resource availability, stress, and disturbance, expressed with three axes of primary components: C ('Competitors'), S ('Stress tolerators'), and R ('Ruderals'). Each species can be classified into one out of 19 classes with different combinations of strategy components, e.g. C, SR/R, or CSR. We retrieved CSR strategies from Hunt et al. [23] (313 species), supplemented by the BioFlor database [24] (35 species). For 11 species, we assigned the CSR strategy according to the method of

Hodgson et al. [25], using seven life-history traits of these species retrieved from the LEDA database [26]. The remaining 13 species, for which we could not attribute CSR strategy due to lack of trait information, were excluded from the calculation of plot-mean CSR values. For each species, scores of each primary component (C, S, and R) were determined from its proportional contribution to a specific strategy (e.g. C scores of C, CS, CSR strategy are 1, 0.5, and 0.33, respectively: cf. [27]). Subsequently, for each site, plot-mean scores of C, S, and R (again, not weighted average by species abundance) were calculated. The trait coverage ranged from 86% to 100% (median 100%).

Statistics

Variation partitioning. Soil fertility measures were correlated, especially strongly within the group of N-related measures and that of P-related measures (table S1). In order to examine the relative contribution of soil N and soil P to plant trait variation among sites, we partitioned the variance of each plant trait t ($\mathbf{T}t$, a vector of n plots) into unique and shared effects of the two groups of predictor variables; i.e. to N-related fertility measures ($\mathbf{N}$, a matrix of n x p_N, where p_N is the number of N-related fertility measures, $p_N = 6$) and P-related fertility measures ($\mathbf{P}$, a matrix of n x p_P, where p_P is the number of P-related fertility measures, $p_P = 5$) [28]. The fraction of variance explained was indicated by the coefficient of determination of linear regression analysis for $\mathbf{T}t$ regressed by $\boldsymbol{X}$, where $\boldsymbol{X}$ is either $\mathbf{N}$, $\mathbf{P}$, or $\mathbf{N\&P}$. In order to correct for the different number of fertility measures within each group, we used adjusted R^2, $R^2_{(\mathbf{T}t|\boldsymbol{X})\ \mathrm{adj}}$, according to [29]. The unique effects of $\mathbf{N}$ were calculated as the fraction of variance in $\mathbf{T}t$ explained by $\mathbf{N\&P}$ minus the effects of $\mathbf{P}$ on $\mathbf{T}t$: $R^2_{(\mathbf{T}t|\mathbf{N\&P})\ \mathrm{adj}} - R^2_{(\mathbf{T}t|\mathbf{P})\ \mathrm{adj}}$. Identically, the unique effects of $\mathbf{P}$ on $\mathbf{T}t$ were calculated as $R^2_{(\mathbf{T}t|\mathbf{N\&P})\ \mathrm{adj}} - R^2_{(\mathbf{T}t|\mathbf{N})\ \mathrm{adj}}$. Shared effects of $\mathbf{N}$ and $\mathbf{P}$ on $\mathbf{T}t$ were calculated as $R^2_{(\mathbf{T}t|\mathbf{N})\ \mathrm{adj}} + R^2_{(\mathbf{T}t|\mathbf{P})\ \mathrm{adj}} - R^2_{(\mathbf{T}t|\mathbf{N\&P})\ \mathrm{adj}}$. The analysis was performed in R [30].

Additionally, we tested if N-related fertility measures explain significantly more variance in plot-mean traits than P-related fertility measures, and vice versa, following the bootstrapping method described in [29] using R [30]. Bootstrapped adjusted R^2 was computed 1000 times, and the difference between the adjusted R^2 between groups, $D_i = R^2_{(\mathbf{T}t|\mathbf{N})\ \mathrm{adj\text{-}boot},i} - R^2_{(\mathbf{T}t|\mathbf{P})\ \mathrm{adj\text{-}boot},i}$, was calculated for each ith bootstrapping. p-values were calculated from the distribution of D_i, as the fraction of D_i that falls below zero (when the median of D_i was positive, i.e. variance explained by $\mathbf{N}$ was larger than that by $\mathbf{P}$) or above zero (when the median of D_i was negative).

Hierarchical partitioning. In order to examine the most relevant timescale of soil fertility for explaining the variation of plant traits, we used the hierarchical partitioning method [31,32]. This method allows, within the hypothetical relationship between trait variance (response variable) and k fertility measures with different timescales (predictor variables), to quantify the independent contribution of a fertility measure S to the explained variance of a trait without being confounded by the other k-1 fertility measures. The hierarchical partitioning method computes the increase in goodness-of-fit when S is added to the model (in our case: a linear multivariate regression model of a trait regressed by fertility measures) compared to the model without S, and averages the increase over all possible models that include S as a predictor variable (i.e. 2^k models). In this way, the variation of a trait explained by S is divided into an independent effect of S and joint effect of S with other fertility measures. Negative values of a joint effect mean that the interactive effects of S and the other fertility measures on the trait are suppressive, rather than enhancing. An advantage of using the hierarchical partitioning over a one-model approach is that the averaging eliminates the problem of multicollinearity among predictor variables [32]. We conducted hierarchical partitioning separately for N-related and P-related fertility measures (i.e. $k = p_N$ or $k = p_P$), so that interactions of N and P do not obscure the effects of timescale. We used R^2 as the goodness-of-fit measure of the models. The analysis was performed in R [30] with the package 'hier.part' [33].

Foe each fertility measures, the statistical significance of their independent contribution to a plant trait was tested by randomizing the pairs of trait and fertility values for 1000 times [32] in R [30]. Z-scores were calculated from the generated distribution of randomized independent contributions, and statistical significance was determined based on the upper 95% confidence limit ($Z \geq 1.65$).

In addition, independent contributions were compared among fertility measures with different timescales by means of bootstrapping in R [30]. 134 plots were randomly selected with replacement, and the 95% confidence interval was computed from the distribution of independent contributions of 1000x bootstrapped 134 plots. Furthermore, the difference in independent effects between two fertility measures was computed for all combinations. When more than 95% of the difference was larger or smaller than zero, we considered that the two fertility measures had significantly ($p<0.05$) different magnitudes of independent effects.

Principal component regression. Since the soil fertility measures were strongly correlated, we extracted the main axes of variation by a Principal Component Analysis (PCA) in R [30], based on a correlation matrix to correct for differences in metrics among variables. We used six soil N measures, five soil P measures, and soil N:P ratio. All these variables were log-transformed prior to the analysis. We extracted the scores of sites along PCA axis 1 and 2 and related them to plot-mean values of each trait using linear regression.

Results

Relative contributions of soil N and P measures to plot-mean plant traits

Bivariate correlations between plant traits and soil fertility measures were often significant (table S2). All soil fertility measures together explained less than half of the among-site variation in leaf nutrient concentration (32.7% for LNC [Fig. 1A] and 42.8% for LPC [Fig. 1B]), in which a large part was attributed to the shared effects of soil N and P measures. There was no significant difference between soil N and P measures in their contribution to the total explained trait variance ($p = 0.42$ for LNC, $p = 0.30$ for LPC). P concentrations at the whole-canopy level were more strongly related to soil fertility measures than those on a leaf level (i.e. 65.9% of variance in WPC was explained, Fig. 1D). In contrast, the relationships between N concentrations and soil fertility were weak both on a whole-canopy level and on a leaf level (i.e. 31.0% of variance in WNC was explained, Fig. 1C). For both WPC and WNC, the shared effects of soil N and P measures were relatively small, and the contribution of soil P measures was larger than that of soil N measures (significant for WPC [$p<0.001$] but not for WNC [$p = 0.42$]).

55.5% of the variance in IV_{nut} was explained by soil fertility measures (Fig. 1E), of which less than half was attributed to the shared effects of soil N and P measures (18.2%). The contribution of soil P and N measures was not significantly different ($p = 0.33$).

Of the three components of the CSR strategy, soil fertility measures explained the variance in the S component to a largest extent (50.7%, Fig. 1G), followed by the C component (33.6%,

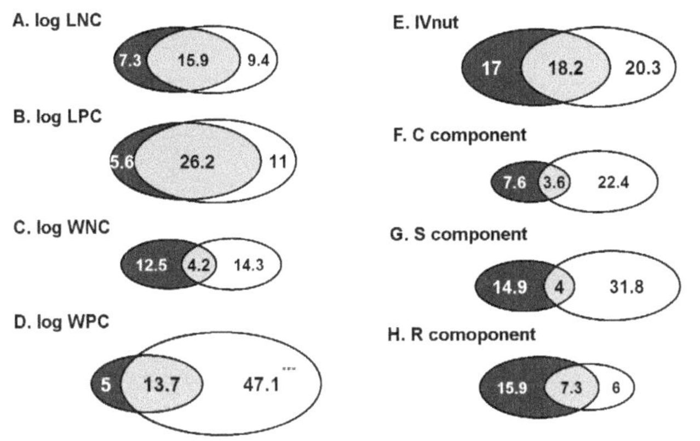

Figure 1. Partitioning of among-site plant trait variation. Trait variations (in percentage of variance) are divided into unique effects of soil N measures (dark grey), unique effects of soil P measures (white), and shared effects of both (light grey). Examined plot-mean plant traits are A: log-transformed leaf N concentration, LNC (mg/g) ($n=51$), B: log-transformed leaf P concentration, LPC (mg/g) ($n=51$), C: log-transformed N concentration of above-ground plant biomass, WNC (mg/g) ($n=82$), D: log-transformed P concentration of above-ground plant biomass, WPC (mg/g) ($n=82$), E: indicator value for nutrients, IV_{nut} ($n=134$), F: C component ($n=134$), G: S component ($n=134$), and H: R component ($n=134$). When the contribution of N or P measures to the total explained variance is significantly larger than the other, this is indicated by asterisks (***: $p<0.001$).

Fig. 1F) and the R component (29.3%, Fig. 1H). Shared effects contributed to a small proportion of the explained variance (4.0% for S, 3.6% for C, and 7.3% for R). For the R component, the contribution of soil N measures was almost significantly ($p=0.057$) larger than that of soil P measures, whereas for the C and S components, the contribution of soil P measures was considerably (but not significantly, $p=0.16$ and $p=0.21$, respectively) larger than that of soil N measures.

Effects of different timescales of soil fertility measures on plot-mean plant traits

Only few fertility measures had significant independent effects on LNC and WNC (i.e. dissolved P and soil total P for LNC [Fig. 2I], soil total N and annual Pmin for WNC [Fig. 2C and 2K]). The independent effects were not significantly ($p>0.05$) different among fertility measures with different timescales (Table S3).

For LPC, independent effects were significant for three out of six soil N measures (Fig. 2B) and for three out of five soil P measures (Fig. 2J). Longer-term measures (e.g. soil total N, soil total P) tended to have larger independent effects, but the differences with short-term measures were not very apparent (e.g. only dissolved N had slightly smaller effects than longer-term measures i.e. annual Nmin, 5yr Nmin, and soil total N, $p<0.05$, Table S3).

For WPC, all soil P measures had significant independent effects (Fig. 2L). Longer-term measures (e.g. 5yr Pmin and soil P:C ratio) tended to have larger independent effects, but the differences were not significant ($p>0.05$) except between 5yr Pmin and annual Pmin ($p<0.01$, Table S3).

For IV_{nut}, independent effects were significant for all fertility measures (Fig. 2E, Fig. 2M). The independent effects were not significantly ($p>0.05$) different among fertility measures with different timescales except between 5yr Pmin and annual Pmin ($p<0.05$, Table S3).

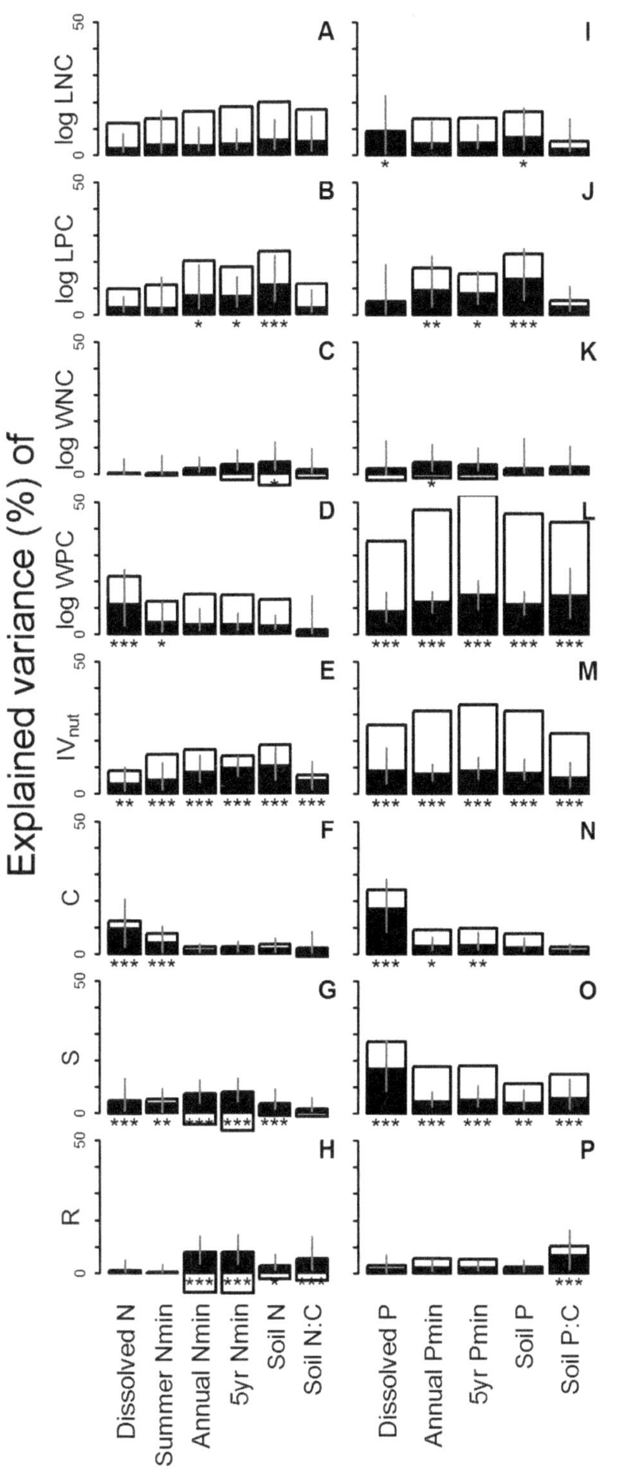

Figure 2. Hierarchical partitioning of among-site plant trait variation. Trait variations are divided into independent effects of a soil fertility measure (black bars) and its joint effects with other measures (white bars). All fertility measures (see Table 1 for specification) were log-transformed prior to the analysis. Examined plot-mean plant traits are A&I: log LNC, B&J: log LPC, C&K: log WNC, D&L: log WPC, E&M: IV_{nut}, F&N: C component, G&O: S component, and H&P: R component. See the caption of Figure 1 for the specification and sample number of each plant trait. Computation was done within each group of fertility measures (i.e. within soil N measures [A–H] and soil P measures [I–P]). Asterisks indicate that the independent effect was significant, based on Z-scores computed with randomization (*: $p<0.05$, **: $p<0.01$, ***:

$p<0.001$). 95% confidence intervals of independent effects, obtained by 1000-time bootstrapping, are shown.

For the C component of the CSR strategy, independent effects were significant and stronger for shorter-term fertility measures (e.g. dissolved N, summer Nmin, dissolved P; Fig. 2F and Fig. 2N). In contrast, for the S component, almost all fertility measures had significant independent effects (Fig. 2G, Fig. 2O), but the difference was less apparent (although dissolved P had stronger independent effects than all other soil P measures, $p<0.05$, Table S3). The strong independent effects of dissolved P on the C and S components could be an artefact of using Olsen extraction methods for some of the acid soils (see Appendix S2 for details). For the R component, soil P measures hardly had significant independent effects, and timescale did not matter ($p>0.05$, Table S3) (Fig. 2P), whereas intermediate-term soil N measures (e.g. annual Nmin, 5yr Nmin) had significant and stronger ($p<0.01$) independent effects than shorter-term measures (e.g. dissolved N, summer Nmin) (Fig. 2H, Table S3).

Relations between integrated soil fertility measures and plot-mean plant traits

PCA analysis extracted major axes of variation in soil fertility measures. The first axis, explaining 59.8% of the variance, was related to overall nutrient availability of a site (i.e. high in both soil N and P), with the negative axis values representing fertile conditions (Fig. 3). The second axis, explaining 18.3% of the variance, separated relatively P-rich (positive axis values) from N-rich (negative axis values) sites (Fig. 3).

Most plant traits, except WNC and the R component, were significantly related to PCA axis 1 (Fig. 4A–H). PCA axis 2 was significantly related only to WPC, IV_{nut}, the S component, and the R component (Fig. 4I–P). Multiple regression analysis showed that both PCA axes had significant ($p<0.05$) effects for WPC, IV_{nut}, and the S component (and almost significant [$p<0.07$] effects for

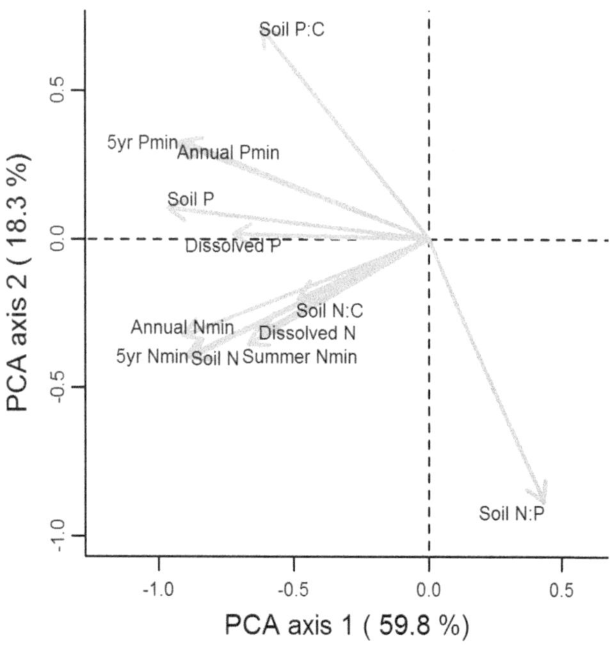

Figure 3. Principal Component Analysis (PCA) of 12 soil fertility measures for 134 plots.

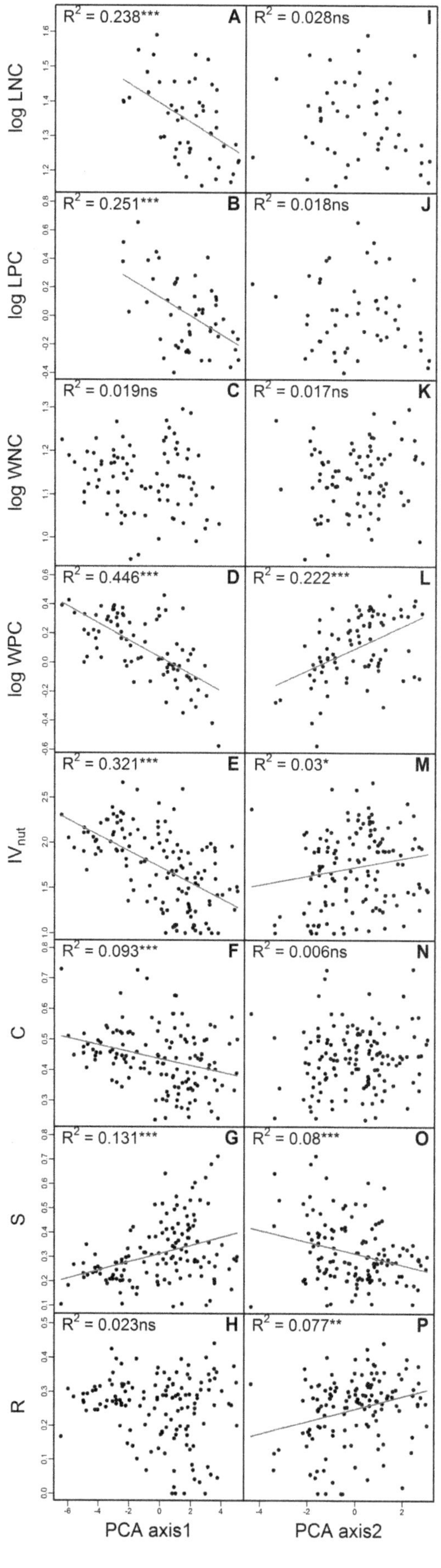

Figure 4. Relationships of the first and second PCA axis with plot-mean plant traits. R^2 and *p*-values of linear regression analysis are shown. Lines represent the regression model (only when $p<0.05$). See the caption of Figure 1 for the specification and sample numbers for each plant traits.

the R component); this indicates that the two axes had complementary rather than redundant effects on these traits.

Discussion

Mutual and stoichiometric control of soil N and P on plant traits

In line with previous studies [2,4,5], we found that fertile soils are associated with plant communities composed of species with higher nutrient concentrations, and include more competitor rather than stress-tolerant types of species. However, we highlight the importance of considering both soil N and soil P concentrations in explaining trait variation among communities. Indeed, we found that soil N and P had a strong mutual (shared) effect on LNC and LPC while other traits tend to be related to unique effect of soil N or soil P.

There are several possible mechanisms that explain why LNC is related to both soil N and soil P (rather than only to soil N) and why LPC is related to both soil N and soil P (rather than only to soil P). Although leaf-level nutrient concentrations change plastically with changing concentrations of that element in the soil [34], LNC and LPC are strongly coupled with a 2/3-power law of scaling [35,36] due to the mutually dependent roles of N and P in photosynthesis. Consequently, rapid growth, which is associated with high LNC and high LPC [35], is realized only when both soil N and P availability for plants is high. This clarifies the observed pattern that the shared effects of soil N and P (i.e. overall fertility effects), rather than the unique effects of soil N and soil P, explain the large variation in LNC and LPC.

In contrast, the nutrient concentrations on a whole-canopy level (WNC and WPC) reflect the availability of single elements more strongly than on a leaf level, because plants store excess elements in non-leaf organs. This is reflected in a much larger variation in N:P ratio in non-leaf organs than in leaves [37]. The predictive power of soil P on WPC was especially strong, because plants can adjust P concentrations much more flexibly than N concentrations to the nutrient availability in the soil [38].

For IV_{nut}, the unique effects of single elements contributed to most of the explained variance, although the shared effect of soil N and P also accounted for a large part of the variance. For the CSR strategy, most of the explained variance was related to the unique effects of single elements, rather than to the shared effects of soil N and P. Here, the effects of N and P were highly different, indicating a stoichiometric control of soil N and P on the CSR strategy of vegetation. The unique effects of soil P were larger than those of soil N for the C and S components, in accordance with the findings of Ceulemans et al. [39] based on Olsen-extracted P availability and KCl-extracted N availability. Variance in the R component was least explained, reflecting that environmental drivers other than nutrient availability (i.e. disturbance) are the primary determinants of the distribution of ruderals [21].

To test whether the above-mentioned trends were merely an artefact of biased distribution of N- and P-limited ecosystems in our study, we conducted the same analyses for N-limited and P-limited plots separately (Appendix S3). When only N-limited plots were considered, contributions of soil N measures to plant trait variance tended to increase slightly (Fig. S3 left); however, other major trends (e.g. a stronger effect of soil P on WPC than that of soil N [$p=0.094$], small shared effects on CSR strategy) remained. When only P-limited plots were considered, the contributions of soil P remained mostly unchanged or even decreased for some traits (Fig. S3 right). It is particularly notable that the stronger determinants for stress tolerators were N-related fertility measures in N-limited plots ($p<0.05$) and P-related fertility measures in P-limited plots ($p=0.079$) (Fig. S3). Furthermore, signs of correlations between soil N measures and integrative plant traits (IV_{nut}, CSR strategy) reversed in P-limited plots compared to N-limited plots for most cases (e.g. stress tolerator increased as soil N availability increased in P-limited plots only; see Table S4 in comparison to Table S5 in Appendix S3). These reversed relationships imply that, under P-poor conditions, high soil N availability results in an extreme imbalance of soil N and P and therefore induces harsher environments for plant species. This emphasizes the importance of explicitly considering stoichiometric effects of nutrients on plant functioning.

Timescale has only minor impact on fertility–trait relationships

Contrary to previous results [2,10], we found no clear indication that the fertility–trait relationship is sensitive to the timescale of the soil fertility measure. Almost all soil fertility measures were closely correlated, and therefore the independent effects of a particular fertility measure were almost never outstandingly stronger than that of others.

It is more difficult to estimate the availability of soil P for plants than that of soil N; this is because various geochemical processes of soil inorganic P, such as adsorption and precipitation, are involved. These processes were not explicitly included in our soil P measures, possibly obscuring the impact of timescale on the fertility–trait relationships. To test this, we examined for a subset of our dataset (36 plots) whether adding extra measurements of soil P availability improved the fertility–trait relationships (Appendix S4). Neither summer gross P mineralization rates (i.e. increase in Olsen-extractable P [both inorganic and organic] in in-situ incubation experiments), oxalate-extractable P (i.e. an estimate of reducible amount of P, which includes P adsorbed on Al- and Fe-hydroxides and oxides [40]), nor the degree of phosphate saturation (i.e. the percentage of oxalate-extracted P over half of oxalate-extracted Al plus Fe, an index for soil capacity to release P [41]) were superior to other P fertility measures in explaining trait variations (Fig. S5 in Appendix S4). So, the virtual absence of timescale impacts on soil fertility–trait relationships was not likely due to the inadequacy of selected P measures.

Use of integrated soil fertility measures to explain community trait composition

There was no single fertility measure that dominantly explained plant trait variations; this indicates that plant traits are mutually controlled by multiple soil fertility measures, suggesting the usefulness of using integrated fertility measures. Indeed, the main axis of variation in fertility measures, overall fertility gradient (PCA axis 1), was related to almost all plant traits. However, the explanatory power of the PCA axis 1 (i.e. R^2 values in Fig. 4) was only slightly higher for LNC and LPC than that of the best single fertility measures (i.e. variance explained by a fertility measure, including both independent and joint effects, in Fig. 2). For the other traits, PCA axis 1 explained less than the best single fertility measures. The other integrated fertility measure, the relative availability of soil N and P (PCA axis 2), had small but complementary effects for some plant traits (i.e. WPC, IV_{nut}, S component, R component). Similarly, including the type of nutrient limitation of plants (i.e. N- or P-limited) improved the relationship between overall fertility and several plant traits (WNC, WPC, IV_{nut}, R component) (Fig. S4 in Appendix S3). This means that the fertility–trait relationships are modulated by the type and magnitude of nutrient limitation, probably because the most influential factor of fertility is not identical for all sites but depends on which nutrient is actually limiting the plant growth of the site.

These findings imply that simultaneous consideration of overall fertility and N:P stoichiometry (either in soil or in plants) is a prerequisite for improving fertility–trait relationships. Note that the N:P stoichiometry effect cannot be assessed by individual fertility measures alone, but it can be explicitly tested by integrated fertility measures (i.e. PCA axis 2 in our case). This suggests the appropriateness of using integrated fertility measures as a starting point to explore which aspect of soil fertility has a relevant effect on the specific plant trait.

Wrong choice of fertility measure, or intrinsically moderate relationships between fertility and plant traits?

As in previous studies [2,4], variance of LNC was less strongly explained by fertility measures than that of LPC in our study, even if various types of fertility measures were considered (i.e. 32.7% explained by all fertility measures together, 19.9% by the best single measure [soil total N] and 23.8% by the best integrated measure [PCA axis 1]). Also, only a minor part of the variation was explained by soil fertility for WNC (31.0%, 3.2%, and 1.9%, respectively) and for R component (29.3%, 6.7%, and 7.7%, respectively). In contrast, several other plant traits could be well explained by fertility measures: 65.9%, 52.6%, and 44.6% for WPC, respectively, and 55.5%, 33.6%, and 32.1% for IV_{nut}, respectively. Reasonably good relationships between IV_{nut} or equivalent (e.g. Ellenberg indicator value for nutrients) and a single soil fertility measure have also been observed before (e.g. 49% explained by annual N mineralization rates [5], 35% explained by 'nitrification degree' [9], 29% explained by oxalate-extractable P [42]).

Thus, the moderate fertility–trait relationships of some plant traits are most likely not because of a wrong choice of fertility measure but because of the intrinsic nature of the relationships for these traits as explained earlier. Furthermore, environmental factors other than soil fertility also influence plant traits. For example, the occurrence of ruderal species (R component of CSR strategy) is primarily associated with disturbance rather than soil fertility [21], and LNC and LPC are weakly but consistently related with drought and oxygen stress [43]. Since plant traits are coordinated through physiological trade-offs and the coordination of the traits is strongly modulated by environmental factors [19,44], simultaneous consideration of multiple environmental factors is necessary to improve the prediction of these traits. Reasonably good relationships between fertility and IV_{nut} encourage the application of this relationship in species-distribution models to predict the functional composition of plant species. Also, other plant traits unexamined in our study, such as specific leaf area, may be considered for examining fertility–trait relationships. Specific leaf area was, however, not better related to soil fertility measures than LPC [4].

In conclusion, our study showed that among-site variations in nutrient-related plant traits are consistently, although moderately for some traits, related to soil fertility measures. Whether a trait has only a moderate relationship depends on the mechanism through which soil fertility and other factors regulate the trait variation. The timescale of the fertility measure has only negligible or minor effects on fertility–trait relationships, whereas the mutual and/or stoichiometric effects of N and P should be considered to improve the relationships. Since the relative importance of soil fertility measures is different among plant traits, a scan of integrated fertility measures will facilitate identification of influential fertility measures (or groups of fertility measures) for each specific plant trait separately.

Supporting Information

Table S1 Correlations among soil fertility measures.

Table S2 Correlations between plot-mean plant traits and soil fertility measures.

Table S3 Comparison of 1000x bootstrapped independent effects among soil fertility measures.

Appendix S1 Testing the effects of using two extraction methods for dissolved P.

Appendix S2 Testing the effects of using Olsen-P extraction for acid soils.

Appendix S3 Fertility–trait relationships in N- and P-limited ecosystems.

Appendix S4 Testing additional soil P measures for a subset of dataset.

Acknowledgments

This research was carried out in the framework of CARE project of the Dutch national research program 'Knowledge for Climate' and the joint research programme of the Dutch Water Utility sector. We thank Staatsbosbeheer, Veluwe national park, and PWN for the permission to sample soils; H. Olde Venterink and J. Ordonez for the permission to use their database; F. Daniels, H. Roelofsen, and J. Runhaar for field sampling; and J.C. Douma and R.P. Bartholomeus for SWAP simulations.

Author Contributions

Conceived and designed the experiments: YF PMvB JPMW. Performed the experiments: YF. Analyzed the data: YF PMvB. Wrote the paper: YF PMvB JPMW.

References

1. Aerts R, Chapin FS, III (2000) The mineral nutrition of wild plants revisited: a re-evaluation of processes and patterns. Advances in Ecological Research 30: 1–67.
2. Ordonez JC, van Bodegom PM, Witte JPM, Wright IJ, Reich PB, et al. (2009) A global study of relationships between leaf traits, climate and soil measures of nutrient fertility. Global Ecology and Biogeography 18: 137–149.
3. Ordonez JC, van Bodegom PM, Witte JPM, Bartholomeus RP, van Hal JR, et al. (2010) Plant Strategies in Relation to Resource Supply in Mesic to Wet Environments: Does Theory Mirror Nature? American Naturalist 175: 225–239.
4. Ordoñez JC, Van Bodegom PM, Witte JPM, Bartholomeus RP, Van Dobben HF, et al. (2010) Leaf habit and woodiness regulate different leaf economy traits at a given nutrient supply. Ecology 91: 3218–3228.

5. Douma JC, Witte J-PM, Aerts R, Bartholomeus RP, Ordoñez JC, et al. (2012) Towards a functional basis for predicting vegetation patterns; incorporating plant traits in habitat distribution models. Ecography 35: 294–305.
6. Sonnier G, Shipley B, Navas M-L (2010) Quantifying relationships between traits and explicitly measured gradients of stress and disturbance in early successional plant communities. Journal of Vegetation science 21: 1014–1024.
7. Fukami T, Dickie IA, Paula Wilkie J, Paulus BC, Park D, et al. (2010) Assembly history dictates ecosystem functioning: evidence from wood decomposer communities. Ecology Letters 13: 675–684.
8. Cavender-Bares J, Kitajima K, Bazzaz FA (2004) Multiple trait associations in relation to habitat differentiation among 17 Floridian oak species. Ecological Monographs 74: 635–662.
9. Schaffers AP, Sykora KV (2000) Reliability of Ellenberg indicator values for moisture, nitrogen and soil reaction: A comparison with field measurements. Journal of Vegetation science 11: 225–244.
10. Van Duren IC, Pegtel DM (2000) Nutrient limitations in wet, drained and rewetted fen meadows: evaluation of methods and results. Plant and Soil 220: 35–47.
11. Olde Venterink H, Pieterse NM, Belgers JDM, Wassen MJ, de Ruiter PC (2002) N, P and K budgets along nutrient availability and productivity gradients in wetlands. Ecological Applications 12: 1010–1026.
12. Schaffers AP (2000) In situ annual nitrogen mineralization predicted by simple soil properties and short-period field incubation. Plant and Soil 221: 205–219.
13. Carmo Horta Md, Roboredo M, Coutinho J, Torrent J (2010) Relationship between Olsen P and ammonium lactate-extractable P in Portuguese acid soils. Communications in Soil Science and Plant Analysis 41: 2358–2370.
14. Fujita Y, van Bodegom PM, Olde Venterink H, Runhaar H, Witte J-PM (2013) Towards a proper integration of hydrology in predicting soil nitrogen mineralization rates along natural moisture gradients. Soil Biology and Biochemistry 58: 302–312.
15. Del Grosso S, Ojima D, Parton W, Mosier A, Peterson G, et al. (2002) Simulated effects of dryland cropping intensification on soil organic matter and greenhouse gas exchanges using the DAYCENT ecosystem model. Environmental pollution 116: S75–S83.
16. Van Dam JC, Groenendijk P, Hendriks RFA, Kroes JG (2008) Advances of modeling water flow in variably saturated soils with SWAP. Vadose Zone Journal 7: 640–653.
17. von Asmuth JR, Maas K, Knotters M, Bierkens MFP, Bakker M, et al. (2012) Software for hydrogeologic time series analysis, interfacing data with physical insight. Environmental Modelling & Software 38: 178–190.
18. Pakeman RJ, Quested HM (2007) Sampling plant functional traits: What proportion of the species need to be measured? Applied Vegetation Science 10: 91–96.
19. Wright IJ, et al. (2004) The worldwide leaf economics spectrum. Nature 428: 821–827.
20. Witte JPM, Wójcik RB, Torfs PJJF, De Haan MWH, Hennekens S (2007) Bayesian classification of vegetation types with Gaussian mixture density fitting to indicator values. Journal of Vegetation science 18: 605–612.
21. Grime JP, Hodgson JG, Hunt R (2007) Comparative plant ecology: a functional approach to common British species. KirkcudbrightshireUK: Castlepoint Press. 748 p.
22. Käfer J, Witte JPM (2004) Cover-weighted averaging of indicator values in vegetation analyses. Journal of Vegetation science 15: 647–652.
23. Hunt R, Hodgson JG, Thompson K, Bungener P, Dunnett NP, et al. (2004) A new practical tool for deriving a functional signiture for herbaceous vegetation. Applied Vegetation Science 7: 163–170.
24. Klotz S, Kühn I, Durka W (2002) BIOLFLOR - Eine Datenbank zu biologisch-ökologischen Merkmalen der Gefäβpflanzen in Deutschland. Schriftenreihe für Vegetationskunde 38. Bonn: Bundesamt für Naturschutz.
25. Hodgson JG, Wilson PJ, Hunt R, Grime JP, Thompson K (1999) Allocating C-S-R plant functional types: a soft approach to hard problem. OIKOS 85: 282–296.
26. Kleyer M, Bekker RM, Knevel IC, Bakker JP, Thompson K, et al. (2008) The LEDA Traitbase: a database of life-history traits of the Northwest European flora. Journal of Ecology 96: 1266–1274.
27. Willby NJ, Pulford ID, Flowers TH (2001) Tissue nutrient signatures predict herbaceous-wetland community responses to nutrient availability. New Phytologist 152: 463–481.
28. Borcard D, Legendre P, Drapeau P (1992) Partialling out the Spatial Component of Ecological Variation. Ecology 73: 1045–1055.
29. Peres-Neto PR, Legendre P, Dray S, Borcard D (2006) Variation partitioning of species data matrices: estimation and comparison of fractions. Ecology 87: 2614–2625.
30. R Development Core Team (2013) R: A language and environment for statistical computing. R Foundation for Statistical Computing, Vienna, Austria. URL http://www.R-project.org/. Accessed 2013 May 16.
31. Chevan A, Sutherland M (1991) Hierarchical Partitioning. The American Statistician 45: 90–96.
32. Mac Nally R (2002) Multiple regression and inference in ecology and conservation biology: further comments on identifying important predictor variables. Biodiversity & Conservation 11: 1397–1401.
33. Walsh C, Mac Nally R (2008) hier.part: Hierarchical Partitioning. R package version 1.0–3.
34. Garnier E, Cheplick GP (1998) Interspecific variation in plasticity of grasses in response to nitrogen supply. Population biology of grasses: Cambridge University Press, Cambridge. pp. 155–182.
35. Reich PB, Oleksyn J, Wright IJ, Niklas KJ, Hedin L, et al. (2010) Evidence of a general 2/3-power law of scaling leaf nitrogen to phosphorus among major plant groups and biomes. Proceedings of the Royal Society B: Biological Sciences 277: 877–883.
36. Niklas KJ, Owens T, Reich PB, Cobb ED (2005) Nitrogen/phosphorus leaf stoichiometry and the scaling of plant growth. Ecology Letters 8: 636–642.
37. Kerkhoff AJ, Fagan WF, Elser JJ, Enquist BJ (2006) Phylogenetic and growth form variation in the scaling of nitrogen and phosphorus in the seed plants. American Naturalist 168: E103–E122.
38. Güsewell S (2004) N:P ratios in terrestrial plants: variation and functional significance. New Phytologist 164: 243–266.
39. Ceulemans T, Merckx R, Hens M, Honnay O (2011) A trait-based analysis of the role of phosphorus vs. nitrogen enrichment in plant species loss across North-west European grasslands. Journal of Applied Ecology 48: 1155–1163.
40. Van Bodegom PM, Van Reeven J, Van Der Gon HACD (2003) Prediction of reducible soil iron content from iron extraction data. Biogeochemistry 64: 231–245.
41. Lookman R, Vandeweert N, Merckx R, Vlassak K (1995) Geostatistical assessment of the regional distribution of phosphate sorption capacity parameters (FeOX and AlOX) in northern Belgium. Geoderma 66: 285–296.
42. Ertsen ACD, Alkemade JRM, Wassen MJ (1998) Calibrating Ellenberg indicator values for moisture, acidity, nutrient availability and salinity in the Netherlands. Plant Ecology 135: 113–124.
43. Douma JC, Bardin V, Bartholomeus RP, van Bodegom PM (2012) Quantifying the functional responses of vegetation to drought and oxygen stress in temperate ecosystems. Functional Ecology 26: 1355–1365.
44. Westoby M, Falster DS, Moles AT, Vesk PA, Wright IJ (2002) Plant ecological strategies: Some leading dimensions of variation between species. Annual Review of Ecology and systematics 33: 125–159.

5

Biotic and Abiotic Properties Mediating Plant Diversity Effects on Soil Microbial Communities in an Experimental Grassland

Markus Lange[1], Maike Habekost[1], Nico Eisenhauer[2], Christiane Roscher[3], Holger Bessler[4], Christof Engels[4], Yvonne Oelmann[5], Stefan Scheu[6], Wolfgang Wilcke[7], Ernst-Detlef Schulze[1], Gerd Gleixner[1]*

1 Max Planck Institute for Biogeochemistry, Jena, Germany, **2** Institute of Ecology, Friedrich- Schiller-University Jena, Jena, Germany, **3** UFZ, Helmholtz Centre for Environmental Research, Department of Community Ecology, Halle, Germany, **4** Department of Plant Nutrition, Faculty of Agriculture and Horticulture, Humboldt-Universität zu Berlin, Berlin, Germany, **5** Geoecology/Geography, Eberhard Karls University of Tübingen, Tübingen, Germany, **6** J.F. Blumenbach Institute of Zoology and Anthropology, Georg August University Göttingen, Göttingen, Germany, **7** Soil Science Group, Geographic Institute, University of Berne, Bern, Switzerland

Abstract

Plant diversity drives changes in the soil microbial community which may result in alterations in ecosystem functions. However, the governing factors between the composition of soil microbial communities and plant diversity are not well understood. We investigated the impact of plant diversity (plant species richness and functional group richness) and plant functional group identity on soil microbial biomass and soil microbial community structure in experimental grassland ecosystems. Total microbial biomass and community structure were determined by phospholipid fatty acid (PLFA) analysis. The diversity gradient covered 1, 2, 4, 8, 16 and 60 plant species and 1, 2, 3 and 4 plant functional groups (grasses, legumes, small herbs and tall herbs). In May 2007, soil samples were taken from experimental plots and from nearby fields and meadows. Beside soil texture, plant species richness was the main driver of soil microbial biomass. Structural equation modeling revealed that the positive plant diversity effect was mainly mediated by higher leaf area index resulting in higher soil moisture in the top soil layer. The fungal-to-bacterial biomass ratio was positively affected by plant functional group richness and negatively by the presence of legumes. Bacteria were more closely related to abiotic differences caused by plant diversity, while fungi were more affected by plant-derived organic matter inputs. We found diverse plant communities promoted faster transition of soil microbial communities typical for arable land towards grassland communities. Although some mechanisms underlying the plant diversity effect on soil microorganisms could be identified, future studies have to determine plant traits shaping soil microbial community structure. We suspect differences in root traits among different plant communities, such as root turnover rates and chemical composition of root exudates, to structure soil microbial communities.

Editor: Gabriele Berg, Graz University of Technology (TU Graz), Austria

Funding: This work (GL 262) was funded by the Deutsche Forschungsgemeinschaft within the scope of the biodiversity research group (FOR 456) with support from the Friedrich Schiller University Jena and the Max Planck Society. The funders had no role in study design, data collection and analysis, decision to publish, or preparation of the manuscript.

Competing Interests: The authors have declared that no competing interests exist.

* E-mail: ggleix@bgc-jena.mpg.de

Introduction

The soil microbial community holds a central position in ecosystem processes like carbon and nitrogen cycling (e.g., [1,2]). The performance and shape of soil microbial communities on one hand depend on soil properties, such as pH, temperature, texture and moisture [3–6], but on the other hand the soil microbial community is closely linked to plant communities through complex interactions [7–12]. Plants affect the soil microbial community through biomass production, litter quality, seasonal variability of litter production, root-shoot carbon allocation and root exudates [13–15]. In turn, soil microbial communities mineralize organic matter and enhance nutrient release by mineral weathering. Both processes increase the availability of nutrients enhancing plant growth [16,17] and consequently accelerate the matter flow between the aboveground and belowground parts of ecosystems.

Plant diversity influences a wide range of ecosystem processes, but the underlying mechanisms are not well understood [18]; for example the link between plant diversity and belowground processes is just fragmentarily explained. Increasing plant diversity modifies resource availability for microbial communities in soil [19,20], which might lead to higher niche differentiation and facilitation of the soil microbial community [21]. Beside species richness, the number of plant functional groups, i.e. species with similar morphological, phenological and physiological traits, impact soil microorganisms [21]. Plant functional groups, such as legumes and grasses, differ in litter quality and the amount of carbon and nitrogen released to the soil [22,23], thereby affecting microbial decomposition processes [24]. Bacteria and fungi form

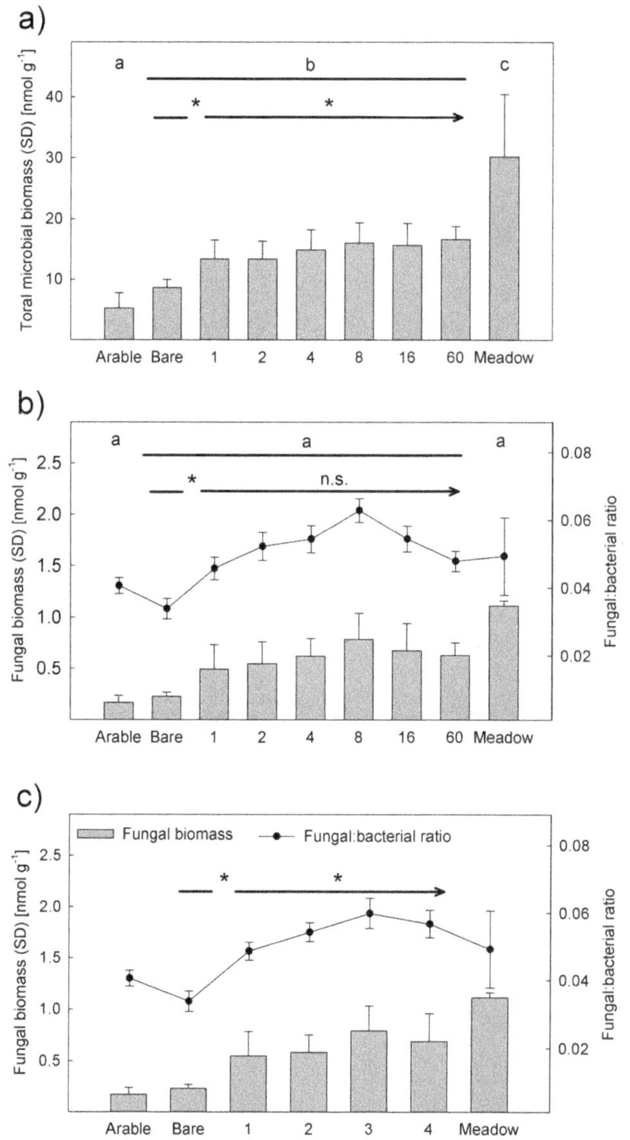

Figure 1. Impact of land use and plant diversity soil microbial community. Differences ($P<0.05$) between experimental plots and control sites (arable land and semi-natural meadows) were analyzed with Tukey's HSD test and indicated by letters. Differences between bare ground vs. vegetated plots and significant effects of plant diversity were tested with ANOVA (Table 1) and indicated by asterisks ($P<0.05$). Figures show effect of plant species richness on (a) total microbial biomass (b) fungal biomass and the fungal-to-bacterial biomass ratio and (c) number plant functional groups effect on fungal biomass and the fungal-to-bacterial biomass ratio.

most of the soil microbial biomass and represent the main drivers of organic matter turnover [25]. Since both groups prefer different qualities of resources they might be differently affected by plant diversity and plant community composition. Fungi are able to decompose litter with high C:N ratios [26]. Therefore, it has been suggested that plant communities providing litter with high C:N ratio favor decomposition by fungi, whereas plant communities producing litter with low C:N ratio favor decomposition by bacteria [27]. Moreover, there are differences among bacteria: Gram-negative bacteria are mainly root-associated and thus decompose organic molecules of low molecular weight [28], whereas Gram-positive bacteria are decomposing more complex materials, such as soil organic matter and litter [29]. As a consequence, the presence of certain plant functional groups is likely to promote distinct microbial groups. Therefore, higher plant diversity, as number of species or number of functional groups, might affect the composition of the soil microbial community by differences in litter input quantity and quality. Most studies investigating effects of plant diversity on soil microbial community focus on plant-originated inputs, often ignoring that differences in diversity and composition of plant communities also affect microclimatic conditions such as soil moisture. Conversely, studies considering plant mediated effects on soil moisture [23,30] usually do not account for root inputs or changes in the soil microbial composition. Identifying the relative importance of drivers changing soil microbial communities is needed to better understand the functioning of soils [31].

We assessed the effect of plant diversity and functional group composition on soil microbial communities using phospholipid fatty acids (PLFAs) [29,32,33]. The study was conducted in the framework of the Jena Experiment, a biodiversity experiment established by sowing different combinations of grassland species

[34] on a fallow agricultural soil. In addition to experimental plots with different levels of plant diversity and vegetation-free bare ground plots, we studied long-term meadows and on-going arable plots as adjacent to the field experiment as control sites to assessed, how the soil microbial community developed five years after establishing the experimental site. We hypothesized that (1) higher plant diversity increases soil microbial biomass, caused by higher amounts of litter input as well as by improved microclimatic conditions for soil microbes, and (2) plant functional group composition drives composition of the soil microbial community, exemplified e.g., by changes of fungal-to-bacterial biomass ratio (F:B ratio). Due to the production of low quality litter, we expected plant communities containing grasses but not legumes to favor fungi, whereas plant communities producing litter of high quality to favor bacteria.

Materials and Methods

Site description and experimental design

The field site of the Jena Experiment is located close to the city of Jena (Germany) in the floodplain of the river Saale (50°55′ N, 11°35′ E, 130 m a.s.l.). No specific permission was required to work on "The Jena Experiment" and no endangered or protected species were involved in this study. The soil (Eutric Fluvisol) has developed from up to 2 m thick fluvial sediments presenting a systematic variation of soil texture. The sand content decreases with distance to the river from 40% to 7%, while the silt and clay content increase (silt: 44% to 69%; clay: 16% to 24%). Experimental plots were arranged in four blocks parallel to the river to account for these differences in soil characteristics. Before the establishment of the Jena Experiment in 2002, the site was used as arable land since the early 1960s and ploughed and fertilized regularly. The Jena Experiment comprises 86 plots (82 vegetated and 4 bare ground plots, each 20 m by 20 m). The experimental design manipulates a gradient in sown plant species richness from 1 to 60 (1, 2, 4, 8, 16, and 60) near-orthogonal with a gradient in plant of functional group numbers from 1 to 4 (1, 2, 3 and 4). All 60 species are typical for Central European mesophilic grasslands [34,35]. They were grouped into four functional groups according to their morphological, phenological and physiological traits. The species pool included 16 grasses, 12 small herbs, 20 tall herbs and 12 legumes [34]. To maintain the diversity levels, all experimental communities have been weeded manually twice a year. Plots are mown twice a year, in June and September and are not fertilized. In addition bare ground plots with four replicates were established. Furthermore, soil microbial community was determined on two adjacent regularly mown non-fertilized meadows and two arable plots on the experimental site. The arable plots were continuously managed according to conventional agricultural procedures, growing cereals.

Soil sampling and phospholipid fatty acid (PLFA) analysis

In early May 2007, six soil samples per plot were taken with a core cutter (inner diameter: 4.8 cm, Eijkelkamp Agrisearch Equipment, Giesbeek, The Netherlands) to a depth of 5 cm, pooled and placed immediately in cooling boxes. Within 48 hours after sampling the soil was kept at 4°C, sieved <2 mm, remains of roots were manually removed and finally the samples were stored at −20°C until further sample processing. Soil samples were shaken with a mixture of chloroform, methanol and 0.05 M phosphate buffer (pH 7.4) to extract soil lipids [29,36]. The lipids were split into neutral lipids, glycolipids and phospholipids by eluting with chloroform, acetone and methanol from a silica-filled solid phase extraction column. Subsequently, phospholipids were hydrolyzed and methylated by a methanolic KOH solution and the PLFA-methyl esters were identified and quantified by gas chromatography with atomic emission detector (GC-AED) (Agilent, Böblingen, Germany) and gas chromatography-mass spectrometry (GC/MS) (Thermo Electron, Dreieich, Germany). Peak areas and the resulting amount of PLFA were calculated relative to the internal standard PLFA 19:0. The sum of all PLFAs (Table S1) was taken as total soil microbial biomass. Furthermore PLFAs were assigned to microbial groups [37,38]. The PLFAs 14:0, 14:0br, 15:0, 16:0, 17:0, 18:0 were used as general microbial markers. All monounsaturated and cyclic fatty acids were grouped as Gram-negative bacteria (Gram-), while all branched PLFAs were grouped as Gram-positive bacteria (Gram+). PLFA 18:2ω6 was used as a fungal biomarker [38,39]. The F:B ratio was calculated using the molar weight of the fungal PLFA marker divided by the sum of molar weights of bacterial PLFA biomarker [40].

Covariables

Fine root standing biomass (termed as 'root biomass' hereafter), leaf area index (LAI) and soil moisture were considered as potentially meaningful covariables. Unfortunately, in 2007, the year of the PLFA sampling, root biomass was not determined, thus we used an average of 2006 and 2008 root biomass measurements. In both years root biomass was sampled to a depth of 30 cm. In addition, the sampling in 2008 was stratified, so that the 0–5 cm depth increment could directly related to the sample of the soil microbes. Based on the ratio of the top increment (0–5 cm) to the total root biomass in 2008, we calculated the specific root biomass in the top soil (0–5 cm). Furthermore, nitrogen concentration of fine roots was determined using root material from ingrowth cores from, sampled between 2007 and 2008 [41]. N concentrations in the biomass were determined with an elemental analyzer (Vario EL Element Analyzer, Elementar, Hanau, Germany). In the course of the PLFA soil sampling soil moisture was determined, too, as the gravimetric soil water content. Leaf area index (LAI) was measured approx. 5 cm above ground level [42] using a LAI-2000 plant canopy analyzer (LI-COR) in late May 2007 (shortly before the first mowing of the year; see experimental design).

Statistical analysis

Using analyses of variance (ANOVA) followed by Tukey's HSD test we assessed microbial biomass and F:B ratio in experimental plots of the Jena Experiment, and their relationship to those from control plots (arable fields and meadows). ANOVA with sequential sum of squares (type I SS) was applied to test for effects of plant diversity on microbial biomass (total, Gram+, Gram− and fungal). The Jena Experiment is based on a factorial design with different combinations of plant species richness and number of functional groups, where all plots are arranged in a block design accounting for differences in soil texture among the blocks [34]. Therefore, 'block effect' was included as random factor and was fitted first. The contrast between bare ground plots vs. sown plots was fitted next, before testing for the effect of richness (log-linear term) and functional groups (linear term) as continuous variables. Finally, the presence of each plant functional group (small herbs, tall herbs, grasses and legumes) was included into the model in a series of alternative models. Furthermore, non-metric multidimensional scaling (NMDS) [43] was used to compare plot-specific patterns of PLFA profiles. The data used in the NMDS was normalized to the peak area of the highest peak (18.1n11) set at 100%. Bray-Curtis was used as dissimilarity index.

To investigate which mechanisms underlie the effects of plant diversity, we used structural equation modelling (SEM, see also

Table 1. Results of ANOVAs on the effect of the experimental variables on microbial community.

		MicBM		F:B ratio		Fungi		Gram+		Gram-	
	Df	**F value**	***P***	**F value**	***P***	**F value**	***P***	**F value**	***P***	**F value**	***P***
Block	3	**9.33**	**<0.001**	**3.54**	**0.019**	**8.20**	**<0.001**	**8.64**	**<0.001**	**6.64**	**<0.001**
Bare grounds	1	**18.83**	**<0.001**	**8.93**	**0.004**	**16.10**	**<0.001**	**15.30**	**<0.001**	**18.19**	**<0.001**
PSR	1	**13.60**	**<0.001**	0.09	0.768	2.65	0.108	**8.70**	**0.004**	**13.84**	**<0.001**
FG	1	0.03	0.858	**6.72**	**0.012**	**11.70**	**<0.001**	0.00	0.968	0.00	0.974
legumes	1	0.00	0.983	**21.74**	**<0.001**	**9.83**	**0.002**	0.01	0.933	0.26	0.609
Grasses	1	1.62	0.207	1.83	0.180	0.08	0.775	**4.83**	**0.031**	0.93	0.339
Tall herbs	1	0.10	0.757	0.97	0.329	0.13	0.725	0.27	0.607	0.09	0.763
Small herbs	1	0.47	0.496	2.72	0.103	0.41	0.524	1.33	0.252	0.45	0.506

Impact of plant diversity (plant species richness (PSR, log transformed) and number plant functional groups (FG)) and presence of distinct plant functional groups (legumes, grasses, small herbs, tall herbs) on total microbial biomass (MicBM), Gram-positive (Gram+), Gram-negative bacteria (Gram−), fungal biomass (Fungi) and the composition of the soil community, characterized by the fungal-to-bacterial biomass ration (F:B ratio). Numbers in bold display p values < 0.05 and numbers in italic display p values <0.1.

Table S2) with observed variables [44]. In SEM, all diversity levels, except bare ground, were considered. For every group of PLFAs assigned to specific microbial taxa, a full model was set up (Figure S1) including all experimental variables that were significant in the preceding ANOVA. As possible means by which the effect of plant diversity might be manifest, we included root biomass as measure of belowground plant input, root nitrogen concentration as measure of litter quality and LAI as a measure of plant community influence on evaporation and thus the micro-climatic conditions (e.g., soil moisture and temperature). The categorical variable 'block' was substituted by the continuous variable 'clay' content of soil. We considered aboveground plant inputs as negligible, because all above ground biomass was harvested twice a year. The minimal parsimonious models were identified using specification search, based on the Bayes information criterion (BIC) [45]. The adequacy of the model was tested with Chi-squared tests (χ^2 tests) and root mean square error of approximation (RMSEA) [44].

ANOVA and Tukey's HSD test were performed using R 2.15.2 [46] and structural equation modeling was performed using AMOS 18.0 [47].

Results

Soil microbial biomass

The mean of the total PLFA concentration, henceforth termed total microbial biomass, was14.4±3.5 nmol g^{-1} soil dry weight (mean ± sd) on experimental plots (vegetated plots and bare ground). This was significantly higher than measured on plots of arable land (5.3±2.5 nmol g^{-1}) and significantly lower than on meadows (30.2±10.3 nmol g^{-1}; Figure 1a).

ANOVA revealed block as a significant predictor of the total microbial biomass (Table 1). Furthermore, total microbial biomass was significantly lower on bare ground plots (8.7±1.3 nmol g^{-1}) than on vegetated plots (14.7±3.4 nmol g^{-1}). Plant species richness had a significant positive effect on the total microbial biomass on vegetated plots (Table 1; Figure 1a). The presence of individual plant functional groups did not affect total microbial biomass.

Structural equation modeling (SEM) showed that block (represented by the continuous variable clay content of soil) and plant richness, the design variables with significant influence indirectly affected total microbial biomass (Figure 2a). The minimal parsimonious model ($\chi^2_{13} = 21.74$, $P = 0.060$; RMSEA = 0.093, $P = 0.147$) explained 45% of the variance of total microbial biomass. Total microbial biomass was mainly explained by its positive relationship to soil moisture. Soil moisture increased with increasing plant richness and higher clay content. The major effect on soil moisture was attributed to increasing leaf area index (LAI), which itself was strongly correlated to plant richness. The negatively influence of root nitrogen concentration on total microbial biomass was driven by higher root biomass, which itself was increased at higher plant richness.

Soil microbial community structure

The F:B ratio did not differ among experimental plots, arable plots and meadow plots (experimental plots: 0.052±0.015; arable land: 0.041±0.003; meadows: 0.049±0.016; Figure 1b). In contrast, the F:B ratio was significantly lower on bare ground (0.034±0.006) than on vegetated plots of the biodiversity experiment (0.053±0.015). F:B ratio was positively affected by an increasing number of plant functional groups and negatively by the presence of legumes (Table 1; Figure 1c). Plant richness did not significantly affect the F:B ratio. However, the F:B ratio increased

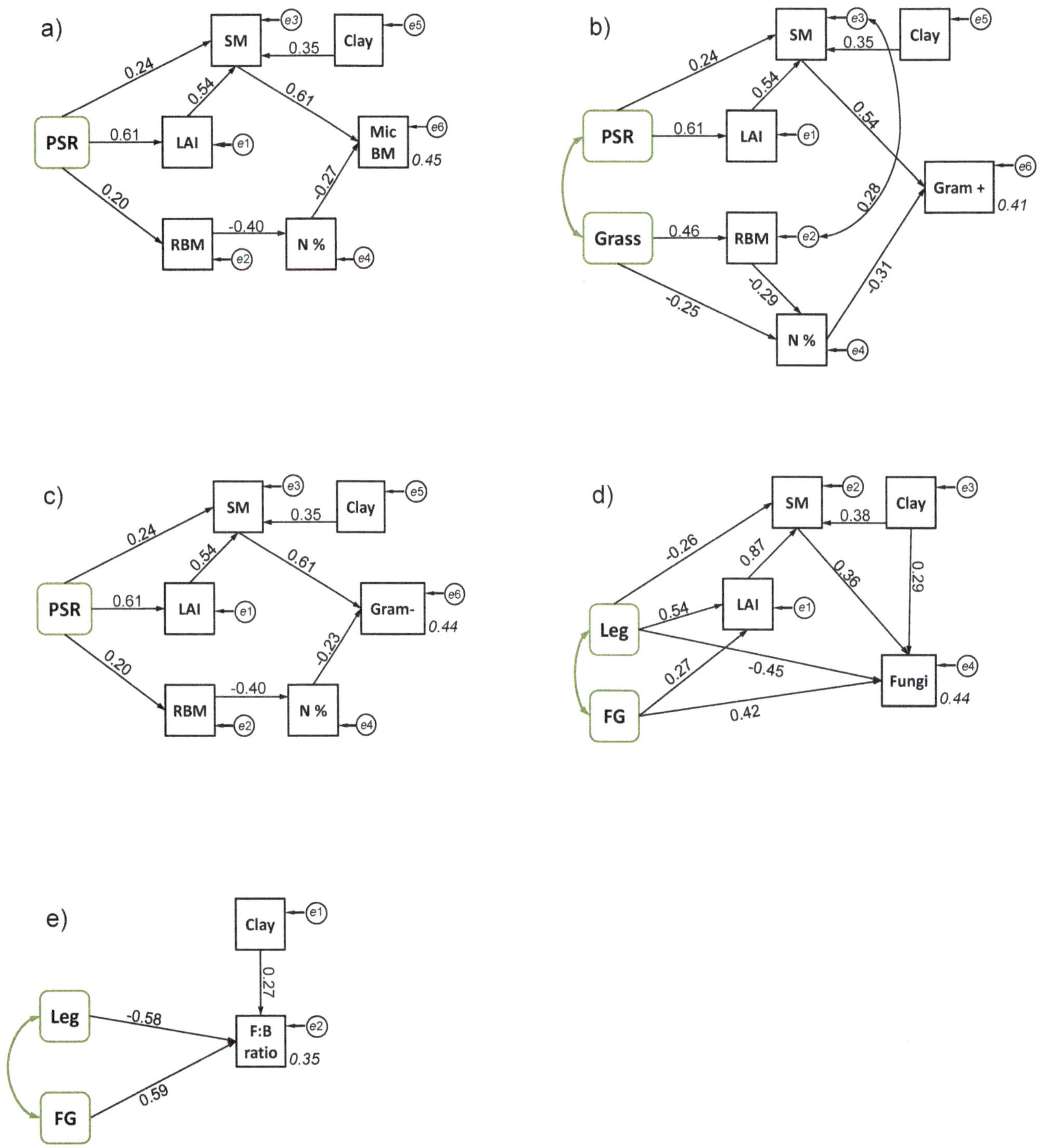

Figure 2. Minimal parsimonious models, testing direct and indirect effects of plant diversity on soil microbial community. Minimal SEM for a) total soil microbial biomass (MicBM), b) biomass of Gram positive bacteria (Gram+), c) biomass Gram negative bacteria (Gram−), d) fungal biomass (Fungi), and e) fungal-to-bacterial biomass ratio (F:B ratio). Arrows show significant relationships between variables. Numbers next to arrows show standardized parameter estimates (i.e., standardized regression weights). Circles (e1–e6) indicate error terms, and double-headed arrows indicate significant correlations between the error terms. Squared multiple correlations (R^2) for the dependent soil microbial biomass are given next to the box of the dependent variable. See the non-standardized estimates of the regression weights in Table S3a-e. Abbreviations are PSR: plant species richness, FG: plant functional group richness, LEG: presence of legumes, GRASS: presence of grasses, RBM: fine root standing biomass, N%: nitrogen concentration of fine roots, LAI: leaf area index, SM: soil moisture, Clay: clay content of soil

from low plant richness plots to medium ones with eight plant species, and decreased again in plots with high plant richness. Regression analyses of the relationship between F:B ratio and both, fungal and bacterial biomass revealed that the F:B ratio was more related to fungi ($R^2 = 0.67$, $P<0.001$) than to bacterial biomass ($R^2 = 0.043$, $P<0.066$).

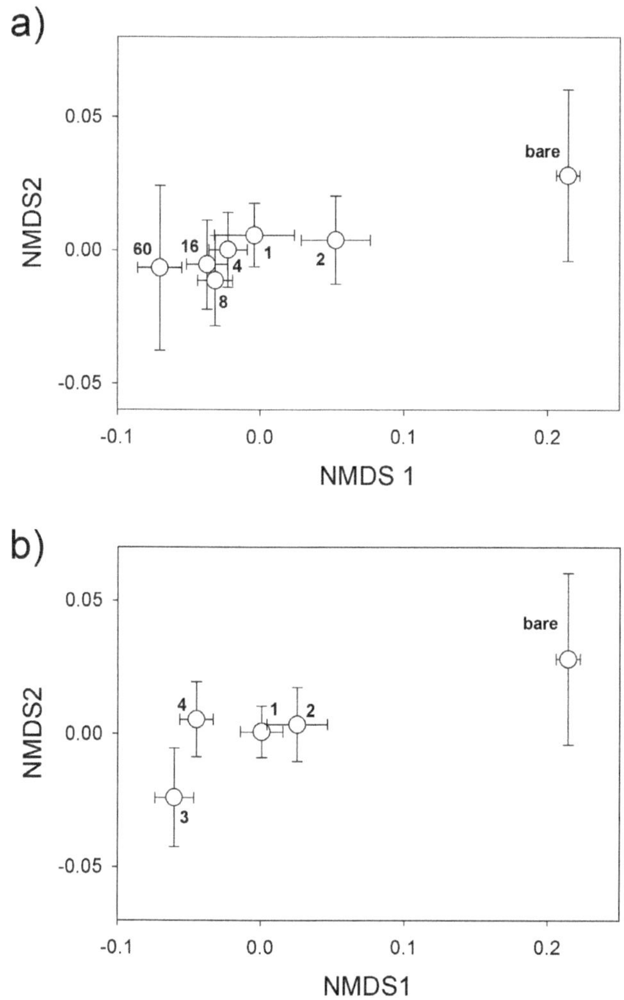

Figure 3. Summary of non-metric multidimensional scaling (NMDS) of the PLFAs. Differences among bare ground plots and different plant diversity levels are shown for (a) 1 to 60 sown plant species and (b) 1 to 4 functional groups. Bray-Curtis was used as dissimilarity index.

Considering biomass of Gram+, Gram- and fungi separately, all groups differed significantly among blocks and between bare ground and vegetated plots (Table 1). Plant diversity positively affected both bacterial groups. Gram+ as well as Gram- bacteria were reduced on bare ground plots (Gram+ = 2.3±0.4; Gram- = 4.6±0.6) compared to vegetation plots (Gram+ = 3.9±1.0; Gram- = 7.7±1.7) and biomass of both bacterial groups were increased with increasing plant richness (Table 1). However, the ANOVA also revealed differences: while Gram+ were positively influenced by the presence of grasses, Gram- were not affected by any of the plant functional groups. Fungal biomass was positively affected by the number of functional groups present, and negatively by the presence of legumes.

The results of the NMDS (Figure 3), based on the PLFA composition, confirmed the strong dissimilarity between bare ground plots and vegetated plots. The dissimilarity between the vegetated plots was relatively small, though we found a clear effect of plant diversity, i.e. the higher the plant diversity on the plot the more different were the microbial communities compared to low diverse plots.

SEM for Gram+ ($\chi^2_{17} = 24.877$, $P = 0.098$; RMSEA = 0.078, $P = 0.231$) and Gram- bacteria ($\chi^2_{13} = 19.80$, $P = 0.100$; RMSEA = 0.082, $P = 0.217$) explained to 41% and 44% of variance, respectively, and revealed high analogy between the groups (Figure 2b, 2c). Both bacterial groups were mainly driven by soil moisture, which was mostly affect by LAI. Furthermore, both groups had a negative relationship with nitrogen concentration of fine roots, mediated by increased root biomass. The minimal parsimonious model explains 44% the variation in fungal biomass ($\chi^2_5 = 5.43$, $P = 0.365$; RMSEA = 0.034, $P = 0.475$). In contrast to the bacterial groups, fungal biomass was neither affected by the amount of root biomass nor by its quality (Figure 2d). Although there was a positive indirect pathway from functional groups and legumes via LAI and soil moisture to fungal biomass, strong direct paths from functional groups and legumes remained in the minimal parsimonious model. These direct paths indicate that the diversity effect was driven by mechanisms other than soil moisture or quantity and quality of root biomass. The F:B ratio was explained to 35% by the minimal model ($\chi^2_2 = 2.27$, $P = 0.322$; RMSEA = 0.041, $P = 0.390$). In contrast to all other models, only direct paths connected the experimental variables to F:B ratio (Figure 2e): it strongly decreased in the presence of legumes, but increased with increasing number of plant functional groups to almost the same extent. These relationships could not be explained by our measured covariables.

Discussion

In the framework of the Jena Experiment we investigated how soil microbial communities are affected by plant diversity and the underlying mechanisms of these effects. In grassland with manipulated plant species richness and number of plant functional groups we showed that the soil microbial communities are strongly linked to plant diversity. Corresponding to hypothesis 1, the positive plant diversity effect on total microbial biomass was mainly driven by improved microclimatic conditions, while we found only a minor influence of the amount of root litter inputs on the soil microbes. Furthermore, number of plant functional groups and the plant functional composition, in particular the presence of legumes, highly impact the microbial community composition, referring to hypothesis 2. Below, we will discuss in detail, how plant diversity drives the soil microbial community.

Five years after conversion from arable land to grasslands, increased soil microbial biomass indicates that the microbial community performs better. In addition to the growth of the soil microbial community, it has been reported that the community also has a higher metabolic activity compared to the initial conditions [21]. Lower microbial biomass on arable land probably is due to soil disturbance by tillage and the tillage-induced changes of soil properties [48]. The lower organic carbon concentration in arable soils is attributed to faster decomposition of soil organic matter, which in turn reduces the microbial biomass in the long term (e.g., [49,50]). However, even in plots with highest plant diversity, i.e., 60 species and 4 functional groups, microbial biomass was lower than in adjacent meadows. This indicates that the time since conversion of our study area from arable use to grassland was not sufficient to reach the state of microbial biomass of permanent meadows. However, as total microbial biomass significantly increased with increasing plant richness, higher plant diversity promotes the development towards the stage of permanent meadows (Figure 1a).

Confirming hypothesis 1, plant richness as well as clay content of the soil indirectly increased total soil microbial biomass. Interestingly, structural equation modeling suggests that this was mediated via soil moisture. Soil moisture itself holds a central position in the interplay between plant diversity, abiotic soil conditions and microbial biomass (Figure 2a). The strong influence of soil texture on soil moisture is well known: with smaller particle size soil water holding capacity increases. Results of the present study suggest, however, that the positive effect of plant diversity on soil microbial biomass may exceed that soil texture via changes in soil moisture. Higher plant diversity increased canopy density of the plant stands, measured as LAI, which presumably reduced evaporation from the soil surface [23,51]. Plant richness also affected soil microbial biomass via root inputs, namely via root nitrogen concentration. The detrimental effect of nitrogen concentration on microbial biomass was closely related to increased root biomass; with increasing plant richness root biomass increased, while at the same time nitrogen concentration decreased, which confirms earlier findings [22,41].

Results of NMDS showed that the composition of PLFAs differed mainly between bare ground and vegetated plots, while in vegetated plots (1-60 plant species, 1-4 plant functional groups) the composition of PLFAs was similar. However, the dissimilarity of the microbial community composition was more pronounced in plots with different diversity levels, i.e., low diverse plots differed most from high diverse plots. Higher diversity in plant communities leads to more diverse organic matter input in quantity, quality and timing [9,13], and this likely is responsible for the observed changes in microbial communities along diversity levels. The plant diversity effect on microbial community composition was also reflected in the F:B ratio, but in contrast to total microbial biomass, the F:B ratio was more affected by functional groups than by species richness, supporting hypothesis 2. Moreover, this relation to plant functional groups reflects the stronger dependency of F:B ratios from changes in fungal biomass than in bacterial biomass [52]. We further found legumes to be a strong predictor of F:B ratio, which is in line with previous findings [15,24,53]. However, neither the underlying mechanisms of the positive effect of functional groups nor of the negative legume effect on the F:B ratio was mediated by the considered covariables. Similar results have been reported by Lamb *et al.* [54], who studied the effect of plant species richness and evenness on soil microbial communities in a pot experiment. The lack of relationships between F:B ratio and root litter quantity and quality as well as soil moisture indicates that both microbial groups are similarly affected by these variables. The strong direct link between plant diversity and F:B ratio, however, points to other plant resources as major drivers of soil microorganisms, such as root exudates. Indeed, root exudates were reported to strongly influence soil microbial communities [14,55]. In more diverse plant mixtures resource supply for microorganisms may be assumed to be higher and more diverse, while resource supply in monocultures is expected to be more one-sided and temporally limited. Furthermore, it is known, that the number of plant functional groups and presence of legumes may be related to turnover rates and decomposition of fine roots [24,56,57], which might cause changes in microbial community structure.

Although bacteria and fungi were similarly affected by plant diversity, we found bacteria more related to plant diversity-controlled abiotic soil properties, while and fungi were more affected by the input of organic materials. As shown by de Vries *et al.* [58] fungal-based soil food webs are more resistant to disturbances, while bacterial communities are more resilient due to their fast life cycle. This might explain why the bacterial community was in our study more related to fast changing abiotic conditions, such as soil moisture.

Conclusion

We identified changes in microclimatic conditions, in particular increased soil moisture, as a main mechanism how plant diversity affects soil microbial biomass in the topsoil. Furthermore, the results indicate that shifts in the microbial community composition, namely in the F:B ratio, heavily rely on differences in the quality and quantity of root exudates. Changes in soil microbial biomass with plant diversity suggest that microbial communities of the established grassland systems develop towards permanent meadows, but that reaching the state of these meadows takes decades. Notably, however, differences in microbial biomass indicate that high diverse plant communities promote faster transition towards permanent meadows indicating that plant diversity is a key factor for restoring functional grassland systems on former arable land.

Supporting Information

Figure S1 Scheme of the full model used in the structural equation modelling. The full model included simultaneously all measures of plant diversity with significant impact (potentially plant species richness, number of plant functional groups and the presence/absence of legumes, grasses, small herbs and tall herbs) on microbes. Measurements of plant inputs (fine root biomass (RBM), and nitrogen content of fine roots (N%) leaf area index (LAI), soil moisture (SM) and clay content of soil (Clay) have also been included in the model to explain the underlying mechanisms of the diversity effect.

Table S1 Mean concentrations of identified PLFAs in nmol g^{-1} dw (dry weight).

Table S2 Correlation matrix of the predictors. Abbreviations are: Block numeric (Bl_num), logarithmic plant species richness (log_PSR), number of plant functional groups (FG), presence/absence of grasses (gras), presence/absence of small herbs (sherb), presence/absence of tall herbs (therb), presence/absence of legumes (leg), soil content of clay [%] (clay), leaf area index (LAI), fine root standing biomass [$g\ m^{-2}$] (RBM), nitrogen content of fine root biomass [%] (N%).

Table S3 Estimates of the minimal adequate structural equation models (maximum likelihood) for a) Total microbial biomass (MicMB b) Gram positive bacteria (Gram+), c) Gram negative bacteria (Gram-), d) Fungal-to-bacterial ratio (F:B ratio) and e) Fungi.

Acknowledgments

We thank Janine Seyfferth, Sibylle Steinbeiss and Steffen Rühlow for their help during sampling and for helping with the PLFA extraction, Alexandra Weigelt for providing the LAI data and Francesca Hopkins for her helpful comments on the manuscript. We gratefully acknowledge all the people that were involved in planning, set up and maintaining of the experiment and the German Research Foundation (DFG) for financial support of FOR 456 and Gl262-14.

Author Contributions

Conceived and designed the experiments: GG SS WW EDS CR. Performed the experiments: MH GG HB. Analyzed the data: ML MH CR NE. Contributed reagents/materials/analysis tools: GG. Wrote the paper: ML MH NE CR HB CE YO SS WW EDS GG.

References

1. Balser TC, Firestone MK (2005) Linking microbial community composition and soil processes in a California annual grassland and mixed-conifer forest. Biogeochemistry 73: 395–415.
2. de Vries FT, Bloem J, Quirk H, Stevens CJ, Bol R, et al. (2012) Extensive Management Promotes Plant and Microbial Nitrogen Retention in Temperate Grassland. Plos One 7.
3. Griffiths RI, Thomson BC, James P, Bell T, Bailey M, et al. (2011) The bacterial biogeography of British soils. Environmental Microbiology 13: 1642–1654.
4. Papatheodorou EM, Argyropoulou MD, Stamou GP (2004) The effects of large- and small-scale differences in soil temperature and moisture on bacterial functional diversity and the community of bacterivorous nematodes. Applied Soil Ecology 25: 37–49.
5. Steenwerth KL, Drenovsky RE, Lambert JJ, Kluepfel DA, Scow KM, et al. (2008) Soil morphology, depth and grapevine root frequency influence microbial communities in a Pinot noir vineyard. Soil Biology & Biochemistry 40: 1330–1340.
6. Thoms C, Gattinger A, Jacob M, Thomas FM, Gleixner G (2010) Direct and indirect effects of tree diversity drive soil microbial diversity in temperate deciduous forest. Soil Biology & Biochemistry 42: 1558–1565.
7. Gleixner G, Kramer C, Hahn V, Sachse D (2005) The Effect of Biodiversity on Carbon Storage in Soils. In: Scherer-Lorenzen M, Körner C, Schulze E-D, editors. Forest Diversity and Function: Springer Berlin Heidelberg. pp. 165–183.
8. Habekost M, Eisenhauer N, Scheu S, Steinbeiss S, Weigelt A, et al. (2008) Seasonal changes in the soil microbial community in a grassland plant diversity gradient four years after establishment. Soil Biology and Biochemistry 40: 2588–2595.
9. Hooper DU, Bignell DE, Brown VK, Brussaard L, Dangerfield JM, et al. (2000) Interactions between aboveground and belowground biodiversity in terrestrial ecosystems: Patterns, mechanisms, and feedbacks. Bioscience 50: 1049–1061.
10. Latz E, Eisenhauer N, Rall BC, Allan E, Roscher C, et al. (2012) Plant diversity improves protection against soil-borne pathogens by fostering antagonistic bacterial communities. Journal of Ecology 100: 597–604.
11. Potthoff M, Steenwerth KL, Jackson LE, Drenovsky RE, Scow KM, et al. (2006) Soil microbial community composition as affected by restoration practices in California grassland. Soil Biology & Biochemistry 38: 1851–1860.
12. Steinbeiss S, Bessler H, Engels C, Temperton VM, Buchmann N, et al. (2008) Plant diversity positively affects short-term soil carbon storage in experimental grasslands. Global Change Biology 14: 2937–2949.
13. Angers DA, Caron J (1998) Plant-induced changes in soil structure: Processes and feedbacks. Biogeochemistry 42: 55–72.
14. Bais HP, Weir TL, Perry LG, Gilroy S, Vivanco JM (2006) The role of root exudates in rhizosphere interations with plants and other organisms. Annual Review of Plant Biology 57: 233–266.
15. Bezemer TM, Lawson CS, Hedlund K, Edwards AR, Brook AJ, et al. (2006) Plant species and functional group effects on abiotic and microbial soil properties and plant-soil feedback responses in two grasslands. Journal of Ecology 94: 893–904.
16. Porazinska DL, Bardgett RD, Blaauw MB, Hunt HW, Parsons AN, et al. (2003) Relationships at the aboveground-belowground interface: Plants, soil biota, and soil processes. Ecological Monographs 73: 377–395.
17. van der Heijden MGA, Bardgett RD, van Straalen NM (2008) The unseen majority: soil microbes as drivers of plant diversity and productivity in terrestrial ecosystems. Ecology Letters 11: 296–310.
18. Cardinale BJ, Duffy JE, Gonzalez A, Hooper DU, Perrings C, et al. (2012) Biodiversity loss and its impact on humanity. Nature 486: 59–67.
19. De Deyn GB, Quirk H, Yi Z, Oakley S, Ostle NJ, et al. (2009) Vegetation composition promotes carbon and nitrogen storage in model grassland communities of contrasting soil fertility. Journal of Ecology 97: 864–875.
20. Zak DR, Holmes WE, White DC, Aaron DP, Tilman D (2003) Plant Diversity, Soil Microbial Communities, and Ecosystem Function: Are There Any Links? Ecology 84: 2042–2050.
21. Eisenhauer N, Bessler H, Engels C, Gleixner G, Habekost M, et al. (2010) Plant diversity effects on soil microorganisms support the singular hypothesis. Ecology 91: 485–496.
22. Abbas M, Ebeling A, Oelmann Y, Ptacnik R, Roscher C, et al. (2013) Biodiversity Effects on Plant Stoichiometry. Plos One 8.
23. Rosenkranz S, Wilcke W, Eisenhauer N, Oelmann Y (2012) Net ammonification as influenced by plant diversity in experimental grasslands. Soil Biology & Biochemistry 48: 78–87.
24. Breulmann M, Schulz E, Weisshuhn K, Buscot F (2012) Impact of the plant community composition on labile soil organic carbon, soil microbial activity and community structure in semi-natural grassland ecosystems of different productivity. Plant and Soil 352: 253–265.
25. Baath E, Anderson TH (2003) Comparison of soil fungal/bacterial ratios in a pH gradient using physiological and PLFA-based techniques. Soil Biology & Biochemistry 35: 955–963.
26. Bowen RM, Harper SHT (1990) Decomposition of Wheat Straw and Related-Compounds by Fungi Isolated from Straw in Arable Soils. Soil Biology & Biochemistry 22: 393–399.
27. Wardle DA, Walker LR, Bardgett RD (2004) Ecosystem properties and forest decline in contrasting long-term chronosequences. Science 305: 509–513.
28. Griffiths BS, Bonkowski M, Dobson G, Caul S (1999) Changes in soil microbial community structure in the presence of microbial-feeding nematodes and protozoa. Pedobiologia 43: 297–304.
29. Kramer C, Gleixner G (2006) Variable use of plant- and soil-derived carbon by microorganisms in agricultural soils. Soil Biology and Biochemistry 38: 3267–3278.
30. Vogel A, Eisenhauer N, Weigelt A, Scherer-Lorenzen M (2013) Plant diversity does not buffer drought effects on early-stage litter mass loss rates and microbial properties. Global Change Biology 19: 2795–2803.
31. O'Donnell AG, Seasman M, Macrae A, Waite I, Davies JT (2001) Plants and fertilisers as drivers of change in microbial community structure and function in soils. Plant and Soil 232: 135–145.
32. Ruess L, Chamberlain PM (2010) The fat that matters: Soil food web analysis using fatty acids and their carbon stable isotope signature. Soil Biology & Biochemistry 42: 1898–1910.
33. Zelles L, Bai QY (1993) Fractionation of Fatty-Acids Derived from Soil Lipids by Solid-Phase Extraction and Their Quantitative-Analysis by Gc-Ms. Soil Biology & Biochemistry 25: 495–507.
34. Roscher C, Schumacher J, Baade J, Wilcke W, Gleixner G, et al. (2004) The role of biodiversity for element cycling and trophic interactions: an experimental approach in a grassland community. Basic & Applied Ecology 5: 107–121.
35. Ellenberg H (1988) Vegetation Ecology of Central Europe: Cambridge University Press.
36. Bligh EG, Dyer WJ (1959) A Rapid Method of Total Lipid Extraction and Purification. Canadian Journal of Biochemistry and Physiology 37: 911–917.
37. Seifert A-G, Trumbore S, Xu X, Zhang D, Kothe E, et al. (2011) Variable effects of labile carbon on the carbon use of different microbial groups in black slate degradation. Geochimica Et Cosmochimica Acta 75: 2557–2570.
38. Zelles L (1999) Fatty acid patterns of phospholipids and lipopolysaccharides in the characterisation of microbial communities in soil: a review. Biology and Fertility of Soils 29: 111–129.
39. Frostegard A, Baath E, Tunlid A (1993) Shifts in the structure of soil microbial communities in limed forests as revealed by phospholipid fatty acid analysis. Soil Biology & Biochemistry 25: 723–730.
40. Frostegard A, Baath E (1996) The use of phospholipid fatty acid analysis to estimate bacterial and fungal biomass in soil. Biology and Fertility of Soils 22: 59–65.
41. Bessler H, Oelmann Y, Roscher C, Buchmann N, Scherer-Lorenzen M, et al. (2012) Nitrogen uptake by grassland communities: contribution of N-2 fixation, facilitation, complementarity, and species dominance. Plant and Soil 358: 301–322.
42. Weigelt A, Marquard E, Temperton VM, Roscher C, Scherber C, et al. (2010) The Jena Experiment: six years of data from a grassland biodiversity experiment. Ecology 91: 930–931.
43. Oksanen J, Blanchet FG, Kindt R, Legendre P, O'Hara RB, et al. (2010) Community Ecology Package. 1.17-2 ed. pp. Ordination methods, diversity analysis and other functions for community and vegetation ecologists.
44. Grace JB (2006) Structural Equation Modeling and Natural Systems: Cambridge University Press.
45. Arbuckle JL (2009) Amos 18 User's Guide. Crawfordville, FL, USA: SPSS Incorporated.
46. RDevelopmentCoreTeam (2012) R: A language and environment for statistical computing. R Foundation for Statistical Computing, Vienna, Austria. URL.
47. AmosDevelopmentCorporation (2009). Crawfordville, FL 32327 USA.
48. Cookson WR, Murphy DV, Roper MM (2008) Characterizing the relationships between soil organic matter components and microbial function and composition along a tillage disturbance gradient. Soil Biology & Biochemistry 40: 763–777.
49. Hamelinck CN, Faaij APC (2006) Outlook for advanced biofuels. Energy Policy 34: 3268–3283.
50. Ravit B, Ehenfeld JG, Haggblom MM (2006) Effects of vegetation on root-associated microbial communities: A comparison of disturbed versus undisturbed estuarine sediments. Soil Biology & Biochemistry 38: 2359–2371.
51. Kreutziger Y (2006) Rückkopplungseffekte verschieden diverser Grünlandökosysteme auf die Komponenten des Bodenwasserhaushaltes an einem Auestandort der Saale - Ergebnisse des Jenaer Biodiversitätsexperiments. [Dissertation]. Jena: Friedrich-Schiller-University.
52. Bardgett RD, Hobbs PJ, Frostegard A (1996) Changes in soil fungal:bacterial biomass ratios following reductions in the intensity of management of an upland grassland. Biology and Fertility of Soils 22: 261–264.
53. Zhang C-B, Liu W-L, Wang J, Chen T, Yuan Q-Q, et al. (2011) Plant functional group richness-affected microbial community structure and function in a full-scale constructed wetland. Ecological Engineering 37: 1360–1368.
54. Lamb EG, Kennedy N, Siciliano SD (2011) Effects of plant species richness and evenness on soil microbial community diversity and function. Plant and Soil 338: 483–495.
55. Baudoin E, Benizri E, Guckert A (2003) Impact of artificial root exudates on the bacterial community structure in bulk soil and maize rhizosphere. Soil Biology & Biochemistry 35: 1183–1192.

56. Goebel M, Hobbie SE, Bulaj B, Zadworny M, Archibald DD, et al. (2011) Decomposition of the finest root branching orders: linking belowground dynamics to fine-root function and structure. Ecological Monographs 81: 89–102.
57. Leigh MB, Fletcher JS, Fu XO, Schmitz FJ (2002) Root turnover: An important source of microbial substrates in rhizosphere remediation of recalcitrant contaminants. Environmental Science & Technology 36: 1579–1583.
58. de Vries FT, Manning P, Tallowin JRB, Mortimer SR, Pilgrim ES, et al. (2012) Abiotic drivers and plant traits explain landscape-scale patterns in soil microbial communities. Ecology Letters 15: 1230–1239.

6

Tillage, Mulch and N Fertilizer Affect Emissions of CO_2 under the Rain Fed Condition

Sikander Khan Tanveer, Xiaoxia Wen, Xing Li Lu, Junli Zhang, Yuncheng Liao*

College of Agronomy, Northwest A&F University Yangling, Shaanxi, P.R. China

Abstract

A two year (2010–2012) study was conducted to assess the effects of different agronomic management practices on the emissions of CO_2 from a field of non-irrigated wheat planted on China's Loess Plateau. Management practices included four tillage methods i.e. T_1: (chisel plow tillage), T_2: (zero-tillage), T_3: (rotary tillage) and T_4: (mold board plow tillage), 2 mulch levels i.e., M_0 (no corn residue mulch) and M_1 (application of corn residue mulch) and 5 levels of N fertilizer (0, 80, 160, 240, 320 kg N/ha). A factorial experiment having a strip split-split arrangement, with tillage methods in the main plots, mulch levels in the sub plots and N-fertilizer levels in the sub-sub plots with three replicates, was used for this study. The CO_2 data were recorded three times per week using a portable GXH-3010E1 gas analyzer. The highest CO_2 emissions were recorded following rotary tillage, compared to the lowest emissions from the zero tillage planting method. The lowest emissions were recorded at the 160 kg N/ha, fertilizer level. Higher CO_2 emissions were recorded during the cropping year 2010–11 relative to the year 2011–12. During cropping year 2010–11, applications of corn residue mulch significantly increased CO_2 emissions in comparison to the non-mulched treatments, and during the year 2011–12, equal emissions were recorded for both types of mulch treatments. Higher CO_2 emissions were recorded immediately after the tillage operations. Different environmental factors, i.e., rain, air temperatures, soil temperatures and soil moistures, had significant effects on the CO_2 emissions. We conclude that conservation tillage practices, i.e., zero tillage, the use of corn residue mulch and optimum N fertilizer use, can reduce CO_2 emissions, give better yields and provide environmentally friendly options.

Editor: Ben Bond-Lamberty, DOE Pacific Northwest National Laboratory, United States of America

Funding: The study was supported by the National Natural Science foundation of China (31071375 and 31171506). The funders had no role in study design, data collection and analysis, decision to publish, or preparation of the manuscript.

Competing Interests: The authors have declared that no competing interests exist.

* E-mail: yunchengliao@163.com or yunchengliao@nwsuaf.edu.cn

Introduction

Studies regarding soil CO_2 emissions have attracted significant attention because the concentration of CO_2 in the atmosphere is increasing very rapidly as a consequence of fossil fuel combustion and deforestation. The past two centuries of human activities have reportedly contributed as much as approximately half of the increase in CO_2 emissions [1], [2]. Global terrestrial ecosystems absorbed carbon at the rate of 1–4 Pg yr^{-1}, during 1980s and 1990s, which made up approximately 10–60% of the fossil fuel emissions [3], [4]. Currently, significant attention is given to CO_2 emissions from soils because this source significantly affects the global carbon cycle and the function of the terrestrial ecosystem [5]. Fluxes of greenhouse gases (CO_2, N_2O and CH_4) between the atmosphere and agricultural soils considerably influence the stock of anthropogenic greenhouse gases [6]. Agriculture is an important source of emissions for these different gases, and its contribution to climate change is approximately 20% on an annual basis [7]. It has been reported that soils have already contributed approximately 50 Pg of anthropogenic CO_2 to the atmosphere in the past, through cultivation processes [8].

Tillage is an integral part of agriculture which not only significantly affects crop production but is also considered one of the leading factors in soil degradation. This technique is a fundamental operation that has affected both the soil and the environment and is considered one of the most important sources of CO_2 emissions into the atmosphere [9] because humans have tilled the soil for crop production for thousands of years [10] and approximately 23–44% of total CO_2, is emitted into the atmosphere through soil preparation-related operations [11]. Approximately 30–50% of soil C has already been lost through the adaptations of intensive tillage practices [12], and major C losses from, soils in the form of CO_2 occur immediately after the tillage operations [13].

Agricultural management practices affect different soil processes (i.e., soil temperature, soil moisture and soil pH), and other ongoing soil decomposition processes, which ultimately result in the conversions of plant-derived C to soil organic matter and CO_2 [14]. Applications of inorganic as well as organic fertilizers [15] and different degrees of soil moisture and temperature strongly affect the fluxes of soil CO_2 [16], [17] & [18]. Similarly, the application of N fertilizer also affects soil CO_2 emissions [19]. Instead of burning crops residues, farmers, applications of inorganic fertilizers and use of green manures as well as organic manures can be of great use in maintaining soil fertility [20]. These practices can provide essential nutrients to crops and reductions in the burning of crops can reduce CO_2 emissions into the atmosphere [21].

Agricultural tillage practices can be helpful in the sequestering of atmospheric CO_2 [22], [23] and [24]. Conservation tillage has the potential to increase soil C and N [25] and other types of

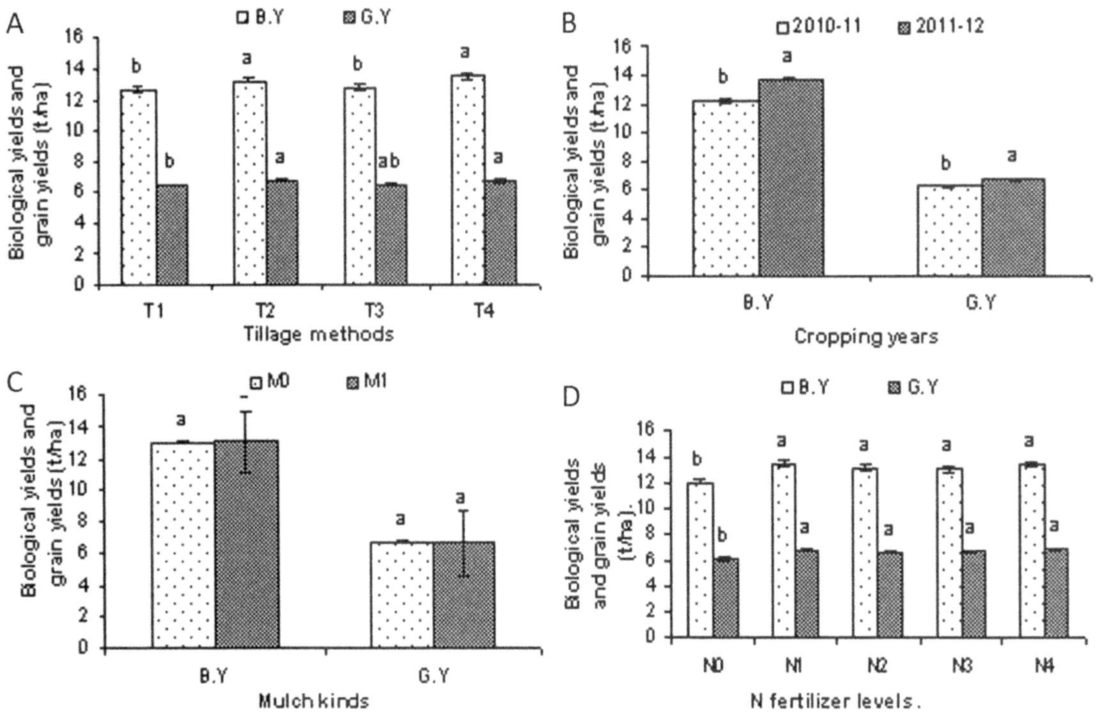

Figure 1. Wheat crop biological and grain yields as affected by different tillage methods, different mulch kinds and different N fertilizer levels during different cropping years (2010–12). (A), Wheat crop biological and grain yields as affected by different tillage methods i.e. T_1, chisel plow tillage., T_2, zero tillage., T_3, rotary tillage and T_4, mold board plow tillage., (B), Mean wheat crop biological and grain yields as affected by different tillage methods, mulch kinds and different N fertilizer levels during two cropping years (2010–12). (C), Wheat crop biological and grain yields as affected by different mulch kinds i.e. M_0, No mulch and M_1, corn residue mulch., (D), Wheat crop biological and grain yields as affected by different N fertilizer levels during two cropping years (2010–12) including, N_0, 0 kg N/ha., N_1, 80 kg N/ha, N_2, 160 kg N/ha., N_3, 240 kg N/ha and N_4, 320 kg N/ha.

conservation practices can be helpful in reducing the loss of soil organic carbon from the soils [26], [27]. Similarly, the retention of crop residue, nitrogen fertilization and no-tillage are generally supposed to enhance the soil organic carbon (SOC) stocks in the soil [28] because these farms, management practices not only increase crops biomass, but are also considered very important for the microbial decomposition of crop residues [29]. As far as N fertilization is concerned, some scientists have reported that increased N fertilization can depress CO_2 emissions [30], [31] however, others [32] have reported that N fertilization has no effect on SOC, while some other scientists [33] have reported that higher N fertilization improves the SOC of the soil.

The Loess Plateau has an area of approximately 63, 5000 km^2. It covers many provinces in China, and is home to millions of people. It is one of the most highly eroded areas of the world, and traditional agriculture, i.e., intensive tillage, is considered one of the leading man-made factors responsible for this erosion. However many crops residues are produced in this region. A small portion of these residues is used for forage or fuel consumption and the remaining residues are generally burned. Mold board plowing followed by harrowing is commonly used for the tillage operations in this region [34]. Few studies showing CO_2 emissions to the adaptation of different agronomic management practices have been previously reported from this region of China. In this area intensive tillage methods i.e. rotary tillage and mold board plow tillage methods are commonly used for land preparation. Commonly higher levels of N fertilizers are applied and crop residues are removed from the fields at the time of soil preparation.

The main aim of this two year study was to identify the effects of different tillage methods i.e. chisel plow and zero tillage in comparison with intensive tillage practices i.e. rotary tillage and mold board plow tillage methods, different N fertilizer levels and the application of corn residue mulch on CO_2 emissions. The results from this study can be of great help in improving the management of soils not only in this area of China but also in other regions of the world.

Results

Wheat crop yields

Significant variations in wheat crops biomass and grain yields were recorded during both study years (2010–12). When compared to the other tillage methods, there were better yields overall with the zero tillage planting method. No grain yields differences were recorded under different mulch treatments and similarly low yields were recorded for N_0 nitrogen fertilizer levels. Statistically equal grain yields were recorded for all of the other higher tested levels of N fertilizer (Fig. 1).

Soil CO_2 flux

All tested treatments had significant effects on the CO_2 emissions (Figs. 2, 3). The details are given below.

Tillage method effects on CO_2 emissions

Two years of combined data show that the rotary tillage and mold board plow tillage methods had their highest and statistically equal CO_2 emissions during the first week of planting the wheat

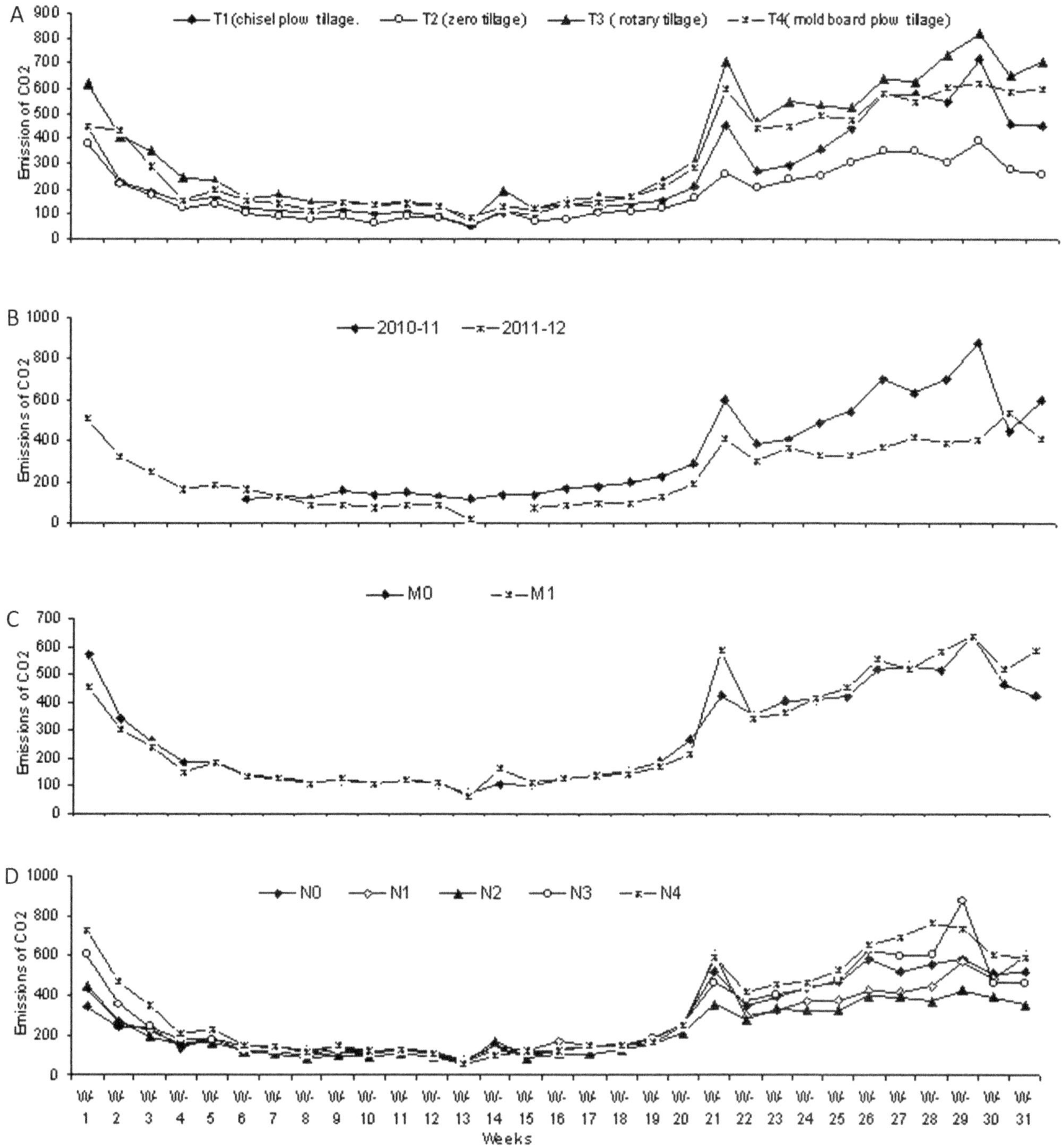

Figure 2. CO_2 emission trends as affected by different tillage methods, mulch kinds and N fertilizer levels during two cropping years (2010–12). (A). Emissions trends of CO_2 from different tillage methods i.e. T_1, chisel plow tillage, T_2, zero tillage, T_3, rotary tillage and T_4, mold board plow tillage., (B).CO_2 emission trends as affected by different kinds of tillage methods, different types of mulch and different N fertilizer levels, during two cropping years (2010–12). (C). CO_2 emission trends due to different mulch kinds i.e. M_0, no mulch and M_1, corn residue mulch., (D). CO_2 emissions trends due to different N fertilizer levels during two cropping years (2010–12) including, N_0, 0 kg N/ha., N_1, 80 kg N/ha, N_2, 160 kg N/ha, N_3, 240 kg N/ha and N_4, 320 kg N/ha.

crop, compared to the chisel plow tillage and zero tillage planting methods. When compared to all the other tillage methods, the lowest CO_2 emissions were recorded for the zero tillage planting method (Fig. 2; Fig. 3). Emission trends recorded during the whole wheat crop seasons show that although there were variations in the CO_2 emissions in response to different tillage methods, the overall highest emissions were recorded for the rotary tillage planting method, followed by mold board plow tillage. Although higher CO_2 emissions were recorded for the chisel plow tillage method, the lowest CO_2 emissions were generally recorded for zero tillage planting method (Fig. 2; Fig. 3).

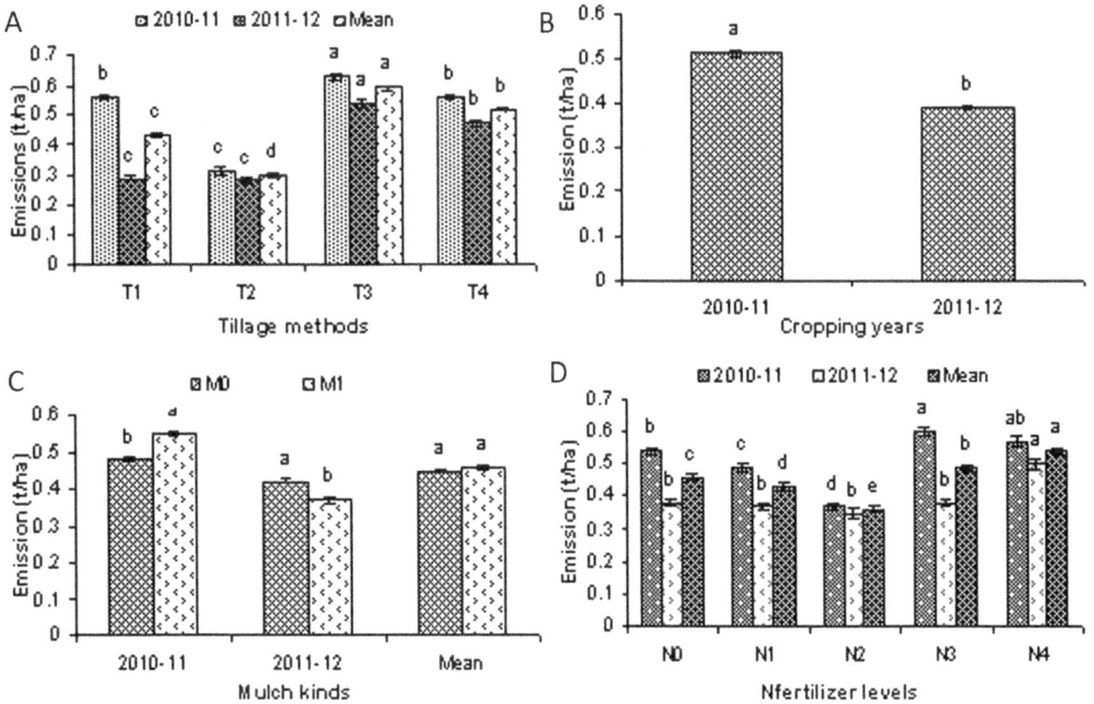

Figure 3. Total/mean emissions of CO_2 as affected by different tillage methods, mulch kinds, and N fertilizer levels during two cropping years (2010–12). (A).Emissions of CO_2 from different tillage methods i.e. T_1, chisel plow tillage., T_2, zero tillage.,T_3, rotary tillage and T_4., mold board plow tillage., (B). Total emissions of CO_2 as affected by different kinds of tillage methods, different kinds of mulch and different levels of N fertilizer, during two cropping years (2010–12)., (C), Emissions of CO_2 from the different mulch kinds i.e. M_0, no mulch and M_1, corn residue mulch., (D).Total/mean emissions of CO_2 as affected by different levels of N fertilizer, during two cropping years (2010–12) including, N_0, 0 kg N/ha., N_1, 80 kg N/ha., N_2,160 kg N/ha., N_3, 240 kg N/ha and N_4, 320 kg N/ha.

Effects of cropping years on CO_2 emissions

With the exception of one week (i.e., week 23), the highest weekly emissions of CO_2 were recorded during cropping year 2010–11 in comparison to the weekly emissions of CO_2 during cropping year 2011–12 (Figs. 2, 3).

Mulch effects on CO_2 emissions

Weekly CO_2 emissions during both wheat crops growing seasons (2010–11 and 2011–12) and the mean of the two years (2010–12) of data (Figs. 2, 3) show that there were more CO_2 emissions recorded from the corn residue-mulched treatments during cropping year 2010–11 than from the non-residue mulched treatments (Figs. 2, 3). However, fewer CO_2 emissions were recorded from the corn residue mulched treatments during cropping year 2011–12 and in comparison to the non residue-mulched treatments, over all less emissions of CO_2 were recorded from the corn residue mulched treatments (Figs. 2, 3). However the two year mean data show mixed types of emissions were recorded on a weekly basis following the applications of corn residue mulch or no mulch (Fig. 2).

N fertilizer level effects on CO_2 emissions

CO_2 emissions data recorded during both wheat cropping seasons, i.e., 2010–11 and 2011–12 and the two year (2010–12)) mean data show that there were significant differences in the weekly CO_2 emissions in response to different N fertilizer levels (Fig. 2; Fig. 3). This finding also indicates that the lowest CO_2 emissions were recorded for the N_0, N fertilizer level at the start of the wheat crop growing seasons compared to all the other higher N fertilizer levels (Fig. 2). During the winter months the CO_2 emissions decreased under all treatments but when the temperatures rose again, higher CO_2 emissions were recorded (Fig. 2). However, when compared to all the other N fertilizer level treatments, the lowest overall CO_2 emissions were recorded for 160 kg N/ha (Fig. 2). CO_2 emission fluxes increased with the increase in crop growth and temperatures, so during the last weeks of wheat crop growth equal CO_2 emissions were recorded for all of the N fertilizer treatments (Fig. 2).

Cumulative CO_2 emissions

Two years (2010–12) of CO_2 emissions data show that on a mean cumulative basis, except in the case of corn residue mulch treatments, significant differences in the emissions of CO_2 were recorded for all of the different tillage methods and different N fertilizer level treatments (Fig. 3). Statistically significant variations in the total and mean CO_2 emissions were recorded for all of the tillage methods, and the emissions trend for the different tillage methods was $T_3>T_4>T_1>T_2$ (Fig. 3). Different N fertilizer levels had significant effects on the total and mean CO_2 emissions. The emissions trend for the different nitrogen fertilizer levels was $N_4>N_3>N_0>N_1>N_2$ (Fig. 3). On a cumulative basis, more CO_2 emissions were recorded during the cropping year 2010–11 than during cropping year 2011–12 (Fig. 3), and on the whole approximately 30% more CO_2 emissions were recorded during cropping year 2010–11 than during 2011–12 (Fig. 3).

CO_2 emissions varied for all of the tillage methods and N fertilizer levels. For the chisel plow tillage treatment, using 160 kg N/ha reduced the emissions of CO_2 (data not shown) and the CO_2

emissions trend in case of the chisel plow tillage method and different N fertilizer levels was $N_4>N_3>N_0>N_1>N_2$ (data not shown) (Table 1). For the Zero tillage planting method and N fertilizer level interactions, the CO_2 emission varied but the emissions trend was $N_0>N_3>N_2>N_1>N_4$ (data not shown) (Table 1). For the rotary tillage planting method and N fertilizer level interactions, the CO_2 emissions trend was $N_4>N_3>N_1>N_0>N_2$ (data not shown) (Table 1). Similarly for the mold board plow tillage method and N fertilizer level interactions, the CO_2 emissions trend was $N_0>N_4>N_1>N_3>N_2$ (data not shown) (Table 1).

With the exception of the mold board plow tillage method, corn residue mulch applications increased the CO_2 emissions in all of the other three tillage methods. Lower CO_2 emissions were recorded for all the tillage methods during cropping year 2011–12 in comparison to cropping year 2010–11 and throughout cropping year 2010–11, corn residue mulch applications increased the CO_2 emissions to 16.5% compared to the non residue mulched treatments. Similarly during cropping year, 2011–12, 12.6% fewer CO_2 emissions were recorded for the corn residue mulched treatments in comparison to the non-residue mulched treatments (data not shown) (Table 1).

For the N_0 (0 kg N/ha), N_1 (80 kg N/ha), N_2 (160 kg N/ha) and N_4 (320 kg N/ha), treatments, the use of corn residue mulch increased the CO_2 emissions by approximately 6.2%, 5.5%, 0.6% and 35.6%, respectively, and for N_4 (240 kg N/ha), the application of corn residue mulch reduced the CO_2 emissions by approximately 35.6% (data not shown) (Table 1).

Soil temperatures versus CO_2 emissions

Temperature changes for the top 5 cm of soil from the different treatments during the two years (2010–12) of study are given in Fig. 4. Although with the passage of time, there were variations in soil temperatures for the different tillage methods, however intermediary types of temperature changes were recorded following zero tillage planting in comparison to the other three tillage methods (Fig. 4). Similarly, although statistically comparable temperatures were recorded for both types of mulch treatments, slightly increased temperatures were recorded in the corn residue mulched treatments relative to the non-residue mulched treatments (Fig. 4). However, no soil temperature differences were recorded for the different N fertilizer level treatments (Fig. 4). Generally speaking, higher soil temperatures were recorded during the cropping year 2010–11 than during cropping year 2011–12 (Fig. 4). CO_2 emission trends showed that CO_2 emissions increased with an increase in soil temperatures and vice versa (Figs. 2, 4).

Soil moisture versus CO_2 emissions

When compared to all of the other tillage methods, the highest soil water contents were recorded in response to the rotary tillage planting method (Fig. 5). Similarly, higher water contents were recorded during the cropping year 2011–12 for the different crop growth stages relative to cropping year 2010–11 (Fig. 5). Corn residue mulch increased the soil moisture contents of the crop revival stage and on the booting stage compared to the non-residue mulched treatments (Fig. 5). Lower soil moisture contents were recorded for the 80 kg N/ha (N_1) and 160 kg N/ha (N_2), treatments compared to all of the other N fertilizer level treatments (Fig. 5). Two years of mean data show that the CO_2 emissions were lower in those tillage methods or N fertilizer levels treatments that had lower water contents (Fig. 2; Fig. 4).

Table 1. ANOVA (Mean Square Values) of biological yields, grain yields, soil organic carbon, and cumulative emissions of CO_2 during two cropping years (2010–12).

Source	D.F	B.Y	G.Y	SOC	CO_2
Tillage methods	3	6593271.9***	122260.74**	32.12145866***	9605737865***
Planting years	1	13496603.4***	29504910.64***	97.90534734***	8211670776***
Mulch kinds	1	132977.5NS	20786.14NS	3.74523078***	93546720 NS
N Fertilizer levels	4	1470235.5***	4549170.27***	3.78932599***	21321766841***
Tillage methods X Mulch kinds	3	6780289.1***	1690508.29***	26.55366456***	1455896530***
Tillage methods X N Fertilizer levels	12	1443071.8 NS	302485.89NS	5.42559025***	1418530119***
Tillage methods X planting years	3	3443481.8NS	1276931.92*	12.85179871***	1455478300***
Planting years X Mulch kinds	1	66692.7NS	418751.64NS	0.67995212 NS	2547564904***
Planting years X N Fertilizer levels	4	10758689.1***	3983648.62***	3.45638215***	720140735***
Mulch kinds X N Fertilizer levels	4	975872.1NS	461889.16NS	3.23681715***	3518560905***
Tillage methods X Planting years X Mulch kinds	3	1162751.4NS	190004.99NS	12.61344035***	406313591***
Tillage methods X Planting years X N fertilizer levels	12	12108232.1NS	369168.67NS	4.50848299***	959919429***
Tillage methods X Mulch kinds X N fertilizer levels	12	1785389.2NS	420989.67NS	6.20244131***	2881830882***
Tillage methods X Mulch kinds X Planting years X N fertilizer levels	16	2029198.1 NS	782603.91***	4.05394235***	1326809061***

*Significant at 0.05 probability levels.
**Significant at 0.01 probability levels.
***Significant at 0.001 probability levels.
B.Y, Biological yields., G.Y, Grain yields., SOC, Soil organic carbon.
X, indicates interactions between different factors i.e. Tillage's X Mulches indicates interactions between different tillage methods and mulch kinds.

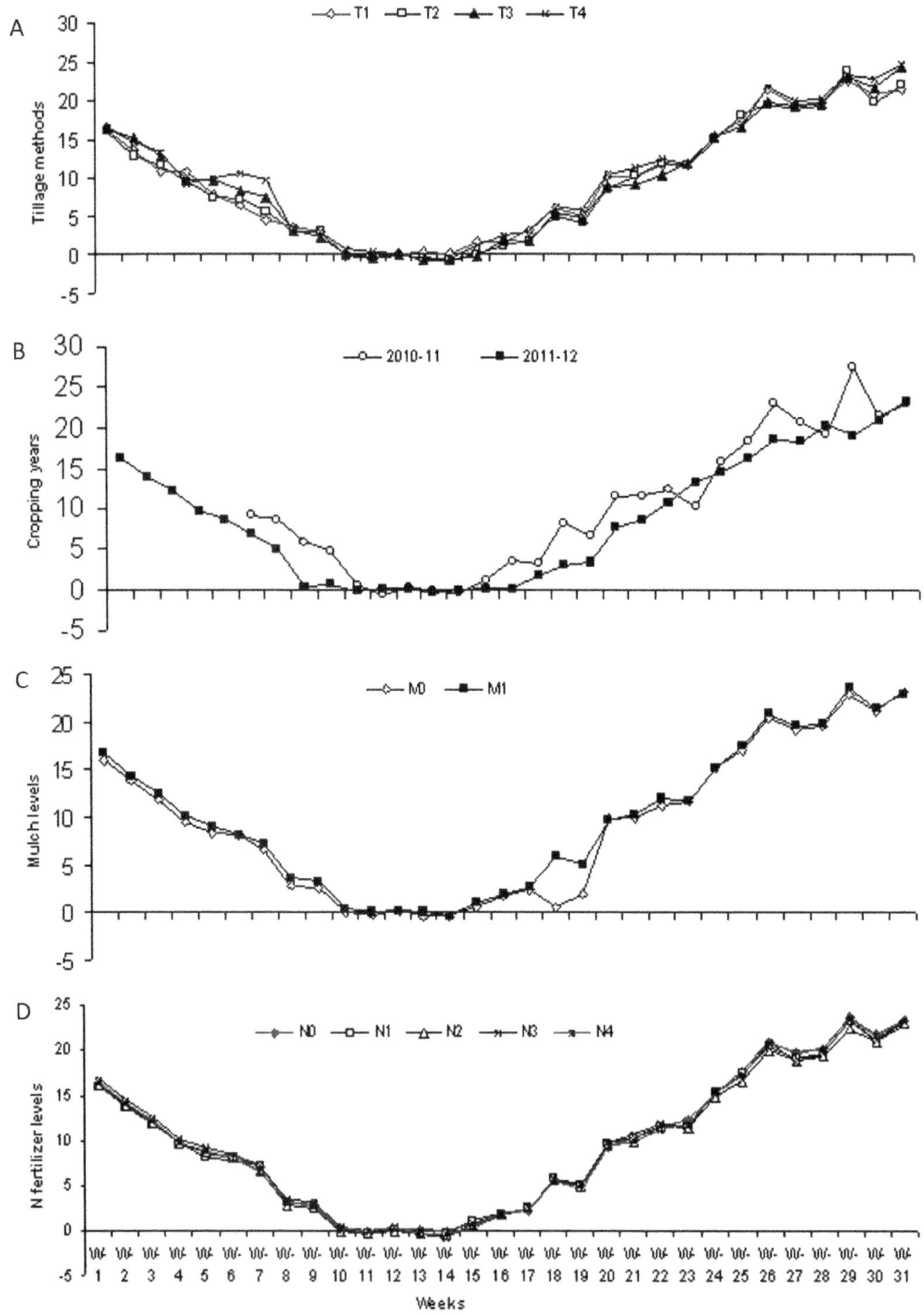

Figure 4. Changes in soil temperatures (0–5 cm depth) due to different tillage methods, corn residue mulch and N fertilizer levels during two cropping years (2010–12). (A). Changes in soil temperatures (0–5 cm depth) due to different tillage methods i.e. T_1, chisel plow tillage, T_2, zero tillage, T_3, rotary tillage and T_4, mold board plow tillage method., (B) Changes in soil temperatures (0–5 cm depth), as affected by different tillage methods, different mulch kinds and different N fertilizer levels during two cropping years (2010–12)., (C), Changes in soil temperatures (0–5 cm depth) during two cropping years (2010–12) due to different mulch kinds i.e. M_0, no mulch and M_1, corn residue mulch., (D). Changes in soil temperatures (0–5 cm depth) due to different N fertilizer levels during two cropping years (2010–12) including, N_0, 0 kg N/ha., N_1, 80 kg N/ha., N_2, 160 kg N/ha., N_3, 240 kg N/ha and N_4, 320 kg N/ha.

Soil organic carbon versus CO_2 emissions

Two years of mean data show that the tillage methods had significant effects on the SOC from 0–10 cm soil depths and that compared to all of the tillage methods, the highest SOC was recorded following chisel Plow tillage (Fig. 6). Higher SOC contents were recorded for all of treatments during cropping

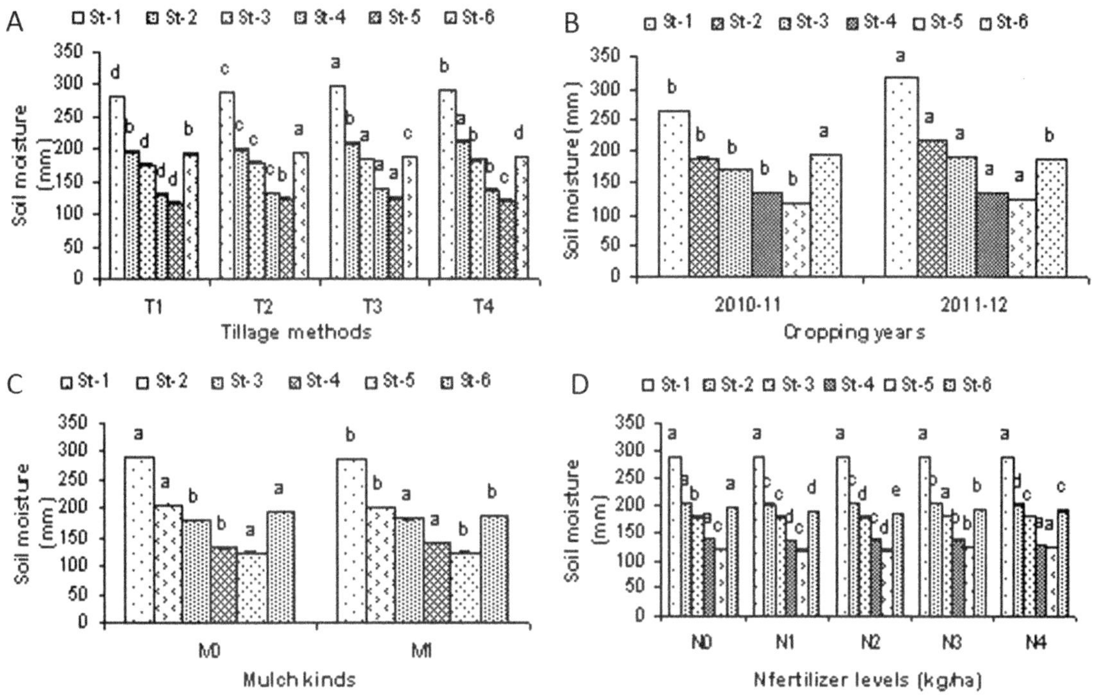

Figure 5. Total soil moisture contents (0–100 cm soil depth) as affected by different tillage methods, mulch kinds, and N fertilizer levels during two cropping years (2010–12). (A). Total soil moisture contents (0–100 cm soil depth) due to different tillage methods i.e. T_1, chisel plow tillage., T_2, zero tillage., T_3, rotary tillage and T_4, mold board plow tillage method., (B).Total soil moisture contents (0–100 cm soil depth) as affected by different tillage methods, different mulch kinds and different N fertilizer levels, during two cropping years (2010–12).(C), Total soil moisture contents (0–100 cm soil depth) as affected by different mulch kinds during the two cropping years (2010–12) i.e. M_0, no mulch and M_1, corn residue mulch., (D). Total soil moisture contents (0–100 cm soil depth) as affected by different N fertilizer levels, during two cropping years (2010–12) including, N_0, 0 kg N/ha., N_1, 80 kg N/ha, N_2, 160 kg N/ha., N_3, 240 kg N/ha and N_4, 320 kg N/ha., (*) Stage-1, (Crop sowing stage), Stage-2, (Crop revival stage), Stage-3, (Stem elongation stage), Stage-4, (Booting stage), Stage-5, (Grain formation stage) and Stage-6 (Crop harvesting stage).

year 2011–12 in comparison to cropping year 2010–11 (Fig. 6). The use of corn residue mulch increased the SOC compared to the non-mulched treatments (Fig. 6), and higher over all SOC contents were recorded in the cases of N_1 (80 kg N/ha) and N_2 (160 kg N/ha) in comparison to all of the other N fertilizer levels treatments (Fig. 6). The data show that CO_2 emissions were lower in those treatments that had higher SOC contents (Fig. 3; Fig. 6).

Effects of seasonal variations on CO_2 emissions

Seasonal temperature variations had significant effects on the CO_2 emissions. Because there were normal temperatures when the wheat crops were sown, higher CO_2 emissions were recorded, but reduced emissions were recorded with the decline in temperature during the winter seasons. With increased of crop growth and ascending of seasonal temperatures higher CO_2 emissions were recorded during both years (Figs. 2, 3, 4, 7, 8,).

Discussion

Soil CO_2 fluxes

Variations in seasonal temperatures had significant effects on soil temperatures, which ultimately affected the CO_2 emissions (Figs. 2, 3, 4, 7, 8). Higher CO_2 fluxes were recorded immediately after the tillage operations, which continued for a few days, and these emissions decreased with the passage of time (Fig. 2). Our results are in agreement with other findings [35]. Other investigators have reported changes in CO_2 emissions with seasonal variations. According to these investigators, seasonal variations in CO_2 emissions are controlled not only by the soil temperatures and soil moistures but also by the tillage practices. Changes in CO_2 emissions from seasonal variations have been reported for almost all ecosystems. These emissions mainly depend both on the type of climate and the ecosystem [36].

Similar findings regarding variations in CO_2 fluxes with changes in soil temperatures and crop growth stages have been reported for rice crops [37]. In our study, the air and soil temperatures had significant effects on CO_2 emissions. Thus with the decrease in soil temperatures during the winter months, the CO_2 emissions decreased, and the CO_2 emissions again increased with the rise in soil temperatures during the summer months (Figs. 2, 3, 4, 7, 8).

Generally, the seasonal CO_2 emissions variations found in our experiment were similar to other findings [38]. Other investigators reported that these emission variations might be related to variations in autotrophic and in heterotrophic respiration because both are involved in soil CO_2 emissions. In addition a large amount of CO_2 is released from plant roots, during the continuation of the plant-energy system. Microbial and root respiration can also significantly contribute to CO_2 emissions.

Tillage method effects on CO_2 emissions

Higher CO_2 emissions were recorded immediately after the tillage operations, which continued for a few days (Fig. 2). Our results are in agreement with findings from other researchers, who reported that CO_2 emissions following tillage increased up to 2–15 times [39], [40], [41], [42] and [43]. According to these findings [39], instead of microbial activity, the basic reason for higher CO_2 emissions immediately after the tillage was actually the release of entrapped CO_2 from the soil pores as a result of physical

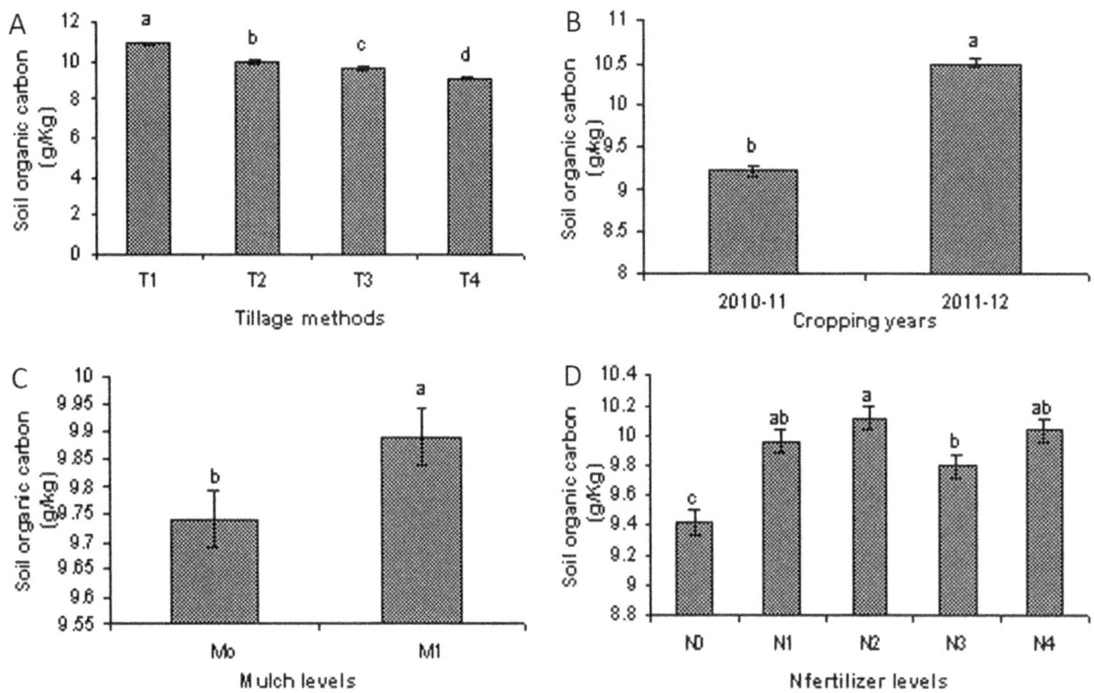

Figure 6. Soil organic carbon (SOC) contents (g/kg) as affected by different tillage methods, different mulch kinds and different N fertilizer levels during different cropping years (2010–12). (A), Soil organic carbon (SOC) contents as affected by different tillage methods i.e. T_1, chisel plow tillage., T_2, zero tillage., T_3, rotary tillage and T_4, mold board plow tillage method., (B), Mean soil organic carbon (SOC) contents as affected by different tillage methods, mulch kinds and different N fertilizer levels during two cropping years (2010–12)., (C), Soil organic carbon (SOC) contents as affected by different mulch kinds i.e. M_0, no corn residue mulch and M_1, corn residue mulch., (D), Soil organic carbon (SOC) contents as affected by different N fertilizer levels during two cropping years (2010–12) including, N_0, 0 kg N/ha., N_1, 80 kg N/ha., N_2, 160 kg N/ha., N_3, 240 kg N/ha and N_4, 320 kg N/ha.

operations. The other reasons for these higher emissions might be that (1) tillage operations break soil aggregates and expose their organic matter to microbial attack [44], [45]; (2) tillage operations encourage the mineralization of soil organic matter by incorporating crops residues into the soil [46]; and (3) tillage operations enhance soil aeration [43]. In our study, the tillage methods had significant effects on the CO_2 emissions and, overall, the rotary tillage and mold board plow tillage methods led to higher CO_2 emissions compared to the chisel plow and zero tillage planting methods (Fig. 2 and Fig. 3). Similar findings under different tillage systems have been previously reported [13]. These researchers reported significantly more CO_2 emission fluxes from the fields tilled by a mold board plow relative to the fields prepared by the chisel plow methods. According to these researchers, the basic reason for the greater emission fluxes from the mold board plow tillage method compared to the chisel plow tillage method was the depth and extent to which the soil was disturbed by using the different tillage implements. In our experiment, higher soil temperatures in the top 5 cm depth (Fig. 4) and generally higher moisture contents (Fig. 5) in the rotary tillage and mold board plow methods might be responsible for the higher emissions, in addition to the soil preparation depths. The tillage depths also resulted in a reduction of the SOC in the top 0–10 cm soil layers for the rotary tillage and mold board plow methods, compared to the chisel plow tillage and zero tillage planting methods (Fig. 6).

Corn residue mulch effects on CO_2 emissions

Weekly CO_2 emissions data (Figs. 2, 4) show that during cropping year 2010–11, the application of corn residue mulch caused an overall increase in CO_2 emissions compared to the non-residue mulched treatments, but during the cropping year 2011–12, fewer CO_2 emissions were recorded in response to the application of corn residue mulch relative to the non-residue mulched treatments (Fig. 2 and Fig. 4). Although more CO_2 emissions fluxes were recorded in the corn residue mulched treatments in comparison to the non-residue mulched treatments, these might be due to the more microbial activities in the corn residue mulched treatments, which might have increased the SOC with the passage of time (Fig. 6). However, the two year mean data show that statistically non significant differences were recorded for CO_2 emissions following the use of the corn residue mulched or non mulched treatments. These findings might be explained by noting that both years CO_2 collecting chambers were fixed before the application of corn residue mulch. As a result there were fewer corn residues within the CO_2 collecting chambers from the corn residue mulched treatments. This smaller amount might be the reason why no CO_2 emission differences were recorded following the applications of different corn residue mulch treatments. The other reason might be that during the cropping year 2010–11, the application of corn residue mulch increased the CO_2 emissions but this mulch also increased the SOC contents of the corn residue mulched treatments, possibly resulting in the lower emission of CO_2 from the corn residue mulched treatments during the year 2011–12 (Fig. 3). A modeling study reported [47] that instead of applying higher rates of fertilizers, the use of crop residues or manure amendments would mitigate GHG emissions more efficiently. Similarly, it has been reported [48] that applications of straw increased the SOC sequestration in the soil which ultimately influenced the temporal patterns of CO_2 emissions from the soil.

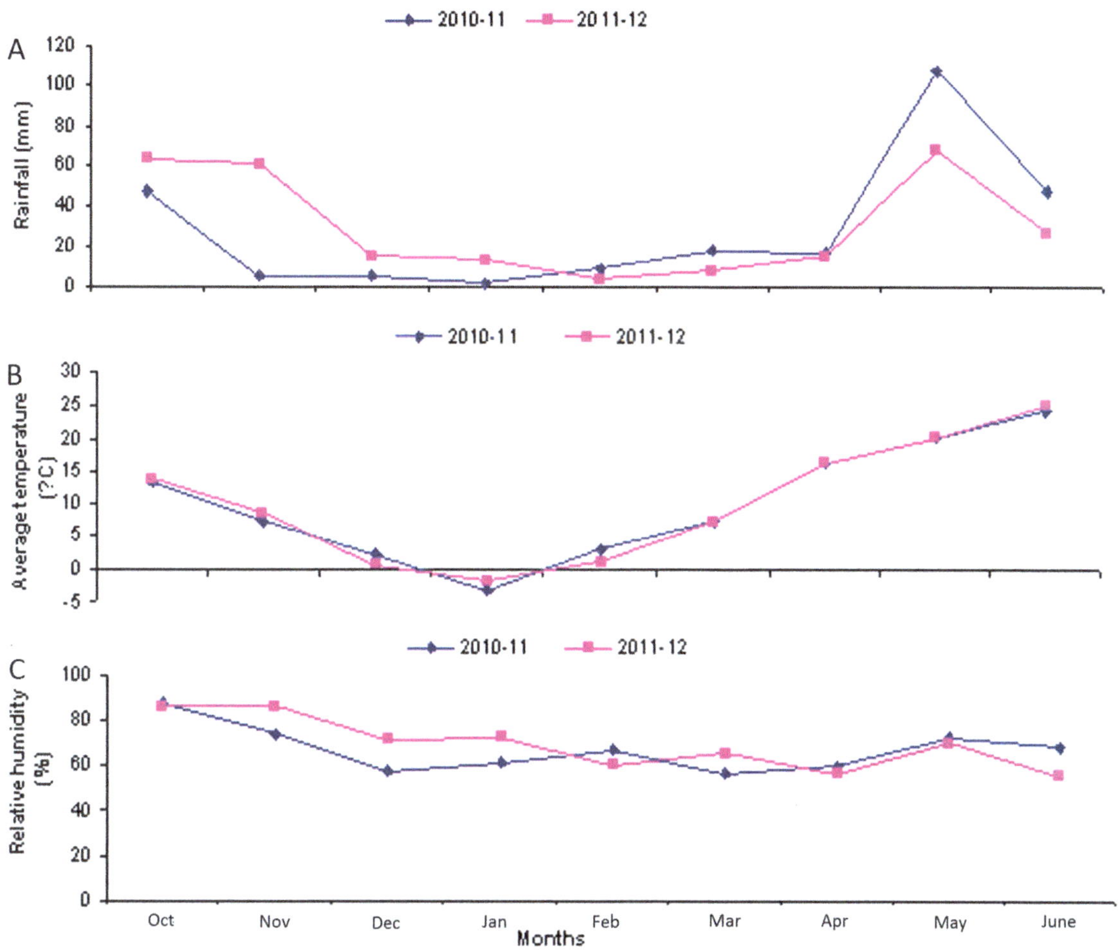

Figure 7. Monthly rainfalls, average temperatures and average relative humidity's of the study area during two wheat crop growing seasons (2010–12). (A).Rainfalls of the study area during two wheat crop growing seasons (2010–12), (B) Average temperatures of the study area during two wheat crop growing seasons (2010–12) (C) Average relative humidities of the study area during two wheat crop growing seasons (2010–12). (*). R.F-1, Rainfalls during cropping year 2010–11., R.F-2, rainfalls during cropping year 2011–12., Av.temp-1, Monthly average temperatures during the cropping year 2010–11., Av.temp-2, Monthly average temperatures during the cropping year 2011–12., R.H (%)-1, Average monthly relative humidities during the cropping year 2010–11., R.H (%)-2, Average monthly relative humidities during the cropping year 2011–12.

N fertilizer level effects on CO_2 emissions

Very few studies regarding CO_2 emissions in relation to the different tillage methods, corn residue mulch and N fertilizer levels have previously been reported in this region of China. It is expected that the application of inorganic N fertilizers along with organic materials will affect the mineralization of soil organic matter and crop productions, which will ultimately affect CO_2 emissions [8].

However variations in CO_2 emissions following fertilizer applications have been reported for different areas of China. Some scientists [15] have reported that the fertilizer applications suppresses CO_2 emissions and others [49] have reported that fertilizer application enhances CO_2 emissions. Moreover, some other scientists [19] have reported that fertilizer applications have no effects on CO_2 emissions.

Our study shows that the use of 80 kg and 160 kg N/ha, suppressed CO_2 emissions when compared with the 0 N fertilizer level, but that further increases in N fertilizer application rates enhanced CO_2 emissions. Higher emissions from the N_0, nitrogen fertilizer treatments might be explained by noting that plants under unfertilized N treatments are considered to respond to a relative shortage of N by increasing the plant's carbon allocation to its structures and functions, which are responsible for N acquisition [50]. In our study, the use of different levels of N fertilizer relative to the no nitrogen fertilizer level significantly increased crop yields (Fig. 1).This result shows that the higher CO_2 emission fluxes in response to higher levels of N fertilizer might be explained by the increased use of C for microbial growth [51] and it might also be explained by the less efficient use of carbon by the microbial biomass, which resulted in a greater proportion of carbon loss in the form of CO_2 fluxes [52].

In our case, the use of corn residue mulch in combination with different levels of N fertilizer increased CO_2 emissions following the chisel plow, zero tillage and rotary tillage planting methods (Table 1). Our results regarding CO_2 emissions following the combination of N fertilizer and organic amendments are in agreement with the results of other scientists [53], [38] and [54]. These investigators reported a higher CO_2 emission flux from the treatments that utilized both fertilizers and organic manures. This finding shows that up to a certain extent the use of higher N fertilizer levels suppresses CO_2 emissions but in our case, higher levels of N fertilizer (i.e., 240 and 320 kg N/ha) enhanced CO_2 emissions. These results are contrary to the findings of other

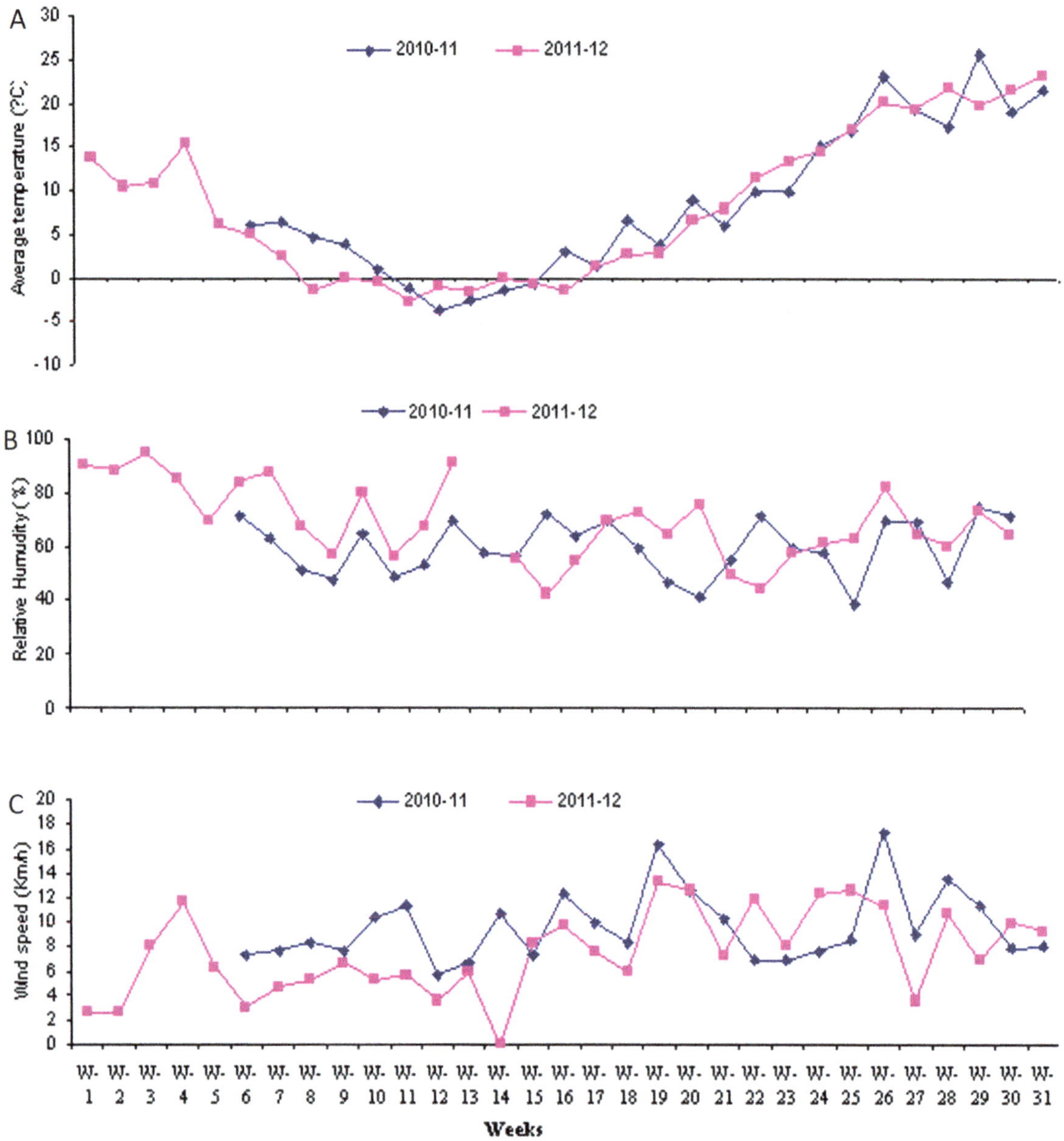

Figure 8. Average weekly temperatures, relative humidity's and wind speeds during the two wheat crop growing seasons (2010–12). (A) Average weekly temperatures of the study area during the two wheat crop growing seasons (2010–12), (B).Average weekly relative humidity's of the study area during the two wheat crop growing seasons (2010–12), (C) Average weekly wind speeds of the study area during the two wheat crop growing seasons (2010–12).

scientists [30] and [31]. These investigators reported no clear reasons for the CO_2 emission reductions.

Cropping year effects on CO_2 emissions

Two years of data on a weekly as well as on a cumulative basis show that there were more CO_2 emissions from all of the treatments during cropping year 2010–11 compared to the CO_2 emissions during cropping year 2011–12 (Figs. 2, 3). The main reasons for these lower emissions during cropping year 2011–12 were, on the whole, lower air temperatures, higher relative humidities and higher over all rainfall concentrations during the early periods of wheat crop growth relative to cropping year 2010–11 (Fig. 7). These factors all ultimately reduced the soil temperatures, which resulted in reduced CO_2 emissions during the cropping year 2011–12 compared to cropping year 2010–11 (Figs. 4, 7, 8). Another reason for the lower emissions might be an increase in the SOC from the corn residue mulched treatments, which might have ultimately helped to reduce the CO_2 emissions during cropping year 2011–12 in comparison to cropping year 2010–11 (Figs. 2, 3, 6).

Effects of soil temperatures, soil moisture and soil organic carbon on CO_2 emissions

Respiration of ecosystem mainly depends on both the heterotrophic (microbe) and autotrophic (plant) activities and both of these factors are controlled by the prevailing environmental conditions (basically temperature and water availability), availability of carbohydrates and substrates and others [55], [56] and

Table 2. Q10 values of the different treatments during the different cropping years (2010–12).

Treatments	chisel plow		zero tillage		rotary tillage		mold board plow tillage	
	2010–11	2011–12	2010–11	2011–12	2010–11	2011–12	2010–11	2011–12
	$Y = 69.220e^{0.0516x}$	$Y = 47.81\ e^{0.0898x}$	$Y = 73.371e^{0.0736x}$	$Y = 74.836e^{0.807x}$	$Y = 140.49e^{0.0427x}$	$Y = 67.628\ ^{0.0749x}$	$Y = 287.91e^{0.0707x}$	$Y = 78.941e^{0.1001x}$
N_0	$R^2 = 0.7076$	$R^2 = 0.8064$	$R^2 = 0.0823$	$R^2 = 0.7213$	$R^2 = 0.7472$	$R^2 = 0.7372$	$R^2 = 0.6076$	$R^2 = 0.7439$
	Q10 = 1.68	Q10 = 2.43	Q10 = 2.09	Q10 = 1.90	Q10 = 2.54	Q10 = 2.55	$Y = 287.91e^{0.0707x}$	Q10 = 2.72
	$Y = 79.76e^{0.0744x}$	$Y = 73.979e^{0.0816x}$	$Y = 57.341e^{0.0845x}$	$Y = 62.738e^{0.0996x}$	$Y = 142.39e^{0.0691x}$	$Y = 93.744e^{0.092x}$	$Y = 247.36e^{0.0385x}$	$Y = 129.37e^{0.0818x}$
N_1	$R^2 = 0.6672$	$R^2 = 0.7158$	$R^2 = 0.0869$	$R^2 = 0.752$	$R^2 = 0.8187$	$R^2 = 0.8074$	$R^2 = 0.5688$	$R^2 = 0.736$
	Q10 = 2.10	Q10 = 2.26	Q10 = 2.33	Q10 = 2.71	Q10 = 1.63	Q10 = 2.11	Q10 = 2.03	Q10 = 2.72
	$Y = 80.442e^{0.0827x}$	$Y = 0.787e^{0.1012x}$	$Y = 114.12e^{0.0509x}$	$Y = 57.382e^{0.0891x}$	$Y = 90.813e^{0.0616x}$	$Y = 122.76e^{0.0669x}$	$Y = 108.97e^{0.0634x}$	$Y = 79.756e^{0.0873x}$
N_2	$R^2 = 0.5944$	$R^2 = 0.8201$	$R^2 = 0.528$	$R^2 = 0.7701$	$R^2 = 0.790$	$R^2 = 0.6534$	$R^2 = 0.8134$	$R^2 = 0.7563$
	Q10 = 2.29	Q10 = 2.75	Q10 = 1.66	Q10 = 2.70	Q10 = 2.00	Q10 = 2.51	Q10 = 1.47	Q10 = 2.27
	$Y = 125.25e^{0.1085x}$	$Y = 70.222e^{0.0931x}$	$Y = 82.409e^{0.0478x}$	$Y = 55.391e^{0.1006x}$	$Y = 176.74e^{0.097x}$	$Y = 149.95e^{0.0859x}$	$Y = 103.92e^{0.0614x}$	$Y = 69.581e^{0.0843x}$
N_3	$R^2 = 0.7977$	$R^2 = 0.7054$	$R^2 = 0.7439$	$R^2 = 0.7937$	$R^2 = 0.6429$	$R^2 = 0.7876$	$R^2 = 0.852$	$R^2 = 0.6457$
	Q10 = 2.96	Q10 = 2.47	Q10 = 2.33	Q10 = 2.02	Q10 = 1.85	Q10 = 1.95	Q10 = 1.89	Q10 = 2.32
	$Y = 55.139e^{0.0653x}$	$Y = 61.6e\ ^{0.0904x}$	$Y = 43.884e^{0.0903x}$	$Y = 60.312e^{0.8051x}$	$Y = 90.419e^{0.058x}$	$Y = 221.26e^{0.0624x}$	$Y = 116.85e^{0.0613x}$	$Y = 96.28e^{0.1086x}$
N_4	$R^2 = 0.6443$	$R^2 = 0.7961$	$R^2 = 0.7996$	$R^2 = 0.9015$	$R^2 = 0.6951$	$R^2 = 0.6654$	$R^2 = 0.7312$	$R^2 = 0.7969$
	Q10 = 1.02	Q10 = 2.18	Q10 = 1.61	Q10 = 2.16	Q10 = 2.64	Q10 = 2.32	Q10 = 1.85	Q10 = 2.96

N_0, 0 kg N/ha., N_1, 80 kg N/ha., N_2, 160 kg N/ha., N_3, 240 kg N/ha and N_4, 320 kg N/ha.

Table 3. Q10 values of the different treatments during the different cropping years (2010–12).

Treatments	chisel plow		zero tillage		rotary tillage		mold board plow tillage	
	2010–11	2011–12	2010–11	2011–12	2010–11	2011–12	2010–11	2011–12
	$Y = 43.137e^{0.1316x}$	$Y = 87.49e^{0.0778x}$	$Y = 163.72e^{0.0363x}$	$Y = 61.295e^{0.09x}$	$Y = 62.946e^{0.1111x}$	$Y = 99.78e^{0.0961x}$	$Y = 183.66e^{0.0518x}$	$Y = 55.625e^{0.0947x}$
N_0+M	$R^2 = 0.8849$	$R^2 = 0.6632$	$R^2 = 0.4067$	$R^2 = 0.8372$	$R^2 = 0.7113$	$R^2 = 0.8096$	$R^2 = 0.753$	$R^2 = 0.8216$
	Q10 = 3.72	Q10 = 2.69	Q10 = 2.47	Q10 = 2.10	Q10 = 1.79	Q10 = 2.96	Q10 = 1.85	Q10 = 2.58
	$Y = 254.21e^{0.0484x}$	$Y = 0.0988e^{0.0516x}$	$Y = 84.882e^{0.0413x}$	$Y = 49.272e^{0.0965x}$	$Y = 175.25e^{0.0557x}$	$Y = 76.273e^{0.0858x}$	$Y = 112.84e^{0.0703x}$	$Y = 51.454e^{0.0957x}$
N_1+M	$R^2 = 0.7689$	$R^2 = 0.8367$	$R^2 = 0.7109$	$R^2 = 0.7881$	$R^2 = 0.5183$	$R^2 = 0.8149$	$R^2 = 0.695$	$R^2 = 0.7376$
	Q10 = 1.62	Q10 = 2.52	Q10 = 1.44	Q10 = 2.22	Q10 = 3.04	Q10 = 2.58	Q10 = 1.68	Q10 = 2.60
	$Y = 72.653\ ^{0.0809x}$	$Y = 45.736e^{0.0926x}$	$Y = 80.339e^{0.0722x}$	$Y = 61.262e^{0.0907x}$	$Y = 163.02e^{0.0519x}$	$Y = 104.06e^{0.083x}$	$Y = 93.668e^{0.0666x}$	$Y = 46.012e^{0.1166x}$
N_2+M	$R^2 = 0.8237$	$R^2 = 0.7309$	$R^2 = 0.8477$	$R^2 = 0.7992$	$R^2 = 0.4896$	$R^2 = 0.7173$	$R^2 = 0.6503$	$R^2 = 0.816$
	Q10 = 2.25	Q10 = 2.87	Q10 = 1.52	Q10 = 2.18	Q10 = 1.95	Q10 = 2.36	Q10 = 2.02	Q10 = 3.17
	$Y = 1.015e^{0.0898x}$	$Y = 46.994e^{0.1054x}$	$Y = 164.85e^{0.0502x}$	$Y = 30.192e^{0.1197x}$	$Y = 163.9e^{0.405x}$	$Y = 40.179e^{0.0983x}$	$Y = 75.02e^{0.0742x}$	$Y = 72.535e^{0.1075x}$
N_3+M	$R^2 = 0.8604$	$R^2 = 0.8214$	$R^2 = 0.4050$	$R^2 = 0.8784$	$R^2 = 0.6464$	$R^2 = 0.84$	$R^2 = 0.7363$	$R^2 = 0.8221$
	Q10 = 2.45	Q10 = 1.11	Q10 = 2.06	Q10 = 2.18	Q10 = 1.68	Q10 = 2.29	Q10 = 1.95	Q10 = 2.93
	$Y = 8.69e^{0.10179x}$	$Y = 69.932e^{0.0859x}$	$Y = 51.822e^{0.0641x}$	$Y = 36.672e^{0.1073x}$	$Y = 287.97e^{0.0937x}$	$Y = 121e^{0.1050x}$	$Y = 127.88e^{0.0702x}$	$Y = 73.455e^{0.1177x}$
N_4+M	$R^2 = 0.8342$	$R^2 = 0.7264$	$R^2 = 0.7247$	$R^2 = 0.8364$	$R^2 = 0.6465$	$R^2 = 0.854$	$R^2 = 0.7656$	$R^2 = 0.8821$
	Q10 = 2.76	Q10 = 2.36	Q10 = 1.65	Q10 = 2.74	Q10 = 1.50	Q10 = 2.67	Q10 = 2.10	Q10 = 3.24

N_0+M, 0 kg N/ha + corn residue mulch., N_1+M, 80 kg N/ha + corn residue mulch., N_2+M, 160 kg N/ha + corn residue mulch., N_3+M, 240 kg N/ha + corn residue mulch., N_4+M, 320 kg N/ha + corn residue mulch.

[57]. Many studies have shown that seasonal variations in CO_2 emissions were mainly caused by the soil temperature, soil moisture or the combination of both these factors [58], [36].

Our study also indicates that soil temperature was an important driving force for the increased CO_2 emissions, which is also supported by the Q10 values given in Table 2 and Table 3. It has also been reported that the CO_2 evolution rate significantly increases with an increase in temperature and moisture [37]. Our results are also in agreement with many other field studies, which have shown strong relationships between soil temperatures and CO_2 flux rates [59], [60], [17] and [18]. Additionally, our results agree with the findings of [61], who reported a stronger polynomial for temperature and moisture interaction ($r^2 = 0.89$) than for temperature alone ($r^2 = 0.47$). Many previous studies have reported that changes in crop management practices, i.e., the appropriate use of tillage operations, proper fertilization, crop residue applications and crop rotations can be helpful for managing soil organic matter, e.g.,. [62].

Our study also shows that the results at the end of cropping year 2010–11 revealed more SOC in the top 0–10 cm of soil than that at the end of cropping year 2011–12. This result shows that the adaptation of different management practices, i.e., the application of crop residues, increased the SOC contents, especially following chisel plow tillage and zero tillage planting compared with rotary tillage and mold board plow tillage. SOC sequestration is a long term processes and various results have been previously reported, i.e. [63] and [64], have reported that conservation tillage is a recommended management practice for agricultural ecosystems that can enhance the pool of soil organic carbon (SOC), in the soil. Following an analysis of global data, NT reportedly sequestered carbon at an average rate of 0.57 Mg C ha^{-1} compared with the mold board plow [65]. It has also been reported that increases in the SOC pools can be credited to either reductions in the CO_2 efflux from the soil or to increases in the C inputs [65]. When comparing the soil surface across different tillage systems, conservation tillage systems retain more crop residues, which ultimately result in the formation of more SOC [66] and [67]. In addition to this finding, the decomposition process of surface applied plants residues as a part of conservation tillage is slow compared to conventional tillage systems because of lower contact with the soil.

In our study a negative but highly significant correlation coefficient (r) value i.e. −0.19403** was recorded between the SOC and cumulative CO_2 emission. This finding might help in the reduction of CO_2 emissions from the chisel plow tillage and zero tillage planting methods compared to the rotary tillage and mold board plow methods (Figs. 2 and 3). In our experiment, increases in the SOC contents from different tillage methods, especially chisel plow tillage and zero tillage, might be related to the lower disturbance of the soil, retention of more crop residues on the soil surface and reductions in the efflux of CO_2 from the soil. Similar results have been reported by [65], [66] and [67].

Conclusions

Intensive tillage and higher N fertilization are not only detrimental to the soil but are also destructive for the entire environment. Adaptations of appropriate tillage methods, crop residue applications and proper fertilization are beneficial for the soil as well as for the environment. These practices are also economically beneficial for resource-poor farmers. The findings from our study clearly indicate that the tillage methods significantly affected CO_2 emissions and the zero tillage planting method emitted the lowest CO_2 compared with the other three tillage methods. No significant differences in CO_2 emissions were recorded for the applications of corn residue mulch, but the applications of corn mulch significantly improved the soil organic carbon (SOC) contents of the soil for all of the tillage systems. In addition, corn residue mulch application reduced the weed infestation by up to 40% (data not shown). Therefore, the application of corn residue mulch can be helpful for reducing the use of herbicides, which will also be helpful in establishing a healthy environment. Applications of different N fertilizer levels also significantly affected CO_2 emissions and, overall, the lowest emissions were recorded for the 160 kg N/ha treatment. Higher CO_2 emissions were recorded immediately after the tillage operations. This study also indicated that both soil temperatures and moistures strongly affected CO_2 emissions and that compared with the other tillage methods, zero tillage planting gave better grain yields. The lowest N fertilizer use gave equal yields, for the two year mean, as did the application of higher N fertilizer levels. These results clearly indicate that proper changes in farm management practices i.e., the adoption of zero tillage, crops residue application and optimum use of N fertilizers can reduce CO_2 emissions from soils. Therefore this type of long term study can be further helpful in reducing the emissions of CO_2 from soils, which will be helpful in reducing the use of inorganic fertilizers. These practices will be helpful in reducing production costs and will be beneficial for the entire environment.

Materials and Methods

Experimental site

A two-year (2010–12)), field study was conducted at the experimental area of Northwest A&F University, Shaanxi Province, northwestern China (latitude of 34°20° N, longitude of 108° 04°E and elevation of 466.7 m above sea level) on the Eum-Orthrosols (Chinese soil Taxonomy) soil, with a mean bulk density of approximately 1.29 g cm^3. The soil in the top 40 cm had an SOC of approximately 14.26 g/kg, total nitrogen 0.74 g/kg and the pH was approximately 7.85. This area is under the corn-wheat rotation system. During both years, the wheat crop was planted after harvesting the corn crop. Both fertilizers, i.e., phosphorous in the form of calcium phosphate (Ca_2 $(PO_4)_3$ with 16% P and nitrogen in the form of urea with N≥46%, were applied to the corn crop at the rate of 750 kg/ha and 375 kg/ha, respectively. The total rainfall during the wheat crop growing season (October-June) was 231.6 mm and 242.7 mm during, cropping years 2010–11 and 2011–12 respectively (Fig. 7).

Experimental design and treatments

A factorial experiment having a strip split-split arrangement, with tillage methods in the main plots, mulch levels in the sub plots and N-fertilizer levels in the sub-sub plots with three replicates, was used for this study. Different tillage methods, i.e., chisel plow tillage (T_1), zero tillage (T_2), rotary tillage (T_3) and mold board plow tillage (T_4) methods, were kept in the main plots, different mulch kinds, i.e., M_0 (no residue mulch) and M_1 (corn residue mulch), in the sub plots, while different nitrogen fertilizer rates, i.e., 0, 80, 160, 240 and 320 kg N/ha, were kept in the sub-sub plots. The area was uniform in terms of fertility. The total experimental plot area (3300 m^2) was equally divided into four main tillage treatments. The area of each tillage treatment (33 m×25 m) was further sub-divided into sub-plots for mulch treatments, and finally the sub plots were further divided into sub-sub plots, and each sub-sub individual plot had an area of 3 m×25 m. Treatments were randomized within each sub-plot.

Chisel plow tillage (T_1), was performed using a chisel plow. Following fertilizer applications, a chisel plow with a shank spacing of approximately 40 cm apart and 30–35 cm deep was used once. Later on, fertilizers were mixed by using the rotavator for up to a 5 cm depth. For zero tillage (T_2) fertilizers were mixed by using the rotavator up to a 5 cm depth due to the lack of availability of a proper zero tillage drill. For rotary tillage (T_3), the seed bed was prepared by using the rotavator up to a depth of 15–20 cm, and in the case of the mold board. plow tillage method (T_4), the soil was plowed up to a 20–30 cm depth by using the mold board plow, followed by the rotavator for the final seed bed preparation.

Urea fertilizer with N≥46%, was used as the source of the nitrogen, and phosphorous (P) fertilizer in the form of calcium phosphate ($Ca_2\ (PO_4)_3$ with 16% P was equally applied to all of the treatments at a rate of 750 kg/ha at the time of soil preparation. The treatments arrangements were kept the same during both years (2010–12) of study. Previously harvested air-dried corn crops residues were used as the source of corn residue mulch during both years. When the wheat crop was at the 3–4 leaf stage, mulch was applied at a rate of 750 g/m^2. The field was flat with a uniform topography. This area is rain fed, and a wheat-corn rotation is the main cropping system. No irrigation was applied to either crop. No changes were made to the areas with different tillage treatments, and a corn crop was planted after the wheat crop harvest by using the same tillage methods. The wheat crop was harvested using a combine harvester, and after harvesting the wheat crop, the corn crop was planted using a corn planter.

Winter wheat (C.V Shaan mai −139) was planted on October 17, 2010 and October 18, 2011 by using wheat drills. The line to line distance was maintained at approximately 16 cm apart. The seed had a 13% moisture contents and a 85% germination rate during both years. During cropping season 2010–11, a seed rate of about @ 190–200 kg/ha was used, while during cropping season 2011–12, a seed rate of approximately 205–210 kg/ha was employed. Experimental treatments were separated from each other by making boundaries between the treatments. Both years an herbicide application i.e. carfentrazone-ethyl ($C_{15}H_{14}C_{12}F_3N_3O_3$), was used to control the weeds. At physiological maturity, which occurred on June 8, 2011 and on June 10, 2012, three samples were randomly selected from each treatment and manually harvested by using the 1 m^2 quadrants to calculate the grain yields. Finally the wheat crop was harvested using a combine harvester.

Measurements

Meteorological factors. Meteorological data for the study area are given in Fig. 7 and Fig. 8, which indicate that during cropping year 2011–12, more rains were concentrated during the early period of crop growth relative to 2010–11, during which more rains were concentrated in the later crop growth stages. Higher average seasonal temperatures were recorded during the year 2010–11 than in 2011–12. However, more relative humidity was recorded during 2011–12 than in 2010–11. Weekly meteorological data showed higher, average temperatures during the year 2010–11 than during 2011–12. On the other hand, higher weekly relative humidities were recorded during the cropping year 2011–12 than during 2010–11. Similarly, higher weekly basis more wind speeds were recorded during the cropping year 2010–11 than during 2011–12.

Monitoring CO_2 emissions. Because of the high number of treatments, both years of CO_2 emissions data were recorded three times per week and one day of CO_2 emissions data was used as one replicate for statistical analysis. CO_2 emissions data were recorded using a GXH-3010E1 portable gas analyzer. This gas analyzer is made by the Beijing Huayun Yiqi Company, and the CO_2 emission was recorded by using the method described by Gao et al [68]. During both years (2010–12) CO_2 emissions data from all of treatments was recorded 6840 times, which included 3120 times during the cropping year 2010–11 and 3720 times during cropping year 2011–12.

One round PVC chamber (21 cm in diameter and 13.5 cm in height), having total area of approximately 0.0047 m^3 was permanently fixed in the center of each treatment plot. The chamber was completely fixed in the soil up to a 4.5 cm depth. The plants growing with in the chambers were removed. As a consequence of some technical problems with the Gas analyzer, during wheat growing season 2010–11, the CO_2 emissions data were recorded after the wheat had been planted for one month, and during cropping season, 2011–12, CO_2 emissions were recorded starting during the first week of wheat planting. However, due to severe snow fall, data were not recorded on the 14th week during this year. Taking data from wheat planting until harvest for 2010–11, the CO_2 emissions data were recorded for 26 weeks, and during cropping year 2011–12, these data were recorded for 31 weeks. Every week, the data were collected 3 times depending upon the environmental condition, i.e., if the field was too wet from rainfall, then the data recording was stopped. Each time the data collection started at 900 a.m., and each sequence of CO_2 flux measurements took at least 4 hours. Due to the large number of treatments the data were randomly collected from different treatments. The main purpose of this randomization was to minimize the effects of different days as well as changing soil temperatures on the emissions of CO_2. The GXH-3010E1 gas analyzer was attached to the data collector chambers with intake and an outtake tubes. These tubes were made up of soft plastic pipes and each one was approximately one meter in length. At the time of data recording, first the CO_2 data, i.e., X_1 were recorded without closing/covering the chamber, and then the chamber top was tightly closed with a cover that had a fixed small fan. The gas within the chamber was mixed for three minutes with this fan. After this CO_2 emission, data (X_2) were recorded using the gas analyzer.

Chambers tops were closed for only three minutes at the time of data recording. To avoid any chemical change in the soil, these chambers were kept open for the whole remaining time. Along with CO_2 data, soil temperatures data were also recorded from each treatment from depths of 5 cm. For this purpose, thermometers were permanently fixed in each plot each year for the whole crop growth period. The CO_2 emission rate was calculated by using equation (1) as described by Gao et al. [68].

$$F = k(X_2\text{-}X_1)H/\Delta t \quad (1)$$

Where F is the CO_2 emission in mg/(m^2.h); K is a constant with a value of 1.80 (25°C) and X_1 and X_2 are the CO_2 emissions rates from the chambers before and after covering of the chambers. H is the height of the chambers in meters and Δt is the time in hours (h). The cumulative emission of CO_2 was calculated using the following relationship, as described by Wilson H.M and Alkaisi W.W (2008) [69],

$$CO_2 - C\left(kgha^{-1}\right) = \sum_{i=first}^{n=last} X_i + X_{i+1} * N + X_{i+2} * N + \ldots + X_{i+n} * N \quad (2)$$

where (i) is the first week of the growing season when the first CO2 emission rate was taken, (n) is the last week of the growing season when the last CO2 emission rate was taken, X is the CO2 rate (Kg ha−1 day−1), and N is the number of days between two consecutive CO2 emission rates measurements. Finally, these CO2 emission rates for the whole wheat crop growing period (taken between 9.00 A.M until approximately 1.00 P.M) were converted into tons/ha.

Soil moisture measurements. Both years soil moisture contents were measured during different crop growth stages, i.e., before planting the wheat crops, at the 5 leaf stage (Zadoks stage 1.0–1.9), stem elongation stage (Zadoks stage 3.0–3.9), booting stage (Zadoks stage 4.0–4.9), grain formation stage (Zadoks stage 7.0–7.9) and harvesting stage. For this purpose, soil samples were collected from each treatment from 0–100 cm soil depths with three replicates, with increments of 0–10 cm, 10–20 cm, 20–30 cm. 30–40 cm, 40–60 cm, 60–80 cm and 80–100 cm. Soil samples were collected in aluminum boxes using a hand auger, and fresh soil samples were immediately transported to the laboratory. After recording the fresh weights, these samples were dried in an oven at 105°C for at least 48 hours, and the soil water contents were then measured using equation (3) given below.

$$\text{Soil water content (\%)} = \frac{\text{fresh weight of soil} - \text{dry weight of soil}}{\text{dry weight of soil}} \times 100 \quad (3)$$

Soil water contents (mm) were calculated as gravimetric moisture contents using the equation (soil water ×B.D× thickness of soil layer). Soil bulk densities from the different tillage treatments and from the different soil depths were measured using the core method. Total soil water contents in the 0–100 cm soil depths of the different treatments were measured on the basis of the bulk densities of these different soil layers.

Soil organic carbon (SOC) measurements. Both year soil SOC samples were collected from the top 0–10 cm soil depth after the wheat harvest. Soil samples were collected from two randomly chosen points from each plot using a hand augur. The samples from each treatment were then mixed together to make a composite sample of each treatment. These samples were then air-dried at room temperature, crushed gently and passed through a 2 mm sieve for further chemical analysis. Soil organic carbon was determined by the oxidation method with $K_2 Cr_2 O_7$ -H_2SO_4. For chemical analysis, 0.5 grams of soil was digested with 5 mL of 1 M $K_2 Cr_2 O_7$ and 5 mL of concentrated H_2SO_4 and was heated at 175°C for 5 minutes followed by titration of the digests with $FeSO_4$ [70].

Statistics

Annual data collected for the CO_2 emission rates and for other related parameters over the whole 2-year period were subjected to an analysis of variance (ANOVA) by using the factorial experiment with the strip-split-split arrangement having the tillage methods in the main plots, mulch in the sub plots and N-fertilizer levels in the sub-sub plots. The SAS analytical software package GLM (8.01) was used for the analyses. Mean values and standard errors (SE) were calculated for each treatment, and an ANOVA was used to assess the treatment effects on the measured variables. Means were declared statistically significant at a 0.05 probability level, or P≤(0.05), using the DUNCAN test (DNMRT).

Author Contributions

Conceived and designed the experiments: SKT YL. Performed the experiments: SKT XLL. Analyzed the data: SKT JLZ. Wrote the paper: SKT XXW YL.

References

1. Post WM, Peng TH, Emanuel WR, King AW, Dale VH, et al (1990) The global carbon cycle. Am Sci 78: 310–326.
2. Houghton RA, Skole DL (1990) Carbon, In: Turner BL, Clark WC, Kates RW, Richards JF, Mathews JT, Meyer WB, editors. The earth as Transformed by Human Action. Cambridge University Press. 393–408.
3. Houghton RA (2007) Balancing the global carbon budget. Ann Rev Earth Planet Sci 35: 313–47.
4. Denman KL, Brasseur G, Chidthaisong A, Ciais P, Cox PM, et al. (2007) In Climate Change 2007: The Physical Science Basis. Contribution of Working Group I to the Fourth Assessment Report of the International Panel on Climate Change. Solomon, S., et al, editors. Cambridge Univ. Press. 499–587.
5. Valentini R, Matteucci G, Dolman AJ, Schulze ED, Rebmann C, et al. (2000) Respiration as the main determinant of carbon balance in European forests. Nature 404: 862–864.
6. IPCC (2007) Climate Change 2007: The Physical Science Basis. Cambridge University Press. 996.
7. Cole CV, Duxbury J, Freney J, Heinemeyer O, Minami K, et al. (1997) Global estimates of potential mitigation of greenhouse gas emissions by agriculture. Nut. Cycl. Agroecosyst. 49: 221–228.
8. Paustian K, Collins HP, Paul EA (1997) Management controls on soil carbon. In: Paul EA, Paustian K, Elliot ET, Cole CV, editors. Soil Organic Matter in Temperate Agro ecosystems (Long-term Experiments in North America). CRC Press, Boca Raton, FL. 15–49.
9. Lal R (1997) Long-term tillage and maize monoculture effects on a tropical alfisol in western Nigeria.1. Crop yield and soil physical properties. Soil and Tillage Res 42: 145–160.
10. Ahmadi H, Mollazade K (2009) Effect of plowing depth and soil moisture content on reduced secondary tillage. Agricultural Engineering International: The CIGR E Journal 11: 1–9.
11. Koga N, Tsuruta H, Tsuji H, Nakanoa H (2003) Fuel consumption – derived CO_2 emission under conventional and reduced tillage cropping systems in northern Japan, Agriculture Ecosystems and Environment 99: 213–219.
12. Schlesinger WH (1986) Changes in soil carbon storage and associated properties with disturbance and recovery. In: Trabalka JR, Reichle DE, editors. The changing Carbon Cycle- A global analysis. Springer Verlag, New York. 194–220.
13. Reicosky DC, Lindstrom MJ (1993) Fall tillage method: Effect on short – term carbon dioxide flux from soil. Agron J 85(6): 1237–1243.
14. Franzluebbers AJ, Hons FM, Zubberer DA (1995) Tillage induced seasonal changes in soil physical properties affecting soil CO_2 evolution under intensive cropping. Soil Tillage Res 34: 41–60.
15. Ding W, Cai Y, Cai Z, Zheng X (2006) Diel pattern of soil respiration in N-amended soil under maize cultivation. Atmos Environ 40: 3294–3305.
16. Ren X, Wang Q, Tong C, Wu J, Wang K, et al. (2007) Estimation of soil respiration in a paddy ecosystem in the subtropical region of China. Chin Sci Bull 52 (19): 2722–2730.
17. Iqbal J, Ronggui H, Lijun D, Lan L, Shan L, et al. (2008) Differences in soil CO_2 flux between different land use types in mid-subtropical China. Soil Bio Biochem 40: 2324–2333.
18. Liu H, Zhao P, Lu P, Wang YS, Lin YB, et al. (2008) Greenhouse gas fluxes from soils of different land-use types in a hilly area of south China. Agric Ecosyst Environ 124(1–2): 125–135.
19. Lee DK, Doolittle JJ, Owens VN (2007) Soil carbon dioxide fluxes in established switch grass land managed for biomass production. Soil Biol Biochem 39: 178–186.
20. Ladd JN, Amato M, Zhou LK, Schultz JE (1994) Differential effects of rotation, plant residue and nitrogen fertilizer on microbial biomass and organic matter in an Australian alfisol. Soil Biol Biochem 26: 821–831.
21. Edmeades DC (2003) The long-term effects of manures and fertilizers on soil productivity: a review. Nutr Cycl Agroecosys 66: 165–180.
22. Kern JS, Johnson MG (1993) Conservation tillage impacts on natural soil and atmospheric carbon levels. Soil Sci SOC Am J 57: 200–210.
23. Reeves DW (1997) The role of soil organic matter in maintaining soil quality in continuous cropping system. Soil Tillage Res 43: 131–167.
24. Smith P, Powlson DS, Glendining MJ, Smith JU (1998) Preliminary estimates of potential for carbon mitigation in European soils through no-till farming. Global Chang Biol 4: 679–685.
25. Schlesinger WH (1999) Carbon sequestration in soils. Science 248: 2095.
26. Johnson MG, Levine ER, Kern JS (1995) Soil organic matter, distribution, genesis, and management to reduce greenhouse gas emissions. Water Air Soil Pollut 82: 593–615.
27. Lal R, Follett RF, Kimble J, Cole CV (1999) Managing US crop land to sequester carbon in soil. J. Soil Water Conserv 54: 374–381.

28. Paustian K, Andren O, Janzen H, Lal R, Smith P, et al. (1997) Agricultural soils as a C sink to offset CO_2 emissions. Soil Use Manage 13: 230–244.
29. Green CJ, Blackmer AM, Horton R (1995) Nitrogen effects on conservation of carbon during corn residue decomposition in soil. Soil Sci Soc Am J 59: 453–459.
30. Kowallenko CG, Ivarson KG, Cameron DR (1978) Effect of moisture content, temperature and nitrogen fertilization on carbon dioxide evolution from field soils. Soil Biol Biochem J 10: 417–423.
31. Fogg K (1988) The effect of added nitrogen on the rate of organic matter decomposition. Biol Rev 63: 433–472.
32. Halvorson AD, Wienhold BJ, Black AL (2002) Tillage, nitrogen, and cropping system effects on soil carbon sequestration. Soil Sci Soc Am J 63: 906–912.
33. Liang BC, Mackenzie AF (1992) Changes in the soil organic carbon and nitrogen after six years of corn production. Soil Sci 153: 307–313.
34. Wang XB, Cai DX, Hoogmoed WB, Oenema O, Perdok UD (2007) Developments in conservation tillage in rainfed regions of North China. Soil & Tillage Res 93: 239–250.
35. Zhang H, Wang X, Feng Z, Pang J, Lu F, et al. (2011) Soil temperature and moisture sensitivities of soil CO_2 efflux before and after tillage in a wheat field of Loess Plateau, China. J Environ Sci 23: 79–86.
36. Lou YZ, Zhou X (2006) Soil respiration and the environment. Academic, San Diego. 110–111.
37. Iqbal J, Hu R, Lin S, Hatano R, Feng M, et al. (2009) CO_2 emission in a subtropical red paddy soil (Ultisol) as affected by straw and N- fertilizer applications: A case study in Southern China. Agriculture Ecosyst and Environ 131: 292–302.
38. Zou J, Huang Y, Zheng X, Wang Y, Chen Y (2004) Static opaque chamber – based technique for determination of net exchange of CO_2 between terrestrial ecosystem and atmosphere. Chin Sci Bull 49 (4): 381–388.
39. Reiccosky DC, Dugas WA, Torbert HA (1997) Tillage- induced soil carbon dioxide loss from different cropping system. Soil Tillage Res 41: 105–118.
40. Calderon FJ, Jackson LE, Scow KM, Rolston DE (2001) Short term dynamics of nitrogen, microbial activity, and phospholipid fatty acids after tillage. Soil Sci Soc Am J 65: 118–126.
41. Reiccosky DC (2002) Long term effect of moldboard plowing on tillage-induced CO_2 loss. In: Kimble JM. Lal R, Follett RF editors. Agricultural practices and policies for carbon sequestration in soil. CRC Press, Boca, 87–96.
42. La Scala N Jr, Lopes A, Panosso AR, Camara FT, Periera GT (2005) Soil CO_2 efflux following rotary tillage of a tropical soil. Soil Tillage Res 84: 222–225.
43. Alvaro-Fuentes J, Cantero-Martinez C, Lopez MV, Arrue JL (2007) Soil carbon dioxide fluxes following tillage in semiarid Mediterranean agro-ecosystems. Soil Tillage Res 96: 331–341.
44. Beare MH, Cabrera ML, Hendrix PF, Coleman DC (1994) Aggregate–protected and unprotected organic matter pools in conventional and no-tillage soils. Soil Sci Soc Am J 58: 787–795.
45. Kladivko EJ (2001) Tillage systems and soil ecology. Soil Tillage Res 61: 61–76.
46. Zhang HX, Zhou X, Lu F, Pang J, Feng Z, et al (2011) Seasonal dynamics of soil CO_2 efflux in a conventional tilled wheat field of the Loess Plateau, China. Ecol Res. 26: 735–743. DOI 10. 1007/s11284-011-0832-5.
47. Li H, Qiu J, Wang L, Tang H, Li C, et al. (2010) Modelling impacts of alternative farming management practices on greenhouse emissions from a winter wheat – maize rotation system in China. Agriculture, Ecosyst and Environ 135: 24–33.
48. Jacinthe PA, Lal R, Kimble JM (2002) Carbon budget and seasonal carbon dioxide emission from a Central Ohio Luvisol as influenced by wheat residue amendment. Soil & Tillage Res 67: 147–157.
49. Xiao Y, Xie G, Lu G, Ding X, Lu Y (2005) The value of gas exchange as a service by rice paddies in suburban Shanghai, PR China. Agric Ecosyst Environ 109: 273–283.
50. Chapin FS III (1991) Effects of multiple environmental stress upon nutrient availability In: Mooney HA, Winner WA, Pell EJ, editors. Integrated Responses of Plants to Environmental Stress. Academic New York. 67–86.
51. Fisk MC, Fahey TJ (2001) Microbial biomass and nitrogen cycling responses to fertilization and litter removal in young northern hardwood forests. Biogeochemistry 53: 201–223.
52. Anderson TH (1994) Physical analysis of microbial communities in soil: applications and limitations. In: Ritz K, Dighton J, Giller KE, editors. Beyond the Biomass: Compositional and Functional Analysis of Soil Microbial Communities. John Wiley and Sons. New York. 67–76.
53. Galantini J, Rosell R (2006) Long-term fertilization effects on soil organic matter quality and dynamics under different production systems in semiarid Pampean soils. Soil Tillage Res 34: 41–60.
54. Ding W, Meng L, Yin Y, Cai Z, Zheng X (2007) CO_2 emission in an intensively cultivated loam as affected by long-term application of organic manure and nitrogen fertilizers. Soil Biol Biochem 39: 669–679.
55. Raich JW, Schlesinger WH (1992) Modeling soil organic matter in organic amended and N-fertilized, long-term plots. Soil Sci Soc Am J 56: 476–488.
56. Davidson EA, Belk E, Boone RD (1998) Soil water content and temperature as independent or confounded factors controlling soil respiration in a temperate mixed hardwood forest. Glob Chang Biol 4: 217–227.
57. Reichstein M, Tenhunen JD, Roupsard O, Ourcival JM, Rambal S, et al. (2002) Ecosystem respiration in two Mediterranean evergreen Holm oak forests: drought effects and decomposition dynamics. Funct Ecol 16: 27–39.
58. Buchmann N (2000) Biotic and abiotic factors controlling soil respiration rates in *Picea abies* stands. Soil Biol Biochem 32: 1625–1635.
59. Saiz G, Black K, Reidy B, Lopez S, Farrell EP (2007) Assessment of soil CO_2 efflux and its components using a process-based model in a young temperate forest site. Geoderma 139: 79–89.
60. Bauer J, Herbst M, Huisman JA, Weihermuller L, Vereecken H (2008) Sensitivity of simulated soil heterotrophic respiration to temperature and moisture reduction functions. Geoderma 145: 17–27.
61. Bowden RD, Newkirk KM, Rullo GM (1998) Carbon dioxide and methane fluxes by a forest soil under laboratory-controlled moisture and temperature conditions. Soil Biol. Biochem 30: 1591–1597.
62. Haynes RJ (2000) Labile organic matter as an indicator of organic matter quality in arable and pastoral soils in New Zealand. Soil Biol Biochem 32: 211–219.
63. Lal R, Kimble JM (1997) Conservation tillage for carbon sequestration. Nutr. Cycling Agroecosyst 49: 243–253.
64. Lal R, Kimble JM, Follet RF, Cole CV (1998) The potential for US crop land to Sequester Carbon and Mitigate the Greenhouse Effect. Chelsea: Ann Arbor Science.
65. West TO, Post WM (2002) Soil organic carbon sequestration rates by tillage and crop rotation: a global data analysis. Soil Sci. Soc. Am J 66: 1930–1946.
66. Drury CF, Tan CS, Welacky TW, Oloya TO, Hamill AS, et al. (1999) Red clover and tillage influence on soil temperature, water content and corn emergence. Agron J 91: 101–108.
67. Hutchinson JI, Campbell CA, Desjardins RL (2007) Some perspectives on carbon sequestration in agriculture. Agric Meteorol 142: 288–302.
68. Gao C, Sun X, Cao J, Luan Y, Hao H, et al. (2008) A method and apparatus of measurement of carbon dioxide flux from soil surface from situ [J]. Journal of Beijing Forestry University 30 (2): 102–105. (In Chinese with English abstract).
69. Wilson HM, Al-kaisi MM (2008) Crop rotation and nitrogen fertilization effect on soil CO_2 emissions in Central Lova. Applied Soil Ecology 39: 264–270.
70. Bao SD (2008) Soil Agricultural Chemistry Analysis [M]. Beijing: China Agricultural press (In Chinese).

7

A Mechanistic Study of Plant and Microbial Controls over R* for Nitrogen in an Annual Grassland

Stephanie G. Yelenik[9¤a], **Benjamin P. Colman***[9¤b], **Jonathan M. Levine**[¤c], **Janneke HilleRisLambers**[¤d]

Department of Ecology, Evolution and Marine Biology, University of California Santa Barbara, Santa Barbara, California, United States of America

Abstract

Differences in species' abilities to capture resources can drive competitive hierarchies, successional dynamics, community diversity, and invasions. To investigate mechanisms of resource competition within a nitrogen (N) limited California grassland community, we established a manipulative experiment using an R* framework. R* theory holds that better competitors within a N limited community should better depress available N in monoculture plots and obtain higher abundance in mixture plots. We asked whether (1) plant uptake or (2) plant species influences on microbial dynamics were the primary drivers of available soil N levels in this system where N structures plant communities. To disentangle the relative roles of plant uptake and microbially-mediated processes in resource competition, we quantified soil N dynamics as well as N pools in plant and microbial biomass in monoculture plots of 11 native or exotic annual grassland plants over one growing season. We found a negative correlation between plant N content and soil dissolved inorganic nitrogen (DIN, our measure of R*), suggesting that plant uptake drives R*. In contrast, we found no relationship between microbial biomass N or potential net N mineralization and DIN. We conclude that while plant-microbial interactions may have altered the overall quantity of N that plants take up, the relationship between species' abundance and available N in monoculture was largely driven by plant N uptake in this first year of growth.

Editor: Maarja Öpik, University of Tartu, Estonia

Funding: This research was supported by a grant from the David and Lucile Packard Foundation (JML and JHRL), and from the National Research Initiative of the United States Department of Agriculture Cooperative State Research, Education and Extension Service, grant 2005-35320-16273 (SGY). The funders had no role in study design, data collection and analysis, decision to publish, or preparation of the manuscript.

Competing Interests: The authors have declared that no competing interests exist.

* Email: colmanb@gmail.com

9 These authors contributed equally to this work.

¤a Current address: Pacific Island Ecosystem Research Center, United States Geological Survey, Kilauea Field Station, Hawaii National Park, Hawaii, United States of America
¤b Current address: Department of Biology, Duke University, Durham, North Carolina, United States of America
¤c Current address: Institute of Integrative Biology, Eidgenössische Technische Hochschule Zurich, Zurich, Switzerland
¤d Current address: Biology Department, University of Washington, Seattle, Washington, United States of America

Introduction

Plant communities are known to be shaped by competition for resources such as light, space, nutrients, and water [1–3]. Differences in species' abilities to capture or alter these resources across space and time can drive competitive hierarchies, successional dynamics, community diversity, and invasions [4–6]. One of the best known attempts to describe these complex interactions with a simple set of rules is R* resource competition theory [5,7,8]. In this theory, Tilman [5] posits that plant species differ in their ability to deplete a limiting resource, and the species that maintains a positive growth rate at the lowest level of this resource will emerge as the competitive dominant. Following from Tilman [5], R* is the equilibrium concentration of resources necessary for the consumer (plant) species to maintain a stable population. Although seemingly simplistic, this theory has successfully predicted the outcome of plant competition in various ecosystems, including California grasslands [9–11], alpine meadows [12], and Midwestern prairies [13,14].

Part of R* theory's success stems from providing clear predictions and testable hypotheses for ecologists [8,15]. For example, it is relatively simple to quantify differences in resource depletion among species by planting monocultures and measuring resource concentrations after biomass stabilizes (R*). Determining whether R* values predict competitive hierarchies and structure communities can be accomplished by comparing these monoculture R* values with the relative abundance of those species in mixtures (i.e. after competition). A negative relationship between species' R* and their relative biomass in mixture supports R* theory, since the species depleting resources to the lowest levels are the competitive dominants [5,13].

An open question is, what drives species-specific R* values in the first place? Differences in species R* levels for a soil nutrient may be due to differences in plant uptake, differences in how plants affect microbial nutrient cycling and sequestration, or both. However, the relative contribution of plants and microbes to the observed resource depletion remains largely untested. R* theory in terrestrial ecosystems has largely been tested in perennial communities or perennial communities subjected to annual invaders [9,10,12–14]. In such systems it is difficult to disentangle over multi-year timescales the relative contributions to R* of plant uptake, microbial nutrient cycling, and the indirect effects of plant traits such as litter quality and other plant-soil feedbacks [16].

Plants are often assumed to be the major drivers of R*. For example, in systems limited by nitrogen (N), depletion has largely been assumed to be due to plant N uptake and sequestration in plant tissues coupled with litter-mediated feedbacks[11,14,17]. Plant species can also alter microbial community composition, size, and activity, and thus alter available soil N concentrations through differences in plant traits such as quality of root exudates or root and leaf litter [18–21]. These effects of plant species on microbial N cycling rates can occur over short time scales [22,23]. Carbon-rich root exudates from grasses, for example, have been shown to increase microbial respiration and N turnover in rhizosphere soils within 24 hours, resulting in higher N uptake by grasses [24].

However, microbes are also strong competitors for N in soils [25–28], given their high surface area [29]. Microbes are integral in the cycling of N in soils by breaking down large organic N molecules into smaller compounds (e.g., breaking proteins into amino acids and ammonium), and converting ammonium into nitrate [30]. While microbial biomass is typically not large, it turns over rapidly. As microbial biomass turns over, N from microbial "necromass" can become less available to plants and microbes as it is incorporated into soil organic matter and is protected from decomposition by a range of chemical, physical, and biological mechanisms [31,32].

To test the extent to which plant or microbial controls on N availability underlie the R* theory of plant competition, we established an experiment in a California grassland ecosystem. We planted annual grass and forb species in monoculture and mixture plots and followed them over one growing season. Our focus on an establishing annual system is very different than the majority of work examining R* theory, most of which has been done over multiple years in perennial systems and does not explicitly examine the extent to which plants or microbes are driving the patterns [10,33]. Although it has been previously established that plant-microbial feedbacks mediated by litter are associated with species' R* after several years in perennial systems [33], this has not been studied mechanistically (i.e., by explicitly examining the plant and microbial N pools and fluxes). Moreover, in the first year of community establishment in an annual system, such feedbacks are unlikely to be important drivers. Thus, examining an annual system over a single growing season allows for a detailed examination of short-term impacts of plants on microbial activity, and microbial activity on plant growth and nutrient acquisition, all in the relative absence of litter mediated feedbacks. As such, this study represents a first step in describing the mechanisms that drive patterns of resource competition in R* theory.

We have previously reported that the available soil N levels in monocultures in this experiment were correlated with the competitive hierarchy among plant species [11], and thus, that R* for N (and therefore, competition for nitrogen) strongly influences community dynamics in this system. The importance of competition for N is not surprising, given that grasslands in this locality have been shown previously to be N-limited [10,34]. Here we examine the degree to which plant and microbial controls drive N levels in soils, and directly or indirectly influence plant-plant competitive dynamics.

A range of different relationships could emerge between plants, microbes, and R*. If differences in R* are driven primarily by plant N uptake, we would expect to see a strong negative relationship between plant N uptake and dissolved inorganic N (DIN), our metric of R*. Furthermore, we would expect no relationship between microbial N cycling and DIN (Figure 1A).

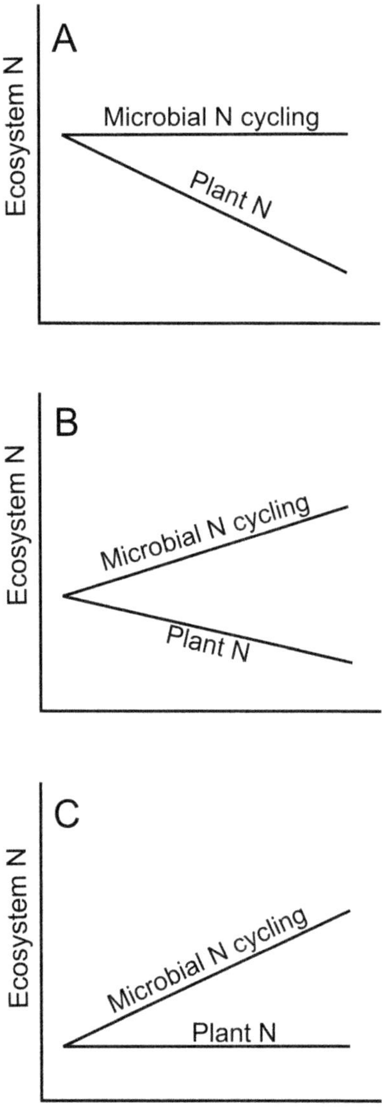

Figure 1. Conceptual models of plant and microbial controls of soil DIN. Our measure of R*, soil DIN, could be affected by **A** plant uptake, **B** plant uptake and microbial N cycling, or **C** microbial N cycling. See Introduction for details.

However, if differences in R* are driven primarily by changes in microbial N cycling rates (*e.g.*, net N mineralization), we would expect that microbial N cycling would be positively correlated to DIN, while plant N uptake would have no relationship with DIN (Figure 1C). If differences are driven by both differences in plant uptake and in microbial N cycling, we would expect a pattern somewhere in between the plant dominated and microbe dominated patterns with a slight positive correlation between microbial N cycling and DIN, and a slight negative relationship between plant N uptake and DIN (Figure 1B).

While we expected that both direct plant uptake of DIN and microbial N cycling would contribute to R*, we were interested in where on the continuum of plant to microbial control this annual ecosystem would fall (Figure 1A–C). Across the different monoculture and bare plots, we measured: 1) pool sizes of total N in plants; 2) DIN at the end of the growing season (our measure of R*) and over time; 3) microbial pools and process rates (microbial

N, microbial biomass, net N mineralization, gross NO_3^- production, and nitrification potential). To tease out drivers we then examined relationships in bare and monoculture plots between DIN and: plant biomass; microbial biomass and biomass N; and microbial process rates.

Materials and Methods

Site and Experimental Design

Experimental plots were established in a pasture/grassland with permission from the owner, the Midland School, in Los Olivos, California. The climate is Mediterranean, with hot, dry summers and cool, wet winters and an average annual rainfall of 550 mm. The land had not been plowed or tilled since 1940, but these soils were not undisturbed; gopher activity causes extensive physical turnover of surface soils in California grasslands, and has been estimated to give complete turnover of surface soils every 3–5 years [35]. Plots have also been grazed by cattle annually, presumably also causing soil disturbance. The vegetation is a mixture of annual and perennial herbaceous forbs and grasses growing with occasional oak trees. Soils are typic Argixerolls, with gravelly fine sandy loam texture.

Study species included three native annual forbs (*Amsinkia menziesii, Calandrinia ciliata, Clarkia purpurea*), three native annual grasses (*Muhlenbergia microsperma, Vulpia microstachys,* and *Vulpia octoflora*), and six exotic annual grasses (*Avena barbata, Bromus hordeaceous, Hordeum murinum, Lamarkia aurea, Polypogon monspielensis* and *Vulpia myorus*). All species were present in nearby grasslands (Stan Harpole, personal communication) and thus our study plants reflect a realistic representation of current California annual species. All data were collected in winter, spring, and summer of 2006.

Five 'blocks' were established within the grassland and were separated by 50 to 500 m. Vegetation was cleared by spraying with Roundup in fall 2005 and roto-tilling with a tractor two weeks after plants had died. Within each block, fourteen 0.64 m^2 plots were established separated by 1 m buffers, and plots were randomly assigned treatments (species monocultures, mixtures, or bare plots). There was one full replicate of the experimental design within each block, consisting of one bare, one mixture (with all 12 species), and 12 monoculture plots, giving a total of five replicate plots for each plot type between the five blocks. Each block was fenced to exclude cattle grazing (see [11] for further details). Data from mixture plots was used to establish the R* relationship in HilleRisLambers et al. [11], by relating relative biomass in mixture to DIN in monoculture plots. Because we were trying to determine what drives DIN concentrations in the monoculture plots used to test R* theory, we focused on monoculture plots only for this study.

To establish plots, we collected seeds locally (*A. barbata, B. hordeaceous, H. murinum, L. aurea*, and *V. myorus*) or used seed from local seed companies (*A. menziesii, C. ciliata, C. purpurea, M. microsperma, P. monspielensis, V. microstachys*, and *V. octoflora*). We added a total of 15 grams of seed/m^2 to each plot, with that weight being divided equally among the species in mixture plots. Seeds were added at the beginning of the growing season (late November 2005) and plots were watered with a volume equivalent to 3 inches of rainfall to encourage germination and establishment. We weeded plots twice to remove nontarget species, once soon after germination and once midway through the growing season. Seeding amounts in monoculture and mixture plots were sufficiently high to generate plots with little bare ground visible for all but two species (JHRL, personal observation), including *M. microsperma*, which failed to germinate. *M. microsperma* thus served as a type of bare plot, and may be considered as a more appropriate control because it had an initial N input from seed, similar to other planted plots. We refer to it as bare (seeded) in the rest of the manuscript and figures.

Soil N Pools and Dynamics

We collected all soils immediately after plant biomass harvest. We sampled soil by extracting two 5 cm diameter, 10 cm depth soil cores from each plot. From each of the five replicate plots for each species, duplicate cores were sieved (2 mm mesh size) prior to soil and microbial analyses (see *Microbial N Pools and Process rates*, below). Dissolved inorganic N (DIN hereafter) was quantified by extracting ammonium and nitrate (NH_4^+ and NO_3^-) with 2M KCl which were then analyzed on a Lachat flow injection autoanalyzer (Lachat Instruments; Loveland, USA), while soil moisture was determined gravimetrically after six days drying at 60°C. DIN was our measure of R* [11]. Total soil N pool (which includes DIN, as well as organic N) was measured on dried soils after they were ground in a ball mill (Wig-L-Bug amalgamator, Crescent Dental, Lyons IL), after which Carbon (C) and N content were measured on a Carlo Erba NA 1500 CHN analyzer (Fisons Instruments, Beverly, MA).

Soil bulk density was measured in June by collecting five soil cores randomly placed within plots in each block (4 cm diameter, 10 cm deep). These were returned to the lab, dried for 48 h at 60°C, rocks were removed, and remaining soil weighed. Block-specific bulk densities were used to convert all soil measures to g m^{-2} unit values to enable comparison of N budgets between soils, plants, and microbes. Outcomes of our statistical analyses on soil N pools did not differ given different N measurement units (i.e., g N m^{-2} versus g N kg^{-1} soil).

To determine whether our R* measure (DIN pools in May) was representative of N dynamics over the growing season, we used resin stakes (PRS Probes; Western Ag Innovations, Saskatoon, Canada) to track inorganic N levels in soils throughout the experiment [36]. Plastic stakes with ion exchange resin surfaces were placed in 0–5 cm soils in mid-month February 2006, and switched each month thereafter until mid-month June 2006. Thus the May sample date represents resin-available N sampled between mid-April and mid-May. Excess soil was washed off of probes with DI water. Probes were then extracted in 0.5 M HCl, and extracts were analyzed for NH_4^+ and NO_3^- as described above. Regenerated stakes were placed into the same slots in the soil as previous stakes.

Plant N Pools

Plant biomass was harvested in mid-May for all species; this timing was chosen based on close monitoring of the diverse phenologies of the species in the experiment. At this time, all species had reached peak biomass and set seed, but had not senesced. Aboveground plant biomass was sub-sampled in each plot by clipping all vegetation within one 10×50 cm quadrat. Roots were collected from duplicate soil cores collected within the clip plots by sieving soils (see above in *Soil N Pools and Dynamics*) in each of the five replicate plots for each vegetation type. All plant material was dried at 60°C for six days and weighed. Plant material was ground with a Wiley mill (Thomas Scientific, Philadelphia, PA) followed by a ball mill. Plant samples were then analyzed for C and N on the CHN analyzer. Total plant N (g N/m^2) in monoculture plots was then quantified as (aboveground biomass/m^2 × shoot % N/100) + (belowground biomass/m^2 × root % N/100).

Microbial N Pools and Process Rates

Prior to examining microbial biomass and N cycling, sieved soils were adjusted to 35% water holding capacity (WHC) to eliminate the confounding effects of variable water content on process rates. Soils were allowed to equilibrate for 7 days to allow the effects of sieving and drying/rewetting to pass, after which soils were weighed out for all analyses. Lab replicates were not included given that soils were sieved, there were five field replicates per treatment, and past work using these same techniques in soils from this area showed tight agreement among lab replicates [37].

To examine how microbial biomass N relates to R*, we used a chloroform slurry extraction to examine the flush of N released from microbial cell lysis and extraction [38]. Briefly, 25 mL of 0.5 M K_2SO_4 was added to each glass tube containing a 4 g soil sample. To this, 0.5 mL of EtOH-free chloroform was added. Tubes were sealed with PTFE lined caps, and shaken for 4 h at 150 rev min^{-1} on an orbital shaker. Tubes were allowed to settle for 10 minutes allowing the bulk of the chloroform and soil slurry to separate. The top 10 mL of extract was filtered through a Pall A/E glass fiber filter (Pall Corporation, Port Washington, NY, USA), then bubbled with air for 20–30 minutes to remove any residual chloroform. Chloroform slurries were compared to slurries extracted without chloroform.

To test the influence of microbial abundance on R* we used substrate induced respiration (SIR), an index of microbial biomass [39]. Briefly, 10 mL of a 12 g L^{-1} Difco Yeast Extract (Becton, Dickinson and Company, Franklin Lakes, NJ, USA) was added to 4 g soil. Soils were capped and shaken for four hours with 1 mL headspace CO_2 subsamples withdrawn at 20 minutes, two hours, and four hours. Gas samples were analyzed on a LI-6262 infrared gas analyzer (LI-COR, Lincoln, NE, USA). The SIR biomass was then calculated by applying a linear fit to the change in CO_2 over time (average $r^2 = 0.99$).

Because microbial biomass turnover rates are faster than those of annual plants, microbes may have driven inorganic N availability in soils via process rates that are not reflected in microbial biomass pools. Thus, we also quantified potential net N mineralization, gross NO_3^- production, and nitrification potential. Net N mineralization is the net accumulation of extractable NH_4^+ and NO_3^- in soil, and represents the balance between microbial production and consumption. Gross nitrification separates microbial production and consumption of NO_3^-, and offers insight into nitrifier abundance and ambient substrate (NH_4^+) availability. In contrast to gross rates, nitrification potential assays add NH_4^+ in excess and thus rates are driven by—and serve as an index of—the abundance and potential activity of ammonia oxidizers. Net accumulation of inorganic N is associated with conditions where heterotrophic microbes are not limited by N availability, while high gross rates of nitrification and high nitrification potential are associated with high availability of NH_4^+ and NO_2^- to autotrophic nitrifiers.

For net N mineralization, two sets of 4 g samples were used, with one set extracted on the first day of incubation for one hour with 25 ml of 0.5 M K_2SO_4, while the other set was incubated for 60 days at 20°C prior to extraction [37]. Net N mineralization was then calculated as the difference in extractable NH_4^+ and NO_3^- in soils between the initial and final time points. These rates can be seen as potential rates because they were held at optimal moisture and temperature for microbial activity. Our method has the advantage of controlling for microclimate variability that may affect mineralization rates, but is not expected to perfectly match field rates.

To quantify gross nitrification, we took duplicate 4 g samples of soil at 35% WHC and added 0.25 ml of 25.5 mg N/L (98% enriched $^{15}N\text{-}NO_3^-$) stirring with the pipette tip after addition. Samples were then extracted for one hour with 25 ml of 0.5 M K_2SO_4, with one sample extracted at 15 min, and the other sample at 24 h. After NO_3^- was measured, extracts were prepared for isotope ratio mass spectrometry. First, NH_4^+ was converted to NH_3 with MgO and driven out of solution. Then, NO_3^- was converted to NH_3 by Devarda's Alloy in the presence of MgO, and the NH_3 was then captured on acidified filter disks in teflon packets using the methods of Sørensen and Jensen [40]. Filter packs were removed and placed in a desiccator with DrieRite and concentrated H_2SO_4 to dry filters and trap any free NH_3. Filters were folded into tin capsules, and ^{15}N enrichment was measured at UCSB's Marine Science Institute Analytical Lab using a Thermo-Finnigan MAT Delta+ Advantage (Thermo Fisher GmBH, Dreieich, Germany). Gross nitrification rates were calculated following Hart et al. [41]. We aimed to measure gross N mineralization in addition to gross nitrification, however NH_4^+ pools in soils were too low, and consumption too high, such that it precluded quantifying gross N mineralization.

Nitrification potential was measured by adding 35 ml of a stock solution containing 8 mL 0.2 M K_2HPO_4, 1 ml 0.2 M KH_2PO_4, 5 ml 0.2 M $(NH_4)_2SO_4$, and 20 ml of 1 M $NaClO_3$ (added to block the activity of nitrite oxidizers; [42]). The rate of appearance of nitrite was measured at three time points over 4 h using a modified Griess-Illosvay reagent [43], with 1 ml of color reagent added to 4 ml of sample.

Statistical Analyses

We asked what the patterns in DIN were over the course of the growing season under our different monoculture types (i.e. plant community composition, time) by fitting the resin N data with five linear generalized mixed effects models (with block as random effect in all), including: a null model; a model with only species identity as explanatory variable; a model with only time as the explanatory variable; a model with both time and species identity as explanatory variables; and a model with the main effects of time and species identity as well as their interaction. Specifying block as a random effect with time as an explanatory variable is analogous to a repeated measures ANOVA. The log of resin N values was used as our response variable to ensure normality. We used AIC's to determine the best-fitting of the five models, and the appropriate likelihood ratio tests to determine the significance of explanatory variables retained in the best fitting model. We also used resin-N data to test whether our one time measure of soil DIN represented soil dynamics across the season, by correlating species-specific cumulative resin N values for each species (over the entire growing season) to species-specific DIN values. We used Kendall's tau rather than Pearson's r for this correlation because both DIN values were non-normally distributed.

We also asked whether DIN values were greater in plots without any vegetation, as we would expect if plants influence available soil N. We did this using a generalized mixed effects model with DIN (log transformed) as the response variable and the categorical explanatory variable 'vegetated' vs. 'bare'.

We then asked how species-specific DIN related to N pools and fluxes. First we used 1- way ANOVAs with species as a fixed effect and block as a random effect to ask whether plant and microbial parameters were different between species (Table 2). We then quantified the correlation between DIN and either: total plant N; Microbial N; Microbial biomass (substrate induced respiration, SIR); Net N mineralization; Nitrification potential; and Gross nitrification (we use the term "microbial biomass/process rates" to encompass these measurements of biomass or process rates hereafter). Kendall's tau was used to assess the relationship

Table 1. Soil (0–10 cm), microbial, and plant N pools.

Nitrogen Pool	g N/m^2	% of Total Soil N
Total Soil N	154±4	
Soil NH_4^++NO_3^-	0.47±0.05	0.3
Microbial biomass N	0.47±0.03	0.3
Plant N (above + below)	4.43±0.33	3.0

*Values represent averages across all treatments ± standard error.

between these variables since many were non-normally distributed. If plant uptake was driving N dynamics, we expected to find a negative relationship between DIN and plant N, and no clear relationship between DIN and other N pools and fluxes (Figure 1A). By contrast, if microbial processes were driving R*, then we would expect to find positive correlations between microbial biomass/process rates and DIN (our proxy for R*), but no clear relationship between plant biomass N and DIN (Figure 1C). If a combination of microbial cycling and plant uptake were driving DIN, we would expect a modest positive correlation between microbial biomass/process rates and DIN, and a modest negative correlation between plant biomass N and DIN (Figure 1B).

While examining the longer term effects of litter mediated plant-soil feedbacks is beyond the scope of this study, we used linear regression to examine the relationship between plant C:N and plant biomass to provide some insight into potential relationships. A positive correlation between biomass and litter C:N would reinforce the short-term R* patterns, with plants that reach high biomass and draw N levels to the lowest levels further depleting soil N with C rich and N poor litter. A negative correlation could suggest that litter-feedbacks might drive the system in the opposite direction, with plants that achieve lower biomass and have a higher R* potentially slowing decomposition over longer time scales with high C:N litter.

Finally, we asked how plant productivity (as measured by total plant biomass per m^2) related to total available N pools (summed over soils, microbes and plant biomass). We used Kendall's tau for these correlations. All analyses were performed using R version 2.13.1 [44].

Results

N Pools and Dynamics

In examining N pools in this annual grassland, total soil N accounted for 154±4 g N/m^2. Microbial biomass N and DIN pools (NO_3^-+NH_4^+) comprised a small fraction of soil N pools, with each accounting for about 0.3% of the total soil N pool (Table 1). In contrast to microbes, plants were a much larger N pool, about ten-fold higher than microbial biomass N (Table 1).

Our indicator of growing season N dynamics, resin N, varied by species over time for all monoculture plots. A model including species identity, time, and their interaction gave the best fit (Figure 2A), as indicated by AIC values (data not shown). Resin available N increased in May in all plots, with a pronounced increase in both seeded bare plots as well as plots with low plant biomass (*Vulpia octoflora*, Vo; *Hordeum murinum*, Hm; and *Calandrinia ciliata*, Cc). In contrast, this response was tempered in monoculture plots with high plant biomass (e.g., *Clarkia purpurea*, Cp).

The five-month sum of resin-available N was significantly correlated with DIN pools measured in May (Figure 2B). This shows that our R* measure (May DIN pools) was correlated with the soil N dynamics through the growing season. The temporal trends in the resin N data showed that plots with high plant biomass maintained lower DIN. Consistent with these data, we found that vegetated plots had lower DIN (R*) values than bare plots (Figure 2C).

Plant N Uptake

Monoculture plant biomass N was significantly and negatively correlated with soil DIN (Figure 3A), and gave a similar trend with resin-available N. Monoculture plots with high plant N (e.g., *Clarkia purpurea*) consistently exhibited lower levels of resin-available N than bare plots (Figure 2A). When comparing N pools of plants and microbes within monoculture plots, even for those plants that took up very little N (*e.g., C. ciliata*), plant biomass was still a larger N pool than microbial biomass. This difference became greater with increasing aboveground plant biomass (Figure 4).

We found that plant species differed in their uptake of N into plant biomass. Differences in plant biomass N were related to differences in total plant biomass yielding a strong positive correlation between the two ($r^2 = 0.81$, $p<0.001$, $n = 54$), whereas plant % N showed a slight negative correlation with N plant biomass N ($r^2 = 0.10$, $p = 0.02$, $n = 54$). Plant species that obtained higher biomass in monoculture plots also had higher C:N ratios (C:N aboveground biomass versus biomass in monoculture: $r^2 = 0.67$, $p = 0.007$, data not shown).

Microbial N Uptake and Process Rates

Microbial biomass N and SIR biomass did not differ in soils across plant monoculture plots (Table 2), and had no relationship with DIN (Figs. 3B, C). Similarly, there was no relationship between DIN and our indicator of microbial N cycling rates, net N mineralization (Figure 3D). Our two metrics of the conversion of NH_4^+ to NO_3^-, nitrification potential and gross nitrate production, did show significant positive correlations with DIN (Figs. 3E, F).

Discussion

While a number of studies have shown that a plant species' depletion of soil available N can be predictive of its competitive ability in mixture, the extent to which plants or microbes drive the N depletion has remained largely unexplored. Set in the context of an N-limited annual grassland ecosystem [10,34], our study suggests that plant N uptake was more important than species-specific effects on microbial processes in driving plant species' differences in their R* (Figure 1A), and thus uptake determined their competitive dominance in the system [11]. Three lines of

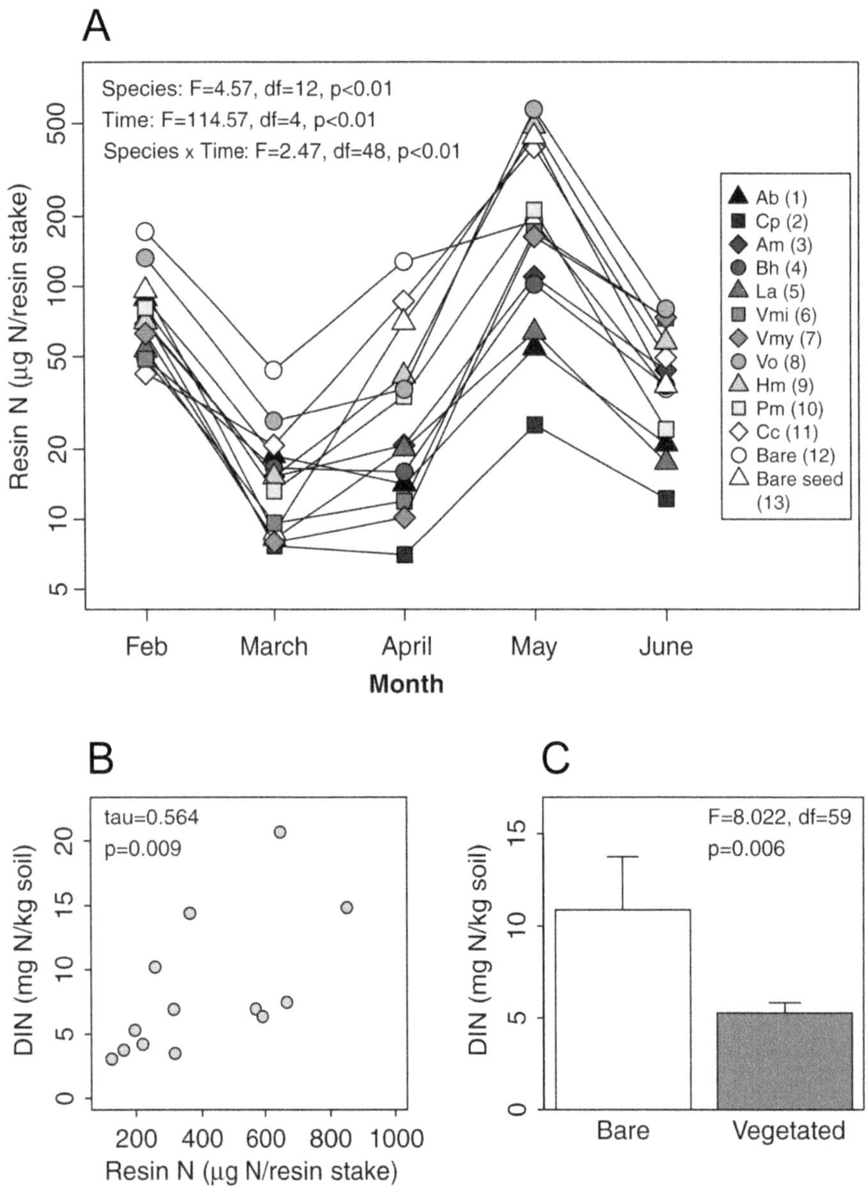

Figure 2. Resin available N in soil during growing season. A Point measures over time in monoculture and bare plots. Species are listed in order of monoculture aboveground biomass and this rank is noted in parenthses (1 = greatest biomass, 12 = lowest biomass) and denoted with increasing grayscale values for symbols indicating greater plant biomass. **B** Relationship between cumulative resin-available N (summed over time) and final DIN levels (R*) in monoculture and bare plots. **C** Final DIN in bare and planted plots. Data were log transformed to account for non-normality prior to analyses, and bars show backtransformed coefficients (±1SE) reflecting bare and planted DIN means. Species abbreviations: *Amsinkia menziesii* (Am), *Avena barbata* (Ab), *Bromus hordeaceous* (Bh), *Calandrinia ciliate* (Cc), *Clarkia purpurea* (Cp), *Hordeum murinum* (Hm), *Lamarkia aurea* (La), *Polypogon monspielensis* (Pm), *Vulpia microstachys* (Vmi), *Vulpia myorus* (Vmy) and *Vulpia octoflora* (Vo). Bare seeded plots received seed addition, but did not germinate (see methods).

evidence support this interpretation: 1) plant N and soil DIN were negatively correlated across the monocultures of different plant species (Figure 3A); 2) microbial N was not correlated with soil DIN (Figure 3B); and 3) net N mineralization was not correlated with soil DIN (Figure 3D).

Evidence that Plants Wear the Pants

The negative correlation between plant N and soil DIN (Figure 3A) suggests that plant uptake was largely responsible for the observed species level differences in R* at the end of the growing season. Additionally, throughout the growing season most of the monocultures—especially those species that achieved high biomass—maintained lower levels of resin-available N than bare plots, where N levels were primarily a function of microbial processes (Figure 2C).

We found that neither microbial biomass N nor net N mineralization was correlated with soil DIN at the end of the growing season (Figure 3B–D) This suggests that microbes were not driving differences in soil DIN via direct uptake and immobilization of available N into microbial biomass, and that microbial processing of organic N was not driving differences in DIN levels across plant species. It has been suggested in the

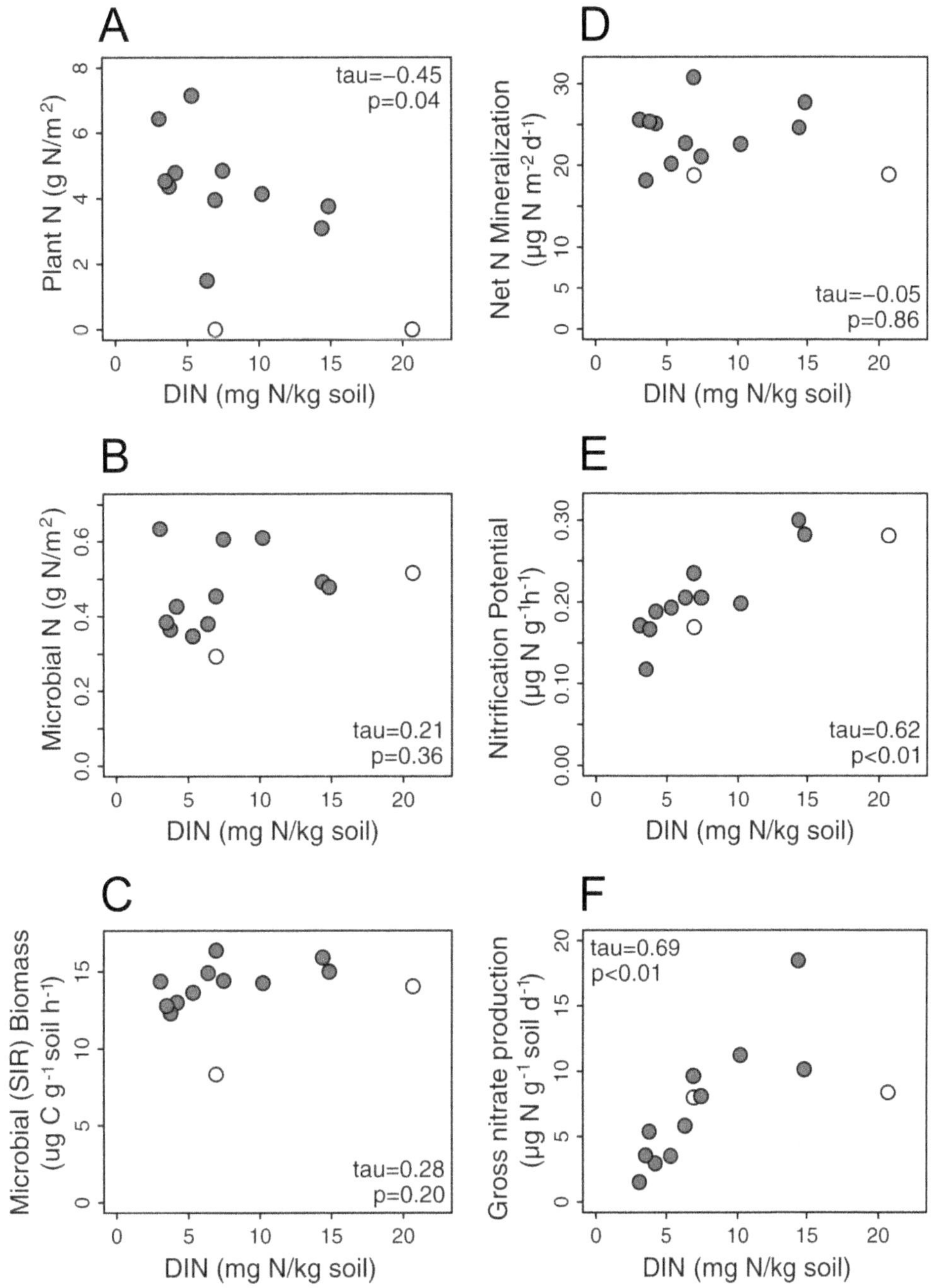

Figure 3. Comparison of soil DIN and plant and microbial metrics. **A** total plant N, **B** Microbial N, **C** Microbial biomass measured by substrate induced respiration (SIR), **D** Potential net N mineralization, **E** Nitrification potential, and **F** Gross nitrification for monocultures and bare plots. White dots represent bare and bare seeded plots, grey dots are vegetated plots.

literature that, while microbes do not drive available N via direct uptake, they do regulate N via processing and plants take up the "leftovers" [27]. If this were the case in the present study net N mineralization would be expected to positively correlate with plant N uptake [27], which is a relationship we did not find in our system (net N mineralization versus plant N uptake: $r^2 = 0.05$, $p = 0.47$).

In fact, nitrification potential and gross nitrification were the only measured microbial parameters that were correlated with DIN (Figs. 3E, F). Nitrification is a dissimilatory process and thus only alters N form, not DIN pool size. Since nitrifier abundance (nitrification potential) and process rates (gross nitrification) correlated to substrate pools (DIN, Figs. 3E, F), we hypothesize that they were limited in their activity by the size of their substrate pools of NH_4^+ and NO_2^- [45,46]. This is consistent with the current understanding of these chemoautotrophic organisms, which are generally considered poorer competitors for soil N than heterotrophic microbes or plant roots [46].

In contrast, the heterotrophic microbes which drive N mineralization and immobilization were likely not limited by N availability. Rather, they functioned similarly across monoculture plots regardless of N uptake by plants and DIN concentrations

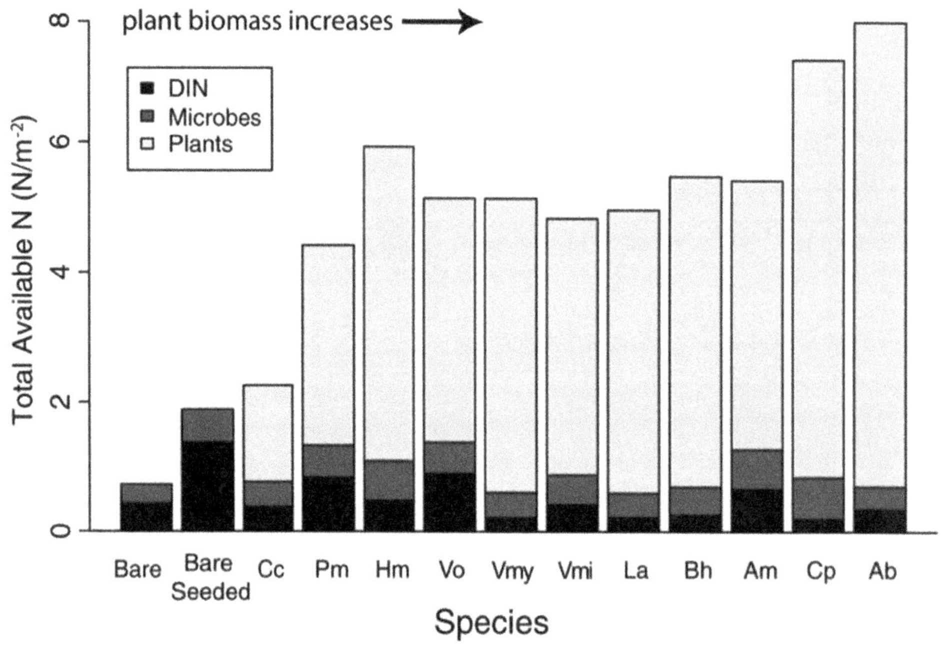

Figure 4. Sum of plant N, microbial N, and final soil DIN to find a total available N in each monoculture plot. Species are listed left to right in order from lowest to highest ranking of aboveground plant biomass in monoculture plots. Kendall's tau between total available N per m^2 (sum of microbial, plant and soil) and above ground plant biomass is 0.68, 2-sided $p = 0.001$. Species abbreviations are as in Figure 1.

(Figs. 3B–D). This is consistent with low soil C:N values, which averaged 7.3 (±0.1), indicating that heterotrophic microbes were more limited by C than N [47]. C-limitation in turn may explain why heterotrophic microbes did not seem to drive final DIN levels.

Evidence of plant-microbial interactions

Our data suggest that plant N uptake drove DIN pool sizes in this annual system, and thus drove R*. This is consistent with ^{15}N tracer experiments which postulate that plants immobilize a greater amount of N into plant biomass over an entire growing season, and thus drive the longer-term patterns of N fate in ecosystems [25,26,29,48]. However, these same studies make it clear that microbes can outcompete plants for available soil N in the short term, with some of the best evidence for this coming from California grasslands [25,26].

Our goal was to test the mechanisms underlying R* by determining the role of plant uptake as compared to changes in microbial biomass and microbially mediated N cycling rates. Understanding such mechanisms is particularly important in such ecosystems where competition outcomes among annuals in the first year of community assembly are likely to be important for long-term community trajectories. By constraining our study to one year in an annual system, we explicitly removed the confounding effect of plant-microbe feedbacks as mediated by litter, which are more likely to develop over several years [20]. Short-term competition stands to be particularly important in these California grasslands, where disturbance plays a large role in exotic annual grass invasion [10,11] and the dominant competitors are annual.

While our data suggest that plant N uptake drove soil DIN pools, another story emerges when the focus is shifted from looking at comparisons of individual pool sizes and fluxes (Figure 3) to looking at the total N in plants, microbes, and DIN for each set of monoculture plots (Figure 4). Examining the sum of N in plants, microbes, and DIN reveals that there were larger pools of biological and/or bioavailable N in plots with higher plant biomass (Figure 4). This suggests that microbes released more plant available N over the course of the growing season in those plots with higher plant biomass and lower R*, even though our measure of microbial N turnover, net N mineralization, did not differ between plots. Comparing planted vs. bare plots, there was more N in these pools (plant N, microbial N, and DIN) in planted (5.4 ± 0.3 g N/m^2) than bare plots (1.3 ± 0.7 g N/m^2; $F_{1,62} = 26.7$, $p < 0.001$) and total N in these pools increased with increasing plant biomass (Figure 4) and decreasing R*. Thus, while plant N uptake was working in a manner consistent with our conceptual models, microbial N cycling may have actually been responding in the opposite direction of our predictions. In other words, microbial N cycling in the field may have been negatively, rather than uncorrelated or positively correlated with DIN (Figure 5); in plots with lower R* and higher plant N uptake, microbes may have generated more plant available N than in plots with higher R* and lower plant N uptake.

It is possible that this pattern—where the sum of plant N, microbial N, and DIN was higher in plots with lower R*—could be explained by large losses of available N from bare plots due to denitrification or leaching. Several lines of evidence suggest that these explanations are unlikely. Denitrification depends on labile C, which would have led to higher potential for N losses due to denitrification in planted plots with roots exuding labile C as opposed to those that were bare or had sparse cover. Thus denitrification would be expected to yield more loss of N with increasing biomass, which would lead to a pattern opposite to that observed in the total N pools (Figure 4). While leaching losses would have been higher in plots with lower biomass due to decreased evapotranspiration, we suggest it is unlikely to have been the main driver of the observed patterns. Over 94% of the annual rainfall had fallen by mid-April, prior to peak plant growth and strong differences in monoculture soil resin available N

Table 2. Effect of plant species on soil N pools and process rates.

Response Variable	F	df	P
Total Soil N (g N m^{-2})	1.41	11,49	0.20
DIN (g N m^{-2})*	2.63	11, 49	0.01
Microbial Biomass N (g N m^{-2})	0.77	11, 49	0.67
Microbial (SIR) Biomass (ug CO_2 g^{-1} soil hr^{-1})	1.30	11, 49	0.25
Potential Net N Mineralization (µg N m^{-2} d^{-1})	1.90	11, 49	0.06
Nitrification Potential (µg N g^{-1} soil h^{-1})*	1.90	11, 49	0.06
Gross Nitrification (µg N g^{-1} soil d^{-1})*	2.56	11,40	0.01
Plant N (g N m^{-2})	3.14	10,43	<0.01

(ANOVA with species as fixed effect and block as a random effect).
*log transformed for normality.

(Figure 2A). The two remaining rainfall events were small enough that they would not have caused significant leaching.

While we cannot rule out denitrification and leaching, we hypothesize that plant 'priming' of microbes led to differences in microbial N cycling in the field, which in turn altered plant uptake of DIN, and thus the total distribution of N in different ecosystem compartments. Priming is the stimulation of microbial activity via labile C-rich root exudates. The low C:N ratio of these soils makes it likely that microorganisms were limited more by C than by N [47]. Plant roots releasing labile C into rhizosphere soil may have led to localized increases in microbial activity [49] leading to higher N availability in rhizosphere soil [21,50]. Because we did not separate rhizosphere from bulk soil, localized increases in microbial activity, such as net N mineralization, were less likely to be observed [50,51]. Priming has been shown to increase plant N uptake in various ecosystems [24,50–52] including California grasslands, where it was shown to increase gross N mineralization rates by ten-fold [51]. Thus plant-microbe interactions may have increased N cycling in this current experiment in plots with lower R* (Figure 5), but our measures of N cycling were not sensitive to those differences given several factors, including: the localized nature of this interaction; the removal of roots and fresh C inputs which would have been the drivers of this interaction; and the short preincubation following sieving and adjustment of water content.

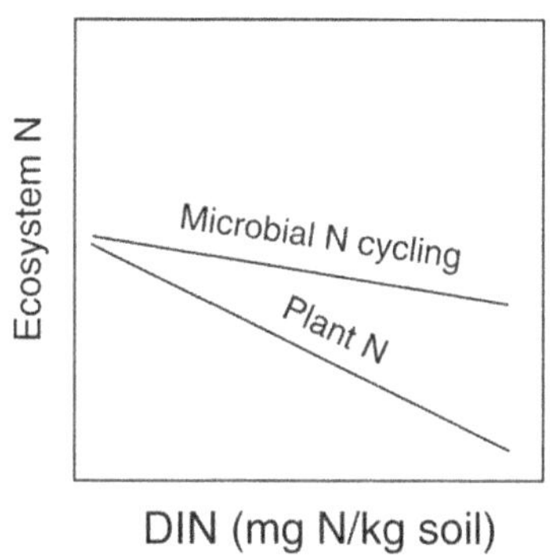

Figure 5. Revised conceptual model plant and microbial contributions to driving DIN (our indicator of R*) in this annual system. In contrast to models shown in Figure 1, in this model, DIN decreases as plant N increases due to plant uptake (Figure 3A), similar to earlier predictions (Figure 1A,B). In addition, greater plant biomass N may have been obtained through stimulation of microbial N cycling (Figure 4), in contrast to earlier predictions (Figure 1B,C).

While our results were consistent with plants driving patterns in R* through uptake of N (Figure 3A), and showed no evidence of microbial processes or biomass driving R* (Figure 3A–C), when ecosystem N pools are taken as a whole (Figure 4) those plots that accumulated more N in microbial, plant, and DIN pools likely had increased cycling and availability of N to plants. This suggests that even over the course of one growing season, plant-microbial interactions (*e.g.*, priming) may be important to ecosystem N cycling and ultimately, may influence plant competition.

It remains an open question as to whether these patterns would persist over multiple growing seasons, and what kind of impact litter mediated feedbacks would have in this annual system. In our experiment, plant species that obtained higher biomass in monoculture plots also had lower quality biomass (higher C:N in aboveground biomass), suggesting that the patterns observed in this study would be reinforced over multiple years by litter mediated feedbacks. It is likely that monoculture plots of annual species with high N use efficiency and low N litter would depress N mineralization, helping drive low R* (Figure 1B,C) and the magnitude of this effect could increase with time as feedbacks become better established. Thus for annuals growing in continuous monoculture, we would expect to see similar patterns to perennial systems [12,33], where the persistence of plants allows for strong litter feedbacks that help determine plant species' R* for nitrogen.

Acknowledgments

We thank the Midland School (Los Olivos, CA) for letting us perform this experiment on their property. Ben Munger and Rick Skillin provided invaluable help in setting up the experiment, and Peter Adler, Anna Oliver, Lilly Hayden and Rebecca Harris assisted in data collection. We also thank Carla D'Antonio, Steve Perakis, Josh Schimel, Alexia Kelley, Bonnie McGill, Marissa Lee, Erin Mordecai, Louie Yang, and three anonymous reviewers.

Author Contributions

Conceived and designed the experiments: SGY BPC JHRL JML. Performed the experiments: BPC SGY JHRL. Analyzed the data: JHRL SGY BPC. Contributed reagents/materials/analysis tools: SGY BPC JHRL JML. Wrote the paper: SGY BPC.

References

1. Grime JP (1977) Evidence for existence of three primary strategies in plants and its relevance to ecological and evolutionary theory. American Naturalist 111: 1169–1194.
2. Connell JH (1983) On the prevalence and relative importance of interspecific competition: evidence from field experiments. American Naturalist 122: 661–696.
3. Pacala SW, Tilman D (1994) Limiting similarity in mechanistic and spatial models of plant competition in heterogeneous environments. American Naturalist 143: 222–257.
4. Connell JH, Slatyer RO (1977) Mechanisms of succession in natural communities and their role in community stability and organization. American Naturalist 111: 1119–1144.
5. Tilman D (1982) Resource competition and community structure. Princeton, NJ: Princeton University Press.
6. Goldberg DE (1990) Components of resource competition in plant communities. In: Grace J, Tilman D, editors.Perspectives on plant competition.San Diego: Academic Press. pp. 27–49.
7. Tilman D (1985) The resource-ratio hypothesis of plant succession. American Naturalist 125: 827–852.
8. Miller TE, Burns JH, Munguia P, Walters EL, Kneitel JM, et al. (2005) A critical review of twenty years' use of the resource-ratio theory. American Naturalist 165: 439–448.
9. Dukes JS (2001) Biodiversity and invasibility in grassland microcosms. Oecologia 126: 563–568.
10. Seabloom EW, Harpole WS, Reichman OJ, Tilman D (2003) Invasion, competitive dominance, and resource use by exotic and native California grassland species. Proceedings Of The National Academy Of Sciences 100: 13384–13389.
11. HilleRisLambers J, Yelenik SG, Colman BP, Levine JM (2010) California annual grass invaders: the drivers or passengers of change? Journal Of Ecology 98: 1147–1156.
12. Suding KN, Larson JR, Thorsos E, Steltzer H, Bowman WD (2004) Species effects on resource supply rates: do they influence competitive interactions? Plant Ecology 175: 47–58.
13. Fargione J, Tilman D (2006) Plant species traits and capacity for resource reduction predict yield and abundance under competition in nitrogen-limited grassland. Functional Ecology 20: 533–540.
14. Harpole SW, Tilman D (2006) Non neutral patterns of species abundance in grassland communities. Ecology Letters 9: 15–23.
15. Fox JW (2002) Testing a simple rule for dominance in resource competition. American Naturalist 159: 305–319.
16. Hodge A, Fitter AH (2013) Microbial mediation of plant competition and community structure. Functional Ecology 27: 865–875.
17. Tilman D, Wedin D (1991) Plant traits and resource reduction for five grasses growing on a nitrogen gradient. Ecology 72: 685–700.
18. Eviner VT (2004) Plant traits that influence ecosystem processes vary independently among species. Ecology 85: 2215–2229.
19. Hawkes CV, Wren IF, Herman DJ, Firestone MK (2005) Plant invasion alters nitrogen cycling by modifying the soil nitrifying community. Ecology Letters 8: 976–985.
20. Yelenik SG, Levine JM (2010) Native shrub re-establishment in exotic annual grasslands: do ecosystem processes recover? Ecological Applications 20: 716–727.
21. Brzostek ER, Greco A, Drake JE, Finzi AC (2013) Root carbon inputs to the rhizosphere stimulate extracellular enzyme activity and increase nitrogen availability in temperate forest soils. Biogeochemistry 115: 65–76.
22. Curiel Yuste J, Baldocchi DD, Gershenson A, Goldstein A, Misson L, et al. (2007) Microbial soil respiration and its dependency on carbon inputs, soil temperature and moisture. Global Change Biology 13: 2018–2035.
23. Fornara DA, Tilman D, Hobbie SE (2009) Linkages between plant functional composition, fine root processes and potential soil N mineralization rates. Journal of Ecology 97: 48–56.
24. Hamilton EW, Frank DA (2001) Can plants stimulate soil microbes and their own nutrient supply? Evidence from a grazing tolerant grass. Ecology 82: 2397–2402.
25. Jackson LE, Schimel JP, Firestone MK (1989) Short-term partitioning of ammonium and nitrate between plants and microbes in an annual grassland. Soil Biology & Biochemistry 21: 409–415.
26. Schimel JP, Jackson LE, Firestone MK (1989) Spatial and temporal effects on plant-microbial competition for inorganic nitrogen in a California annual grassland. Soil Biology & Biochemistry 21: 1059–1066.
27. Kaye JP, Hart SC (1997) Competition for nitrogen between plants and soil microorganisms. Trends in Ecology & Evolution 12: 139–143.
28. Schmidt SK, Costello EK, Nemergut DR, Cleveland CC, Reed SC, et al. (2007) Biogeochemical consequences of rapid microbial turnover and seasonal succession in soil. Ecology 88: 1379–1385.
29. Kuzyakov Y, Xu X (2013) Competition between roots and microorganisms for nitrogen: mechanisms and ecological relevance. New Phytologist 198: 656–669.
30. Schimel JP, Bennett J (2004) Nitrogen mineralization: challenges of a changing paradigm. Ecology 85: 591–602.
31. Schmidt M, Torn MS, Abiven S, Dittmar T, Guggenberger G, et al. (2011) Persistence of soil organic matter as an ecosystem property. Nature 478: 49–56.
32. Kleber M, Sollins P, Sutton R (2007) A conceptual model of organo-mineral interactions in soils: self-assembly of organic molecular fragments into zonal structures on mineral surfaces. Biogeochemistry 85: 9–24.
33. Wedin DA, Tilman D (1990) Species effects on nitrogen cycling: a test with perennial grasses. Oecologia 84: 433–441.
34. Harpole WS, Goldstein L, Aicher RJ (2007) Resource limitation. In: Stromberg MR, Corbin JD, D'Antonio CM, editors.California Grasslands.Berkeley, CA: University of California Press. pp. 119–127.
35. Harrison S, Viers J (2007) Serpentine grasslands. California grasslands: ecology and management University of California Press, Berkeley: 145–155.
36. Johnson DW, Verburg PSJ, Arnone JA (2005) Soil extraction, ion exchange resin, and ion exchange membrane measures of soil mineral nitrogen during incubation of a tallgrass prairie soil. Soil Science Society of America Journal 69: 260–265.
37. Colman BP, Schimel JP (2013) Drivers of microbial respiration and net N mineralization at the continental scale. Soil Biology and Biochemistry 60: 65–76.
38. Fierer N, Schimel JP (2002) Effects of drying-rewetting frequency on soil carbon and nitrogen transformations. Soil Biology and Biochemistry 34: 777–787.
39. Fierer N, Schimel JP, Holden PA (2003) Variations in microbial community composition through two soil depth profiles. Soil Biology & Biochemistry 35: 167–176.
40. Sørensen P, Jensen ES (1991) Sequential diffusion of ammonium and nitrate from soil extracts to a polytetrafluoroethylene trap for ^{15}N determination. Analytica Chimica Acta 252: 201–203.
41. Hart SC, Nason GE, Myrold DD, Perry DA (1994) Dynamics of gross nitrogen transformations in an old-growth forest - the carbon connection. Ecology 75: 880–891.
42. Belser LW, Mays EL (1980) Specific inhibition of nitrite oxidation by chlorate and its use in assessing nitrification in soils and sediments. Applied and Environmental Microbiology 39: 505.
43. QuickChemMethod10-107-04-1-A (1995) Nitrate/nitrite, nitrite in surface water, waste water, 0.2 to 20 mg/L. In: QuickChem Automated Ion Analyzer Methods Manual: Lachat Instruments, Milwaukee, U.S.A.
44. R Development Core Team (2008) R: A language and environment for statistical computing. R Foundation for Statistical Computing, Vienna, Austria. ISBN 3-900051-07-0, URL http://www.R-project.org.
45. Davidson EA, Stark JM, Firestone MK (1990) Microbial production and consumption of nitrate in an annual grassland. Ecology 71: 1968–1975.
46. Verhagen FJM, Laanbroek HJ, Woldendorp JW (1995) Competition for ammonium between plant roots and nitrifying and heterotrophic bacteria and the effects of protozoan grazing. Plant and Soil 170: 241–250.
47. Cabrera ML, Kissel DE, Vigil MF (2005) Nitrogen mineralization from organic residues: research opportunities. Journal of Environmental Quality 34: 75–79.
48. Perakis SS, Hedin LO (2001) Fluxes and fates of nitrogen in soil of an unpolluted old-growth temperate forest, southern Chile. Ecology 82: 2245–2260.
49. Blagodatskaya EV, Blagodatsky SA, Anderson TH, Kuzyakov Y (2009) Contrasting effects of glucose, living roots and maize straw on microbial growth kinetics and substrate availability in soil. European Journal of Soil Science 60: 186–197.
50. Phillips RP, Finzi AC, Bernhardt ES (2010) Enhanced root exudation induces microbial feedbacks to N cycling in a pine forest under long-term CO2 fumigation. Ecology Letters 14: 187–194.
51. Herman DJ, Jaeger KK, Schwartz CH, Firestone E (2006) Root influence on nitrogen mineralization and nitrification in rhizosphere soil. Soil Science Society of America Journal 70: 1504.
52. Frank DA, Groffman PM (2009) Plant rhizospheric N processes: what we don't know and why we should care. Ecology 90: 1512–1519.

Estimating Seasonal Nitrogen Removal and Biomass Yield by Annuals with the Extended Logistic Model

Richard V. Scholtz III*, Allen R. Overman

Agricultural & Biological Engineering Department, University of Florida, Gainesville, Florida, United States of America

Abstract

The Extended Logistic Model (ELM) has been previously shown to adequately describe seasonal biomass production and N removal with respect to applied N for several types of annuals and perennials. In this analysis, data from a corn (*Zea mays* L.) study with variable applied N were analyzed to test hypotheses that certain parameters in the ELM are invariant with respect to site specific attributes, like environmental conditions and soil type. Invariance to environmental conditions suggests such parameters may be functions of the crop characteristics and certain other management practices alone (like plant population, planting date, harvest date). The first parameter analyzed was Δb, the difference between the N uptake shifting parameter and the biomass shifting parameter. The second parameter tested was N_{cm}, the maximum N concentration. Both parameters were shown to be statistically invariant, despite soil and site differences. This was determined using analysis of variance with normalized nonlinear regression of the ELM on the data from the study. This analysis lends further evidence that there are common parameters involved in the ELM that do not rely on site-specific or situation-specific factors. More insight into the derivation of, definition of, and logic behind the various parameters involved in the model are also given in this paper.

Editor: Manuel Reigosa, University of Vigo, Spain

Funding: The funding of this analysis was provided by the Florida Experiment Station. The funders had no role in study design, data collection and analysis, decision to publish, or preparation of the manuscript.

Competing Interests: The authors have declared that no competing interests exist.

* E-mail: rscholtz@ufl.edu

Introduction

Effective water and nutrient management plays an essential role in future attempts at sustainable agricultural production. As the world's population continues to grow, the potable water supply is limited and must be guarded from unnecessary withdrawals and contamination from excessive nutrient loads. Strict monitoring and exercises in groundwater modeling of all agricultural operations is cost prohibitive. It therefore becomes necessary to investigate crop nutrient removal from their environment, and to adopt management procedures and rules that are based on a sound scientific foundation.

Overman et al. first proposed the logistic model as a nutrient management tool to describe seasonal biomass yield dependence of forage grasses on applied N [1]. The original application of the logistic model to plant biomass production was based on inductive reasoning, a process where inferences are made from "real world" observations [2]. While inductive reasoning is innately a more empirical method of model development, all models, no matter the complexity, have some element of empiricism [3]. It is because of the application to the "real world," that engineering and the applied sciences are, for practical reasons, inherently more empirical. The logistic model was extended to include seasonal plant N uptake (removal from the environment) dependence on applied N by forage grasses [4] and then for annuals, like corn [5]. The ELM is a five parameter, non-linear, parametric model that is capable of describing the seasonal biomass yields, N uptake, and N concentration with respect to applied N. Work conducted over the years has indicated that the ELM can effectively describe both annual [6–11] and perennial [8,12–20] crops, for a wide range of nutrient inputs.

The ELM, begins with the simple logistic expression that relates N uptake, N_u, to the N applied, N, which is given by the following relationship:

$$N_u = \frac{A_N}{1+\exp(b_N - cN)} \qquad (1)$$

where A_N is the relative maximum N uptake in kg ha^{-1}, b_N is the dimensionless N uptake shifting parameter, and c is the applied N response parameter given in ha kg^{-1}. The phase relationship between biomass production, Y, and N uptake, N_u, is given by

$$Y = \frac{Y_m N_u}{k_N + N_u} \qquad (2)$$

where Ym is the maximum potential biomass production in Mg ha^{-1}, and k_N is the N uptake response parameter in kg ha^{-1}. The transformation of Eq. (1) by using Eq. (2) yields the following logistic expression:

$$Y = \frac{A}{1+\exp(b - cN)} \qquad (3)$$

that relates biomass production to applied N, where A is the relative maximum biomass production in Mg ha^{-1}, b is the biomass yield shifting parameter, and c is the same applied N response parameter that applies to the N uptake logistic equation. The parameters A_N, A, b_N, b, and c are currently key parameters used with the ELM, and are the easiest to determine from regression analysis.

From the transformation of Eq. (1) into Eq. (3), the relative maximum biomass yield parameter can be written in terms of the maximum potential biomass yield parameter, the relative maximum N uptake parameter, and the N uptake response parameter.

$$A = \frac{Y_m A_N}{k_N + A_N} \quad (4)$$

The biomass shifting parameter can be written in terms of the N uptake shifting parameter, the relative maximum N uptake parameter, and the N uptake response coefficient.

$$b = b_N + \ln\left(\frac{k_N}{k_N + A_N}\right) \quad (5)$$

The difference between the shifting parameter for N uptake and the biomass yield can be written as the following:

$$\Delta b = b_N - b = -\ln\left(\frac{k_N}{k_N + A_N}\right) \quad (6)$$

N concentration is simply defined as the ratio of N uptake to biomass production. This leads to the following relationship:

$$N_c = \frac{N_u}{Y} = N_{cm}\frac{1+\exp(b-cN)}{1+\exp(b_N-cN)} \quad (7)$$

The N concentration model suggests that as N applied is increased to exorbitant levels, there is a maximum limit to the N concentration, N_{cm}. This maximum limit is simply the ratio of the relative maximum N uptake with respect to applied N, A_N, to the relative maximum biomass production with respect to applied N, A.

$$\lim_{N\to+\infty} N_c = \frac{A_N}{A} = N_{cm} \quad (8)$$

It seems logical to suggest that as the background amount of N present in the soil is decreased, the N concentration would be reduced to some prescribed lower limit, as there would be lower limit on the percent of proteins present in a given crop to sustain any growth. Mathematically, the model suggests that this lower limit of N concentration, N_{cl}, is a function of the maximum concentration and the difference between the N uptake shifting parameter and the biomass shifting parameter.

$$\lim_{N\to-\infty} N_c = N_{cm}\exp(-\Delta b) = N_{cl} \quad (9)$$

Also, this lower limit of N concentration, N_{cl}, can be found by taking the ratio of the N uptake response coefficient to the maximum potential biomass production parameter.

$$N_{cl} = \frac{k_N}{Y_m} \quad (10)$$

From the phase relationship between biomass production and N uptake, Eq. (2), N concentration can be found from the following equation with respect to N uptake:

$$N_c = \frac{k_N}{Y_m} + \frac{1}{Y_m}N_u \quad (11)$$

This predicts a linear relationship between N concentration, N_c, and N uptake, N_u. The line should have a slope equal to the inverse of the maximum potential biomass production and an intercept that equals the ratio of the N uptake response parameter to maximum potential biomass production. As this is a phase relationship, this is a functional segment that is bounded between N uptake values from 0 to the peak of A_N, and between N concentration values between N_{cl} and N_{cm}.

From the earlier work of Overman et al. [5], it has been shown that for a given site the applied N response parameter, c, and the N uptake intercept parameter, b_N, and the biomass intercept parameter, b, are not unique to the ELM when applied to grain or the whole plant. Meaning that the harvest index is constant for a given site. Their analysis showed that the only differences between the grain and the whole plant appear in the relative maximum N uptake parameter, A_N, and the relative maximum biomass production parameter, A [5]. Because the b_N, b, and c parameters were shown to be constant for both grain fraction and total biomass production, all the differences in both grain N uptake and biomass production, and the differences in the total plant N uptake and biomass production can be estimated with seven model parameters and a value for the seasonal amount of N applied. This is a comparative reduction of three parameters when b_N, b, and c are not held constant between grain and total plant biomass production.

The goal of this work is to continue to elevate the Extended Logistic Model (ELM) beyond the empiricism of it nascent beginnings and achieve a balance between what can be measured and what should be modeled, as called for by Montieth [21]. The intent is to shed new light on the significance of parameters used in the ELM and to contribute to the search for commonality among parameters. Normalized non-linear regression and analysis of variance (ANOVA) were used to show the invariance of two model parameters with respect to environmental differences, namely soil type and water availability.

Methods

Data Set

This analysis uses data collected by Eugene Kamprath from a corn (Pioneer 3320) N-rate field study that was conducted at three regional research stations in North Carolina from 1981 to 1984. A detailed explanation of the field experiment has been previously reported [22]. Supplemental irrigation was provided at the Clayton experiment station for the well-drained Dothan loamy fine sand (*fine-loamy, siliceous, thermic Plinthic Kandiudults*), at a rate of 10 to 12 cm a season, except for 1982 when no additional water was supplied. No irrigation was provided at the Kinston station for the well-drained Goldsboro sandy loam (*fine-loamy, siliceous, thermic Aquic Paleudults*). At the Plymouth experiment station, no irrigation was provided for the poorly-drained Portsmouth very fine sandy

Table 1. Analysis of variance for model parameters for corn grain and total plant biomass production and for corn grain and total plant N uptake, grown on three different soils.

Mode	Parameters Estimated	Degrees of Freedom	Normalized Residual Sum of Squares	Normalized Mean Sum of Squares	*F* Value
I	21	39	0.0247	0.000633	----
II, Common Δb.	19	41	0.0259	0.000631	----
II–I	--	2	0.00119	0.000595	0.940
III, Common Δb, & N_{cm}.	15	45	0.0267	0.000594	----
III–I	--	6	0.00202	0.000337	0.533

loam (*fine-loamy over sandy or sandy-skeletal mixed, thermic Typic Umbraquults*). The experiments at each station were set up as a RCB design, with four replications. Both total plant and grain fraction biomass were sampled, and every year the experiment was conducted at a new location within the same soil type at each station. This was to limit the impact on the experiment of any residual N in the soil from the previous year. The fertilizer treatments were applied in the form of NH_4NO_3 at rates of 0, 56, 112, 168, and 224 kg ha^{-1} of N. Average values over the four year period were combined for each of the different treatments and the model parameters were evaluated based on those combined averages.

Normalization

Parameters A_N and A for grain, A_N and A for total plant, and b_N, b and c for each site are determined simultaneously, using Newton-Raphson non-linear regression. A detailed description of Newton-Raphson non-linear regression of logistic equation can be found in Overman and Scholtz [8]. The attempt of this methodology is to consistently distribute the standard error amongst all those parameters for further analysis. Because of the unit and an order of magnitude difference between biomass and N uptake parameter values, as well as a subsequent order of magnitude difference between grain and total plant parameter values, a normalization routine is also employed. The error sum of squares for each individual site is initially written as

$$ESS_{NORM} = \underbrace{\sum\left(\frac{Y_g}{A_g} - \frac{1}{1+\exp(b-cN)}\right)^2 + \sum\left(\frac{N_{ug}}{A_{Ng}} - \frac{1}{1+\exp(b_N-cN)}\right)^2}_{grain\ fraction} + \underbrace{\sum\left(\frac{Y_t}{A_t} - \frac{1}{1+\exp(b-cN)}\right)^2 + \sum\left(\frac{N_{ut}}{A_{Nt}} - \frac{1}{1+\exp(b_N-cN)}\right)^2}_{total\ plant} \quad (12)$$

where the total normalized error is resultant from the sum of the normalized error from the three sites. For this study the initial Hessian matrix is 21 by 21 elements and paired with a 21 element Jacobian vector. As a result of the normalization procedure, performing the Newton-Raphson procedure can diverge more readily than a non-normalized procedure. Because of this, it is important to establish a reasonable initial guess for each parameter. First, individual logistic response are evaluated for each data set from all sites, for grain biomass, grain N uptake, for total plant biomass, and total plant N uptake. This results in six A values, six A_N values, six b values, six b_N values, and 12 values for the c parameter. Using the fact that it can be shown for two straight lines

$$y_1 = m_1 x + b_1 \quad (13)$$

and

$$y_2 = m_2 x + b_2, \quad (14)$$

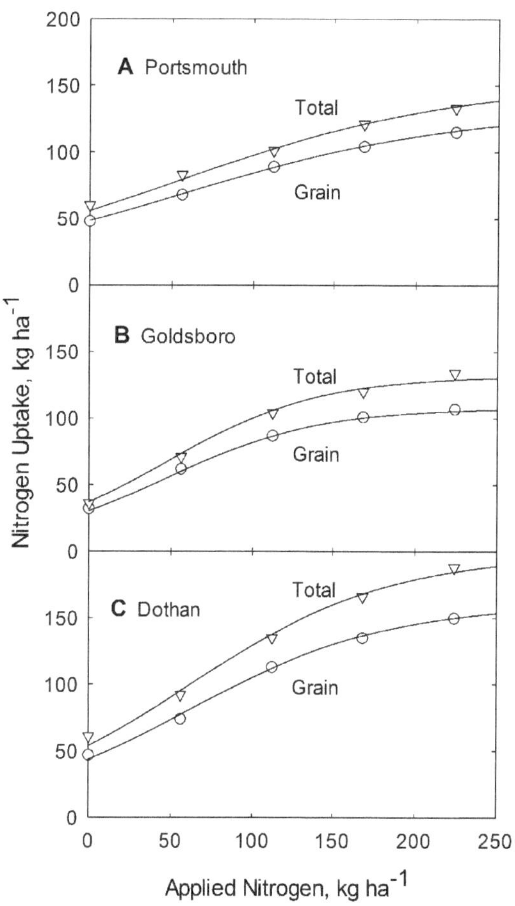

Figure 1. Dependence of grain and total plant N uptake on applied N for corn grown at the Plymouth (A), Kinston (B), and Clayton (C) experiment stations in North Carolina. Curves are constructed from Eq. 1 and from parameters listed in Table 2.

Table 2. Standard logistic model parameters invariant to corn grain and total plant biomass production and for corn grain and total plant N uptake, grown on three different soils.

Soil	Component	Parameter	Estimate
Dothan	Biomass	b	0.319
	N Uptake	b_N	0.978
	Both	c, ha kg^{-1}	0.0161
Goldsboro	Biomass	b	0.282
	N Uptake	b_N	0.941
	Both	c, ha kg^{-1}	0.0209
Portsmouth	Biomass	b	−0.105
	N Uptake	b_N	0.555
	Both	c, ha kg^{-1}	0.0111

who share the same sampling of the independent variable, the best fit single slope shared between them is given by

$$\hat{m}=\frac{m_1+m_2}{2} \tag{15}$$

and the corresponding intercepts become

$$\hat{b}_1=b_1+\frac{1}{2}(m_1-m_2)\bar{x} \tag{16}$$

and

$$\hat{b}_2=b_2+\frac{1}{2}(m_2-m_1)\bar{x}, \tag{17}$$

the initial guess for the c parameter is the average of all 12 values, and the initial guess for each value of bN can be found from

$$\hat{b}_N=b_N+(c-\bar{c})\bar{N} \tag{18}$$

and for each value of b can be found from

$$\hat{b}=b+(c-\bar{c})\bar{N}. \tag{19}$$

The problem is bounded between the maximum and minimum values of the c parameter and each value of b_N is bounded between

$$\hat{b}_{Nmin}=b_N+(c-c_{\max})\bar{N} \tag{20}$$

and

$$\hat{b}_{Nmax}=b_N+(c-c_{\min})\bar{N} \tag{21}$$

and each value of b is bounded between

$$\hat{b}_{min}=b+(c-c_{\max})\bar{N} \tag{22}$$

and

$$\hat{b}_{max}=b+(c-c_{\min})\bar{N}. \tag{23}$$

Analysis

The first hypothesis of this analysis is that the difference between N uptake intercept parameter and the biomass intercept parameter, Δb, is invariant with respect to the differences in soil type and water availability for a given variety of an annual crop. Note that there is no attempt in this work to identify the effects of water availability or site characteristics on the ELM parameters, but to determine which are invariant to those characteristics. For this analysis, the same genetic line of corn is propagated by seeding and harvested at the same relative age. The second hypothesis is that maximum N concentration, N_{cm}, is also invariant with respect to the differences in soil type and water availability. A consequence of both hypotheses being affirmed is that the lower limit to the N concentration, N_{cl}, in the same annual crop is also invariant with respect to soil type and water availability. Parameters were estimated by minimization of the normalized error sum of squares, and analysis of variance (ANOVA) was used to determine the validity of both hypotheses.

For the analysis of variance, three scenarios or modes were used, each with a targeted reduction in the number of parameters used in the ELM to describe the corn data in the Kamprath study. Mode I had 21 separate parameters that were estimated by minimization of the normalized error sum of squares. In Mode I, there are individual values for A, and A_N, for both grain and for total plant, and corresponding values for b, b_N, and c at each of the three sites. For Mode II, the number of parameters estimated dropped to 19, because the Δb parameter was held constant across the three sites. For Mode III, the Δb and the N_{cm} parameters were both held constant across the three sites, reducing the number of estimated parameters to 15.

Nonlinear Coefficients of Determination [23] (Nash-Sutcliffe Model Efficiency Coefficient [24]) will be provided for grain and total plant N uptake and for grain and total plant biomass production just as a relative comparison of fit.

Results

Table 1 contains the summary of the analysis of variance test. The comparison between Modes I and II leads to an increase the *degrees of freedom* to 41 and results in a variance ratio of 0.940. Because the critical F(2,39,95%) value is 3.24, it is concluded that there is no significant difference between the two modes. Thus in

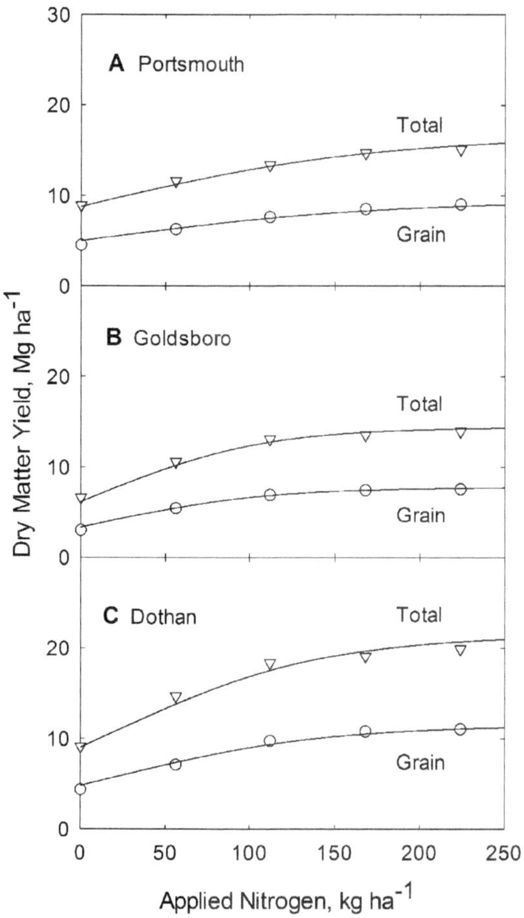

Figure 2. Dependence of grain and total plant biomass production on applied N for corn grown at the Plymouth (A), Kinston (B), and Clayton (C) experiment stations in North Carolina. Curves are constructed from Eq. 3 and from parameters listed in Table 2.

this study the Δb parameter is invariant to all soil and site differences.

Also from Table 1, the comparison between Modes I and III results in an increase the *degrees of freedom* to 45, and in a variance ratio of 0.533. With a critical F(6,39,95%) value of 2.34, not only is the Δb parameter is invariant, but so are the total plant and grain N_{cm} parameters. Thus, the soil, the field conditions, the environmental constraints, and even water availability play no role in either is the Δb or the two N_{cm} parameters. This leads to an invariance in the total plant and grain N_{cl} parameters, by virtue of Eq. (9).

The dependence of grain and whole plant N uptake on applied N at harvest is represented by Figure 1 for the three soil types. In general there is good agreement between the model line and the data. The resulting N uptake model lines (depicted in Figure 1), are generated from Eq. (1), using parameter values for b_N and c from Table 2 and values for A_{Ng} and A_{Nt} found in Table 3. Equation specific Non-linear Coefficient of Determination values and Error Sum of Squares are provided in Table 4.

Grain and whole plant biomass production versus applied N is shown in Figure 2 for all three soil types. In general there is good agreement between the model line and the data. The resulting biomass model lines (depicted in Figure 2), are generated from Eq. (3), using parameter values for b and c from Table 2 and values for A_g and A_t found in both Table 3. Equation specific Non-linear Coefficient of Determination values and Error Sum of Squares are provided in Table 4.

N concentration dependence on applied N is shown in Figure 3 for all three soil types. The resulting N concentration model lines (depicted in Figure 3), are generated from Eq. (7), using parameter values for b, b_N and c from Table 2 and values for $N_{cm\ g}$ and $N_{cm\ t}$ from Table 6.

The phase relationship between biomass production and N uptake for the corn grain and the whole plant is represented by Figure 4 for each of the three soils. The resulting biomass – N uptake phase model lines (depicted in Figure 4), are generated from Eq. (2), using parameter values for k_{Ng}, k_{Nt}, Y_{mg} and Y_{mt} found in Table 5.

The phase relationship between N Concentration and N uptake for the grain and the whole plant is represented by Figure 4 for each of the three soils. The resulting between N Concentration –

Table 3. Standard logistic model parameters specific to corn grain and total plant biomass production and for corn grain and total plant N uptake, grown on three different soils.

Soil	Component	Parameter	Plant Fraction	Estimate
Dothan	Biomass	A_g, Mg ha^{-1}	Grain	11.5
		A_t, Mg ha^{-1}	Total Plant	21.5
	N Uptake	A_{Ng}, kg ha^{-1}	Grain	161
		A_{Nt}, kg ha^{-1}	Total Plant	198
Goldsboro	Biomass	A_g, Mg ha^{-1}	Grain	7.76
		A_t, Mg ha^{-1}	Total Plant	14.4
	N Uptake	A_{Ng}, kg ha^{-1}	Grain	108
		A_{Nt}, kg ha^{-1}	Total Plant	132
Portsmouth	Biomass	A_g, Mg ha^{-1}	Grain	9.50
		A_t, Mg ha^{-1}	Total Plant	16.7
	N Uptake	A_{Ng}, kg ha^{-1}	Grain	133
		A_{Nt}, kg ha^{-1}	Total Plant	154

Table 4. Standard statistical measures of fit, based on specific component and plant fraction data.

Soil	Component	Plant Fraction	Coefficient of Determination	Specific Error Sum of Squares
Dothan	Biomass	Grain	0.984	0.523
		Total Plant	0.969	2.361
	N Uptake	Grain	0.997	24.5
		Total Plant	0.993	83.9
Goldsboro	Biomass	Grain	0.992	0.145
		Total Plant	0.983	0.468
	N Uptake	Grain	0.998	7.81
		Total Plant	0.993	35.2
Portsmouth	Biomass	Grain	0.977	0.169
		Total Plant	0.985	0.592
	N Uptake	Grain	0.999	5.01
		Total Plant	0.991	19.4

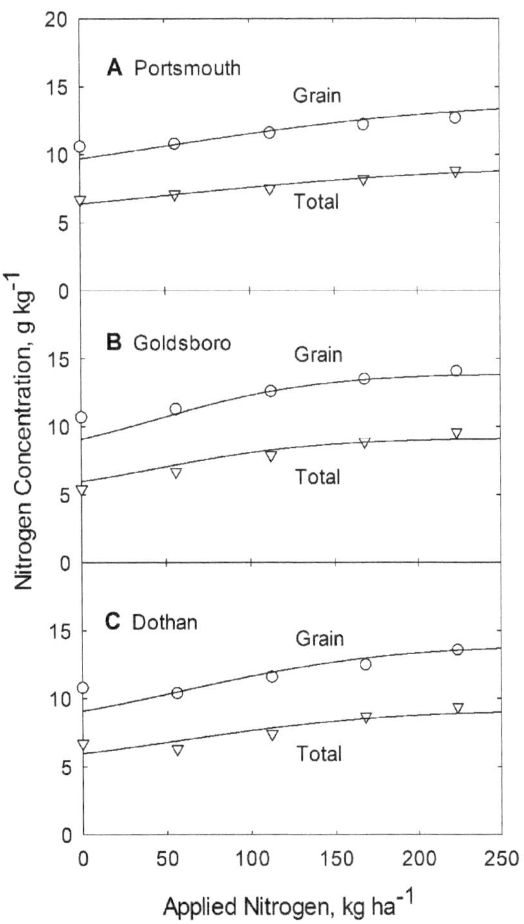

Figure 3. Dependence of grain and total plant N concentration on applied N for corn grown at the Plymouth (A), Kinston (B), and Clayton (C) experiment stations in North Carolina. Curves are constructed from Eq. 7 and from parameters listed in Table 2.

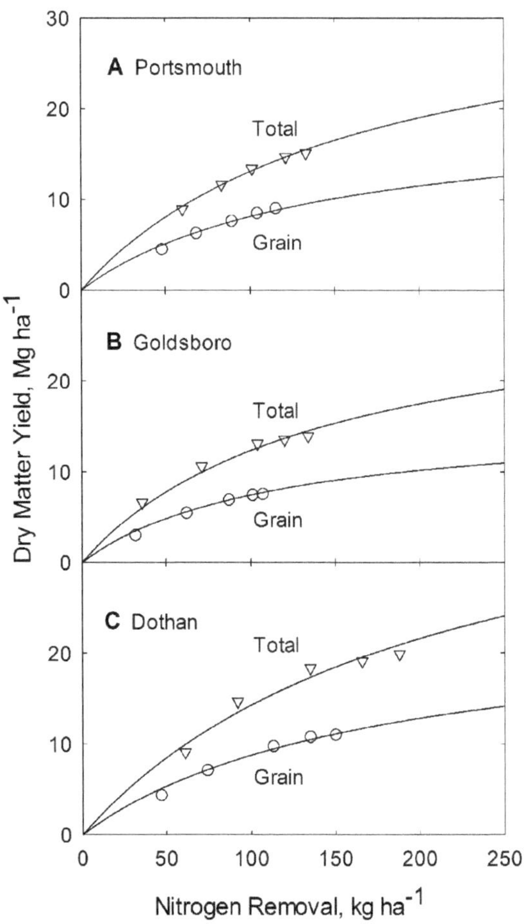

Figure 4. Dependence of grain and total plant biomass production on N uptake for corn grown at the Plymouth (A), Kinston (B), and Clayton (C) experiment stations in North Carolina. Curves are constructed from Eq. 2 and from parameters listed in Table 3.

Table 5. Standard phase model parameters for corn grain and total plant biomass production and for corn grain and total plant N uptake.

Soil	Parameter	Plant Fraction	Estimate
Dothan	k_{Ng}, kg ha^{-1}	Grain	157
	k_{Nt}, kg ha^{-1}	Total Plant	193
	Y_{mg}, Mg ha^{-1}	Grain	22.7
	Y_{mt}, Mg ha^{-1}	Total Plant	42.6
Goldsboro	k_{Ng}, kg ha^{-1}	Grain	104
	k_{Nt}, kg ha^{-1}	Total Plant	125
	Y_{mg}, Mg ha^{-1}	Grain	15.1
	Y_{mt}, Mg ha^{-1}	Total Plant	27.6
Portsmouth	k_{Ng}, kg ha^{-1}	Grain	122
	k_{Nt}, kg ha^{-1}	Total Plant	142
	Y_{mg}, Mg ha^{-1}	Grain	17.7
	Y_{mt}, Mg ha^{-1}	Total Plant	31.4

N uptake phase model lines (depicted in Figure 5), are generated from Eq. (11), using parameter values for k_{Ng}, k_{Nt}, Y_{mg} and Y_{mt} found in Table 5.

Discussion

From this analysis it is concluded that there are aspects of the ELM that are invariant with respect to both soil type and water availability for a given variety of annual crop propagated by seeding and harvested at the same relative age. This analysis has shown, for the Kamprath N-rate study conducted on corn in North Carolina [22] that both the difference between N uptake intercept parameter and the biomass intercept parameter, Δb, and the maximum N concentration, N_{cm}, are in fact invariant with respect to the crop's surrounding environmental conditions. From the model, these facts lead to the conclusion that both the upper limit N concentration, N_{cm}, and the lower limit N concentration, N_{cl}, are both invariant with respect to soil type and water availability in the study analyzed. This further suggests that the N_{cm} and N_{cl} parameters are of more importance to the model. While other parameters, such as

$$A=\frac{A_N}{N_{cm}} \tag{24}$$

$$\Delta b=\ln\left(\frac{N_{cm}}{N_{cl}}\right) \tag{25}$$

$$b=b_N-\Delta b=b_N-\ln\left(\frac{N_{cm}}{N_{cl}}\right) \tag{26}$$

$$Y_m=\frac{A_N}{N_{cm}-N_{cl}} \tag{27}$$

and

$$k_N=N_{cl}Y_m=\frac{N_{cl}A_N}{N_{cm}-N_{cl}} \tag{28}$$

can be rewritten, to show the significance that upper and lower limit concentrations have in each parameter and ultimately seasonal plant response to nutrient application. Having upper

Table 6. Parametric factors invariant to site attributes, including soil type.

Plant Fraction	Parameter	Estimate
Both	Δb, kg ha^{-1}	0.660
Grain	$N_{cm\ g}$, g kg^{-1}	14.0
Total Plant	$N_{cm\ t}$, g kg^{-1}	9.18
Grain	$N_{cl\ g}$, g kg^{-1}	7.23
Total Plant	$N_{cl\ t}$, g kg^{-1}	4.75
Grain	$N_{cm\ t}-N_{cl\ t}=A_{Nt}/Y_{mt}$, g kg^{-1}	6.75
Total Plant	$N_{cm\ g}-N_{cl\ g}=A_{Ng}/Y_{mg}$, g kg^{-1}	4.43

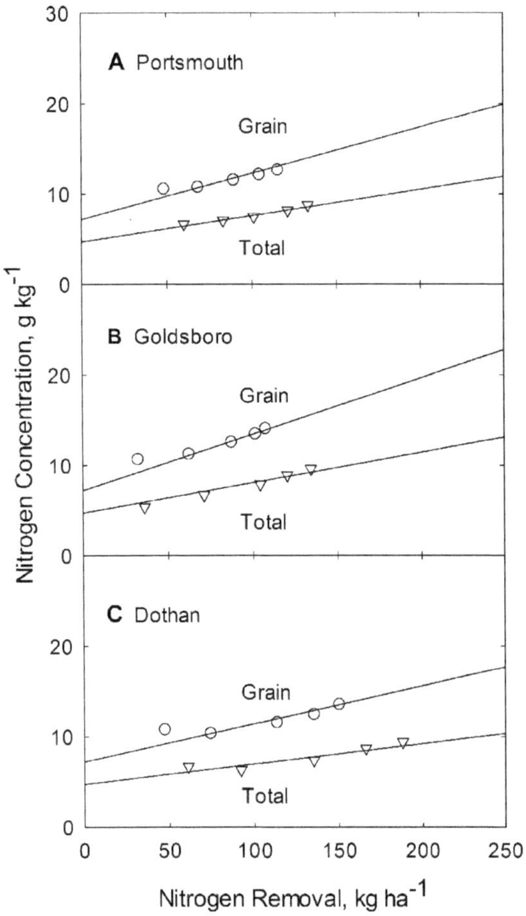

Figure 5. Dependence of grain and total plant N concentration on N uptake for corn grown at the Plymouth (A), Kinston (B), and Clayton (C) experiment stations in North Carolina. Curves are constructed from Eq. 11 and from parameters listed in Table 3.

and lower limits to plant nutrient concentration corresponds with plant physiology. Without a minimum level of a given required nutrient, there can be no yield, seasonal or otherwise. There should also be a maximum concentration that can be approached, as there should be diminishing yield increases as higher agronomic rates are applied, or there would be unbounded growth.

If this invariance with respect to soil type and water availability holds for all crops propagated by seeding, the model could be written in terms of parameters that have measurable physiological significance and could give further insight into relationships that govern plant development and nutrient removal. Initial evidence appears promising that perennial crops, such as ryegrass (*Lolium perenne* L.), when held to comparably the same seasonal management practices also exhibit very nearly the same conclusions with regard to both constant values of the N_{cm} and N_{cl} parameters [25].

Given that N uptake and biomass production can be described by five parameters, and if two are invariant to all but crop type and season length, the model reduces to three parameters (A_N, b_N, and c) when a crop and season length are chosen. From Overman & Scholtz [26] the logistic response originates within the soil's buffering capacity for P and for K, and the c parameter remains the same from the plant extractable logistic response to nutrient uptake logistic response, and to biomass production. It is here assumed that the c parameter for applied N also originates as the rate response parameter for the soil's buffering capacity of N. The c parameter can be modified by plant population [11]. A Future step should be to analyze various field studies to catalogue soil physical and chemical characteristics and the resulting impact on the c parameter. Mathematically b_N represents shifting parameter which in conjunction with the c parameter as

$$N_{0.5} = \frac{b_N}{c} \tag{29}$$

$N_{0.5}$ represents the effective level of N necessary to achieve peak N uptake efficiency [10,16,17]. Ultimately, for environmental considerations, setting applied rate of N to the peak uptake level will result in the most N removed per unit N applied. Plus, provided the difference between b and b_N is greater than 0, then the yield will be on the upper portion of the logistic biomass curve to yield

$$Y = \frac{A_N}{N_{cm}[1+\exp(-\Delta b)]}. \tag{30}$$

The b_N parameter is affected by changes in plant population [11] and is also influenced by crop type [8,25]. The remaining parameter, A_N, is a linear parameter that is affected by the various environmental conditions, the crop type, the soil type, and various management practices [5–20]. Given Eq. (30), knowing the invariant Δb value for a given crop, and having a reasonable estimate for the background level of N already present in the soil, represented by Eq. (29), exists the beginning of a framework for a more reasonable and more sustainable nutrient management guide. Further analyses are being conducted to verify these findings with other annual propagated by seeding and with perennial crops.

Acknowledgments

The authors would like to acknowledge the hard work of field researchers, like Dr. Kamprath, whose tireless efforts have helped bring about greater understanding of the complexities of agricultural production.

Author Contributions

Conceived and designed the experiments: RVS ARO. Performed the experiments: RVS. Analyzed the data: RVS. Contributed reagents/materials/analysis tools: RVS. Wrote the paper: RVS. Mentorship and review: ARO.

References

1. Overman AR, Martin FG, Wilkinson SR (1990) A logistic equation for yield response of forage grass to nitrogen. Commun Soil Sci Plan 21: 595–609.
2. Ferris T (1989) Coming of age in the milky way. New York: Bantam Doubleday Dell Publishing Group. 495 p.
3. France J, Thornley JHM (1984) Mathematical models in agriculture. London: Butterworth and Company. 352 p.
4. Overman AR, Wilkinson SR, Wilson DM (1994) An extended model of forage grass response to applied nitrogen. Agron J 86: 617–620.
5. Overman AR, Wilson DM, Kamprath EJ (1994) Estimation of yield and nitrogen removal by corn. Agron J 86: 1012–1016.
6. Reck WR, Overman AR (1996) Estimation of corn response to water and applied nitrogen. J Plant Nutr 19: 201-214.

7. Overman AR (1999) Model for accumulation of dry matter and plant nutrient elements by tobacco. J Plant Nutr 22: 1, 81–92.
8. Overman AR, Scholtz RV (2002) Mathematical models of crop growth and yield. New York: Marcel Decker. 342 p.
9. Overman AR, Scholtz RV (2002) Corn response to irrigation and applied nitrogen. Commun Soil Sci Plan 33: 3609–3619.
10. Overman AR, Brock KH (2003) Model analysis of corn response to applied nitrogen and tillage. Commun Soil Sci Plan 34: 2177–2191.
11. Overman AR, Scholtz RV, Brock KH (2006) Model analysis of corn response to applied nitrogen and plant population density. Commun Soil Sci Plan 37: 1157–1172.
12. Overman AR, Howard JC (2012) Model analysis of response to applied nitrogen by cotton. J Plant Nutr 35: 2118–2123.
13. Overman AR, Wilkinson SR (1995) Extended logistic model of forage grass response to applied nitrogen, phosphorus, and potassium. T ASAE 38: 103–108.
14. Overman AR, Stanley RL (1998) Bahiagrass responses to applied nitrogen and harvest interval. Commun Soil Sci Plan 29: 237–244.
15. Overman AR, Wilkinson SR (2003) Extended logistic model of forage grass response to applied nitrogen as affected by soil erosion. T ASAE 46: 1375–1380.
16. Overman AR, Brock KH (2003) Model comparison of coastal bermudagrass and pensacola bahiagrass response to applied nitrogen. Commun Soil Sci Plan 34: 2163–2176.
17. Overman AR, Scholtz RV, Taliaferro CM (2003) Model analysis of response of bermudagrass to applied nitrogen. Commun Soil Sci Plan 34: 1303–1310.
18. Overman AR, Rhoads FM, Brock KH (2005) Model analysis of response of pensacola bahiagrass to applied nitrogen, phosphorus, and potassium. I. Seasonal dry matter. J Plant Nutr 27: 1747–1756.
19. Overman AR, Rhoads FM, Brock KH (2005) Model analysis of response of pensacola bahiagrass to applied nitrogen, phosphorus, and potassium. II. Seasonal P and K uptake. J Plant Nutr 27: 1757–1777.
20. Overman AR, Scholtz RV (2005) Model analysis for response of dwarf elephantgrass to applied nitrogen and rainfall. Commun Soil Sci Plan 35: 2485–2494.
21. Monteith JL (1996) The quest for balance in crop modeling. Agron J 88: 695-697.
22. Kamprath EJ (1986) Nitrogen studies with corn on coastal plain soils, Technical Bulletin 282. Raleigh: North Carolina Agricultural Research Service, North Carolina State University. 15 p.
23. Draper NR, Smith H (1998) Applied regression analysis.3rd ed. New York: John Wiley and Sons. 736 p.
24. Nash JE, Sutcliffe JV (1970) River flow forecasting through conceptual models part I - A discussion of principles. J Hydrol 10: 282–290.
25. Scholtz RV (2002) Mathematical modeling of agronomic crops: analysis of nutrient removal and dry matter accumulation. Doctor of Philosophy Dissertation. Gainesville: University of Florida. 139 p.
26. Overman AR, Scholtz RV (2012) A memoir on: A model of yield and phosphorus uptake in response to applied phosphorus by potato. Gainesville: University of Florida. 16p. Available: http://ufdc.ufl.edu/IR00001234/00001 Accessed 11 December 2.

9

Effect of Abandonment on Diversity and Abundance of Free-Living Nitrogen-Fixing Bacteria and Total Bacteria in the Cropland Soils of Hulun Buir, Inner Mongolia

Huhe[1¤], Shinchilelt Borjigin[1], Yunxiang Cheng[2], Nobukiko Nomura[1], Toshiaki Nakajima[1], Toru Nakamura[1], Hiroo Uchiyama[1*]

1 Graduate School of Life and Environmental Sciences, University of Tsukuba, Tsukuba, Ibaraki, Japan, **2** State Key Laboratory of Grassland Agro-Ecosystems, College of Pastoral Agriculture Science and Technology, Lanzhou University, Lanzhou, China

Abstract

In Inner Mongolia, steppe grasslands face desertification or degradation because of human over activity. One of the reasons for this condition is that croplands have been abandoned after inappropriate agricultural management. The soils in these croplands present heterogeneous environments in which conditions affecting microbial growth and diversity fluctuate widely in space and time. In this study, we assessed the molecular ecology of total and free-living nitrogen-fixing bacterial communities in soils from steppe grasslands and croplands that were abandoned for different periods (1, 5, and 25 years) and compared the degree of recovery. The abandoned croplands included in the study were natural restoration areas without human activity. Denaturing gradient gel electrophoresis and quantitative PCR (qPCR) were used to analyze the *nifH* and 16S rRNA genes to study free-living diazotrophs and the total bacterial community, respectively. The diversities of free-living nitrogen fixers and total bacteria were significantly different between each site ($P<0.001$). Neither the total bacteria nor *nifH* gene community structure of a cropland abandoned for 25 years was significantly different from those of steppe grasslands. In contrast, results of qPCR analysis of free-living nitrogen fixers and total bacteria showed significantly high abundance levels in steppe grassland ($P<0.01$ and $P<0.03$, respectively). In this study, the microbial communities and their gene abundances were assessed in croplands that had been abandoned for different periods. An understanding of how environmental factors and changes in microbial communities affect abandoned croplands could aid in appropriate soil management to optimize the structures of soil microorganisms.

Editor: Bas E. Dutilh, Radboud University Medical Centre, NCMLS, Netherlands

Funding: This work was supported by a Grant-in-Aid for Scientific Research (B) (No. 21310050 to HU) and CREST program of Japan Science and Technology Corporation (JST; http://www.jst.go.jp/EN/index.html). The funders had no role in study design, data collection and analysis, decision to publish, or preparation of the manuscript.

Competing Interests: The authors have declared that no competing interests exist.

* Email: uchiyama.hiroo.fw@u.tsukuba.ac.jp

¤ Current address: Institute of Soil and Fertilizer and Save Water Agricultural, Gansu Academy of Agricultural Sciences, Anning District, Lanzhou, Gansu, China

Introduction

Steppe grasslands are distributed over vast areas in the arid and semiarid regions of the Eurasian continent [1]. The Inner Mongolia steppe is an important part of Eurasia and has been used by pastoral nomads for long periods of time. However, vast areas were converted to cropland, and farmers have increased the size of these croplands during the past 40 years because of a rapid increase in the human population. Subsequently, many of the croplands were abandoned because of soil degradation and desertification caused by inappropriate agricultural management [2,3]. Further, some croplands were abandoned to restore natural vegetation, as in the case where the Chinese government proposed to restore farmlands to grasslands or forests [4]. Some plant species, including grasses, shrubs, and trees, are introduced into the abandoned croplands to restore the vegetation in some locations [5–8]. In most areas, the plant community of abandoned cropland is likely to be restored naturally by controlling human activities [9]. Many researchers have evaluated the abandoned cropland ecosystem changes by studying the vegetation changes [9–12]. There are interactions between plants and microorganisms. Plants exude diverse compounds, such as organic acids, enzymes, and polysaccharides, from the roots. Further, plants can recognize microbe-derived compounds and adjust their defense and growth responses according to the type of microorganisms encountered. Conversely, microorganisms can detect suitable plant hosts and initiate their colonization strategies in the rhizosphere by producing canonical plant growth-regulating substances such as auxins or cytokinins [13]. However, most studies have not evaluated changes in soil microbial community and its gene abundance in croplands that have been abandoned for different periods of time.

Cultivated soil or grassland soil contains an estimated 2×10^9 prokaryotic cells per gram [14]. Soil microbial communities are important factors of agriculturally managed systems because they are responsible for most nutrient transformations in soil and influence the aboveground plant diversity and productivity [15]. Next to water, nitrogen is the second-most limiting factor for plant growth [16]. Nitrogen cycling in natural ecosystems and during traditional agricultural production rely on nitrogen fixation of diazotrophic bacteria [17]. Diazotrophs are highly diverse and widely distributed across bacterial and archaeal taxa [18]. Approximately 80% of biological nitrogen fixation is performed by diazotrophs in symbiosis with legumes [19]. However, under specific conditions, the free-living bacteria in soil (e.g., cyanobacteria, *Pseudomonas*, *Azospirillum*, and *Azotobacter*), may fix significant amounts of nitrogen (0–60 $kg{\cdot}N{\cdot}ha^{-1}{\cdot}year^{-1}$) [20,21]. This may be particularly important in abandoned field soils, where legume plants have not been cultivated and there are very few symbiotic plants (e.g., Leguminosae, *Azolla*, *Myrica*, and *Alnus*.).

In this study, the diazotrophic population was monitored by PCR-denaturing gradient gel electrophoresis (DGGE) exploiting the *nifH* gene. The *nifH* gene is the most conserved gene in the *nif* operon and encodes the Fe subunit of the nitrogenase enzyme [22]. Because of the conserved nature of the *nifH* gene, it has been possible to identify primer sets that can be used to analyze nitrogen fixers so that this community can be analyzed by a PCR-DGGE–based technique [20,23–25]. We have tested the diversity and abundances of free-living nitrogen fixers and the total bacterial population changes that occurred over time in soils belonging to artificially disrupted environments (abandoned cropland soils).

Materials and Methods

Ethics statement

No specific permissions are required for our conducting field survey in this area, since land in China belongs to the public and our field studies did not involve any endangered or protected plant species within.

Site description and sample collection

The study area is located in the Hulun Buir grassland (115°31′–126°04′ E, 47°05′–53°20′ N) in northeastern Inner Mongolia, China (Figure 1). The Hulun Buir grassland area is about $2.6{\times}10^5$ km^2, with a west to east distribution of arid steppe, semi-arid steppe, and meadow steppe. The study area is located in the semi-arid areas.

We established 4 sites, 3 abandoned croplands and a light-grazing steppe grassland (LGSG) (Figure 1) that had an intensity of about 1.4 sheep ha^{-1}. The 3 croplands were abandoned for 1, 5, and 25 years (Y1, Y5, and Y25, respectively), and the control area was a LGSG. In the Y1, Y5, and Y25 sites, *Zea mays*, *Helianthus annuus*, and *Elymus cylindricus* were rotated for approximately 40 years. The sites were subsequently abandoned because of land degradation; soil fertility including both organic C and total N had decreased by approximately 70%.

Plant surveying and soil sampling were conducted in August 2010. All of the sites were selected for their similar topography (flat). Each site contained 5 replicates in a randomized plot (1×1 m) design: Y1 (site 1), plots 1–5; Y5 (site 2), plots 6–10; Y25 (site 3), plots 11–15; and LGSG (site 4), plots 16–20. The coordinates and elevations of the sampled sites are as follows: Y1, 48° 38′ 43″ N, 116° 57′ 56″ E, 545 m; Y5, 48° 38′ 50″ N, 117° 00′ 48″ E, 550 m; Y25, 48° 38′ 45″ N, 117° 01′ 56″ E, 545 m and LGSG, 48° 32′ 00″ N, 116° 40′ 18″ E, 568 m. The mean temperature and precipitation from 2000 to 2009 for each site [26] were as follows: LGSG, Y1, Y5, and Y25, 1.6°C and 213 mm. In each plot, the species composition was recorded. Plant communities were classified on the basis of their differential species [27,28]; all species were identified and measured for cover, height, and density, and Shannon-Wiener diversity index was calculated. Soil moisture was measured with a TRIME-FM (Ettlingen, Germany). Above-ground plant biomass was also determined by clipping the plants at ground level, sorting by species, drying at 60°C for 48 h, and weighing the samples (Table 1 and Table S1).

In each plot, the soil samples were collected from 5 randomly selected points (0 to 10 cm deep) and mixed into 1 sample. After carefully removing the surface organic materials and fine roots, each mixed sample was divided into 2 parts. One part was air-dried for the analysis of soil physicochemical properties. The other was sifted through a 2-mm sieve, sealed in sample vials, kept on ice for transport to the laboratory, and stored at −20°C for microbial assays.

The soil texture was determined by mechanical analysis using the pipette method [29], and the soil texture was used to estimate the saturated hydraulic conductivity [30] for each site (Table S2). The soil texture was classified according to the International

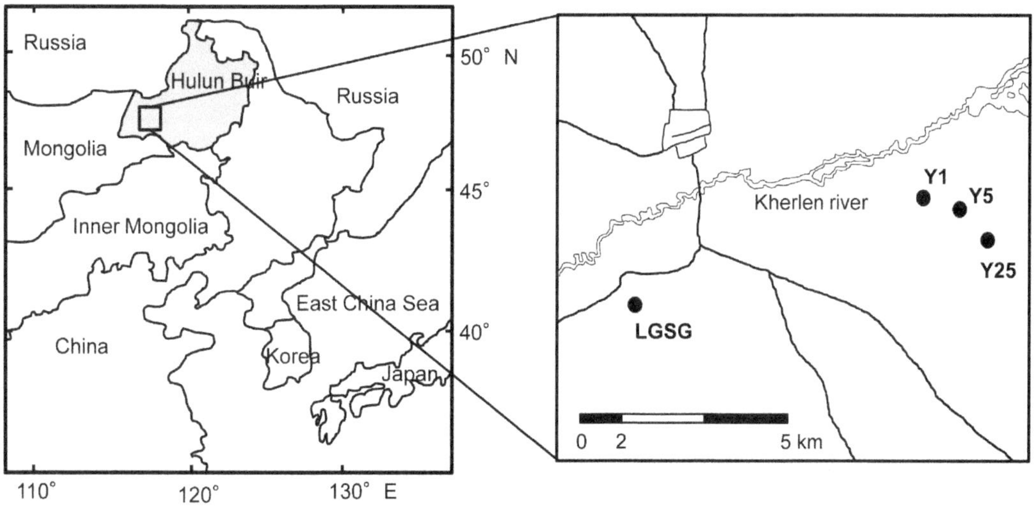

Figure 1. Map of the study area and the 4 study sites (•) in Hulun Buir.

Table 1. Changes to pH, available NO_3-N, available NH_4-N, available P, soluble Fe, organic C, total N, plant diversity (P-*H'*), plant biomass (P-B), soil moisture, and hydraulic conductivity (HC) across the field trial and Pearson's product-moment correlation analysis comparing data to *nifH* and 16S rRNA diversity and gene copies.

Abandoned Cropland or significance parameter	pH	NO_3-N (mg kg^{-1})	NH_4-N (mg kg^{-1})	P (mg kg^{-1})	Fe (g kg^{-1})	Organic C (g kg^{-1})	Total N (g kg^{-1})	H_2O (%)	P-*H'*	P-B	HC (×10^{-3} cm s^{-1})
Abandoned Cropland*											
Y1	7.72±0.13 a	3.35±0.42 a	1.59±0.64 a	12.38±4.09 a	0.43±0.06 a	6.55±0.96 a	0.7±0.09 a	8.0±0 a	1.63±0.34 a	48.5±9.77 a	2.93±0.07 a
Y5	8.57±0.37 b	3.99±0.67 b	0.76±0.04 a	12.92±2.68 a	0.56±0.04 a	9.16±1.51 a	0.94±0.12 b	9.06±0.13 ab	1.71±0.43 a	17.16±4.44 b	2.95±0.03 a
Y25	7.82±0.17 a	3.54±0.53 a	1.09±0.21 a	12.5±1.8 a	0.46±0.04 b	7.23±0.64 a	0.75±0.06 ab	10.7±1.09 b	1.82±0.21 a	79.65±13.5 c	2.79±0.02 b
LGSG	6.21±0.14 c	4.21±0.06 c	4.84±0.37 b	24.48±5.67 b	1.12±0.06 c	24.84±3.09 b	2.3±1.71 c	12.12±1.4 bc	0.94±0.02 b	141.9±5.54 d	2.75±0.02 b
Correlation (*P***)											
with *nifH* DGGE *H'*	+++	++	NS	+++	NS	NS	NS	NS	NS	NS	NS
with *nifH* copy number	NS	NS	++	+++	+++	+++	++	+++	—	++	NS
with 16S rRNA DGGE *H'*	NS	NS	NS	NS	+++	++	++	+++	NS	++	---
with 16S rRNA copy number	NS	NS	+++	NS	NS	++	++	NS	NS	++	NS

*Y1, Y5 and Y25 mean field abandoned for 1 year, 5 years and 25 years, respectively. LGSG is light grazing steppe grassland. The values shown for management factors are means ± standard errors.
***P*, Pearson's product-moment correlation coefficient; NS, not significant; ++/—, significant positive or negative correlation at $P<0.05$; +++/---, significant positive or negative correlation at $P<0.01$. Significant differences are indicated by different letters.

Society of Soil Science (ISSS) classification system. The soils of all 4 sites were sandy loam.

Concentrations of NO_3-N and NH_4-N in the KCl extracts were determined with the zinc reduction-naphthylethylenediamine method for NO_3-N [31] and the indophenol blue colorimetric method for NH_4-N [32]. The soil phosphorus content was determined by the Truog method [33]. Ferrous iron was measured by the *o*-phenanthroline method [34,35]. The organic C and total N were determined by the dry combustion method using an NC analyzer (Sumigraph NC-900; Sumika Chemical Analysis Service, Tokyo, Japan). The soil samples were previously treated with acid to eliminate water and inorganic carbonates [36]. The soil pH was obtained by measuring the equilibrium pH of soil pastes containing 1 g of soil homogenized in 1 mL of H_2O (Table 1).

DNA extraction and PCR

DNA from each of the 20 samples (4 sites×5 replicates) was extracted in 3 subsamples from 0.5 g of soil with the FastDNA Spin Kit for soil (MP Biomedicals, Illkirch, France) according to the manufacturer's protocol. The quality and quantity of the DNA extracts were checked with a SmartSpec Plus spectrophotometer (Bio-Rad Laboratories, United States). The samples were pooled and stored at -20°C until use.

A fragment of the *nifH* gene (approximately 360 bp) was amplified with a nested PCR strategy. First-round reactions were performed with the primers nifH32F (TGAGACAGATAGCTA-TYTAYGGHAA) and nifH623R (GATGTTCGCGCGGCAC-GAADTRNATSA) as described previously [37]. The genomic DNA extract (15–40 ng) was added to PCR mixtures containing 5 μL of 10×ExTaq buffer (Takara, Madison, WI, United States), 4 μL of a mix of deoxynucleoside triphosphates (2.5 mM each), 37.5 μL of water, 0.5 μL of 100 μM nifH32F, 0.5 μL of 100 μM nifH623R, and 0.5 μL of ExTaq DNA polymerase (5 U/μL; Takara, Madison, WI, United States). The reaction mixtures were amplified by 1 denaturation step (5 min at 94°C), followed by 30 cycles of 94°C for 1 min, 50°C for 1 min, and 72°C for 1 min, and 1 final 7 min extension cycle at 72°C. For the second round of the nested amplification, 1 μL of this reaction mixture was used as the template in a 50 μL reaction mixture containing the same reagent mixture described above, but with 39 μL of water, 0.25 μL of 100 μM nifH1-GC (in order to clamp the products for DGGE, the primer nifH1 [CTGYGAYCCNAARGCNGA] was added to the GC-clamp [CGCCCGCCGCGCGCGGCGGGCGGGGCGG-GGGCACGGGGGG] on the 5′ side), and 0.25 μL of 100 μM nifH2 (ADNGCCATCATYTCNCC) [38]. The thermal cycling protocol for the nested reactions was the same as above except that the annealing temperature was raised to 57°C.

Fragments of approximately 200 bp, corresponding to the V3 region of the 16S rRNA gene [39], were amplified using a reaction mixture that contained the same reagent mixture described above, except with 0.25 μL of 100 μM 357F-GC (CCTACGGGAGG-CAGCAG-GC-clamp) and 0.25 μL of 100 μM 518R (GTAT-TACCGCGGCTGG); products were amplified using a touch-down thermocycling program [39]. All of the PCR amplicons were electrophoresed on an agarose gel to ascertain the sizes and purified using the UltraClean PCR Clean-Up Kit (MO BIO Laboratories, Carlsbad, CA, United States).

DGGE

DGGE was performed using the D-Code system (Bio-Rad Laboratories, Hercules, CA, United States) as described by Baxter and Cummings [40]. The polyacrylamide concentration in the gel was 8%, and the linear denaturing gradient was 30% to 60% (100% denaturant corresponds to 7 M urea and 40% deionized formamide). The gel was run at 36 V for 18 h at 60°C in 0.5×TAE buffer. The gel was then stained for 30 min with 1:10,000 (v/v) SYBR Gold, rinsed with 0.5×TAE, and scanned on a transilluminator. Bands were identified, and relative intensities were calculated based on the percentage of intensity of each band in a lane. This was done with an image-analyzing system (Image Master; Amersham Pharmacia Biotech, Uppsala, Sweden). Shannon-Wiener diversity index (H') was calculated by the formula $H' = -\Sigma\, p_i \ln(p_i)$, where p_i is the ratio of relative intensity of band i compared with the relative intensity of the lane. The used of different gradient gel (30% to 70%) to assessed the reproducibility of the DGGE results, similar results were obtained (data not shown).

Real-time PCR assay

Reactions were set up using SYBR green (Bio-Rad Laboratories, the Netherlands) according to Baxter and Cummings [41] with the LightCycler 1.5 system (Roche Applied Sciences, Indianapolis, IN, United States). Reaction mixtures were heated to 95°C for 15 min to denature the DNA before completing 40 cycles of denaturation (95°C for 45 s/15 s [*nifH*/16S rRNA]), annealing (55°C for 45 s/65°C for 15 s [*nifH*/16S rRNA]), and extension (72°C for 45 s/15 s [*nifH*/16S rRNA]). Soil DNA extracts were diluted 1:100 to prevent inhibition of PCR by soil contaminants (e.g., by co-extracted humic substances), and each run included triplicate reactions for each DNA sample, the standard curve, and the no template control. The average copy number was converted into copies of the gene per gram of soil. nifHF (AAAGGYGGWATCGGYAARTCCACCAC) and nifHR (TTGTTSGCSGCR TACATSGCCATCAT) primers [42] were used for *nifH* quantitative PCR (qPCR), and 357F and 518R primers were used for total bacteria qPCR. Dilution series of pGEM-T Easy vector (Promega, Madison, WI, United States) DNA with cloned bacterial *nifH* (*Azospirillum brasilense* ATCC 29729) and 16S rRNA gene (*Pseudomonas aeruginosa* PAO1) fragments were used to generate standard curves ranging from 10^1 to 10^7 gene copies·μL^{-1} for DNA quantification. The specificity of the amplified products was checked by the observation of a single melting peak and the presence of a unique band of the expected size in a 2% agarose gel stained with ethidium bromide. The standard curve produced was linear (r^2>0.98), and the PCR efficiency (Eff = $10^{(-1/\mathrm{slope})}$-1) was>0.90.

Statistical analysis

In all tests, significant effects/interactions were those with a P value that was <0.05. Statistical analysis was performed using SPSS statistical software package (version 19.0; SPSS, Inc., Chicago, IL, United States). Variables of each group must be normally distributed to perform Pearson's product-moment correlations and analysis of variance (ANOVA). The normal distribution of the residuals was evaluated using the Kolmogorov-Smirnov test. If the requirement was not met, data were log-transformed prior to analysis. The homogeneity of the variances was checked by Levene's test. For pairwise comparison of means, Tukey's test was applied. Differences between main effects were tested by ANOVA. Correlations between *nifH* and 16S rRNA diversity and gene copies and environmental factors were tested by Pearson's product-moment correlations.

The choice between a linear or unimodal species response model depends on the underlying gradient length, which is measured in standard deviation units along the first ordination axis and can be estimated by detrended correspondence analysis (DCA). It is recommended to use linear methods when the gradient length is <3, unimodal methods when it is>4, and any method for intermediate gradient lengths [43]. The DCA gradient length for *nifH* gene patterns was 1.77, and that for 16S rRNA patterns was 1.11. Therefore, linear species response models such as partial least-squares regression (PLSR) and redundancy analysis (RDA) were used for multivariate statistical analysis. PLSR is an extension of multiple regression analysis in which the effects of linear combinations of several predictors on a response variable (or multiple response variables) are analyzed. PLSR is especially useful when the number of predictor variables is similar to or higher than the number of observations and/or predictors are highly correlated [44,45]. For PLSR model generation, the software assigns a value known as the variable influence on projection (VIP) to each environmental variable. The VIP indicates the relative importance to the model. Significance was assessed using the VIP parameter (a VIP>1 indicated a significant contribution of the environmental factor to the statistical model) [46,47]. RDA can be considered an extension of principal component analysis (PCA) in which the main components are constrained to be linear combinations of the environmental variables. RDA not only represents the main patterns of species variation as much as they can be explained by the measured environmental variables but also displays correlations between each species and each environmental variable in the data [48]. DCA, RDA, and PCA were performed using the Canoco program for Windows 4.5 (Biometris, Wageningen, the Netherlands). PLSR was performed using the Simca-P 11.0 (Umetrics AB, Umea, Sweden).

Results

Diversity of *nifH*

DGGE gels are shown in Figure S1A. Analysis of the *nifH* DGGE Shannon-Wiener diversity index values for the whole data set indicated that the abandoned period significantly affected the *nifH* diversity ($P<0.001$) by a separate ANOVA. For these reasons, multiple comparisons of the *nifH* DGGE Shannon-Wiener diversity index values were conducted.

Soils of Y1 (2.8±0.02) and Y5 (2.83±0.06) showed significantly higher *nifH* diversity than soils of Y25 (2.72±0.08) and LGSG (2.72±0.1) (Y1×Y25, $P=0.04$; Y1×LGSG, $P=0.02$; Y5×Y25, $P=0.009$; Y5×LGSG, $P=0.025$) (Figure 2A). The Y25 and LGSG soils did not show significant differences in *nifH* diversity. Pearson's product-moment correlation indicated that pH, NO_3-N, and P were positively correlated with the *nifH* Shannon-Wiener diversity index (Table 1).

qPCR of the *nifH* gene

We used qPCR to compare copy numbers of the functional gene (*nifH*) at the 4 sites (Figure 3A). The *nifH* gene copy number in LGSG soil (5.5×10^5 copies·g^{-1} of soil) was the higher compared to the soils of the other 3 abandoned croplands (average number of copies·g^{-1} of soil, 1.9×10^5 in Y1, 3.4×10^5 in Y5, and 2.3×10^5 in Y25). In the abandoned cropland soils, the *nifH* gene copy number did not significantly change over time ($P>0.05$) (Figure 3A).Using Pearson's product-moment correlation, we found that NH_4-N, P, Fe, organic C, total N, soil moisture, and plant biomass were positively correlated, and the Shannon-Wiener diversity index was negatively correlated with the *nifH* copy number (Table 1).

Total bacterial diversity

There are clearly differences in the nitrogen-fixing communities of soil from long and short abandoned periods. In order to ensure that these factors are affecting the nitrogen-fixing community specifically and not the bacterial community as a whole, the 16S rRNA gene diversity and abundance were also analyzed. DGGE gels showing the diversity of total bacteria are shown in Figure S1B. As with the free-living nitrogen-fixing community, ANOVA results for the Shannon-Wiener diversity index of the 16S rRNA gene indicated that the abandoned period ($P<0.001$) was a significant factor. Therefore, each sample was analyzed separately by multiple comparisons. Soils of Y25 and LGSG (both almost equal, 3.17±0.07) showed higher 16S rRNA diversity than soils of Y1 (2.8±0.04) and Y5 (3.06±0.01) (Figure 2B). Pearson's product-moment correlation indicated that organic C, total N, Fe, soil moisture, and plant biomass were positively correlated, and hydraulic conductivity was negatively correlated with the 16S rRNA gene Shannon-Wiener diversity index (Table 1).

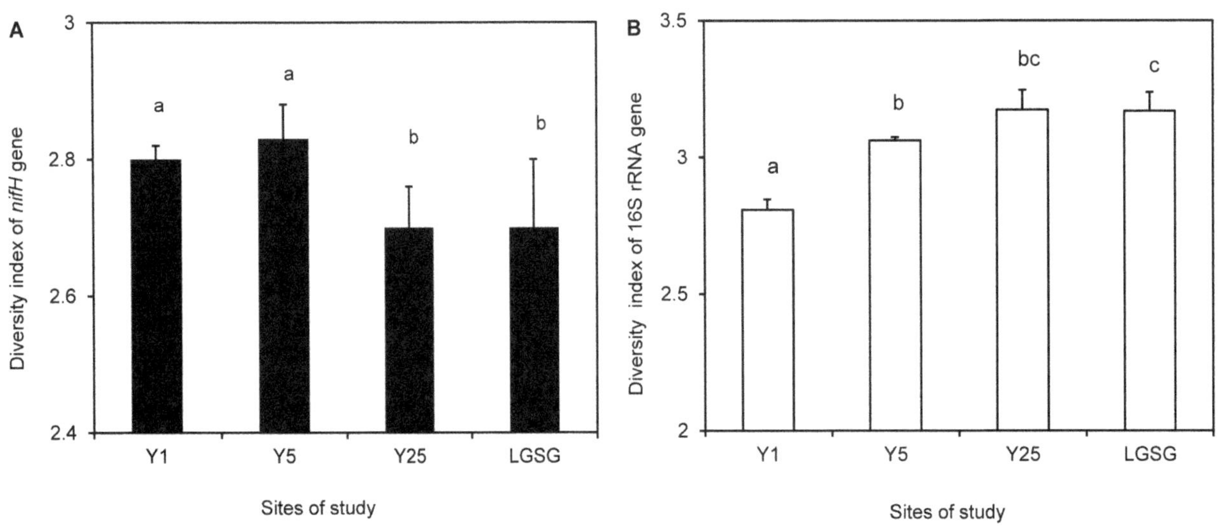

Figure 2. Shannon-Wiener diversity index values for *nifH* (A) and 16S rRNA genes (B). Data sets and results of analysis of variance (ANOVA) in the abandoned cropland (Y1, Y5, and Y25) and light-grazing steppe grassland (LGSG) soils (n = 5; error bars represent standard deviations). Significant differences are indicated by different letters.

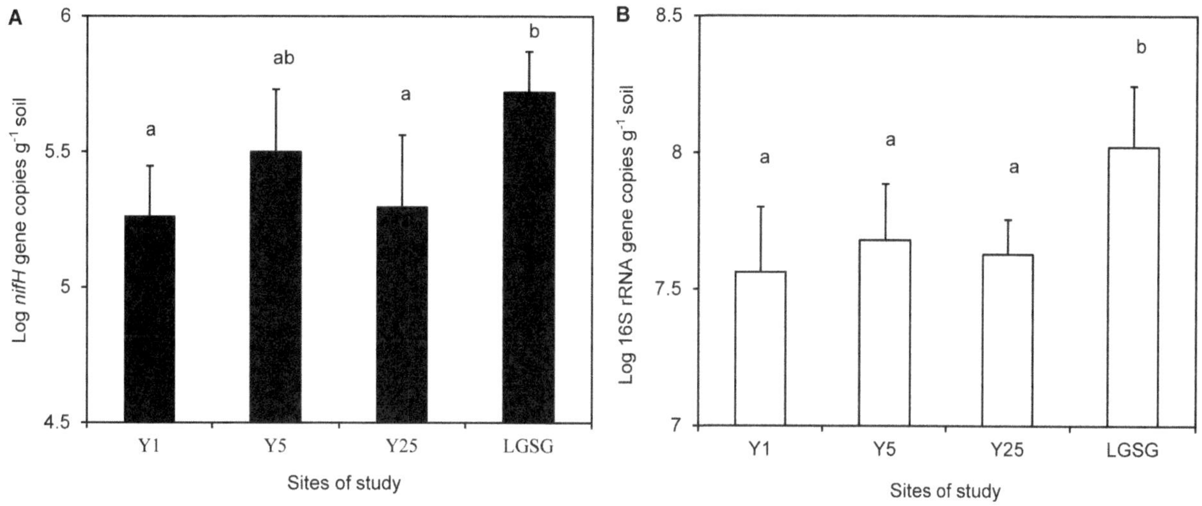

Figure 3. Copy numbers of the *nifH* (A) and 16S rRNA (B) genes. Results of analysis of variance (ANOVA) in the abandoned cropland (Y1, Y5, and Y25) and light-grazing steppe grassland (LGSG) soils (n = 5; error bars represent standard deviations). Significant differences are indicated by different letters.

qPCR of the 16S rRNA gene

By performing the qPCR analysis, we aimed to compare the differences between each site rather than attain an absolute quantification. Similar to *nifH* results presented above, the gene copy numbers of 16S rRNA in LGSG soil (1.2×10^8 copies·g^{-1} of soil) was higher than that of 3 other abandoned cropland soils (average numbers of copies·g^{-1} of soil, 8.2×10^7 in Y1, 5.2×10^7 in Y5, and 4.4×10^7 in Y25) (Figure 3A and B). The 16S rRNA gene copy numbers did not change significantly over time in the abandoned cropland soils ($P>0.05$) (Figure 3B). Pearson's product-moment correlation indicated that NH_4-N, organic C, total N, and plant biomass were positively correlated with the 16S rRNA gene copy number (Table 1).

Effects of environmental variables on the microbial community structure

Because of multicollinearity among environmental variables (Table S3), potential effects of environmental variables on free-living nitrogen-fixing bacteria and the total bacteria community composition were assessed by PLSR (Table 2). Plant biomass, hydraulic conductivity, pH, Fe, and soil moisture, and NH4-N predominantly affected on both free-living nitrogen-fixing bacteria and the total bacteria community composition in this study area. Moreover, C and N also exerted effects on the total bacteria community composition (Table 2).

These correlations of microbial diversity with environmental parameters were also supported by RDA (Figure S2). The ordination plots for the *nifH* and 16S rRNA genes are given in Figure S2A and S2B, respectively. In these figures, the first 2 canonical axes explained 43.6% and 59.1% of the variance of the species data (functional genes) and 51.0% and 63.4% of the variance of the species–environment relationship, respectively. The projection of environmental variables with respect to the free-living nitrogen-fixing community composition revealed that the first canonical axis is positively correlated with soil moisture, Fe, NH4-N, and plant biomass, and it is negatively correlated with pH and hydraulic conductivity (Figure S2A). With respect to the total bacteria community composition, the first canonical axis is positively correlated with Fe, NH4-N, plant biomass, C, and N, and it is negatively correlated with pH and hydraulic conductivity (Figure S2B).

Discussion

This study allowed a detailed analysis of the effects of key components of abandoned cropland with the same land use history (permanent grassland was turned into arable land by cultivation about 40 years ago and had rotated crops of *Zea mayscorn*, *Helianthus annuus* and *E. cylindricus*) and light-grazing steppe grassland systems on soil bacterial and free-living nitrogen-fixing bacterial population structure and gene copy number. Cultivation imparted a strong effect on both total and free-living nitrogen-fixing bacterial population structures (measured by DGGE profiles) and abundances (measured by gene copy numbers).

The copy numbers of *nifH* per gram of LGSG soil was higher than that in Y1, Y5, and Y25 soil. In this study, markedly lower levels of nutrient (e.g., organic C, P, and Fe) in the soil of abandoned cropland (Table 1) may suppress the growth of free-living nitrogen-fixing bacteria. Organic C, P, and Fe and *nifH* copy number have a significant positive Pearson's product-moment correlation (Table 1). The source of carbohydrate is important to allow N_2 fixation activity, which requires large amounts of energy and reducing equivalents [49]. Phosphorus availability can increase nitrogen fixation because nitrogen fixation process requires large amounts of adenosine triphosphate (ATP) [50]. The [4Fe-4S] cluster of the *nifH* gene requires the Fe [22,51]; therefore, a low copy number of *nifH* might be related to the content of organic C, P, and Fe. In addition, Coelho et al. [52,53] found that 30% more free-living diazotrophs could be isolated from soil in the presence of low levels of nitrogen fertilizer than from soil in the presence of high levels of nitrogen fertilizer. In contrast, low copy number of *nifH* was obtained from the abandoned cropland of low nitrogen levels (NO_3-N and NH_4-N) (Table 1). The result may be caused by nutrients content of abandoned cropland soils.

Shannon-Wiener diversity index of *nifH* of LGSG was lower than that of Y1 and Y5 (Figures 2A and 3A). Previous studies indicated that pH is an important factor for the structure of the bacterial community [54,55]. Belnap [56] thought that most

Table 2. Results of partial least squares regression (PLSR) analysis with the explanatory capacity of the first component and the variable influence on projection (VIP) of each predictor within each component to estimate significant predictors.

Gene	***nifH***	**16S rRNA**
Explained variance in fingerprinting pattern (%)	15.2	22.5
Explained variance of component predictors (%)	7.80	15.40
Environment variables*	**Variable Influence on Projection (VIP)**	
P-B	1.29	1.23
HC	1.17	1.07
pH	1.15	1.07
Fe	1.13	1.17
H_2O	1.07	1.00
NH_4-N	1.04	1.02
ELE	0.97	0.95
N	0.94	1.01
C	0.93	1.01
P	0.77	0.73
P-*H′*	0.66	0.75
NO_3-N	0.61	0.70

*All environmental variables are shown and include pH, NO_3-N, NH_4-N, organic carbon (C), total nitrogen (N), available phosphorus (P), soluble iron (Fe), elevation (ELE), hydraulic conductivity (HC), soil moisture (H_2O), plant biomass (P-B), and plant diversity (P-*H′*).

nitrogen-fixing microorganisms have an optimum soil pH of 7 or above. In this study, the samples from abandoned croplands with pH>7 had a higher diversity of free-living nitrogen-fixing microorganisms than the samples from the LGSG with pH<7.

The *nifH* gene profile showed there was a higher variability between replicate samples than that of the 16S rRNA gene profile. Therefore, this result suggested that the free-living nitrogen-fixing bacteria community was susceptible in this study area. An explanation for this may be the local changes at the microsite/aggregate scale because of spatial and temporal variation in exudation along plant roots [57–59]. Similar approaches have demonstrated that there are differences in ammonia oxidizer populations in sediment and soil [60].

The RDA revealed that the distribution of the factors that influence both the free-living nitrogen-fixing bacteria and the broader bacterial community distribution are very similar. This suggests that soil factors that affect the free-living nitrogen fixers are likely to affect the community as a whole in steppe grassland soil. Larkin et al. [61] reported that plant effects are the most important drivers of soil microbial community characteristics within a given site and soil type. In our study, PLSR and RDA showed that plant biomass strongly influenced (Table 2) and was positively correlated with the first axis (Figure S2) in both the free-living nitrogen-fixing bacteria and the broader bacterial community distribution. This is caused by interactions between plants and microorganisms. Above-ground net primary productivity was expected to increase soil carbon input by enhancing the turnover of plant biomass and enhancing root exudation and may therefore influence carbon-limited microbial communities in the soil [62,63]. Concurrently, microorganisms also affect plants by producing canonical plant growth-regulating substances such as auxins or cytokinins [13]. However, the Shannon-Wiener diversity index was not significantly influenced in either the free-living nitrogen-fixing bacteria or the broader bacterial community distribution in this study area. Pearson's product-moment correlations showed that plant diversity was not significantly correlated with *nifH* gene diversity, or 16S rRNA gene diversity and copy number, but did negatively correlate with the copy number of the *nifH* gene (Table 1). Carney and Matson [64] found that plant diversity had a significant effect on the microbial community composition through alterations in microbial abundance rather than community composition. However, several studies report that plant diversity has little direct effect on bacterial community composition [65,66]. Therefore, this effect may depend on the type of plant community examined.

A number of additional factors such as hydraulic conductivity, pH, Fe, soil moisture, and NH4-N significantly influenced both the free-living nitrogen-fixing bacteria and the broader bacterial community distribution (Table 2). In our study, the soil from different sites was the same type, but the water holding capacity function of each research soil was different. The soil moisture content is in the order of LGSG>Y25>Y5>Y1 (Table 1). This suggests that a balance of macropores and micropores has not been fully recovered in the abandoned crop soils. A balance of macropores and micropores in soil influences the permeability and water holding capacity of soil [67]. In addition, the biological decomposition of organic materials produces natural glues, which bind and strengthen soil aggregates [67], and helps soils hold water and nutrients, which may change the balance of macropores and micropores. Organic matter also is a long-term, slow-release storehouse of nitrogen, phosphorus, and sulfur [67]. Accordingly, organic matter significantly influenced the total bacterial community distribution (Table 2), which may have resulted because of the significant difference in soil organic matter content in the abandoned cropland and LGSG ($P<0.05$) (Table 1). Additionally, the soil texture and hydraulic conductivity is closely correlated, and soil texture influences the balance between macropores and micropores [67]. The hydraulic conductivities of sites Y1 and Y5 were significantly different from those of sites Y25 and LGSG (Table 1), and the RDA also indicated that soil moisture was negatively correlated with hydraulic conductivity (Figure S2). Therefore, the hydraulic conductivity of soil also showed that the water holding capacity function of soils from the research sites is different.

The significance of NH4-N indicates that the fixed nitrogen of the free-living nitrogen-fixing bacteria will be the main nitrogen source in the soil of steppe grasslands, where there are few plants such as legumes. Nitrogen cycling in natural ecosystems and traditional agricultural production relies on biological nitrogen fixation primarily by diazotrophic bacteria [16].

In our study, higher copy number/diversity of the *nifH* gene/16S rRNA gene was observed in the LGSG soil, where the Fe content was significantly higher, than in the abandoned cropland soil. According to Pearson's product-moment correlation and PLSR, the Fe content affects the nitrogen-fixing community. This may result because Fe is a required material for the [4Fe-4S] cluster of the *nifH* gene [22,51], and it is possible that the Fe content affects the 16S rRNA gene diversity in other microorganisms with the iron-containing enzyme [68-70] as it does the nitrogen-fixing bacteria.

Gaby and Buckley [71] comprehensively evaluated primers of *nifH* gene and found that *nifH* gene different primer sets amplified different groups in the *nifH* phylogeny. In this study, we used primer sets for DGGE and qPCR analyses of the *nifH* gene that targeted different sequence positions, thus the results obtained were not comparable. However, environmental factors differed greatly between the abandoned cropland and LGSG soils, thus having differential effects on the soil microbial communities. Furthermore, the trends in the changing copy number and diversity of the *nifH* gene were similar to that of the 16S rRNA gene. Therefore, the results we obtained regarding the copy number and diversity of *nifH* gene are able to explain the dynamic change in the trend of free-living nitrogen-fixing bacteria communities in abandoned cropland during different periods. Further research on molecular ecology could provide more detail analysis of microbial diversity, such as by sequencing of DGGE band and high-throughput sequencing technology etc., which would contribute to more accurate insights into the black box of the dynamic change of microbial communities.

We discovered that microbial communities of the abandoned cropland in this study area are strongly influenced by plant biomass, soil moisture, Fe, and NH4-N. Robust information about the mechanisms that regulate the diversity, structure, and composition of natural communities is urgently needed to help conserve ecosystem function and mitigate biodiversity loss from current and future environmental changes. Conversely, the results of this study suggest that advances in desertification may be prevented by adjusting environmental factors of the abandoned cropland, such as the soil moisture content, Fe, and NH4-N, which will enhance the function of a microbial community and possibly increasing plant biomass production. In addition, *nifH* gene copy number had significant positive Pearson's product-moment correlation with soil moisture, organic C, P, and Fe etc., therefore, by adjusting these environmental factors may also increase the abundance of free-living nitrogen-fixing bacteria in abandoned cropland.

Supporting Information

Figure S1 Denaturing gradient gel electrophoresis (DGGE) profiles of the *nifH* (A) and 16S rRNA genes (B). For all images, the numbers refer to the plot numbers in the sample areas.

Figure S2 Redundancy analysis (RDA) of *nifH* (A) and 16S rRNA (B) genes data. Ordination plots of *nifH* (A) and 16S rRNA (B) genes associated with abandoned croplands for different times: Y1 (•), Y5 (▲), and Y25 (■), and *light-grazing* steppe grassland (LGSG, ○). The plots were generated by redundancy analysis (RDA) of the denaturing gradient gel electrophoresis (DGGE) profiles. All environmental variables are shown, including pH, NO_3-N, NH_4-N, organic carbon (C), total nitrogen (N), available phosphorus (P), soluble iron (Fe), elevation (ELE), hydraulic conductivity (HC), soil moisture (H_2O), plant biomass (P-B), and plant species richness (P-*H'*). Values on the axes indicate the percentages of total variation explained by each axis.

Table S1 Total plant cover, plant Shannon's diversity index, plant biomass, plant types, and their coverage in each research plot.

Table S2 Soil texture (clay, silt, and sand) content and saturated hydraulic conductivity of soil samples at each site.

Table S3 Collinearity among environmental parameters as determined by Spearman's rank correlation coefficient rho. Significantly correlated parameters show Spearman's rank correlation coefficient rho$>$0.6 and <-0.6 in bold. All environmental variables are shown, such as pH, NO_3-N, NH_4-N, organic carbon (C), total nitrogen (N), available phosphorus (P), soluble iron (Fe), elevation (ELE), hydraulic conductivity (HC), soil moisture (H_2O), plant biomass (P-B), and plant diversity (P-*H'*).

Acknowledgments

We thank Dr. Kenji Tamura (University of Tsukuba) for his valuable suggestions on the research content and soil sampling, and Takashi Kanda (University of Tsukuba) for excellent technical support in measuring organic C and total N concentrations.

Author Contributions

Conceived and designed the experiments: H SB HU. Performed the experiments: H SB. Analyzed the data: H YXC. Contributed reagents/materials/analysis tools: HU NN T. Nakajima T. Nakamura. Wrote the paper: H.

References

1. Archibold OW (1995) Ecology of World Vegetation. London: Chapman & Hall. pp. 1–522.
2. He C, Zhang Q, Li Y, Li X, Shi P (2005) Zoning grassland protection areas using remote sensing and cellular automata modeling – A case study in Xilingol steppe grassland in northern China. J Arid Environ 63: 814–826.
3. Tong C, Wu J, Yong S, Yang J, Yong W (2004) A landscapescale assessment of steppe degradation in the Xilin River basin, Inner Mongolia, China. J Arid Environ 59: 133–149.
4. Li JH, Fang XW, Jia JJ, Wang G (2007) Effect of legume species introduction to early abandoned field on vegetation development. Plant Ecol 191: 1–9.
5. Van Der Putten WH, Mortimer SR, Hedlund K, Van Dijk C, Brown VK, et al. (2000) Plant species diversity as a driver of early succession in abandoned fields: a multi-site approach. Oecologia 124: 91–99.
6. Xue ZD, Hou QC, Han RL, Wang SQ (2002) Trails and research on ecological restoration by *Sophora viciifolia* in Gullied Rolling Loess Region. J Northwest Forestry Univ 17: 26–29.
7. Zhang JD, Qiu Y, Chai BF, Zheng FY (2000) Succession analysis of plant communities in Yancun low middle hills of Luliang Mountains. J Plant Resour. Environ 9: 34–39.

8. Zou HY, Chen JM, Zhou L, Hongo A (1998) Natural recoverage succession and regulation of the Prairie vegetation on the Loess Plateau. Res. Soil. Water Conserv 5: 126–138.
9. Cheng YX, Nakamura T (2007) Phytosociological study of steppe vegetation in east Kazakhstan. Grassland Sci 53: 172–180.
10. El-Sheikh MA (2005) Plant succession on abandoned fields after 25 years of shifting cultivation in Assuit, Egypt. J Arid Environ 61: 461–481.
11. Prévosto B, Kuiters L, Bernhardt-Römermann M, Dölle M, Schmidt W, et al. (2011) Impacts of land abandonment on vegetation: successional pathways in European habitats. Folia Geobot 46: 303–325.
12. Štolcová J (2002) Secondary succession on an early abandoned field: vegetation composition and production of biomass. Plant Protection Sci 38: 149–154.
13. Ortiz-Castro R, Contreras-Cornejo HA, Macias-Rodriguez L, Lopez-Bucio J (2009) The role of microbial signals in plant growth and development. Plant Signal. Behav 4: 701–712.
14. Daniel R (2005) The metagenomics of soil. Nat Rev Microbiol 3: 470–478.
15. Van Der Heijden M, Bardgett R, van Straalen N (2008) The unseen majority: soil microbes as drivers of plant diversity and productivity in terrestrial ecosystems. Ecol Lett 11: 296–310.
16. Vitousek PM, Aber J, Howarth RW, Likens GE, Matson PA, et al. (1997) Human alteration of the global nitrogen cycle: causes and consequences. Ecol Appl 7: 737–750.
17. Orr CH, James A, Leifert C, Cooper JM, Cummings SP (2011) Diversity and activity of free-living nitrogen-fixing bacteria and total bacteria in organic and conventionally managed soils. Appl Environ Microbiol 77: 911–919.
18. Dixon R, Kahn D (2004) Genetic regulation of biological nitrogen fixation. Nat Rev Microbiol 2: 621–631.
19. Peoples MB, Herridge DF, Ladha JK (1995) Biological nitrogen fixation: an efficient source of nitrogen for sustainable agricultural production? Plant Soil 174: 3–28.
20. Burgmann H, Widmer F, Von Sigler W, Zeyer J (2004) New molecular screening tools for analysis of free-living diazotrophs in soil. Appl Environ Microbiol 70: 240–247.
21. Kahindi JHP, Woomer P, George T, de Souza Moreira FM, Karanja NK, et al. (1997) Agricultural intensification, soil biodiversity and ecosystem function in the tropics: the role of nitrogen-fixing bacteria. Appl Soil Ecol 6: 55–76.
22. Roeselers G, Stal LJ, van Loosdrecht MCM, Muyzer G (2007) Development of a PCR for the detection and identification of cyanobacterial nifD genes. J Microbiol Methods 70: 550–556.
23. Poly F, Monrozier LJ, Bally R (2001) Improvement in the RFLP procedure for studying the diversity of *nifH* genes in communities of nitrogen fixers in soil. Res Microbiol 152: 95–103.
24. Rosado AS, Duarte GF, Seldin L, Van Elsas JD (1998) Genetic diversity of nifH gene sequences in Paenibacillus azotofixans strains and soil samples analyzed by denaturing gradient gel electrophoresis of PCR-amplified gene fragments. Appl Environ Microbiol 64: 2770–2779.
25. Widmer F, Shaffer BT, Porteous LA, Seidler RJ (1999) Analysis of *nifH* gene pool complexity in soil and litter at a Douglas fir forest site in the Oregon Cascade Mountain range. Appl Environ Microbiol 65: 374–380.
26. Matsuura K, Willmott CJ (2009) Terrestrial precipitation: 1900-2010 gridded monthly time series. Available: http://climate.geog.udel.edu/~climate/. Accessed 2014 Aug 27.
27. Braun-Blanquet J (1964) Pflanzensoziologie, 3rd revised edn. New York: Springer-Verlag. pp. 1–865.
28. Mueller-Dombois D, Ellenberg H (1974) Aims and Methods of Vegetation Ecology. New York: John Wiley and Sons. pp. 1–547.
29. Day PR (1965) Particle fraction and particle-size analysis. In: Black CA, editor. Methods of soil analysis. Madison: American Society of Agronomy. pp. 545–566.
30. Siosemarde M, Byzedi M (2011) Studding of number of dataset on precision of estimated saturated hydraulic conductivity. World Academy Sci, Eng Technol 74: 521–524.
31. Leonardo M (2009) Development and validation of a method for determination of residual nitrite/nitrate in foodstuffs and water after zinc reduction. Food Anal Methods 2: 212–220.
32. Motsara MR, Roy RN (2008) Guide to laboratory establishment for plant nutrient analysis. FAO Fertilizer and Plant Nutrition Bulletin No. 19, FAO, Rome. Pp.17–76.
33. Truog E, Meyer AH (1929) Improvements in the deniges colorimetric method for phosphorus and arsenic. Indus, and Engin Chem, Analyt Ed 1: 136–139.
34. Saywell LG, Cunningham BB (1937) Determination of iron: colorimetric o-phenanthroline method. Ind Eng Chem Anal 9: 67–69.
35. Sugio T, Taha T, Kanao T, Takeuchi F (2007) Increase in Fe^{2+}-producing activity during growth of *Acidithiobacillus ferrooxidans* ATCC 23270 on sulfur. Biosci Biotechnol Biochem 71: 2663–2669.
36. Schumacher BA (2002) Methods for the determination of total organic carbon (TOC) in soils and sediments. U.S. Environmental Protection Agency, Washington, Ecological Risk Assesment Support Center.
37. Steward GF, Zehr JP, Jellison RP, Montoya JP, Hollibaugh JT (2004) Vertical distribution of nitrogen-fixing phylotypes in a meromictic, hypersaline lake. Microb Ecol 47: 30–40.
38. Zehr JP, McReynolds LA (1989) Use of degenerate oligonucleotides for amplification of the *nifH* gene from the marine *cyanobacterium Trichodesmium thiebautii*. Appl Environ Microbiol 55: 2522–2526.
39. Muyzer G, De Waal EC, Uitterlinden AG (1993) Profiling of complex microbial populations by denaturing gradient gel electrophoresis analysis of polymerase chain reaction-amplified genes coding for 16S rRNA. Appl Environ Microbiol 59: 695–700.
40. Baxter J, Cummings SP (2006) The impact of bioaugmentation on metal cyanide degradation and soil bacteria community structure. Biodegradation 17: 207–217.
41. Baxter J, Cummings SP (2008) The degradation of the herbicide bromoxynil and its impact on bacterial diversity in a top soil. J Appl Microbiol 104: 1605–1616.
42. Rösch C, Mergel A, Bothe H (2002) Biodiversity of denitrifying and dinitrogen-fixing bacteria in an acid forest soil. Appl Environ Microbiol 68: 3818–3829.
43. ter Braak CJF, Smilauer P (2002) CANOCO reference manual and CanoDraw for Windows user's guide: software for canonical community ordination (version 4.5). Ithaca, USA (url: www.canoco.com): Microcomputer Power. 500 p.
44. Carrascal LM, Galvan I, Gordo O (2009) Partial least squares regression as an alternative to current regression methods used in ecology. Oikos 118: 681–690.
45. Naether A, Foesel BU, Naegele V, Wüst PK, Weinert J, et al. (2012) Environmental factors affect acidobacterial communities below the subgroup level in grassland and forest soils. Appl Environ Microbiol 78: 7398–7406.
46. Umetrics AB (2005) SIMCA-P v. 11 Analysis Advisor. Umeå, Sweden: Umetrics AB.
47. Tremaroli V, Workentine ML, Weljie AM, Vogel HJ, Ceri H, et al. (2009) Metabolomic investigation of the bacterial response to a metal challenge. Appl Environ Microbiol 75: 719–728.
48. Ramette A (2007) Multivariate analyses in microbial ecology. FEMS Microbiol Ecol 62: 142–160.
49. Chan YK, Barraquio WL, Knowles R (1994) N_2-fixing pseudomonas and related soil bacteria. FEMS Microb Rev 13: 95–118.
50. Reed SC, Seastedt TR, Mann CM, Suding KN, Townsend AR, et al. (2007) Phosphorus fertilization stimulates nitrogen fixation and increases inorganic nitrogen concentrations in a restored prairie. Appl Soil Ecol 36: 238–242.
51. Gavini N, Burgess BK (1992) FeMo cofactor synthesis by a *nifH* mutant with altered MgATP reactivity. J Biol Chem 267: 21179–21186.
52. Coelho MRR, de Vos M, Carneiro NP, Marriel IE, Paiva E, et al. (2008) Diversity of *nifH* gene pools in the rhizosphere of two cultivars of sorghum (Sorghum bicolor) treated with contrasting levels of nitrogen fertilizer. FEMS Microbiol Lett 279: 15–22.
53. Coelho MRR, Marriel IE, Jenkins SN, Lanyon CV, Seldin L, et al. (2009) Molecular detection and quantification of nifH gene sequences in the rhizosphere of sorghum (Sorghum bicolor) sown with two levels of nitrogen fertilizer. Appl Soil Ecol 42: 48–53.
54. Noll M, Wellinger M, (2008) Changes of the soil ecosystem along a receding glacier: Testing the correlation between environmental factors and bacterial community structure. Soil Biol Biochem 40: 2611–2619.
55. Wakelin SA, Macdonald LM, Rogers SL, Gregg AL, Bolger TP, et al. (2008) Habitat selective factors influencing the structural composition and functional capacity of microbial communities in agricultural soils. Soil Biol Biochem 40: 803–813.
56. Belnap J (2001) Factors influencing nitrogen fixation and nitrogen release in biological soil crusts. In: Belnap J, Lange OL, editors. Biological soil crusts: structure, function, and management. Berlin: Springer-Verlag. pp. 241–261.
57. Clayton SJ, Clegg CD, Murray PJ, Gregory PJ (2005) Determination of the impact of continuous defoliation of *Lolium perenne* and *Trifolium repens* on bacterial and fungal community structure in rhizosphere soil. Biol Fertil Soils 41: 109–115.
58. Marilley L, Vogt G, Blanc M, Aragno M (1998) Bacterial diversity in the bulk soil and rhizosphere fractions of *Lolium perenne* and *Trifolium repens* as revealed by PCR restriction analysis of 16S rDNA. Plant Soil 198: 219–224.
59. Marschner P, Yang CH, Liebere R, Crowley DE (2001) Soil and plant specific effects on bacterial community composition in the rhizosphere. Soil Biol Biochem 33: 1437–1445.
60. Stephen JR, McCaig AE, Smith Z, Prosser JI, Embley TM (1996) Molecular diversity of soil and marine 16S rRNA gene sequences related to beta-subgroup ammonia-oxidizing bacteria. Appl Environ Microbiol 62: 4147–4154.
61. Larkin RP, Honeycutt CW (2006) Effects of different 3-year cropping systems on soil microbial communities and *Rhizoctonia* diseases of potato. Phytopathology 96: 68–79.
62. Niklaus PA, Alphei J, Ebersberger D, Kampichler C, Kandeler E, et al. (2003) Six years of *in situ* CO_2 enrichment evokes changes in soil structure and soil biota of nutrient-poor grassland. Glob Chang Biol 9: 585–600.
63. Zak DR, Holmes W, White DC, Peacock A, Tilman D (2003) Plant diversity, soil microbial communities, and ecosystem function: are there any link? Ecology 84: 2042–2050.
64. Carney KM, Matson PA (2005) Plant communities, soil microorganisms, and soil carbon cycling: does altering the world belowground matter to ecosystem functioning? Ecosystems 8: 928–940.
65. Kennedy N, Brodie E, Connolly J, Clipson N (2004) Impact of lime, nitrogen and plant species on bacterial community structure in grassland microcosms. Environ Microbiol 6: 1070–1080.
66. Nunan N, Daniell TJ, Singh BK, Papert A, McNicol JW, et al. (2005) Links between plant and rhizoplane bacterial communities in grassland soils, characterized using molecular techniques. Appl Environ Microbiol 71: 6784–6792.

67. Cogger C (2000) Soil Management for Small Farms. EB 1895. Pullman, WA: Washington State University Cooperative Extension. 24 p.
68. Sze IS, Dagley S (1984) Properties of salicylate hydroxylase and hydroxyquinol 1,2-dioxygenase purified from *Trichosporon cutaneum*. J Bacteriol 159: 353–359.
69. Conway T, Ingram LO (1989) Similarity of *Escherichia coli* propanediol oxidoreductase (fucO product) and an unusual alcohol dehydrogenase from *Zymomonas mobilis* and *Saccharomyces cerevisiae*. J Bacteriol 171: 3754–3759.
70. Drennan CL, Heo J, Sintchak MD, Schreiter E, Ludden PW (2001) Life on carbon monoxide: X-ray structure of Rhodospirillum rubrum Ni-Fe-S carbon monoxide dehydrogenase. Proc Natl Acad Sci USA 98: 11973–11978.
71. Gaby JC, Buckley DH (2012) A comprehensive evaluation of PCR primers to amplify the nifH gene of nitrogenase. PLoS ONE 7: e42149.

10

Soil Carbon and Nitrogen Fractions and Crop Yields Affected by Residue Placement and Crop Types

Jun Wang[1], Upendra M. Sainju[2*]

1 College of Urban and Environmental Sciences, Northwest University, Xian, Shaanxi Province, China, **2** U.S. Department of Agriculture, Agricultural Research Service, Northern Plains Agricultural Research Laboratory, Sidney, Montana, United States of America

Abstract

Soil labile C and N fractions can change rapidly in response to management practices compared to non-labile fractions. High variability in soil properties in the field, however, results in nonresponse to management practices on these parameters. We evaluated the effects of residue placement (surface application [or simulated no-tillage] and incorporation into the soil [or simulated conventional tillage]) and crop types (spring wheat [*Triticum aestivum* L.], pea [*Pisum sativum* L.], and fallow) on crop yields and soil C and N fractions at the 0–20 cm depth within a crop growing season in the greenhouse and the field. Soil C and N fractions were soil organic C (SOC), total N (STN), particulate organic C and N (POC and PON), microbial biomass C and N (MBC and MBN), potential C and N mineralization (PCM and PNM), NH_4-N, and NO_3-N concentrations. Yields of both wheat and pea varied with residue placement in the greenhouse as well as in the field. In the greenhouse, SOC, PCM, STN, MBN, and NH_4-N concentrations were greater in surface placement than incorporation of residue and greater under wheat than pea or fallow. In the field, MBN and NH_4-N concentrations were greater in no-tillage than conventional tillage, but the trend reversed for NO_3-N. The PNM was greater under pea or fallow than wheat in the greenhouse and the field. Average SOC, POC, MBC, PON, PNM, MBN, and NO_3-N concentrations across treatments were higher, but STN, PCM and NH_4-N concentrations were lower in the greenhouse than the field. The coefficient of variation for soil parameters ranged from 2.6 to 15.9% in the greenhouse and 8.0 to 36.7% in the field. Although crop yields varied, most soil C and N fractions were greater in surface placement than incorporation of residue and greater under wheat than pea or fallow in the greenhouse than the field within a crop growing season. Short-term management effect on soil C and N fractions were readily obtained with reduced variability under controlled soil and environmental conditions in the greenhouse compared to the field. Changes occurred more in soil labile than non-labile C and N fractions in the greenhouse than the field.

Editor: Raffaella Balestrini, Institute for Sustainable Plant Protection, C.N.R., Italy

Funding: Funding came from the U.S. Department of Agriculture-Agricultural Research Service, Sidney, MT, USA and the National Natural Science Foundation of China (No. 31270484). The funders had no role in study design, data collection and analysis, decision to publish, or preparation of the manuscript.

* Email: upendra.sainju@ars.usda.gov

Introduction

Soil organic matter, as indicated by C and N levels, is an important component of soil quality and productivity. Increasing soil organic matter through enhanced C and N sequestration can also reduce the potentials for global warming by mitigating greenhouse gas emissions and N leaching by increasing N storage in the soil [1,2]. Carbon and N sequestration usually occur when non-harvested crop residues, such as stems, leaves, and roots, are placed at the soil surface due to no-tillage [3,4,5]. Carbon and N sequestration rates, however, depend on the balance between the amounts of plant residue C and N inputs and rates of C and N mineralized in the nonmanured soil [6,7]. Other benefits of increasing C and N storage include enhancement of soil structure and soil water-nutrient-crop productivity relationships [8].

Soil and crop management practices can alter the quantity, quality, and placement of crop residues in the soil, thereby influencing soil C and N storage, microbial biomass and activity, and N mineralization–immobilization [9,10]. Residue placement in the soil under different tillage systems can influence C and N levels by affecting soil aggregation, aeration, and C and N mineralization [9,11]. Crop types can affect the quantity and quality (C/N ratio) of crop residue returned to the soil and therefore on soil C and N levels [9,12]. Legumes, such as pea, because of its higher N concentration and lower C/N ratio, decompose more rapidly in the soil and supply greater amount of N to succeeding crops than nonlegumes [12,13]. As a result, N fertilization rates to crops following pea can be reduced to sustain yields [14,15].

Because of large pool sizes and inherent spatial variability, soil organic C (SOC) and total N (STN) (slow or non-labile fractions) change slowly with management practices [16]. Therefore, measurements of SOC and STN alone may not adequately reflect changes in soil quality and nutrient status [16,17]. Active (or labile) C and N fractions, such as potential C and N mineralization (PCM and PNM) that indicate microbial activity and N mineralization, and microbial biomass C and N (MBC and MBN) that refer to microbial biomass and N immobilization, change seasonally [16,18]. Similarly, particulate organic C and N (POC and PON) that represent coarse organic matter and considered as intermediate C and N levels between slow and active fractions, provide substrates for microbes and influence soil aggregation [19,20].

Available N fractions that influence plant growth and N losses due to leaching, denitrification, or volatilization are NH_4-N and NO_3-N [10,12].

Although active C and N fractions in the soil can change more rapidly than the other fractions, these fractions sometime may not be readily changed within a crop growing season due to high variability in soil properties within a short distance in the field or in regions with limited precipitation, cold weather, and a short growing season [10,12,15]. Under controlled soil and environmental conditions, such as in the greenhouse, it may be possible to detect changes in these fractions more rapidly as affected by management practices than in the field. We hypothesized that surface placement of crop residue (a simulation of no-tillage in the field) under spring wheat can increase soil labile and non-labile C and N fractions and sustain crop yields compared to residue incorporation into the soil (a simulation of conventional tillage) under pea or fallow more in the greenhouse than in the field. Our objectives were to: (1) evaluate the effects of residue placement and crop types on crop yields, residue C and N losses, and soil labile and non-labile C and N fractions within a growing season in the greenhouse and the field and (2) determine if soil C and N fractions change more readily in the greenhouse than the field within a growing season.

Materials and Methods

Greenhouse experiment

The experiment was conducted under controlled soil and environmental conditions in the greenhouse with air temperatures of 25°C in the day and 15°C in the night. Soil samples were collected manually from an area of 5 m^2 using a shovel to a depth of 20 cm under a mixture of crested wheatgrass [*Agropyron cristatum* (L.) Gaertn] and western wheatgrass [*Pascopyrum smithii* (Rydb.) A. Love] from a dryland farm site, 11 km east of Sidney, Montana, USA. The research farm site where soil samples were collected is under the management of USDA, Agricultural Research Service, Sidney, Montana and no endangered or protected species were involved or was negatively impacted by this research. The soil was a Williams loam (fine-loamy, mixed, frigid, Typic Argiborolls [International classification: Luvisols]) with 350 g kg^{-1} sand, 325 g kg^{-1} silt, 325 g kg^{-1} clay, 1.42 Mg m^{-3} bulk density, and 7.2 pH at the 0–20 cm depth. Soil C and N fractions in the sample before the initiation of the experiment are shown in Table 1. Soil was air-dried and sieved to 4.75 mm after discarding coarse organic materials and rock fragments. Eight kilograms of soil was placed in a plastic pot, 25 cm high by 25 cm diameter, above 3 cm of gravel at the bottom.

Treatments consisted of two residue placements (surface placement vs. incorporation into the soil) and three crop types (spring wheat, pea, and fallow [or no crop]) arranged in a completely randomized design with three replications. In order to match the residue and crop type, spring wheat residue was placed under spring wheat and fallow and pea residue under pea. Residues included nine-week old spring wheat and pea plants collected from the field without grains, chopped to 2 cm, and oven-dried at 60°C for 3 d. Fifteen grams of residues per pot (corresponding to 2.6 Mg ha^{-1} of residue found in the field) were either placed uniformly at the soil surface or incorporated into the soil by mixing the residue with the soil by hand. The surface placement of residue corresponded to the simulated no-tillage system in the field, although the soil was disturbed during collection, and incorporated residue to the simulated conventional tillage system. Spring wheat received 0.96 g N pot^{-1} as urea, similar to the recommended N fertilization rate (80 kg N ha^{-1}) in the field, while pea received 0.11 g N pot^{-1} (or 9 kg N ha^{-1}) while applying monoammonium phosphate as the P fertilizer. Half of 0.96 g N pot^{-1} was applied at planting and other half at four weeks later. Both spring wheat and pea also received P fertilizer (monoammonium phosphate) at 0.25 g P pot^{-1} (or 27 kg P ha^{-1}) and K fertilizer (muriate of potash) at 0.50 g K pot^{-1} (or 29 kg K ha^{-1}). No fertilizers were applied to the fallow treatment.

In July 2012, five spring wheat (cultivar Reeder) and pea (cultivar Majoret) seeds were planted per pot, except in the fallow treatment. At a height of 3 cm, seedlings were thinned to two plants per pot. In order to compensate for the water received as rainfall in the field, water was applied to all treatments in the greenhouse experiment to field capacity (0.25 m^3 m^{-3}) [21] at 300 to 500 mL pot^{-1}. Water was applied at planting and at 3 to 7 d intervals thereafter, depending on soil water content (as determined by a soil water probe [TDR 300, Spectrum Technologies Inc., Aurora, IL] installed to a depth of 15 cm). Since measured amount of water was applied according to soil water content and crop demand, only a negligible amount of water was leached below the pot that was not determined. Herbicides and pesticides were applied to plants as needed. At 105 d after planting, shoot biomass including grains was harvested from the pot, washed with water, oven-dried at 60°C for 3 to7 d, and dry matter yield was determined. Because of the small amount of grain production, grains were also included in the shoot biomass. After crop harvest, soil from the entire pot was sieved to 2 mm to separate coarse residue and root fragments, which were picked by hand, washed with water, and oven-dried at 60°C for 3 to7 d to determine dry matter yields. A portion (100 g) of residue and root-free soil sample visible to the naked eye was collected from each pot, air-dried, and used for determinations of C and N fractions. The remaining soil samples were further washed in a nest of 1.0 and 0.5 mm sieves under a continuous stream of water to separate fine roots. Roots left in the sieves were picked using a tweezers, oven-dried at 60°C for 3 to7 d, and dry matter yield was determined. Total root biomass was determined by adding biomass of coarse and fine roots.

Shoot and root biomass and crop residues added to the soil at the initiation of the experiment and those (>2.00 mm) recovered from the soil at the end were ground to 1 mm and C and N concentrations (g kg^{-1}) were determined with a high induction furnace C and N analyzer (LECO, St. Joseph, MI). Amounts of C and N in the residue added and recovered from the soil were determined by multiplying C and N concentrations by the weight of the soil in the pot. Carbon and N losses from the residue were determined as: Residue C and N losses (g kg^{-1}) = (Residue C and N added – Residue C and N recovered) × 1000/Residue C and N added. While determining the amount of C and N recovered in the residue, it was assumed that fine residue (<2.00 mm) was a part of soil organic matter.

Field experiment

The field experiment was conducted using identical treatments, design, and replications as in the greenhouse from April to August 2012 near the place where soil samples were collected for the greenhouse experiment. As a result, soils were similar in both field and greenhouse experiments. The field site has mean monthly air temperature ranging from −8°C in January to 23°C in July and August. The mean annual precipitation (105-yr average) is 340 mm, 80% of which occurs during the crop growing season (April-October). Equivalent amounts of crop residues and fertilizers using the same treatments as in the greenhouse were applied to spring wheat, pea, and fallow in the field. Because the amount of residue applied was similar, the amounts of C and N

Table 1. Average soil organic C (SOC), total N (STN), particulate organic C and N (POC and PON), potential C and N mineralization (PCM and PNM), microbial biomass C and N (MBC and MBN), and NH_4-N and NO_3-N concentrations at the start of the experiment (n = 4).

Parameter	Concentration
SOC (g C kg^{-1})	11.80
POC (g C kg^{-1})	3.18
PCM (mg C kg^{-1})	9.25
MBC (mg C kg^{-1})	117.6
STN (g N kg^{-1})	1.29
PON (g N kg^{-1})	0.34
PNM (mg N kg^{-1})	8.95
MBN (mg N kg^{-1})	69.0
NH_4-N (mg N kg^{-1})	2.86
NO_3-N (mg N kg^{-1})	5.04

added in residue to the soil were also identical in the greenhouse and field. Residues and fertilizers were placed at the soil surface in the no-till system and incorporated to a depth of 10 cm using tillage with a field cultivator in the conventional tillage system. Plot size was 12.2×6.1 m.

Spring wheat and pea were planted in April with a no-till drill at a spacing of 20.3 cm. Growing season weeds were controlled with selective post emergence herbicides appropriate for each crop. Contact herbicides were applied at postharvest and preplanting. Crops were grown under dryland condition receiving only precipitation without irrigation. In August, biomass yield of spring wheat and pea was determined from two 0.5 m^2 areas outside yield rows within each plot and grain yield was determined by harvesting grains from a swath of 1.5 m×12.0 m using a combine harvester. Carbon and nitrogen concentrations in the grain and biomass were determined after oven drying subsamples at 55°C and using the C and N analyzer as above. Carbon and N contents (Mg C or N ha^{-1}) in grain and biomass were determined by multiplying C and N concentrations by grain and biomass yields, respectively. Total aboveground biomass and C and N contents were determined by adding yields and C and N contents of grain and biomass.

Soil samples were collected from five random locations in central rows of the plot to a depth of 20 cm using a truck-mounted hydraulic probe (3.5 cm inside diameter). Samples were composited within a plot, air-dried, ground, and sieved to 2 mm for determining C and N concentrations. No attempts were made to collect the surface residue at soil sampling because of residue loss and contamination with soil and residue from one plot to another due to actions of wind and water. Therefore, residue C and N losses were not determined in the field.

Soil carbon and nitrogen fractions measurements

The SOC concentration in the greenhouse and field soils were determined with a high induction furnace C and N analyzer as above after pretreating the soil with 5% H_2SO_3 to remove inorganic C [22]. The STN concentration was determined by using the analyzer without pretreating the soil with the acid. For determining POC and PON concentrations, 10 g soil sample was dispersed with 30 mL of 5 g L^{-1} sodium hexametaphosphate by shaking for 16 h and the solution was poured through a 0.053 mm sieve [19]. The solution and particles that passed through the sieve and contained mineral-associated and water-soluble C and N were dried at 50°C for 3 to 4 d and SOC and STN concentrations were determined by using the analyzer as above. The POC and PON concentrations were determined by the difference between SOC and STN in the whole-soil and that in the particles that passed through the sieve after correcting for the sand content.

The PCM and PNM concentrations in air-dried soils were determined by the modified method of Haney et al. [23]. Two 10 g soil subsamples were moistened with water at 50% field capacity [21] and placed in a 1 L jar containing beakers with 4 mL of 0.5 mol L^{-1} NaOH to trap evolved CO_2 and 20 mL of water to maintain high humidity. Soils were incubated in the jar at 21°C for 10 d. At 10 d, the beaker containing NaOH was removed from the jar and PCM was determined by measuring CO_2 absorbed in NaOH, which was back-titrated with 1.5 mol L^{-1} $BaCl_2$ and 0.1 mol L^{-1} HCl. One beaker containing soil was removed from the jar and extracted with 100 mL of 2 mol L^{-1} KCl for 1 h. The NH_4-N and NO_3-N concentrations in the extract were determined by using the autoanalyzer (Lachat Instrument, Loveland, CO). The PNM was calculated as the difference between the sum of NH_4-N and NO_3-N concentrations in the soil before and after incubation.

The other beaker containing moist soil and incubated for 10 d (used for PCM determination above) was used for determining MBC and MBN concentrations by the modified fumigation–incubation method for air-dried soils [24]. The moist soil was fumigated with ethanol-free chloroform for 24 h and placed in a 1 L jar containing beakers with 2 mL of 0.5 mol L^{-1} NaOH and 20 mL water. As with PCM, fumigated moist soil was incubated for 10 d and CO_2 absorbed in NaOH was back-titrated with $BaCl_2$ and HCl. The MBC was calculated by dividing the amount of CO_2–C absorbed in NaOH by a factor of 0.41 [25] without subtracting the values from the nonfumigated control [24]. For MBN, the fumigated–incubated sample at 10 d was extracted with 100 mL of 2 mol L^{-1} KCl for 1 h and NH_4-N and NO_3-N concentrations were determined by using the autoanalyzer as above. The MBN was calculated by the difference between the sum of NH_4-N and NO_3-N concentrations in the sample before and after fumigation–incubation and divided by a factor of 0.41 [25,26]. The NH_4-N and NO_3-N concentrations determined in the nonfumigated–nonincubated samples were used as available fractions of N.

Data analysis

Data for C and N contents in crop biomass and residue and soil C and N fractions were analyzed by using the MIXED model of SAS [27]. Treatment was considered as the fixed effect and replication as the random effect. Means were separated by using the least square means test when treatments and interactions were significant [27]. Statistical significance was evaluated at $P \leq 0.05$, unless otherwise stated.

Results

Greenhouse experiment

Shoot and root biomass yields and carbon and nitrogen contents. Shoot and root biomass yields and C and N contents varied among residue placements and crop types (Table 2). Interaction between residue placement and crop types on these parameters was not significant. Shoot and root biomass yields and C and N contents were greater in surface placement than incorporation of residue into the soil. Shoot biomass yield and C and N contents were also greater in wheat than in pea. Because of the negligible amount of roots, root biomass yield and C and N contents in pea were not determined. Absence of plants in the fallow also resulted in non-existence of crop data in this treatment. The coefficient of variation (CV) for crop parameters ranged from 38.2 to 62.5%.

Residue carbon and nitrogen losses. Total amounts of C and N added through residue application and leaf fall and those recovered in coarse fractions (>2 mm) after crop harvest varied with residue placements and crop types, with the significant residue placement × crop type interaction for C and N recovered in the residue (Table 3). Although the amount of residue applied was similar in all treatments (15 g of wheat or pea residue pot^{-1}), differences in C and N concentrations between residues and those added through leaf fall during crop growth varied residue C and N additions among treatments. Averaged across crop types, residue C addition was greater in surface placement than incorporation of residue into the soil. Averaged across residue placements, residue C addition was greater under wheat than pea or fallow, but residue N addition was greater under pea than wheat or fallow. Residue C recovery was greater in surface placement under wheat and fallow than surface placement under pea and incorporation under fallow. Residue N recovery was also greater in surface placement under wheat and fallow than surface placement under pea and incorporation under fallow and wheat. Averaged across crop types, residue N recovery was greater in surface placement than incorporation of residue into the soil. Averaged across residue placements, residue C recovery was greater under wheat than pea. The coefficient of variation for residue C and N addition and recovery varied from 7.2 to 17.8%.

Residue C and N losses also varied with residue placements and crop species, with the significant residue placement × crop species interaction (Table 3). Residue C loss was greater in surface placement under pea and incorporation under fallow than surface placement under fallow. Residue N loss was in the order: surface placement and incorporation under pea > incorporation under wheat and fallow > surface placement under wheat > surface placement under fallow. Averaged across crop types, residue N loss was greater in residue incorporation than surface placement. Averaged across residue placements, residue N loss was greater under pea than under fallow and wheat. The coefficient of variation for residue C and N losses varied from 14.3 to 31.6%.

Soil carbon and nitrogen fractions. The SOC, POC, and PCM concentrations varied among residue placements and crop types (Table 4). Averaged across crop types, SOC and PCM were greater in surface placement than incorporation of the residue into the soil. Averaged across residue placements, SOC was greater under wheat than pea and fallow and POC was greater under wheat than pea. The MBC was not influenced by treatments. The coefficient of variation for soil C fractions ranged from 2.6 to 14.3%.

The STN, PNM, MBN, NH_4-N, and NO_3-N concentrations also varied among residue placements and crop types (Table 4). Averaged across crop types, PNM and NH_4-N were greater in surface placement than incorporation of residue into the soil. Averaged across residue placements, STN was greater under wheat than pea and MBN was greater under wheat than fallow. In contrast, PNM was greater under pea and fallow than wheat and NO_3-N was greater under fallow than wheat. The PON was not influenced by treatments. The coefficient of variation for soil N fractions ranged from 4.6 to 15.9%.

Field experiment

Aboveground total crop biomass yield and C and N contents varied with crop types (Table 5). Averaged across tillage practices, crop biomass yield and C content were greater in wheat than pea, but the trend reversed for N content. Tillage and its interaction with crop type were not significant for crop biomass yield and C and N contents. The coefficient of variation for crop biomass yield and C and N contents ranged from 28.1 to 41.9%.

Soil MBN, NH_4-N, and NO_3-N concentrations varied with tillage practices and MBC and PNM varied with crop types (Table 5). Averaged across crop types, MBN and NH_4-N were greater in no-tillage than conventional tillage, but NO_3-N was greater in conventional tillage than no-tillage. Averaged across tillage practices, MBC was greater under wheat than fallow and PNM was greater under pea than wheat and fallow. Tillage, crop type, and their interaction were not significant for SOC, POC, PCM, STN, and PON. The coefficient of variations for soil C and N fractions ranged from 8.0 to 36.7%.

Discussion

Enhanced soil water conservation due to mulch action of the residue at the soil surface [28] may have increased shoot and root biomass yields and C and N contents in surface placement compared to incorporation of residue into the soil in the greenhouse (Table 2). It has been reported that surface placement of residue in the no-till system increased spring wheat yield compared to residue incorporation in the conventional till system [3,15]. In our field experiment, crop biomass yield and C and N contents, however, were not influenced by tillage (Table 5). It may be possible that wheat and pea residues applied by hand at the soil surface were more uniformly distributed in the greenhouse than in the field where residues were distributed by a machine sprayer. As a result, soil water was probably conserved more, resulting in increased crop yield and C and N contents with the surface placement than incorporation of residue in the greenhouse compared to the field.

Differences in the amount of N fertilizer applied and N fixation capacity may have resulted in variation in crop biomass yields and C contents among crop species in the greenhouse and the field (Tables 2 and 5). Higher amount of N fertilizer application may have increased biomass yield and C and N contents in wheat than pea in the greenhouse. Higher amount of N fertilizer application also may have increased biomass yield and C content in wheat and pea, but greater N fixation may have increased N content in pea than wheat in the field [14,28]. Grain and biomass yields are usually greater in wheat which receives N fertilizer than pea which

Table 2. Effects of residue placement and crop type on crop shoot (grains+leaves+stems) and root biomass C and N contents in the greenhouse.

Residue placement	Crop type	Shoot biomass	Root biomass	Shoot biomass C	Root biomass C	Shoot biomass N	Root biomass N	Total biomass C	Total biomass N
		--------g pot^{-1}------		--------g C pot^{-1}------		--------g Npot^{-1}------		g C pot^{-1}	g N pot^{-1}
Incorporated		4.51b[a]	2.37b	1.83b	0.74b	0.14b	0.05b	2.91b	0.25b
Surface		7.41a	6.14a	3.07a	1.74a	0.24a	0.11a	5.55a	0.44a
	Pea	4.55b	-----[b]	1.91b	-----	0.12b	-----	1.91b	0.12b
	Wheat	7.36a	4.25	3.00a	1.24	0.26a	0.08	4.23a	0.34a
CV (%)[c]		42.4	61.9	42.4	56.4	55.0	62.5	40.4	38.2
Significance									
Residue placement (R)		*	-----	*	*	**	*	*	*
Crop species (C)		*	*	*	-----	**	-----	*	-----
R×C		NS[d]	NS	NS	-----	NS	-----	-----	-----

*Significant at $P = 0.05$.
**Significant at $P = 0.01$.
[a]Numbers followed by different letters within a column in a set are significantly different at $P \leq 0.05$ by the least square means test.
[b]Non-measurable values due to negligible amount of root biomass.
[c]Coefficient of variation.
[d]Not significant.

Table 3. Effects of residue placement and crop type on residue C and N addition, recovered in coarse fragments (>2 mm), and losses during the crop growing period in the greenhouse.

Residue placement	Crop type	Crop residue					
		C added[a]	N added[a]	C recovered	N recovered	C loss	N loss
		g C pot^{-1}	g N pot^{-1}	g C pot^{-1}	g N pot^{-1}	g kg^{-1}	g kg^{-1}
Incorporated	Fallow	7.80a[b]	0.40a	3.84b	0.23b	508a	438b
	Pea	7.80a	0.64a	4.24ab	0.26ab	457ab	588a
	Wheat	8.76a	0.44a	4.89ab	0.24b	430ab	456b
Surface	Fallow	7.80a	0.40a	5.16a	0.31a	339b	213d
	Pea	7.80a	0.64a	3.82b	0.23b	511a	640a
	Wheat	8.92a	0.46a	4.95a	0.31a	446ab	325c
CV (%)[c]		7.2	17.8	15.4	16.5	143	316
Means							
Incorporated		8.07b	0.49a	4.32a	0.24b	465a	494a
Surface		8.17a	0.50a	4.64a	0.28a	432a	393b
	Fallow	7.80b	0.40b	4.50ab	0.27a	423a	326b
	Pea	7.80b	0.64a	4.02b	0.25a	484a	614a
	Wheat	8.76a	0.45b	4.92a	0.28a	438a	390b
Significance							
Residue placement (R)		*	NS[d]	NS	*	NS	**
Crop species (C)		***	***	*	NS	*	***
R×C		NS	NS	*	*	*	**

*Significant at $P=0.05$.
**Significant at $P=0.01$.
*** Significant at $P=0.001$.
[a]Includes C and N added from the residue application and leaf fall.
[b]Numbers followed by different letters within a column in a set are significantly different at $P \leq 0.05$ by the least square means test.
[c]Coefficient of variation.
[d]Not significant.

Table 4. Effects of residue placement and crop type on soil organic C (SOC), total N (STN), particulate organic C and N (POC and PON), potential C and N mineralization (PCM and PNM), microbial biomass C and N (MBC and MBN), and NH_4-N and NO_3-N concentrations in the greenhouse.

Residue placement	Crop type	SOC	POC	PCM	MBC	STN	PON	PNM	MBN	NH_4-N	NO_3-N
		--------- g C kg^{-1}------		-------- mg C kg^{-1}------		------ g N kg^{-1}------		------------------------ mg N kg^{-1}---------------			
Incorporated		12.0b[a]	3.28a	11.6b	128.0a	1.29a	0.33a	5.80b	50.8a	1.14b	11.8a
Surface		12.3a	3.28a	14.4a	143.8a	1.31a	0.31a	9.56a	57.5a	1.65a	16.5a
	Fallow	12.0b	3.26ab	11.4a	118.9a	1.30ab	0.32a	9.36a	46.1b	1.44a	23.7a
	Pea	12.0b	3.18b	13.1a	144.0a	1.26b	0.29a	9.68a	56.3ab	1.38a	12.8ab
	Wheat	12.4a	3.40a	14.5a	145.1a	1.34a	0.34a	5.98b	97.6a	1.36a	6.1b
CV (%)[b]		2.6	11.8	14.3	13.3	4.6	15.9	14.2	14.3	14.8	13.6
Significance											
Residue placement (R)		**	NS[c]	*	NS	NS	NS	*	NS	*	NS
Crop species (C)		**	*	NS	NS	*	NS	*	*	NS	*
R×C		NS	NS	NS	NS	NS	NS	NS	NS	NS	NS

Soil samples were collected at the 0–20 cm depth in the field and used for the greenhouse experiment.
*Significant at $P = 0.05$.
**Significant at $P = 0.01$.
[a]Numbers followed by different letters within a column in a set are significantly different at $P \leq 0.05$ by the least square means test.
[b]Coefficient of variation.
[c]Not significant.

Table 5. Effects of residue placement and crop type on crop aboveground biomass (grains+stems+leaves) yield, C and N contents, and soil organic C (SOC), total N (STN), particulate organic C and N (POC and PON), potential C and N mineralization (PCM and PNM), microbial biomass C and N (MBC and MBN), and NH_4-N and NO_3-N concentrations at the 0–20 cm depth in the field.

Tillage[a]	Crop type	Crop biomass yield	Crop C content	Crop N content	SOC	POC	PCM	MBC	STN	PON	PNM	MBN	NH_4-N	NO_3-N
		Mg ha^{-1}	Mg C ha^{-1}	kg N ha^{-1}	---- g C kg^{-1}----		---- mg C kg^{-1}----		---- g N kg^{-1}----		---------- mg N kg^{-1}----------			
CT		4.91a[b]	2.06a	59.6a	11.0a	2.56a	45.1a	114.4a	1.33a	0.24a	3.21a	12.6b	3.05b	4.54a
NT		5.08a	2.11a	65.3a	110.a	2.56a	56.8a	122.9a	1.37a	0.28a	4.28a	19.6a	3.82a	2.08b
	Fallow	------[c]	------	------	10.6a	2.55a	45.61a	111.4b	1.30a	0.25a	2.85b	14.2a	2.93a	3.36a
	Pea	4.68b	1.87b	72.3a	11.4a	2.56a	51.0a	118.6ab	1.42a	0.27a	5.56a	15.5a	3.43a	2.87a
	Wheat	5.31a	2.18a	52.6b	11.0a	2.57a	56.0a	126.0a	1.34a	0.26a	2.83b	18.6a	3.93a	3.71a
CV (%)		28.1	30.0	41.9	8.0	18.7	27.1	28.9	9.3	18.4	36.4	36.7	27.9	25.0
Significance														
Tillage (T)		NS[d]	NS	NS	NS	NS	NS	NS	NS	NS	NS	*	*	*
Crop species (C)		*	*	**	NS	NS	NS	*	NS	NS	*	NS	NS	NS
T×C		NS	NS	NS	NS	NS	NS	NS	NS	NS	NS	NS	NS	NS

*Significant at $P=0.05$.
**Significant at $P=0.01$.
[a]Tillage are CT, conventional tillage; and NT, no-tillage.
[b]Numbers followed by different letters within a column in a set are significantly different at $P\leq0.05$ by the least square means test.
[c]Crop absent in the fallow.
[d]Not significant.

receives no N fertilizer due to increased water-use efficiency, but higher N concentration due to increased atmospheric N fixation can increase N content in pea than wheat [10,15,28]. The fact that different trends in N content in pea vs. wheat occurred in the field and the greenhouse was probably related to root growing soil volume. It may be possible that roots exploited greater soil volume that resulted in increased N fixation by pea and therefore increased its N content in the field compared to the greenhouse where plants were grown in a limited soil volume in the pot.

Greater residue input due to higher biomass yield may have increased residue C addition in surface placement than incorporation of residue into the soil or increased under wheat than pea or fallow in the greenhouse (Tables 2 and 3). In contrast, higher N concentration may have increased residue N addition under pea than wheat or fallow. Greater C and N recovered in the residue placed at the soil surface under wheat and fallow were probably due to reduced mineralization of wheat residue as a result of its higher C/N ratio than pea residue. While surface placement of residue reduces its contact with soil microorganisms that result in reduced mineralization [29,30], increased mineralization of pea residue due to its lower C/N ratio may have resulted in reduced C and N recovery in the residue placed at the soil surface under pea. Residues of legumes, such as pea with lower C/N ratio, decompose more rapidly than those of nonlegumes, such as wheat with higher C/N ratio [12]. When incorporated into the soil, residue C and N recovery were lower under fallow and wheat. As a result, C and N losses were higher in surface placement of residue under pea or residue incorporation under pea and fallow than the other treatments. It may be possible that some of C and N lost from the residue converted into soil C and N fractions, as discussed below.

Reduced mineralization of residue may have increased SOC, PCM, PNM, and NH_4-N concentrations in surface placement than incorporation of residue into the soil in the greenhouse (Table 4). Similar increases in MBN and NH_4-N concentrations in no-tillage compared to conventional tillage were found in the field (Table 5). Several researchers [5,16,31,32,33] have reported greater SOC, POC, MBC, PCM, PNM, and MBN in surface residue placement in the no-tillage system than residue incorporation into the soil in the conventional tillage system. Increased N mineralization due to residue incorporation, however, may have increased NO_3-N concentration in conventional tillage than no-tillage in the field.

Higher C and N substrate availability due to increased yield probably increased SOC, POC, STN, and MBN under wheat than under pea or fallow in the greenhouse (Table 4) or increased MBC under wheat than fallow in the field (Table 5). Root biomass C, residue C addition (Tables 2 and 3), and amount of applied N fertilizer were greater in wheat than pea or fallow. Similar results probably occurred in the field, since treatments were identical in the greenhouse and the field and crop biomass C was higher in wheat than pea in the field (Table 5). Rhizodeposit C released by roots can increase microbial biomass and activity and soil C storage [34]. Liebig et al. [35] also found higher MBC under spring wheat than under fallow. In contrast, greater PNM and NO_3-N under pea and fallow than wheat in the greenhouse were probably either due to increased mineralization of pea residue as a result of its lower C/N ratio than wheat residue [12] or to greater mineralization of soil and wheat residue as a result of enhanced microbial activity from higher soil temperature and water content and absence of plants to uptake N under fallow [11,13,36]. Since residue N loss was greater under pea than wheat and fallow (Table 3), part of N from pea residue may have contributed to increased PNM and NO_3-N concentrations under pea. Similar result of increased PNM under pea than wheat and fallow was also found in the field, since crop biomass N was greater in pea than wheat (Table 5).

Comparison of soil C and N fractions at the beginning and end of the experiment due to residue placement (Tables 1 and 4) showed that SOC increased by 4.2%, PCM by 55.7%, and PNM by 6.1% with surface residue placement in the greenhouse. Corresponding values in SOC, PCM, and PNM with residue incorporation were 1.7, 25.4, and −35.1%, respectively. In the field, MBN reduced by 71.5% in no-tillage and 81.7% in conventional tillage from the beginning to the end of the experiment. This shows that residue placement at the surface either increased soil C and N fractions in the greenhouse or reduced their losses in the field within a crop growing season compared to residue incorporation. Since soil NH_4-N and NO_3-N concentrations vary seasonally due to N mineralization from crop residue and soil, N fertilization, crop N uptake, and N losses due to leaching, volatilization, and denitrification [10,15], variations in their levels from the beginning to the end of the experiment were not taken into account.

Among crop types, SOC increased by 5.1%, POC by 6.9%, STN by 3.9%, and MBN by 41.4%, but PNM decreased by 33.1% under wheat from the beginning to the end of the experiment in the greenhouse. In the field, MBC increased by 7.1%, but PNM decreased by 68.3% under wheat during this period. The corresponding increases in SOC, POC, STN, and MBN or decrease in PNM during this period were lower under pea and fallow. This suggests that wheat increased more soil C and N fractions, except PNM, than pea or fallow due to increased substrate availability from root and rhizodeposition and/or to slow decomposition of wheat than pea residue due to differences in residue quality (e.g. C/N ratio). The greater PNM under pea than wheat or fallow was due to increased N contribution from its residue (Table 5).

When the greenhouse and field experiments were compared, trends in changes in soil C and N fractions due to treatments within a crop growing season were similar. However, greater changes in labile than nonlabile C and N fractions occurred more in the greenhouse than in the field. Furthermore, the coefficient of variations in soil C and N fractions were lower in the greenhouse (2.6 to 15.9%) than in the field (8.0 to 36.7%) (Tables 4 and 5). This indicates that soil C and N fractions changed more readily but with lower variability with management practices within a crop growing season when soil and environmental conditions are controlled in the greenhouse than in field where soil heterogeneity often results in non-significant differences among treatments in these fractions [16,18,30]. Use of disturbed soil in the greenhouse vs. undisturbed (especially in the no-till system) in the field also may have an influence on differences in changes in soil C and N fractions between the two experiments. The greater changes in labile than nonlabile C and N fractions as influenced by management practices within a short period in the greenhouse and the field suggests that labile C and N fractions are better indicators of changes in soil organic matter quality than nonlabile fractions, a case similar to that reported by various researchers [10,11,13,16,30]. The fact that more changes in labile than nonlabile C and N fractions occurred in the greenhouse than in the field suggests that better measurements of changes in soil organic matter due to management practices within a short period can be observed when soil and environmental conditions are controlled. Greater levels of most soil C and N fractions in the greenhouse than in field was probably a result of increased turnover rate plant C and N into soil C and N, because disturbed soil was used in the greenhouse and environmental condition for microbial transformation was more favorable in the greenhouse than the field.

Greenhouse study provided more information on plant and residue parameters, such as measurement of root biomass and C and N contents and residue C and N losses, which cannot be measured easily in the field. This resulted in the measurement of turnover rate of plant C and N into soil C and N in the greenhouse, a fact that was absent in the field. Because of greater changes in soil C and N fractions, greenhouse study provided a more robust method of evaluating C and N cycling and soil quality within a short period of time as affected by management practices than the field experiment. Such changes can also be measured in the field but it may take longer time. While all results from the greenhouse study may not be readily applied in the field, some information, such as root biomass and residue C and N losses, measured in the greenhouse can be extrapolated to the field condition. The effects of short-term study in the greenhouse can be useful to predict the long-term impact of management practices on soil C and N fractions in the field.

Conclusions

Crop yields, residue C and N losses, and soil C and N fractions varied with residue placement and crop types in the greenhouse and the field. Surface placement of residue increased crop yields, residue C and N losses, and enhanced SOC, PCM, MBN, and NH_4-N concentrations, but residue incorporation increased PNM and NO_3-N concentrations. Similarly, spring wheat had higher yield and increased SOC, POC, MBC, STN, and MBN than pea or fallow, but pea had higher N content and increased PNM than wheat or fallow. Placing nonlegume residue at the soil surface using no-tillage can increase soil C and N sequestration and microbial biomass and activity that can improve soil health and quality. Using this practice, producers can claim for C credit. Incorporation of legume and nonlegume residues into the soil using conventional tillage can increase N mineralization and availability which can reduce N fertilization rate to succeeding crops, but can degrade soil quality due to reduced organic matter and increased erosion. Although soil labile C and N fractions changed more readily than nonlabile fractions within a crop growing season both in the greenhouse and field, greater changes in labile than nonlabile fractions occurred with reduced variability more in the greenhouse than in the field. Results suggest that greenhouse study provided a more robust measurement of crop growth and changes in soil C and N fractions within a short period as influenced by management practices than the field experiment. Longer time will be probably needed in the field to obtain results similar to those in the greenhouse. Additional information, such as root growth, residue C and N losses, turnover of plant C and N to soil C and N, and results of short-term study on soil C and N fractions as influenced by management practices in the greenhouse can be used to predict the long-term impact in the field.

Acknowledgments

We appreciate the excellent support of Joy Barsotti and Thecan Caesar-TonThat for analyzing soil and plant samples in the laboratory.

Author Contributions

Conceived and designed the experiments: UMS. Performed the experiments: JW. Analyzed the data: UMS. Contributed reagents/materials/analysis tools: UMS. Wrote the paper: UMS.

References

1. Lal R, Kimble JM, Stewart BA (1995) World soils as a source or sink for radiatively-active gases. In: Lal R, editor, Soil management and greenhouse effect. Advances in soil science. CRC Press, Boca Raton, FL, pp. 1–8
2. Paustian K, Robertson GP, Elliott ET (1995) Management impacts on carbon storage and gas fluxes in mid-latitudes cropland. In: Lal R, editor, Soils and global climate change. Advances in soil science. CRC Press, Boca Raton, FL, USA, pp. 69–83.
3. Halvorson AD, Peterson GA, Reule CA (2002a) Tillage system and crop rotation effects on dryland crop yields and soil carbon in the central Great Plains. Agron J 94:1429–1436.
4. Sherrod LA, Peterson GA, Westfall DG, Ahuja LR (2003) Cropping intensity enhances soil organic carbon and nitrogen in a no-till agroecosystem. Soil Sci Soc Am J 67:1533–1543.
5. Sainju UM, Caesar-TonThat T, Lenssen AW, Evans RG, Kolberg R (2007) Long-term tillage and cropping sequence effects on dryland residue and soil carbon fractions. Soil Sci Soc Am J 71:1730–1739.
6. Rasmussen PE, Allmaras RR, Rhoade CR, Roager NC Jr (1980) Crop residue influences on soil carbon and nitrogen in a wheat-fallow system. Soil Sci Soc Am J 44:596–600.
7. Peterson GA, Halvorson AD, Havlin JL, Jones OR, Lyon DG, Tanaka DL (1998) Reduced tillage and increasing cropping intensity in the Great Plains conserve soil carbon. Soil Tillage Res 47:207–218.
8. Bauer A, Black AL (1994) Quantification of the effect of soil organic matter content on soil productivity. Soil Sci Soc Am J 58:185–193.
9. Ghidey F, Alberts EE (1993) Residue type and placement effects on decomposition: Field study and model evaluation. Trans ASAE 36:1611–1617.
10. Sainju UM, Lenssen AW, Caesar-Tonthat T, Waddell J (2006b) Tillage and crop rotation effects on dryland soil and residue carbon and nitrogen. Soil Sci Soc Am J 70:668–678.
11. Halvorson AD, Wienhold BJ, Black AL (2002b) Tillage, nitrogen, and cropping system effects on soil carbon sequestration. Soil Sci Soc Am J 66:906–912.
12. Kuo S, Sainju UM, Jellum EJ (1997) Winter cover cropping influence on nitrogen in soil. Soil Sci Soc Am J 61:1392–1399.
13. Sainju UM, Lenssen AW, Caesar-TonThat T, Waddell J (2006a) Carbon sequestration in dryland soils and plant residue as influenced by tillage and crop rotation. J Environ Qual 35:1341–1349.
14. Miller PR, McConkey B, Clayton GW, Brandt SA, Staricka JA, Johnston AM, Lafond GP, Schatz BG, Baltensperger DD, Neill KE (2002) Pulse crop adaptation in the northern Great Plains. Agron J 94:261–272.
15. Sainju UM, Lenssen AW, Caesar-TonThat T, Evans RG (2009) Dryland crop yields and soil organic matter as influenced by long-term tillage and cropping sequence. Agron J 101:243–251.
16. Franzluebbers AJ, Hons FM, Zuberer DA (1995) Soil organic carbon, microbial biomass, and mineralizable carbon and nitrogen in sorghum. Soil Sci Soc Am J 59:460–466.
17. Bezdicek DF, Papendick DF, Lal R (1996) Introduction: Importance of soil quality to health and sustainable land management. In: Doran JW, Jones AJ, editors, Methods of assessing soil quality, Spec. Publ. 49, Soil Science Society of America, Madison, USA, pp. 1–18.
18. Franzluebbers AJ, Arshad MA (1997) Soil microbial biomass and mineralizable carbon of water-stable aggregates. Soil Sci Soc Am J 67:1090–1097.
19. Cambardella CA, Elliott ET (1992) Particulate soil organic matter changes across a grassland cultivation sequence. Soil Sci Soc Am J 56:777–783.
20. Six J, Elliott ET, Paustian K (1999) Aggregate and soil organic matter dynamics under conventional and no-tillage systems. Soil Sci Soc Am J 63:1350–1358.
21. Pikul JL Jr, Aase JK (2003) Water infiltration and storage affected by subsoiling and subsequent tillage. Soil Sci Soc Am J 67:859–866.
22. Nelson DW, Sommers LE (1996) Total carbon, organic carbon, and organic matter. In: Sparks DL, editor, Methods of soil analysis. Part 3. Chemical method. SSSA Book Ser. 5. Soil Science Society of America, Madison, pp. 961–1010.
23. Haney RL, Franzluebbers AJ, Porter EB, Hons FM, Zuberer DA (2004) Soil carbon and nitrogen mineralization: Influence of drying temperature. Soil Sci Soc Am J 68:489–492.
24. Franzluebbers AJ, Haney RL, Hons FM, Zuberer DA (1996) Determination of microbial biomass and nitrogen mineralization following rewetting of dried soil. Soil Sci Soc Am J 60:1133–1139.
25. Voroney RP, Paul EA (1984) Determination of k_C and k_N *in situ* for calibration of the chloroform fumigation-incubation method. Soil Biol Biochem 16:9–14.
26. Brookes PC, Landman A, Pruden G, Jenkinson DJ (1985) Chloroform fumigation and the release of soil nitrogen: A rapid direct-extraction method to measure microbial biomass nitrogen in soil. Soil Biol Biochem 17:937–942.
27. Littell RC, Milliken GA, Stroup WW, Wolfinger RR (1996) SAS system for mixed models. SAS Institute Inc., Cary, NC, USA.
28. Lenssen AW, Johnson GD, Carlson GR (2007) Cropping sequence and tillage system influences annual crop production and water use in semiarid Montana. Field Crops Res 100:32–43.

29. Coppens F, Garnier P, de Gryze P, Merckx R, Recous S (2006) Soil moisture, carbon and nitrogen dynamics following incorporation and surface application of labelled crop residues in soil columns. Europ J Soil Sci 57:894–905.
30. Giacomini SJ, Recous S, Mary B, Aita C (2007) Simulating the effects of nitrogen availability, straw particle size and location in soil on carbon and nitrogen mineralization. Plant Soil 301: 289–301.
31. Malhi SS, Lemke R (2007) Tillage, crop residue and nitrogen fertilizer effects on crop yield, nutrient uptake, soil quality and greenhouse gas emissions in the second 4-yr rotation cycle. Soil Tillage Res 96:269–283.
32. Wright AL, Hons FM, Lemon RG, MacFarland ML, Nichols RL (2008) Microbial activity and soil carbon sequestration for reduced and conventional tillage cotton. Appl Soil Ecol 38:168–173.
33. Lupwayi NZ, Lafond GP, Ziadi N, Grant CA (2012) Soil microbial response to nitrogen fertilizer and tillage in barley and corn. Soil Tillage Res 118:139–146.
34. Lu Y, Watanabe A, Kimura M (2002) Contribution of plant-derived carbon to soil microbial biomass dynamics in a paddy rice microcosm. Biol Fertil Soils 36:136–142.
35. Liebig MA, Tanaka DL, Wienhold BJ (2004) Tillage and cropping effects on soil quality indicators in the northern Great Plains. Soil Tillage Res 78:131–141.
36. Kuzyakov Y, Domanski G (2000) Carbon input by plants into the soil: Review. J. Plant Nutri Soil Sci 163:421–431.

11

Intercropping Competition between Apple Trees and Crops in Agroforestry Systems on the Loess Plateau of China

Lubo Gao[1], Huasen Xu[1], Huaxing Bi[1,2]*, Weimin Xi[3], Biao Bao[1], Xiaoyan Wang[1], Chao Bi[1], Yifang Chang[1]

1 College of Water and Soil Conservation, Beijing Forestry University, Beijing, P.R. China, **2** Key Laboratory of Soil and Water Conservation, Ministry of Education, Beijing, P.R. China, **3** Department of Biological and Health Sciences, Texas A&M University-Kingsville, Kingsville, Texas, United States of America

Abstract

Agroforestry has been widely practiced in the Loess Plateau region of China because of its prominent effects in reducing soil and water losses, improving land-use efficiency and increasing economic returns. However, the agroforestry practices may lead to competition between crops and trees for underground soil moisture and nutrients, and the trees on the canopy layer may also lead to shortage of light for crops. In order to minimize interspecific competition and maximize the benefits of tree-based intercropping systems, we studied photosynthesis, growth and yield of soybean (*Glycine max* L. Merr.) and peanut (*Arachis hypogaea* L.) by measuring photosynthetically active radiation, net photosynthetic rate, soil moisture and soil nutrients in a plantation of apple (*Malus pumila* M.) at a spacing of 4 m × 5 m on the Loess Plateau of China. The results showed that for both intercropping systems in the study region, soil moisture was the primary factor affecting the crop yields followed by light. Deficiency of the soil nutrients also had a significant impact on crop yields. Compared with soybean, peanut was more suitable for intercropping with apple trees to obtain economic benefits in the region. We concluded that apple-soybean and apple-peanut intercropping systems can be practical and beneficial in the region. However, the distance between crops and tree rows should be adjusted to minimize interspecies competition. Agronomic measures such as regular canopy pruning, root barriers, additional irrigation and fertilization also should be applied in the intercropping systems.

Editor: Randall P. Niedz, United States Department of Agriculture, United States of America

Funding: This paper was supported by National Scientific and Technology Program of China (No. 2011BAD38B02) and CFERN & GENE Award Funds on Ecological paper. The funders had no role in study design, data collection and analysis, decision to publish, or preparation of the manuscript.

Competing Interests: The authors have declared that no competing interests exist.

* E-mail: bhx@bjfu.edu.cn

Introduction

The Loess Plateau is the birthplace of China's primitive agriculture. However, because of unsound land use and destruction of forests, the Loess Plateau has suffered serious soil erosion. At the same time, rapid population growth has also brought greater pressure to the environment in the region. The ensuing ecological and environmental problems have slowed down the economic development and living standards of local people. These problems lead to further deterioration of ecological environment, forming a vicious cycle. The local government is facing dual pressures from both economy and ecology.

Agroforestry systems have been considered as an effective practice to alleviate the conflicts between the rapidly growing population and the limited arable land resources [1,2]. In recent years, agroforestry management has been widely applied in the Loess Plateau region for reducing soil erosion and water loss, restoring ecological balance, raising land utilization rate and increasing economic benefits [3,4]. However, in most agroforestry systems, competition for light, moisture and nutrients exists at the interface between trees and crops which can cause a reduction of crop yield [5]. It is a major constraint that has affected stability of the structure and the function of the agricultural ecosystems. The competition between woody tree species and understory crop species not only exists aboveground (competition for light) but also comes from belowground (competition for soil moisture and nutrients), leading to lower crop yield. According to Friday and Fownes, the competition between trees and crops is overwhelmingly for light which is the main reason for the reduction of maize in alley cropping system in Hawii, USA [6]. Similar results were reported by Peng et al. in loess area of Weibei in Shaanxi Province, China [7]. Elsewhere in southern Australia, studies showed that reduced crop yields are associated with the competition for water in windbreak and alley systems [8,9]. Kowalchuk and Jong found that, especially in drought years, competition for water is the principal factor affecting the yield of spring wheat intercropped with shelterbelts in Western Saskatchewan [10]. In some related studies, the results indicated that competition for nutrients does not exist in intercropping systems [11–13]. However, others reported that as one of the main reasons leading to the reduction of crop yield, the competition for soil nutrients does exist in the interface of trees and crops and has a negative impact [14,15]. It is very important to explore the competitive mechanism in intercropping systems, in order to provide optimum management strategies and technologies for managing intercropping system with high-yield, high-efficiency and stabilization.

Table 1. Characteristics of apple trees intercropped with soybean and peanut in the experimental sites in July 2011.

Measurement	Intercropped with soybean	Intercropped with peanut
Tree height (m)	2.4	2.5
DBH (cm)	4.1	4.2
Depth of live crown (m)	1.7	1.8
Mean radius of crown (m)	1.3	1.2

Apple-crop intercropping system is one of the most commonly applied agroforestry systems in the Loess Plateau region owing to its good ecological, social and economic benefits. However, only few studies focused on this intercropping system in the area. In order to explore the biological reasons of the competition in typical intercropping systems and to provide effective management techniques, we report on a study of two apple-crop intercropping systems (apple-soybean, apple-peanut) on the Loess Plateau region in the western portion of Shanxi Province. The objectives of our research were (1) to analyze the interspecies competition relationship between trees and crops; (2) to find the limiting factors in the development of intercropping systems in this area; (3) to offer possible solutions to minimize the interspecies competitions and maximize resource utilization; (4) to enrich the related study and to improve the management of the intercropping systems in this region.

Materials and Methods

Study site

The study site was located in the Baidong Village, Jixian County, Shanxi Province, China (36°06′ N, 110°35′ E, 1025 m a.s.l.). The area is a typical hill and gully region of the Loess Plateau. The annual mean rainfall is about 575 mm, and the mean annual temperature is 10°C (1991–2010). The precipitation is unevenly distributed seasonally, with an average rainfall of 463 mm from June to August (1991–2010), which contributed about 80% of annual precipitation. The parent material of the soil is loess, and the soil properties are uniform. The bulk density, pH, total porosity, $CaCO_3$ content, cation exchange capacity, organic C, total N and available P of the top soil layer (100 cm) were 1.32 $Mg{\bullet}m^{-3}$, 8.24, 50.16%, 18.35%, 18.43 $cmol{\bullet}kg^{-1}$, 6.27 $g{\bullet}kg^{-1}$, 0.39 $g{\bullet}kg^{-1}$ and 4.39 $mg{\bullet}kg^{-1}$, respectively. The main intercropping tree species are Apple (*Malus pumila* M.), Apricot (*Prunus armeniaca* L.), Pear (*Pyrus bretschneideri* R.), Chinese arborvitae (*Platycladus orientalis* (L.) and Franco) and Black locust (*Robinia pseudoacacia* L.).

Ethics Statement

No specific permits were required for the described field studies. The sampling locations were not privately-owned or protected in any way and the field studies did not involve endangered or protected species.

Treatments and Crop Cultivation

Two typical intercropping systems of apple-soybean and apple-peanut were chosen for this study during the crop growing season of 2011 and 2012. The apple trees were planted in an East-West orientation in 2007. The characteristics of the apple trees intercropped with soybeans and peanuts in July 2011 are listed in Table 1. There were four treatments in this study: apple-soybean intercropping treatment (AS), soybean monoculture served as control (CS), apple-peanut intercropping treatment (AP) and peanut monoculture served as control (CP). Each treatment had three replicates. Each replicate of intercropping treatment (AS and AP) was an 8 × 10 m plot that included 12 trees planted in three rows with 4 m between trees and 5 m between rows. Each replicate of control treatment (CS and CP) was the same size of 8 × 10 m. For all treatments, the crops were planted at a spacing of 0.4 m with in rows and 0.5 m between rows and received the same agricultural management practices. Soybean and peanut were grown 0.3 m from an adjacent tree row in the intercropping systems. All plots received 147 kg N ha^{-1}, 30 kg P ha^{-1} and 30 kg K ha^{-1} as basal fertilizer and no additional fertilizer or irrigation in the rest of the year.

Measurements of Plant Photosynthesis, Soil Moisture and Nutrients

For the sampling of plant photosynthesis, soil moisture and soil nutrients, six sampling locations at distances of 0.5 m, 1.5 m and 2.5 m, respectively, from both side of tree row were identified as sampling points in each intercropping plot (Figure 1). The sampling points were further divided into three equal groups and denoted as F0.5, F1.5 and F2.5 based on the distance (0.5 m, 1.5 m and 2.5 m) from the tree row. Measurement parameters of F0.5, F1.5 and F2.5 were used to represent the major locations of 0.5 m, 1.5 m and 2.5 m away from apple tree row. For each control plot, five selected points were established with an S-shaped sampling method.

Photosynthetically active radiation (PAR) and net photosynthetic rate (NPR) of crops were performed by two portable Li-6400 photosynthesis systems which had a 6 cm^2 clamp-on leaf chamber connected to the main engine (Li-6400, Li-Cor Inc., Lincoln, NE, USA) under ambient humidity, temperature and irradiance. One fully expanded leaf from the upper part of the crop canopy in each sampling point was selected and measured five times with 2 h intervals during daytime (0900–1700 h). During each measurement period, all sampling points of intercropping treatment and control treatment were visited. These treatments were measured in mid-August 2011 and again in late August 2012, the typical phenological phases of peanut and soybean. For all measurements, the flow velocity was set at 500 $\mu mol{\bullet}s^{-1}$ and the airstream entering the chambers was kept at the growth CO_2 concentration (370 $\mu mol{\bullet}mol^{-1}$) by a computer-controlled CO_2 injector system supplied with Li-6400. PAR and CO_2/H_2O exchanged by the leaf were measured concurrently with the quantum sensor and the infrared gas analyzer on LI-6400. The data were recorded and calculated automatically with the software in the photosynthesis system.

For soil moisture, the samples were taken at different phenological phases of soybean and peanut: 8 July, 23 August, and 23 September in 2011; and 4 July, 11 August, and 22 September in 2012. A drill was used to remove the soil from 0–100 cm in 20 cm intervals in soil profile. The soil moisture content

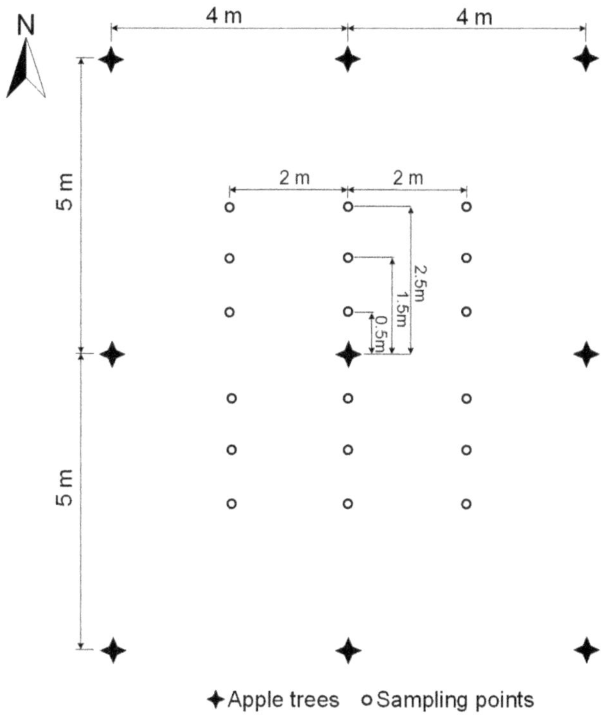

Figure 1. Sampling points of plant photosynthesis, soil moisture and nutrients in the intercropping study sites.

was determined gravimetrically in each layer. The mean soil moisture content of the five layers in all sampling time (2011 and 2012) was calculated and used as the final value of the sampling point.

For the sampling of soil nutrients, the soil samples were taken on 23 August, 2011 and 11 August, 2012, the typical phenological phases of peanut and soybean. The soil samples were collected from a depth of 0–100 cm in soil profile with a drill. Organic matter content was determined by H_2SO_4-$K_2Cr_2O_7$ pyrogenation. Total N was determined using the Kjeldahl method, with a KDY-9830 N Analyzer. Available P was determined by Olsen sodium-bicarbonate extraction. Available K was determined by flame photometer.

Measurements of Crop Growth and Yields

For the sampling of crop growth, we used the same sample locations as for soil moisture. A single crop plant was sampled at each sampling point on 24 August 2011 and 12 August 2012. A total of 69 soybean plants and 69 peanut plants were harvested during each measurement period. In the lab, plant height, hundred leaf dry weight and total above-ground biomass of all plants were measured and recorded.

At the end of the growing season, in each intercropping plot, peanuts and soybeans were harvested from both sides of the tree row in two rectangular areas. The rectangular area was 4.0 m long and 2.7 m wide. As the convenience of the study, the two rectangular areas were divided into three groups: (1) the area of 0.3–1.0 m away from tree row; (2) the area of 1.0–2.0 m away from tree row; and (3) the area of 2.0–3.0 m away from tree row. The yields of the three groups were used to represent the crop yield of F0.5, F1.5 and F2.5, respectively. In the control plots, 2 m × 2 m quadrates of soybean and peanut were harvested to get the grain production. The peanuts and soybeans were dried at 70 °C and then weighed to obtain an average dry weight. Yield values were reported on a per hectare basis.

Data analysis

All parameters (PAR, NPR, soil moisture, soil nutrients content, crop growth and yields) measured for control treatments and three major locations (F0.5, F1.5 and F2.5) of intercropping treatments were described in terms of mean values followed by respective standard deviations. Simple regression analysis was used to examine the relationships between the data of PAR, NPR, soil moisture and the distance from the tree row. Differences among groups for each crop (soybean or peanut) were determined by one-way ANOVA, and the results of the multiple comparisons were performed with least significant difference (LSD) test at $P<0.05$. NPR, total above-ground biomass and yield values of soybean and peanut had a correlated analysis with environmental parameters to decide the effect of apple trees competition on crop growth and productivity via bivariate correlation (Pearson) analysis at $P<0.05$ and $P<0.01$. All the analyses were performed by using the software IBM SPSS Statistics 20.0 for Windows.

Results

Light Interception and Plant Photosynthesis

For both crops, diurnal variation of photosynthetically active radiation (PAR) in the intercropping systems and the monoculture configurations (control treatments) showed a single peak curve with time (Figure 2). The peak of PAR appeared at 13:00 pm and the minimum value appeared in 17:00 pm. Because of reflectance, absorbance and transmittance by the apple tree canopy, the PAR of crops in the intercropping systems were lower than that in the monoculture configurations during the same period. On the horizontal distribution, the general trend was that the closer the crops to the tree rows, the lower the PAR received. The same tendency was found in diurnal variation of net photosynthetic rate (Figure 3).

The daily mean values of PAR showed a clear linear relationship with distance from the apple tree row in both intercropping treatments (Figure 4A). The trend lines of PAR (Y, μmol•s^{-1}•m^{-2}) and distance from trees rows (X, m) were $Y = 78.5\times+865.3$ ($R^2 = 0.999$) in apple-soybean intercropping treatment (AS) and $Y = 82.5\times+881.9$ ($R^2 = 0.873$) in apple-peanut intercropping treatment (AP). The slopes of both regression lines suggested that the PAR in AP treatment had a higher growth than that in AS treatment as the distance from the tree increased. As shown in Figure 4A, PAR reaching the upper parts of the crop canopy in AP treatment also had higher values than that in AS treatment at the same distance away from the tree row. It indicated that peanut canopy could obtain more solar radiation in AP treatment than soybean canopy in AS treatment. At confidence level of 95%, the control treatment PAR mean fell within the confidence intervals of F2.5 in the corresponding intercropping systems. Compared with the corresponding control treatment, PAR at F0.5 and F1.5 showed a reduction of 17.9% and 10.4% in AS treatment, respectively, 17.8% and 5.4% in AP treatment. Similar linear relationships were also obtained through regression analysis of the relationship between NPR and distance from the apple tree row (Figure 4B). The trend lines of NPR (Y, μmol•s^{-1}•m^{-2}) and distance from trees rows (X, m) were $Y = 1.025\times+12.003$ ($R^2 = 0.902$) in AS treatment, and $Y = 0.940\times+10.983$ ($R^2 = 0.951$) in AP treatment. The NPR in AS treatment had higher values and growth than that in AP treatment as the distance from the tree increased which was different from the measurement of PAR. The control treatments

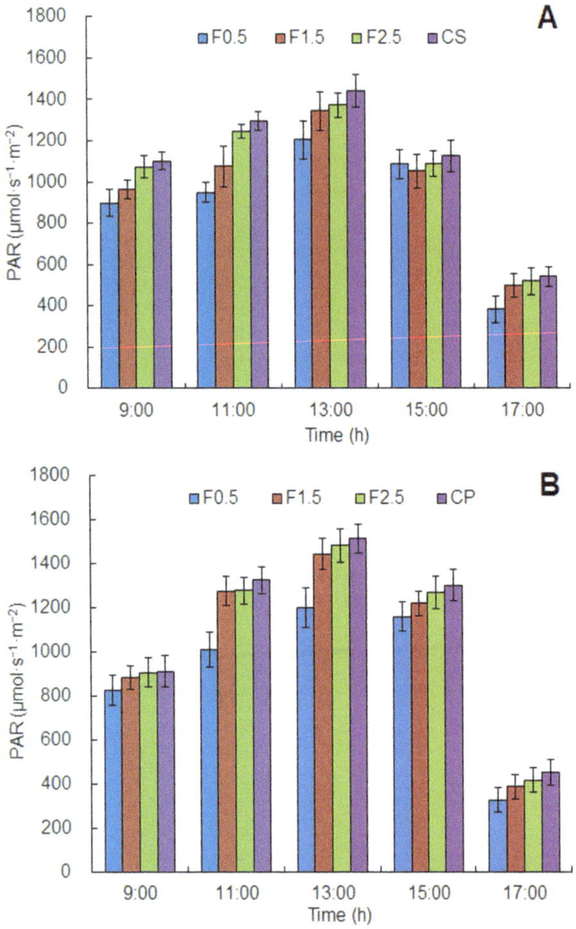

Figure 2. Diurnal variation of photosynthetically active radiation (PAR) for the intercropping systems and its control (A. apple-soybean and B. apple-peanut). F0.5, F1.5 and F2.5 were used to represent the sampling points which had different distance (0.5 m, 1.5 m and 2.5 m) from the tree row. Error bars indicate standard deviation.

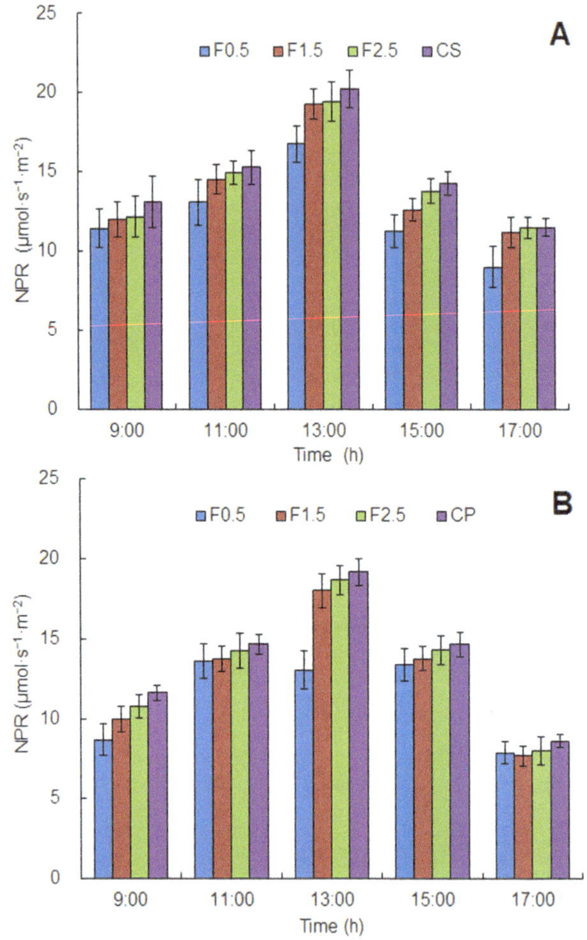

Figure 3. Diurnal variation of net photosynthetic rate (NPR) for the intercropping systems and its control (A. apple-soybean and B. apple-peanut). F0.5, F1.5 and F2.5 were used to represent the sampling points which had different distance (0.5 m, 1.5 m and 2.5 m) from the tree row. Error bars indicate standard deviation.

mean fell within the confidence intervals of F2.5 in the corresponding intercropping systems at confidence level of 95%.

Spatial Distribution of Soil Moisture

Although the soil moisture content in the whole soil profile (0 to 100 cm in depth) in AS was different from AP, the trend of spatial distributions of soil moisture was similar (Figure 5). Soil moisture content in AS was related to distance from the apple tree row and showed a clear linear relationship ($Y = 0.465x + 11.602$, $R^2 = 0.999$), and AP showed the same trend ($Y = 0.590x + 11.002$, $R^2 = 0.900$). Compared with AP, AS had higher values at the same distance away from the tree row. However, with increasing distance from the tree row, soil moisture in AP had a higher growth than that in AS. The lowest soil moisture content was 11.83% in AS and 11.41% in AP, showed a decrease of 10.31% and 11.14% when compared with the corresponding control treatments. The soil moisture at F2.5 in both intercropping systems also had slightly lower values than that in monoculture configurations, however no difference was observed at significance level of 5%, since the control treatments mean fell within the confidence intervals of F2.5 in the corresponding intercropping systems (confidence level 95%). Otherwise, the average soil moisture content in AP was lower than that in AS.

Spatial Distribution of Soil Nutrients

The soil nutrients content in the 0 to 100 cm interval was calculated (Table 2). It represented that organic matter, total N, available P and available K in AS had different degrees of reduction when compared with CS, and showed significant differences ($P<0.05$). Similar results were found between AP and CP, except that no significant difference was observed for total N and available K at the location of F2.5. The average content of organic matter, total N, available P and available K in AS decreased by 30.77%, 63.24%, 56.08% and 27.83% when compared with CS–the monoculture configuration. For AP and CP, the decreased percentages were 18.32%, 21.05%, 36.27% and 7.49% respectively. In addition, except available K, soil nutrients content in AP was higher than that in AS at the same spatial location. With the increasing distance from tree row, the distribution trend of soil nutrients was different from that of PAR, NPR or soil moisture in the same intercropping condition. The lowest content of organic matter, total N, available P and available K in AS were present at the location of F1.5. The similar

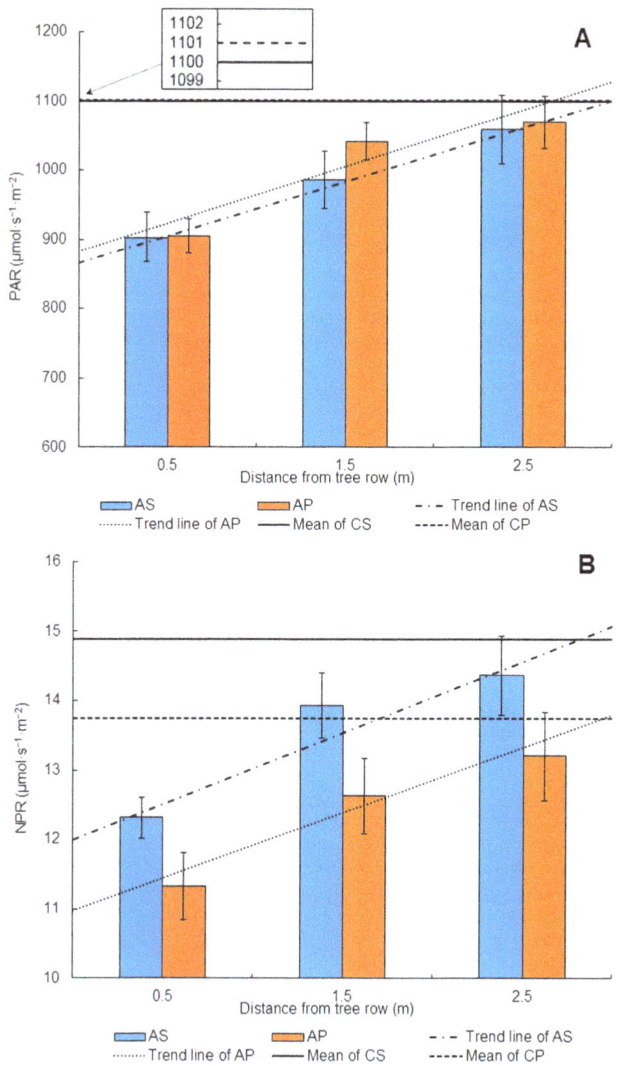

Figure 4. Daily mean of photosynthetically active radiation (PAR) and net photosynthetic rate (NPR) for the intercropping systems and its control (A. PAR and B. NPR). Vertical lines indicate confidence interval at 95% level.

result was found in AP, except that the lowest value of available K was the location of F0.5.

Crop Growth and Yields

Plant height, hundred leaf dry weight and total above-ground biomass in both intercropping systems had lower values when compared with the monoculture configuration ($P<0.05$; Table 3). The locations of F0.5 and F1.5 in all these parameters showed significant differences with corresponding monoculture configuration ($P<0.05$); however, there were no difference observed in the location of F2.5.

The yield of soybean in AS was significantly related to the distance from the row of apple trees ($Y=0.180x+1.400$, $R^2=0.991$), and the yield of peanut in AP showed the same trend ($Y=0.095x+1.554$, $R^2=0.900$) which showed that yield of soybean had greater impacted by distance from the tree row. The yields at F0.5 and F1.5 in AS were lower than that in CS ($P<0.05$), with a reduction of 22.45% and 11.95%, and in AP the yields of reduction were 13.31% and 11.03% when compared with CP. No differences were observed between the locations of F2.5 in both intercropping systems and the corresponding monoculture configuration ($P<0.05$).

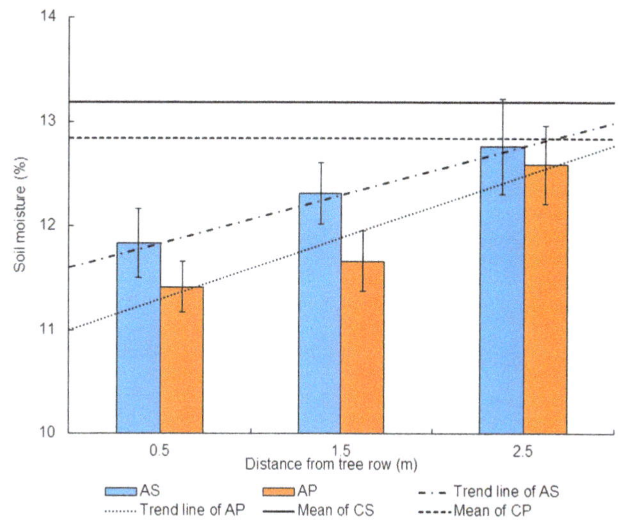

Figure 5. Soil moisture of 0 to 100 cm depth for the intercropping systems and its control. Vertical lines indicate confidence interval at 95% level.

Within plot differences in these parameters were significantly correlated (Table 4). NPR was highly correlated with PAR, and soil moisture. The total above-ground biomass of soybean was highly correlated with PAR, soil moisture and available P. The total above-ground biomass of peanut was correlated with PAR, soil moisture, total N and available P. Yield of soybean was highly correlated with PAR, soil moisture, available P and total N, with a trend of soil moisture >PAR>available P>total N. For peanut, the trend was soil moisture > PAR > total N > available P. It showed that, for both of the intercropping system in the study region, the primary factor affecting the yield is soil moisture, and the secondary factor is photosynthetically active radiation, and soil nutrient also have an impact on crop yield in some depth.

Discussion

Agroforestry system has been studied for a long time and has been widely used in the agricultural production practices in China [3,16,17]. However, there has been little research done on the agroforestry system in the Loess Plateau region. The main intercropping models which have been studied are always walnut-wheat and apple-wheat [18–20]. The types of fruit trees intercropping with economic crops such as soybean and peanut has not been well studied. In fact, compared with wheat, soybean and peanut could bring more economic income to farmers. At the same time, these two crops could be rotated with wheat in order to improving land-use efficiency, and re-establishing the economic viability of the Loess Plateau.

Our study observed a clearly positive linear relationship between distance from the apple tree rows and the daily mean values of PAR and NPR in the intercropping systems. For both apple-crop intercropping systems, the shading of the 4–5 years old apple trees had a significant negative effect on the crops in the range of 1.5 m away from the tree rows and further caused the reduction of crop yield. In other researches of temperate agroforestry systems, the similar results were reported by Reynolds et al. [21] about maize and soybean intercropped with poplar and

Table 2. Soil nutrients for the intercropping systems and control configurations.

Measurement	AS			CS	AP			CP
	F0.5	F1.5	F2.5		F0.5	F1.5	F2.5	
Organic matter (g•kg^{-1})	4.63±0.27a	3.40±0.29b	4.93±0.27a	6.24±0.31c	6.26±0.36a	5.21±0.26b	6.37±0.36a	7.28±0.27c
Total N (g•kg^{-1})	0.28±0.05a	0.22±0.04b	0.25±0.04ab	0.68±0.03c	0.31±0.06a	0.24±0.05b	0.35±0.07ac	0.38±0.04c
Available P (mg•kg^{-1})	2.82±0.56a	2.50±0.54a	3.93±0.31b	7.02±0.22c	5.06±0.61a	3.44±0.44b	4.75±0.42a	6.93±0.33c
Available K (mg•kg^{-1})	97.33±1.77a	84.02±2.09b	95.43±2.97a	127.84±2.75c	87.45±2.81a	90.73±2.45a	98.11±2.42b	99.55±2.19b

Data were given as the means ± SD.
Different lowercase letters within a row of each crop indicate significant differences (LSD, $P<0.05$).

silver maple in Canada and Peng et al. [22] about mungbean and pepper intercropped with walnut and plum in Weibei area, China. For total above-ground biomass and yield of both crops, PAR of soybean had higher correlations than that of peanut, which indicated that soybean is more adversely impacted by tree shading. Within tree-based intercropping systems, many factors such as tree species, tree height, crown shape, tree row orientation and distance between tree rows can influence tree shading of adjoining agricultural crops. Light reduction would depend on the extent and duration of the shade of trees [21]. Regular pruning of fruit trees could reduce light competition within the intercropping system, improving crop yields.

In semiarid and arid regions, it is still a focus of studies whether intercropping system has an overall negative or positive effect on soil moisture [23]. In some related studies, it was considered that the trees can improve soil moisture holistic conditions in intercropping systems [17,24]. In other studies, the opposite results were reported [11,12,25,26]. However, little research has been carried out in this aspect on the Loess Plateau. Our research confirmed that the competition of water between trees and crops do exist, and showed adverse effects in the study site. A clear linear relationship was observed between the distance from the tree row and soil moisture in both of the intercropping systems. The closer to the tree row, the more intense the competition. The lowest soil moisture content in apple-soybean intercropping system and apple-peanut intercropping system showed a reduction of 10.31% and 11.14%, respectively. Only considering competition of water, the mainly affected region of the apple trees was 1.5 m away from the tree rows under the current tree age.

Another key factor of crop growth is soil nutrients in the intercropping systems. Elsewhere, Thomas et al. [27] and Thevathasan et al. [13] have reported that competition for nutrients in intercropping systems does not exist. In our study, it identified that there were competition for soil nutrients between trees and crops in the intercropping systems. The average content of organic matter, total N, available P and available K showed different degrees of reduction in both of the apple-crops intercropping systems than that of the corresponding control treatments. In particular, total N and available P had higher reduction rate than organic matter and available K, and had significantly correlation with yield of crops. As leguminous plants, soybean and peanut could fix nitrogen from the air via a symbiotic relationship with rhizobium bacteria and increase the mineral soil nitrogen content [28,29]. However, the nitrogen coming from biologically fixed N_2 of symbiosis could not meet all the demand of crops growth, and any gaps between N supply by N_2 fixation and crop N demand must be met by N uptake from soil [30]. The deficiency of light and water in the intercropping systems reduced the physiological activity of the crop, and then affected the N fixation capacity, resulting more intense competition for nitrogen between trees and crops. Compared with soybean and peanut, the growth of other non-nitrogen-fixing crop species (i.e. wheat, maize and millet) would be more severely affected because of nitrogen deficiency in the intercropping systems. Different from understory light distribution and soil moisture, soil nutrients had a different variation pattern in both intercropping systems. The main reasons for this phenomenon might be: (1) the crops close to tree row were seriously affected by tree shading, soil moisture stress and human activities, resulting in low physiological activity and low absorption

Table 3. Crop growth, biomass and yield for soybean and peanut intercropped with apple trees and its control.

Measurement	AS			CS	AP			CP
	F0.5	F1.5	F2.5		F0.5	F1.5	F2.5	
Crop height (cm)	43.7±2.8a	44.4±2.3a	47.7±3.2b	50.2±2.6b	19.1±0.7a	21.9±1.0b	22.6±0.9bc	23.5±0.8c
Hundred leaf dry weight (g)	14.12±0.81a	14.14±0.51a	14.95±0.38b	15.37±0.57b	4.39±0.23a	5.79±0.22b	5.90±0.31bc	6.14±0.28c
Total above-ground biomass (g)	54.55±4.66a	66.06±3.81b	76.96±4.30c	79.76±4.22c	50.10±2.62a	51.57±2.45a	79.16±3.28b	79.65±2.46b
Yield (t/ha)	1.48±0.06a	1.69±0.04b	1.84±0.06c	1.91±0.04c	1.62±0.04a	1.66±0.04a	1.81±0.05b	1.86±0.04b

Data were given as the means ± SD.
Different lowercase letters within a row of each crop indicate significant differences (LSD, $P<0.05$).

Table 4. Correlations of soybean and peanut net photosynthetic rate, biomass, and yield with environmental or physiological parameters measured in Jixian, China.

Independent variable	NPR ($\mu mol \cdot s^{-1} \cdot m^{-2}$)	Total above-ground biomass (g)	Yield (kg/ha)
Soybean			
PAR ($\mu mol \cdot s^{-1} \cdot m^{-2}$)	0.973**	0.996**	0.952**
Soil moisture (%)	0.953**	0.977**	0.957**
Organic matter ($g \cdot kg^{-1}$)	0.441	0.575	0.566
Total N ($g \cdot kg^{-1}$)	0.537	0.555	0.601*
Available P ($mg \cdot kg^{-1}$)	0.697*	0.750**	0.763**
Available K ($mg \cdot kg^{-1}$)	0.469	0.538	0.565
Peanut			
PAR ($\mu mol \cdot s^{-1} \cdot m^{-2}$)	0.986**	0.773**	0.816**
Soil moisture (%)	0.926**	0.965**	0.959**
Organic matter ($g \cdot kg^{-1}$)	0.450	0.562	0.583
Total N ($g \cdot kg^{-1}$)	0.513	0.843**	0.770**
Available P ($mg \cdot kg^{-1}$)	0.424	0.628*	0.646*
Available K ($mg \cdot kg^{-1}$)	0.479	0.531	0.446

*Significant at 5% level.
**Significant at 1% level.

of soil nutrients; (2) the decomposition of tree litter leaded to high nutrients content in the area near the tree row; (3) the overlapping of apple tree roots and crop roots resulting in lower nutrient content at F1.5; (4) the tree roots reduced with the increase of the distance from the tree, therefore, the soil nutrients had relatively high content F2.5. Therefore, in the area of 1.5 m away from tree row, strengthen the application of fertilizer (especially nitrogen and phosphorus) would be helpful to alleviate interspecific competition for soil nutrients.

In the apple-crop intercropping systems, the competition of light, water and nutrients resulted in a greater negative impact on crop growth and yields. For the two apple-crop intercropping systems in our study, the primary factor affecting the crop yield was soil moisture, and the secondary factor was light, and deficiency of the soil nutrient also had a negative impact on crop yields. In the same study area, Yun et al. reported a similar research with a different conclusion: the light is the primary limiting factor leading to reduction of crops, followed by soil moisture [31]. In their research, the apple trees had greater crown width, canopy density and root depth due to elder age (9-year-old) and smaller tree spacing (3 m×4 m). Affected by the impact of canopy structure, the obvious microclimate effect inhibited evapotranspiration of soil moisture to some extent [32], in the same time, the low transmittance led to more intense light stress to crops. Furthermore, the effect of hydraulic lift by tree roots also alleviated the interspecific competition for soil water [24]. Combined with these reasons, different results were found. For different intercropping patterns and tree ages, the intensity of competition for resources would be different in the intercropping system. Therefore, a long-term observation should be carried out in this region to obtain more details about the mechanism of interspecific competition in the intercropping systems. In our study, under the current tree age and growth conditions, the influence scope of the apple trees was 1.5 m away from the tree rows. Compared with the corresponding monoculture configuration, the yield of peanut in the intercropping system had a lower reduction than that of soybean. With comprehensive consideration, peanut is more suitable for intercropping with apple trees in this region.

As we have demonstrated in this study, soil moisture, light and soil nutrients were the limiting factors of crop yield. In order to obtain more production, appropriate management measures were needed to minimize competition between trees and crops. Namirembe [33] and Friday [6] have suggested that the competition for light between trees and crops could be alleviated by pruning of trees crown and increasing the intercropping distance. In general, the aboveground competition could be intuitively observed and managed. However, the competition belowground is invisible and easily ignored by farmers or managers. To avoid these yield losses, root barrier in the intercropping interface is considered to be a useful agricultural management practice according to some related studies [12,34,35]. Combined with their research achievements and our experiment result, we offered several specific recommendations to reduce the competition exist in apple-crop intercropping systems: (1) the selection of crop varieties which is more suitable for apple-crop intercropping systems; (2) appropriate distance increase between the crops and apple tree rows; (3) regular pruning of fruit trees, in order to increase canopy light transmittance rate; (4) additional fertilization and irrigation in the key phenological phase of the crops; (5) differences of irrigation and fertilization based on the distance from the apple trees. Management measures such as plastic film and straw mulching have been widely used in agricultural production. Whether these measures have overall positive effects on intercropping system would be one of the focus of our future research work in this region.

Conclusions

As an effective method to increase the efficiency of land use and economic returns, tree-based intercropping systems are particularly important on Loess Plateau. We concluded that the competitions exist both above-ground and below-ground between apple trees and crops. The competition for soil moisture is the primary limiting factor for the crop productivity in this region. Furthermore, the tree shading and the competition for soil nutrients in the interface of trees and crops also have a negative impact on the understory crops. However, it could be minimized by better agricultural technology and management measures.

In summary, our study suggests that there is great potential for intercropping systems in the Loess Plateau. Therefore, in order to relieve the shortage of arable land and promote the sustainable development of natural resources, the intercropping systems would continue to be the hot spot for future research. Canopy structure, roots distribution of trees, the application of different agronomic measures and the role they play in the competition process in the intercropping systems will be the focus of our future research.

Acknowledgments

We are grateful for the support from the Shanxi Jixian Forest Ecosystem Research Station. We also would like to thank the three anonymous reviewers and the editors for their helpful comments.

Author Contributions

Conceived and designed the experiments: LG HX HB BB. Performed the experiments: LG HX HB BB. Analyzed the data: LG HX WX BB XW CB YC. Contributed reagents/materials/analysis tools: XW CB YC. Wrote the paper: LG HB WX.

References

1. Burel F (1996) Hedgerows and their role in agricultural landscapes. Critical Reviews in Plant Sciences 15: 169–190.
2. Gene Garrett HE, Buck L (1997) Agroforestry practice and policy in the United States of America. Forest Ecology and Management 91: 5–15.
3. Li W, Lai S (1994) Agroforestry in China. Beijing: Chinese Science Press. pp. 14–18.
4. Zhu Q, Zhu J (2003) Sustainable management technology for conversion of cropland to forest in loess area. Beijing:Chinese Forestry Press. pp. 160–165.
5. Ong CK, Huxley P (1996) Tree-crop interactions: a physiological approach. Wallingford:CAB International. pp. 386.
6. Friday JB, Fownes JH (2002) Competition for light between hedgerows and maize in an alley cropping system in Hawaii, USA. Agroforestry Systems 55: 125–137.
7. Peng X, Zhang Y, Cai J, Jiang Z, Zhang S (2009) Photosynthesis, growth and yield of soybean and maize in a tree-based agroforestry intercropping system on the Loess Plateau. Agroforestry Systems 76: 569–577.
8. Hall DJM, Sudmeyer RA, McLernon CK, Short RJ (2002) Characterisation of a windbreak system on the south coast of Western Australia. 3. Soil water and hydrology. Australian Journal of Experimental Agriculture 42: 729–738.
9. Unkovich M, Blott K, Knight A, Mock I, Rab A, et al. (2003) Water use, competition, and crop production in low rainfall, alley farming systems of south-eastern Australia. Australian Journal of Agricultural Research 54: 751–762.
10. Kowalchuk TE, Jong E (1995) Shelterbelts and their effect on crop yield. Canadian Journal of Soil Science 75: 543–550.
11. Jose S, Gillespie AR, Seifert JR, Biehle DJ (2000) Defining competition vectors in a temperate alley cropping system in the midwestern USA: 2. Competition for water. Agroforestry Systems 48: 41–59.
12. Miller AW, Pallardy SG (2001) Resource competition across the crop-tree interface in a maize-silver maple temperate alley cropping stand in Missouri. Agroforestry Systems 53: 247–259.
13. Thevathasan NV, Gordon AM, Simpson JA, Reynolds PE, Price G, et al. (2004) Biophysical and ecological interactions in a temperate tree-based intercropping system. Journal of Crop Improvement 12: 339–363.
14. Newman SM, Bennett K, Wu Y (1997) Performance of maize, beans and ginger as intercrops in Paulownia plantations in China. Agroforestry Systems 39: 23–30.
15. Yun L, Bi H, Gao L, Zhu Q, Ma W, et al. (2012) Soil moisture and soil nutrient content in walnut-crop intercropping systems in the Loess Plateau of China. Arid Land Research and Management 26: 285–296.
16. Meng P, Zhang J, Fan W (2003) Research on agroforestry in china. Beijing:Chinese Forestry Press. pp.235.
17. Meng P, Zhang J (2004) Effects of pear-wheat inter-cropping on water and land utilization efficiency. Forest Research 17: 167–171.
18. Zhang J, Meng P, Yin C (2002) Spatial distribution characteristics of apple tree roots in the apple-wheat intercropping. Scientia Silvae Sinicae 38: 30–33.
19. Zhang J, Meng P (2004) Model on wheat potential evapotranspiration in apple-wheat intercropping. Forest Research 17: 284–290.
20. Yun L, Bi H, Ren Y, Ma W, Tian X (2009) Soil moisture distribution at fruit-crop intercropping boundary in the Loess region of Western Shanxi. Journal of Northeast Forestry University 37: 70–78.
21. Reynolds PE, Simpson JA, Thevathasan NV, Gordon AM (2007) Effects of tree competition on corn and soybean photosynthesis, growth, and yield in a temperate tree-based agroforestry intercropping system in southern Ontario, Canada. Ecological Engineering 29: 362–371.
22. Peng X, Cai J, Jiang Z, Zhang Y, Zhang S (2008) Light competition and productivity of agroforestry system in loess area of Weibei in Shaanxi. Chinese Journal of Applied Ecology 19: 2414–2419.
23. Zhang J, Meng P, Yin C, Cui G (2003) Summary on the water ecological characteristics of agroforestry system. World Forestry Research 16: 10–14.
24. Hirota I, Sakuratani T, Sato T, Higuchi H, Nawata E (2004) A split-root apparatus for examining the effects of hydraulic lift by trees on the water status of neighbouring crops. Agroforestry Systems 60: 181–187.
25. Lott JE, Howard SB, Ong CK, Black CR (2000) Long-term productivity of a Grevillea robusta-based overstorey agroforestry system in semi-arid Kenya: II. Crop growth and system performance. Forest Ecology and Management 139: 187–201.
26. Lehmann J, Peter I, Steglich C, Gebauer G, Huwe B, et al. (1998) Below-ground interactions in dryland agroforestry. Forest Ecology and Management 111:157–169
27. Thomas J, Kumar BM, Wahid PA, Kamalam NV, Fisher RF (1998) Root competition for phosphorus between ginger and Ailanthus triphysa in Kerala, India. Agroforestry Systems 41: 293–305.
28. Cheng D (1994) Resource microbiology. Harbin:Northeast Forestry University Press. pp. 32.
29. Wani SP, Rupela OP, Lee KK (1995) Sustainable agriculture in the semi-arid tropics through biological nitrogen fixation in grain legumes. Plant Soil 174: 29–49.
30. Salvagiotti F, Cassman KG, Specht J E, Walters DT, Weiss A, et al. (2008). Nitrogen uptake, fixation and response to fertilizer N in soybeans: A review. Field Crops Research 108:1–13.
31. Yun L, Bi H, Tian X, Cui Z, Zhou H, et al. (2011) Main interspecific competition and land productivity of fruit-crop intercropping in Loess Region of West Shanxi. Chinese Journal of Applied Ecology 22:1225–1232
32. Zhang J, Meng P, Song Z, Gao J (2004) An overview on micro-climatic effects of agro-forestry systems in plain agricultural areas in China. Agricultural Meteorology 25:52–55
33. Namirembe S (1999) Tree management and resource utilization in agroforestry systems with Senna spectabilis in the drylands of Kenya. Bangor:University of Wales. pp. 206.
34. Singh RP, Saharan N, Ong CK (1989) Above and below ground interactions in alley-cropping in semi-arid India. Agroforestry Systems 9: 259–274.
35. Hou Q, Brandle J, Hubbard K, Schoeneberger M, Nieto C, et al. (2003) Alteration of soil water content consequent to root-pruning at a windbreak/crop interface in Nebraska, USA. Agroforestry Systems 57: 137–147.

12

Impacts of *Carpobrotus edulis* (L.) N.E.Br. on the Germination, Establishment and Survival of Native Plants: A Clue for Assessing Its Competitive Strength

Ana Novoa*¤, Luís González

Departamento de Bioloxía Vexetal e Ciencias do Solo, Facultade de Bioloxía, Universidade de Vigo, Vigo, Spain

Abstract

Does *Carpobrotus edulis* have an impact on native plants? How do *C. edulis'* soil residual effects affect the maintenance of native populations? What is the extent of interspecific competition in its invasion process? In order to answer those questions, we established pure and mixed cultures of native species and *C. edulis* on soil collected from invaded and native areas of Mediterranean coastal dunes in the Iberian Peninsula. We examined the impact of the invader on the germination, growth and survival of seeds and adult plants of two native plant species (*Malcolmia littorea* (L.) R.Br, and *Scabiosa atropurpurea* L.) growing with ramets or seeds of *C. edulis*. Residual effects of *C. edulis* on soils affected the germination process and early growth of native plants in different ways, depending on plant species and density. Interspecific competition significantly reduced the germination and early growth of native plants but this result was soil, density, timing and plant species dependent. Also, at any density of adult individuals of *C. edulis*, established native adult plants were not competitive. Moreover, ramets of *C. edulis* had a lethal effect on native plants, which died in a short period of time. Even the presence of *C. edulis* seedlings prevents the recruitment of native species. In conclusion, *C. edulis* have strong negative impacts on the germination, growth and survival of the native species *M. littorea* and *S. atropurpurea*. These impacts were highly depended on the development stages of native and invasive plants. Our findings are crucial for new strategies of biodiversity conservation in coastal habitats.

Editor: Meng-xiang Sun, Wuhan University, China

Funding: The authors received no specific funding for this work.

Competing Interests: The authors have declared that no competing interests exist.

* Email: ANAN@sun.ac.za

¤ Current address: Centre for Invasion Biology, Department of Botany and Zoology, Stellenbosch University, Stellenbosch, South Africa

Introduction

Invasive alien plants are considered as one of the greatest threats to the diversity, structure and functioning of natural ecosystems around the world [1,2]. They can exert significant impacts on many ecological variables [3]. In particular, Mediterranean coastal dune ecosystems are highly sensitive to invasion by exotic plants since the disruption caused by the movement of sand constantly produces open spaces that are susceptible to colonization by alien species [4]. In those open spaces, competitive interactions between invasive and native species are extremely important [5]. As Mediterranean coastal dune ecosystems present a high cultural and ecological value, and support many threatened and endemic species [6], efforts on their alien species management strategies are critically needed.

One of the major invaders of Mediterranean coastal dune ecosystems is *Carpobrotus edulis* (L.) N.E.Br., a perennial clone plant native to South Africa [7–9]. Since its introduction, this invasive succulent now dominates millions of kilometres of Mediterranean dune ecosystems, leading to loss of species and irreversible changes on the substrate [7–10]. However, there is little information about the competitive interactions between this invasive species with native plants, even though it is a crucial aspect for prioritizing Mediterranean coastal dune ecosystems management.

Impacts of invasive species through competition with native plants is a primary ecological process limiting restoration success [11]. The phenologic stage of each species is decisive in these competitive relationships [12]: plants pass through different physiological stages and their development processes and competition occurs not only within species, but also within and between stages of different species [13]. However, most competition studies are focused on plants at the same development stage.

This study deals with the relative competitive ability of native seeds, seedlings and adult plants. It was designed to address the following questions: (i) the presence of *C. edulis* causes changes to soil characteristics (see Novoa et al. (2013a) for further information and detailed data). How do these changes affect the maintenance of native populations during each development stage? And (ii) To what extent are the ecological impacts caused by *C. edulis* based on different development stages of native plants? Despite the fact that the plant invasion process is a result of multiple interacting factors [14,15], to the best of our knowledge this study is the first reporting simultaneous examination of multiple mechanisms: competition, density, timing of sowing, plant developmental stage

Figure 1. Collection of adult plants of *C. edulis*.

and residual soil effects on the limitation of native flora by an invasive plant.

Moreover, since *C. edulis* can reproduce both vegetatively and sexually [16], understanding the competitive relationships established between clones or seeds of *C. edulis* and native species during the germination, establishment and growth processes are crucial for the conservation of the high biodiversity of Mediterranean coastal habitats [17].

Materials and Methods

1. Plant material

The two main dominant native species in the study area were selected: a typical semi-fixed dunes species (*Malcolmia littorea* (L.) R.Br.) and a common fixed-dune and rocky species (*Scabiosa atropurpurea* L.) both of them Chamaephytes. Seeds of the native species *M. littorea* and *S. atropurpurea* and the invasive *C. edulis* were collected between 10 Sep and 10 Oct 2011 from at least 15 plants from 20 different populations of each species located along 20 km of the coast of Pontevedra, Spain (between 42°29′56.17″N 8°52′16.22″O and 42°20′16.22″N 8°49′41.17″O). The land accessed is not privately owned or protected. The seeds were separated from the rest of the fruit and its accessory dispersion parts and stored in the dark at 4°C until assay. Seeds were surface-sterilized for 5 min in 0.1% sodium hypochlorite, rinsed 3 times in distilled water and dried at room temperature.

On the 19 Nov 2010, adult plants of native species (*M. littorea* and *S. atropurpurea*) and apical ramets of the invasive *C. edulis* -as well as a volume of 1 L of sand around each plant- were collected in the same area and immediately transplanted to 1 L sand pots for greenhouse acclimatization, integrating the plant stock. As *C. edulis* presents clonal growth, we could obtain individuals with approximately the same developmental state from the ends of the branches (Figure 1).

2. Soil collection

On the 19 Nov 2010, soil was collected up to 10 cm depth in those dunes from where the seeds, adult plants and ramets had previously been collected. The top soil layer from 20 randomly selected points (1×1 m) was collected in an area invaded by *C. edulis* and in an adjacent native area. The adjacent native area

Table 1. Methodological scheme for greenhouse experiment.

		Ramets of *C. edulis*				
		0	*1*	*2*	*3*	*4*
N° of native species seeds or seedlings	*0*	•				X
	10/1	•			X	
	15/2	•		X		
	20/3	•	X			
	25/4	X				

(X) Represent classic replacement series design. (•) Represent modified replacement series. N = 5.

Table 2. Methodological scheme for growth chamber experiment.

	Day 0	Day 5					Day 10
	5 days before sowing native species	*Sowing date of native seeds*					*5 days after sowing native species*
	Seeds number						
	C. edulis	**native**		*C. edulis*	**native**		*C. edulis*
Assay 1	10	+	10				
	10	+	30				
	30	+	10				
Assay 2		10	+	10			
		30	+	10			
		10	+	30			
Assay 3					10	+	10
					30	+	10
					10	+	30
Monocultures		10		10			
		30		30			

Assay 1: Interspecific competition, *C. edulis* seeds sowed before native seeds. Assay 2: Interspecific competition, *C. edulis* seeds sowed at the same time that native seeds. Assay 3: Interspecific competition, *C. edulis* seeds sowed after native seeds. Monocultures: intraspecific competition for native and *C. edulis* seeds. N = 5.

Table 3. Intraspecific competition effect on the germination indices (Gt: Total germination and AS: Cumulative rate of germination), seedling shoot and root length and survival of the native species *Malcolmia littorea* in both invaded and native soil.

	Invaded soil				Native soil			
Seed number	**10**	**15**	**20**	**25**	**10**	**15**	**20**	**25**
Gt	82.5^{a*}	81.7^{a*}	57.0^{b}	36.0^{c*}	51.2	62.3	53.0	65.6
	(9.8)	(9.6)	(6.8)	(5.7)	(4.0)	(3.4)	(6.4)	(3.4)
AS	105.8^{ab*}	112.9^{a}	84.3^{b*}	47.8^{c*}	67.2	82.7	54.8	88.5
	(14.1)	(10.5)	(6.3)	(3.8)	(8.6)	(10.7)	(12.6)	(43)
Survival (%)	71.0	94.1	82	77.9	72^{b}	81^{b}	79^{b}	92^{a}
	(12.7)	(11.9)	(14.3)	(17.6)	(2.4)	(9.6)	(6.3)	(2.8)
Shoot length (cm)	1.2*	1.5*	1.4*	1.3*	0.9	1.1	0.9	1.1
	(0.1)	(0.2)	(0.1)	(0.1)	(0.1)	(0.1)	(0.1)	(0.1)
Root length (cm)	2.3*	2.3*	2.3*	2.3*	1.4	1.6	1.4	1.5
	(0.2)	(0.3)	(0.2)	(0.2)	(0.1)	(0.1)	(0.1)	(0.1)

Different letters mean significant difference of 5% between treatments (10, 15, 20 or 25 seeds).*: indicates significant difference of 5% between invaded and native soils on each treatment. Numbers in parentheses indicate the standard error. N = 5.

Table 4. Intraspecific competition effect on the germination indices (Gt: Total germination and AS: Cumulative rate of germination), seedling shoot and root length and survival of the native species *Scabiosa atropurpurea* in both invaded and native soil.

	Invaded soil				Native soil			
Seed number	**10**	**15**	**20**	**25**	**10**	**15**	**20**	**25**
Gt	64.0	53.3	43.0	45.6	44.0	49.3	40.0	48.0
	(11.7)	(9.2)	(8.6)	(2.7)	(10.3)	(6.5)	(8.2)	(7.2)
AS	51.0	41.7	33.6	40.1	42.7	39.9	35.8	46.1
	(5.7)	(4.0)	(5.0)	(2.9)	(5.1)	(3.4)	(6.5)	(1.9)
Survival (%)	80.2	87.7	90.2	93.5	68.8	84.7	89.0	94.5
	(11.1)	(4.8)	(4.9)	(1.7)	(12.8)	(4.6)	(3.5)	(2.5)
Shoot length (cm)	0.9^{b}	1.2^{ab}	1.1^{ab}	1.4^{a}	0.79	1.1	1.0	1.2
	(0.1)	(0.2)	(0.2)	(0.2)	(0.1)	(0.2)	(0.1)	(0.2)
Root length (cm)	5.3^{b}	5.6^{b}	5.3^{b}	6.7^{a}	4.5^{b}	5.2^{b}	5.4^{b}	7.1^{a}
	(0.7)	(0.4)	(0.2)	(0.6)	(0.9)	(0.4)	(0.3)	(0.7)

Different letters mean significant difference of 1% between treatments (10, 15, 20 or 25 seeds). Numbers in parentheses indicate the standard error. N = 5.

was sufficiently separated from the invaded area, to affirm that there was no effect of *C. edulis* on the soil. The soil taken from each area (invaded or native) was homogenized (approx. 100 Kg) and reserved for the establishment of the crops, as explained later. See [8] for physicochemical results.

3. Competition between *C. edulis* ramets and native species

In order to look for competitive relationships established between *C. edulis* and native plants in different stages of native plant's development (germination, seedling and adult plant), pot culture experiments were established in both invaded and native soil following the principle of replacement or substitution [18], modified for our purposes (Table 1). Competition experiments were carried out on two soils with different origins (native and invaded soil). To avoid interference in the replacement series due to physico-chemical and biological characteristics of the soil, a concomitant treatment checking intra-specific competition density was established. Four experimental trials were then established to study the competitive interaction between the invasive *C. edulis* and native species: (a) intra-specific competition between native seeds/seedlings (b) ramets of *C. edulis vs* native seeds/seedlings, (c) intra-specific competition between native adult plants and (d) ramets of *C. edulis vs* adult native plants. Cultures were established in 1 L pots filled with soil from native and invaded zones and replicated five times. A total of 320 pots were used.

3.1. Intra-specific competition between native seeds and seedlings. 80 pots with cultures containing 10 15, 20 or 25 seeds of the native species (*M. littorea* or *S. atropurpurea*) were established (Table 1), which we refer to in this paper as "pure seed cultures." Total germination rate (Gt), cumulative rate of germination (AS) [19], survival and early growth were determined. The number of germinated seeds and plant survival were recorded daily for ten weeks.

3.2. Ramets of *C. edulis vs* native seeds and seedlings. In order to check inter-specific competition, mixed cultures were established in 80 pots, combining 3, 2, 1 and 0 ramets of the invasive *C. edulis* with 10, 15, 20 and 25 seeds of native species, referred to as "mixed seed/ramet cultures" (10/3, 15/2, 20/1 and 25/0). Total germination rate (Gt), cumulative rate of germination (AS) [19], survival and early growth were determined. The number of germinated seeds and plant survival were recorded daily for ten weeks.

3.3. Intra-specific competition between native adult plants. In order to take into account intra-specific competition, cultures containing 4, 3, 2 or 1 adult native plants (*M. littorea* or *S. atropurpurea*) were established (Table 1) in 80 pots, referred to as "pure adult cultures". Leaf number, height and survival were recorded every three days for two weeks.

3.4. Ramets of *C. edulis vs* adult native plants. In order to check inter-specific competition, mixed cultures with 3, 2, 1 and 0 ramets of the invasive *C. edulis* combined with 1, 2, 3 or 4 adult native plants were cultivated in 80 pots, referred to as "adult/ramets cultures" (1/3, 2/2, 3/1 and 4/0). Leaf number, height and survival were recorded every three days for two weeks.

4. Competition between *C. edulis* seeds and native seeds

Seeds of *C. edulis* and *M. littorea* or *S. atropurpurea* were sowed at different densities and times, following the scheme proposed by Tielbörger and Prasse [5]. Five replicates of the following seed mixture were established: 10 seeds of each native species plus 10 seeds of *C. edulis*, 10 seeds of each native species plus 30 seeds of *C. edulis*, 30 seeds of each native species plus 10 seeds of *C. edulis* and pure crops of 30 or 10 seeds of each species.

Table 5. Inter-specific competition effect on the germination indices (Gt: Total germination and AS: Cumulative rate of germination), seedling shoot and root length and survival of the native species *Malcolmia littorea* on both invaded and native soil.

	Invaded soil				Native soil			
Seed N°/ramets	10/3	15/2	20/1	25/0	10/3	15/2	20/1	25/0
Gt	46.0*	33.3*	37.0	36.0*	15.8^{c}	26.7^{bc}	45^{ab}	65.6^{a}
	(13.3)	(9.7)	(12.9)	(5.7)	(5.5)	(3.8)	(13.5)	(3.4)
AS	51.5*	42.1	38.0	47.8*	12.8^{c}	23.6^{bc}	44.9^{b}	88.5^{a}
	(10.8)	(8.6)	(10.5)	(3.8)	(7.4)	(5.8)	(15.2)	(43)
Survival (%)	32.5^{b}*	18.6^{b}	51.8^{ab}*	77.9^{a}	0^{b}	0^{b}	92^{a}	(2.8)
	(12.8)	(9.9)	(16.0)	(17.6)				
Shoot length (cm)	-	-	-	-	-	-	-	-
Root length (cm)	-	-	-	-	-	-	-	-

Different letters mean significant difference of 5% between treatments (10/3, 15/2, 20/1 or 25/0 seeds/ramets).*: indicates significant difference of 5% between invaded and native soils on each treatment. Numbers in parentheses indicate the standard error. N = 5.

Table 6. Inter-specific competition effect on the germination indices (Gt: Total germination and AS: Cumulative rate of germination), seedling shoot and root length and survival of the native species *Scabiosa atropurpurea* on both invaded and native soil.

	Invaded soil				Native soil			
Seed N°/ramets	**10/3**	**15/2**	**20/1**	**25/0**	**10/3**	**15/2**	**20/1**	**25/0**
Gt	30.0	41.3	37.1	45.6	14.0	24.0	37.0	48.0
	(8.9)	(3.9)	(3.7)	(2.7)	(9.8)	(7.5)	(3.0)	(7.2)
AS	23.2*	28.3	32.8	40.1*	3.8	19.8	30.4	46.1
	(7.0)	(2.7)	(4.7)	(2.9)	(7.9)	(6.1)	(1.7)	(1.9)
Survival (%)	31.1*	68.3	79.8	93.5	70.0	51.4	51.7	94.5
	(18.1)	(18.1)	(7.5)	(1.7)	(20.0)	(16.4)	(21.5)	(2.5)
Shoot length (cm)	1.2	1.5	1.1	1.4	1.7	1.2	1.2	1.3
	(0.2)	(0.5)	(0.2)	(0.2)	(0.3)	(0.2)	(0.2)	(0.2)
Root length (cm)	4.1	5.0	5.8	6.4	5.4	4.4	4.9	6.3
	(1.1)	(1.4)	(0.7)	(1.4)	(0.5)	(0.9)	(0.3)	(0.9)

Different letters mean significant difference of 5% between treatments (10/3, 15/2, 20/1 or 25/0 seeds/ramets).*: indicates significant difference of 5% between invaded and native soils on each treatment. Numbers in parentheses indicate the standard error. N = 5. None of the post-hoc ANOVA results were significant.

With the aim of testing the effect of time, the seeds of *C. edulis* were sowed at different date ranges (5 days before the native species, at the same time, or 5 days later than the native species). Competition conditions were established in Petri dishes filled with soil from invaded and from native areas. A total of 240 petri dishes were used. The experimental design (Table 2) therefore had 5 independent factors: neighbour density (10 vs 30), target species (two native species), neighbour species (one invasive species), soil type (native or invaded) and timing of sowing (5 days before, at the same time or 5 days latter).

The Petri dishes were incubated in germination chambers with periods of 12 hours of light and 25°C/15°C, temperatures and light regimes similar to those in the field. Substrate moisture in sandy soils is one of the most limiting factors of plant growth [20]. Therefore, all of the seeds were watered every two days, as previous trials have indicated that this procedure permits maximum germination despite the limited amount of substrate. Percolation of the water through holes in the bottom of the dishes was allowed, avoiding the formation of a salt crust. The number of germinated seeds and plant survival were recorded daily for ten weeks. Total germination rate (Gt), and the cumulative rate of germination (As) were determined [19]. After approximately ten weeks of watering, no further germination was observed and the length of leaf, stem and roots of 7 seedlings per plate and species were measured.

5. Statistical analysis

The Kolmogorov–Smirnov test and Levene's test were used to ensure the normality assumption and the homogeneity of variances, respectively. Two-way analysis of variance (ANOVA) was performed to assess the significance of the effects of soil characteristics and density as well as of their interaction on the studied parameters of germination, growth and survival. Tukey's test was applied for all post-hoc comparisons between groups.

A two-way ANOVA involving soil characteristics and density as factors was carried out to detect significant differences between treatments for each native species. All statistical analyses were performed using the IBM SPSS Statistic 19.0 (IBM, Armonk, NY, USA) software package.

Results

1. Competition of *C. edulis* ramets *vs* native species

1.1. Intra-specific competition between native seeds and seedlings. In invaded soil, pure seed cultures of *M. littorea* (10, 15, 20 and 25 seeds) showed a significant decrease in the total germination percentage (up to 56.4%) and cumulative rate of germination (up to 55.1%) with the increase in density (Table 3). However, in native soil there were no significant differences between treatments (seed number). It is quite remarkable that at the highest density (25 seeds), the total germination and cumulative rate of germination were significantly greater on native soil compared with invaded soil ($P \leq 0.05$), while at low seed densities (10, 15 and 20 seeds) we observed the opposite trend (Table 3). The survival percentage of seedlings of *M. littorea* showed a significant increase (21.74%) at high densities in native soil. The shoot and root length of *M.* seedlings in both soils showed no significant effect of intra-specific competition between treatments (number of individuals). Despite this, the growth values from invaded soil are significantly higher than the lengths in the native soil ($P \leq 0.05$).

No significant effect of intra-specific competition was found between *S. atropurpurea* seedlings affecting the growth, survival or germination rates (Table 4).

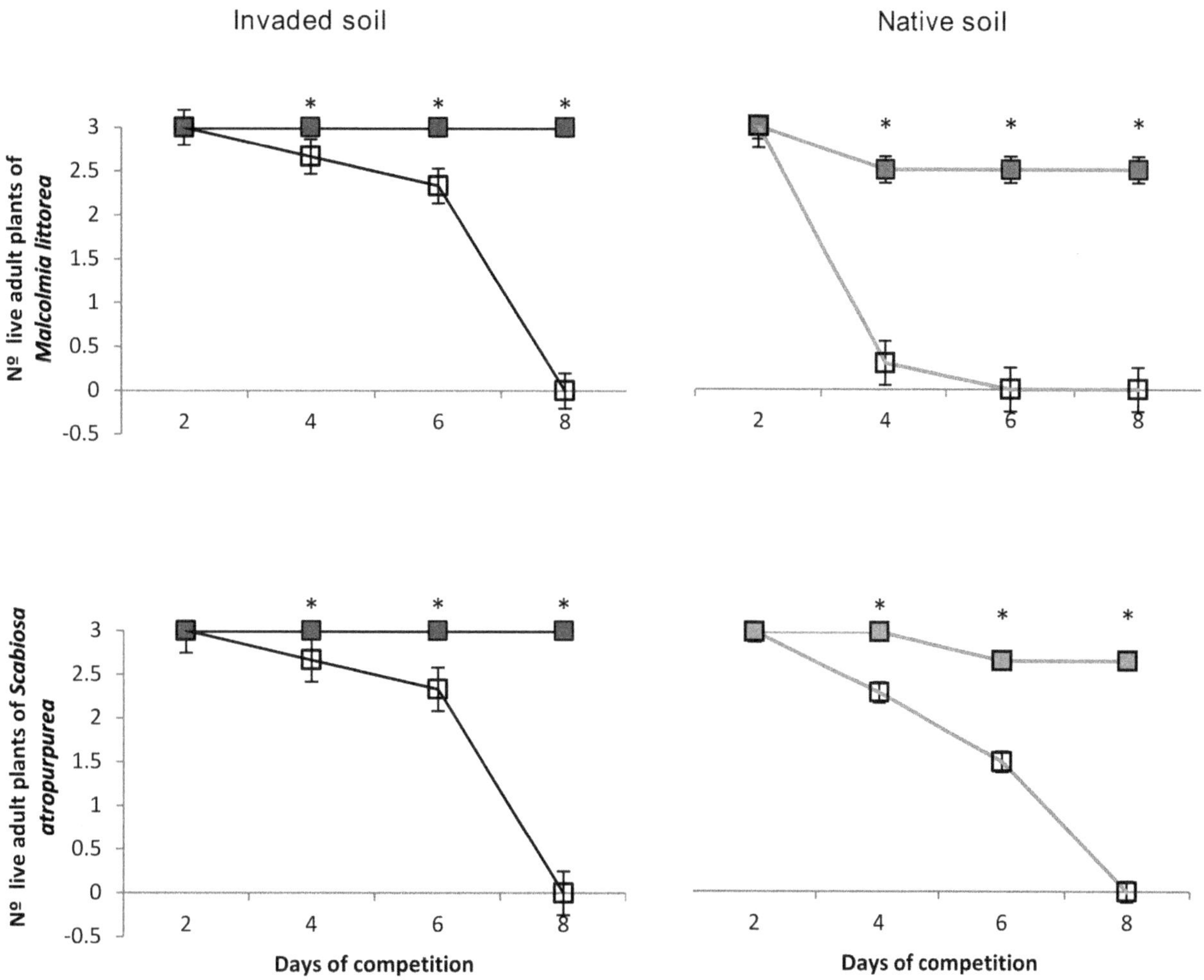

Figure 2. Survival of the native plants *Malcolmia littorea* and *Scabiosa atropurpurea* in pure adult cultures (3 adult native plants) and mixed adult/ramets cultures (3 adult native plants against 1 ramet of *C. edulis*) on invaded and native soil. *: indicate differences ($P \leq 0.05$) between monocultures (filled squares) and mixed (empty squares).

1.2. Ramets of *C. edulis vs* native seeds and seedlings. The increase in density of *C. edulis* per pot did not affect the germination process of *M. littorea* in invaded soil. However, in native soil, the higher density of *C.* caused a decrease in the percentage and speed of germination of *M. littorea* (78% and 87% respectively). The survival of *M. littorea* seedlings was lower as the density of *C. edulis* increased, and was null in native soil (Table 5). Most of the treatments with a high density of *C. edulis* showed higher germination rates and survival percentages in invaded than in native soil. The opposite situation was shown for germination rates when *C. edulis* was not present. It was impossible to test the early seedling response of *M. littorea* experimentally, due to the low or null survival percentage.

The germination process and early growth of *S. atropurpurea* were not affected by the increase in density of *C. edulis* and slightly affected by the type of soil (Table 6).

The interaction between the dependent factors (soil origin and density) was studied by a two-way ANOVA. The interaction was not significant, indicating that the internal relationship between soil origin and density in the germination process and early growth of native plants is very small, or does not exist.

1.3. Intra-specific competition between native adult plants. In native pure adult cultures there were no significant differences in leaf number, height and survival between plant densities or any soil effect on the plant traits recorded (data not shown).

1.4. Ramets of *C. edulis vs* adult native plants. Mixed adult/ramet cultures revealed that the presence of the invasive plant *C. edulis* had a lethal effect on the adult plants of *M. littorea* and *S. atropurpurea*. All native plants died within 8 days of competition, even when the ratio for the invasive plant *C. edulis* was minimal (3 adult native plants against 1 ramet of *C. edulis*) (Figure 2).

2. *C. edulis* seeds *vs* native seeds

2.1. Intra-specific competition and soil effects. Considering pure seed cultures, we found no density effect ($P \leq 0.05$) and almost not soil effect (Table 7). However, the Gt and AS indices of were in general higher in Petri dishes filled with invaded soil than in those filled with native soil while the shoot growth seemed higher on non-invaded soils (Table 7).

2.2. Inter-specific competition and soil effects. In relation to time of sowing, the germination indices for *M. littorea* seeds were generally significantly lower when the seeds of *C. edulis* were sown 5 days before. Values were generally similar in assay 2 and 3 and in monocultures.

Table 7. Effect of soil on the germination and early growth of the invader *C. edulis* and the native species *Malcolmia littorea* and *Scabiosa atropurpurea* in monocultures.

		Gt		AS		Shoot growth		Radicle growth	
	N° seeds	**I**	**N**	**I**	**N**	**I**	**N**	**I**	**N**
C. edulis	10	30.0	23.3	0.19	0.16	0.49*	0.30	1.7*	2.6
		(10.0)	(8.8)	(0.09)	(0.09)	(0.08)	(0.07)	(0.1)	(0.3)
	30	38.9	22.2	0.21	0.12	0.53	0.38	2.1	2.6
		(4.0)	(5.9)	(0.04)	(0.01)	(0.07)	(0.04)	(0.2)	(0.38)
Malcolmia littorea	10	60.0	53.3	0.78	0.53	0.24	0.24	1.79	1.9
		(10.0)	(8.8)	(0.20)	(0.10)	(0.02)	(0.04)	(0.2)	(0.1)
	30	65.6	56.7	1.05*	0.54	0.25	0.22	2.1	2.5
		(1.9)	(10.0)	(0.02)	(0.09)	(0.02)	(0.01)	(0.1)	(0.4)
Scabiosa atropurpurea	10	46.7	73.3	0.73	0.99	0.35	0.33	8.1*	10.2
		(12.0)	(13.3)	(0.24)	(0.19)	(0.02)	(0.02)	(1.2)	(0.8)
	30	65.6	70.0	0.94	0.90	0.44*	0.29	12.0	10.4
		(5.9)	(5.1)	(0.09)	(0.10)	(0.04)	(0.01)	(0.6)	(0.7)

(Gt: Total germination and AS: Cumulative rate of germination). *indicate significant differences at 5% level between the seeds sown in Petri dishes filled with invaded soil and those sowed in native soil. Numbers in parentheses indicate the standard error. I: Invaded soil, N: Native soil. N = 5.

Table 8. Effect of the timing of sowing on the germination of the invader *C. edulis* and the native species *Malcolmia littorea* and *Scabiosa atropurpurea littorea.*

		Gt				AS			
		A1	A2	A3	M	A1	A2	A3	M
CM	I	36.7^{a}	26.7^{b}	20.0^{b}	30.0^{a}	0.50	0.24	0.14	0.19
		(3.3)	(3.3)	(10.0)	(10.0)	(0.04)	(0.09)	(0.09)	(0.09)
	N	30.0^{a}	3.3^{b}	3.3^{b}	23.3^{a}	0.43^{a}	0.02^{b}	0.01^{b}	0.16^{a}
		(0.0)	(3.3)	(3.3)	(8.8)	(0.06)	(0.02)	(0.01)	(0.09)
CS	I	43.3	23.3	30.0	30.0	0.39	0.14	0.07	0.19
		(3.3)	(3.3)	(5.8)	(10.0)	(0.01)	(0.07)	(0.01)	(0.09)
	N	20.0	13.3	10.0	23.3	0.22	0.09	0.05	0.16
		(5.8)	(3.3)	(5.8)	(8.8)	(0.11)	(0.05)	(0.02)	(0.09)
ML	I	33.3^{b}	70.0^{a}	66.7^{a}	60.0^{a}	0.36^{b}	1.01^{a}	0.86^{a}	0.78^{a}
		(8.8)	(17.3)	(12.0)	(10.0)	(0.15)	(0.14)	(0.15)	(0.20)
	N	20.0^{b}	30.0^{a}	40.0^{a}	53.3^{a}	0.13^{b}	0.30^{a}	0.46^{a}	0.53^{a}
		(5.8)	(2.8)	(7.5)	(8.8)	(0.03)	(0.05)	(0.09)	(0.10)
SA	I	46.7	63.3	66.7	46.7	0.59	0.89	0.98	0.73
		(6.7)	(3.3)	(6.7)	(12.0)	(0.07)	(0.01)	(0.06)	(0.24)
	N	70.0	70.0	66.7	73.3	0.77	1.05	0.92	0.99
		(15.3)	(15.3)	(12.0)	(13.3)	(0.19)	(0.20)	(0.10)	(0.19)

Different letters means significant differences at 5% level among treatments (A1, A2, A3, M).
A1: Assay 1, *C. edulis* seeds sown before native seeds. A2: Assay 2: *C. edulis* seeds sowed at the same time that native seeds. A3: Assay 3: *C. edulis* seeds sown after native seeds. M: native monocultures. I: Invaded soil. N: Native soil. CM: *C. edulis* + *M. littorea*. CS: *C. edulis* + *S. atropurpurea*. ML: *M. littorea*. SA: *S. atropurpurea*. Numbers in parentheses indicate the standard error. N = 5.

Table 9. The effect of timing of sowing on the early growth of the native species *Malcolmia littorea* and *Scabiosa atropurpurea*.

		Shoot growth				Radicle growth			
		A1	**A2**	**A3**	**M**	**A1**	**A2**	**A3**	**M**
ML	I	0.20[b]	0.20[b]	0.17[b]	0.24[a]	2.1	1.79	1.7	1.79
		(0.00)	(0.00)	(0.02)	(0.02)	(0.5)	(0.2)	(0.2)	(0.2)
	N	0.24[b]	0.23[b]	0.17[b]	0.30[a]	1.9	1.9	1.7	1.9
		(0.00)	(0.03)	(0.02)	(0.04)	(0.1)	(0.3)	(0.3)	(0.1)
SA	I	0.34	0.35	0.40	0.35	10.9	10.0	10.5	8.1
		(0.03)	(0.02)	(0.03)	(0.02)	(1.3)	(0.6)	(1.0)	(1.2)
	N	0.35	0.32	0.28	0.33	9.0	11.1	10.7	10.2
		(0.03)	(0.02)	(0.03)	(0.02)	(1.1)	(1.1)	(1.0)	(0.8)

Different letters means significant differences at 5% level among treatments (A1, A2, A3, M).
A1: Assay 1, *C. edulis* seeds sown before native seeds. A2: Assay 2: *C. edulis* seeds sowed at the same time that native seeds. A3: Assay 3: *C. edulis* seeds sown after native seeds. M: native monocultures. I: Invaded soil. N: Native soil.
ML: *M. littorea*. SA: *S. atropurpurea*. Numbers in parentheses indicate the standard error. N = 5.

C. edulis seeds reacted the same way in the presence of *M. littorea* seeds in native soils. The germination indices were generally significantly lower when the seeds of *M. littorea* were sown 5 days before (80–90%) but not in the presence of *S. atropurpurea* (Table 8). The germination process of *S.* did not appear to be influenced by any of the treatments (Table 8).

The growth results in relation to the time of sowing factor only refer to native species, as *C. edulis* had different growth periods depending on the treatment (assay 1, 2 or 3). *M. littorea* seedlings showed a decrease in shoot growth with the presence of *C. edulis* seeds, especially in native soils, although the radicle growth of *M. littorea* showed no differences (Table 9). No effects were observed in *S. atropurpurea* growth in relation to the timing factor (Table 9).

Despite the effects of the time of sowing, there were no significant differences ($P \leq 0.05$) between pure and mixed cultures in any treatment (results not shown).

Discussion

1. Competition between ramets of *C. edulis* and native species

1.1. Intra-specific competition between native seeds and seedlings. Soil from habitats invaded by *C. edulis* had a markedly species-dependent effect. *M. littorea*, a typical semi-fixed dunes species, is a specialist plant of poor and slightly saline soils [21]. Its germination and growth process is significantly influenced by soil changes induced by *C. edulis* [7,8]. In general, we found a reduction in the germination process as the density increased in invaded soil. The advantage of this behaviour of density-dependent germination in invaded soils is difficult to understand. It is possible that for these plants there is an advantage of not germinating under competitive conditions, so they may exploit more favourable conditions in later years [5]. Shoot and root length was always greater in invaded soils, probably due to a greater availability of nutrients in these soils [9,22] and salt concentration [23] that stimulate growth. In native soil, the survival percentage increased at the highest density, which may be due to the nurse plant phenomenon [20] taking into account not differences between life forms.

Germination was not density dependent for *S. atropurpurea* in either invaded or native soils. As previously mentioned *S. atropurpurea* presented a greater plasticity to adapt to different soil conditions and seems to be less of a specialist than *M. littorea*.

1.2. Ramets of *C. edulis vs* native seeds and native seedlings. Inter-specific competition is species-dependent and determines what species can coexist [24]. *C. edulis* differentially affected germination of the selected native plants. Soil characteristics became as one of the factors that directly affect the competitive potential of the exotic [25].

M. littorea showed a significant decrease in the germination process caused by the presence of *C. edulis* adult plants. However, the most dramatic effect found on *M. littorea* is in the survivorship, which is soil dependent. In invaded soil, the survival percentage decreases with the proportion of *C. edulis* adult plants in the mixed cultures, while in native soils, none of the germinated seedlings survived at the end of the bioassay. Inter-specific competition seems to be responsible for these results [26,27]. Thus, although barriers to native plant germination could be overcome, *M. littorea* seedlings would not establish viable populations in the presence of *C. edulis* adult plants.

Soil characteristics are not determinant in the response of *S. artropurpurea* to the presence of *C. edulis* (except to mixed seed/ramets cultures 10/3), once again probably due to its plasticity.

Although *C. edulis* threatens the establishment of *S. atropurpurea* in both soils in different ways, seedlings are phenotypically plastic in their allocation of biomass into roots and shoots [28].

1.3. Intra-specific competition between natives adult plants. Our results indicate that the residual effect of *C. edulis* on dune soil does not affect the development of native adult plants. Adult native plants seem to be better adapted to soil changing conditions than seedlings, suffering stress from the residual effect in the soil.

1.4. Ramets of *C. edulis vs* adult native plants. The presence of *C. edulis* in the replacement series had a deleterious effect on native plants independently of density and soil type. Different authors have indicated that allelopathy is a fairly common invasion mechanism (29). Ens et al. [30] proposed that the eventual dominance of invasive species could be explained by direct or indirect chemical inhibition of the establishment of indigenous plants, which was confirmed by Novoa et al. [7] for *C. edulis*.

2. Seed competition of *C. edulis vs* seeds of native species

2.1. Soil effects. Dune species germinate in autumn, the rainy season, when in addition to having more water in the soil, the salt content decreases [20]. Water softens the seed coat so that the root can emerge more easily and also solubilizes nutrients [31]. A high salt content can block the germination process by the osmotic effect, drawing water from seeds [32]. During the assay, all of the seeds were watered every two days. However, invaded soils have a higher level of organic matter than native soils [7,8], so they can hold water for a longer period of time. As a result, the Gt and AS indices of *M.* were higher in Petri dishes filled with invaded soil than in those filled with native soil. Also, in the presence of *C. edulis*, *M. littorea* has greater shoot growth in native soils. This could be due to an allelopathic effect of *C. edulis* seeds on *M. littorea* [7].

Once again *S. atropurpurea* did not show any differences in the germination process between soils. However, the radicle growth of *S. atropurpurea* seedlings seemed to be stimulated by native soil, while shoot growth seemed to be stimulated by invaded soil. The nutrient content of sand (especially nitrogen and phosphorus) positively or negatively affects the growth of dune species depending on the species [20]. The radical growth of *S. atropurpurea* decreases and its shoot growth increases with an increase in the nutrient level [8]. This could explain the differences observed in the growth of *S. atropurpurea* between invaded and native soil.

Finally, when *C. edulis* invades coastal habitats, it modifies the conditions of the substrate and suffers from difficulties as a result of tissue decomposition [33]. This feature could have evolved as a mechanism to facilitate recolonization when the clones die, and it influences the germination process of the invasive species. As a result, the germination process of *C. edulis* depends on the nutrient level of the soil [7]. Therefore, the Gt and AS indices of *C. edulis* were higher in Petri dishes filled with invaded soil than in those filled with native soil. Also, the radicle of *C. edulis* grows more at high pH levels [7], and so, it grew more in native than in invaded soils.

2.2. Timing of sowing. We observed that the timing of sowing affects the establishment of both *C. edulis* and *M. littorea*, although there are no competitive interactions if they are sown at the same time. Therefore, they could have some mechanisms that allow their seeds to evaluate the conditions of their neighbours prior to emergence and to plastically respond to them [5]. *C. edulis* could also have an allelopathic effect on *M. littorea* and vice versa, which should be explored in future assays. These are important results to be taken into account for restoration actions; but also in conservation actions. If *C. edulis* is a new invader arriving from seeds into a native area and *M. littorea* is present, according to our results it appears that M. littorea might have the ability to supress *C. edulis* germination.

Conclusion

Our results show that the impacts of *C. edulis* on native plants are highly dependent on its development stages as well as on the development stages of native plants. These findings are crucial for new strategies of biodiversity conservation in coastal habitats.

Acknowledgments

We thank Paula Lorenzo for valuable comments and Jorge Tabares for technical assistance.

Author Contributions

Conceived and designed the experiments: AN LG. Performed the experiments: AN. Analyzed the data: AN. Contributed reagents/materials/analysis tools: LG. Contributed to the writing of the manuscript: AN LG.

References

1. Kowarik I (2003) Biologische Invasionen-Neophyten und Neozoen in Mitteleuropa. Stuttgart: Ulmer.
2. Simberloff D, Parker IM, Windle PN (2005) Introduced species policy, management, and future research needs. Frontiers in Ecology and the Environment 3: 12–20.
3. Vilà M, Espinar JL, Hejda M, Hulme PE, Jarošík V, et al. (2011) Ecological impacts of invasive alien plants: a meta-analysis of their effects on species, communities and ecosystems. Ecology letters 14: 702–708. Available: http://www.ncbi.nlm.nih.gov/pubmed/21592274. Accessed 25 May 2014.
4. Ley C, Gallego J, Vidal C (2007) Manual de restauración de dunas costeras. Ministerio de medio ambiente.
5. Tielbörger K, Prasse R (2009) Do seeds sense each other? Testing for density-dependent germination in desert perennial plants. Oikos 118: 792–800. Available: http://doi.wiley.com/10.1111/j.1600-0706.2008.17175.x. Accessed 6 November 2012.
6. Council Directive 92/43/EEC (1992): 7.
7. Novoa A, González L, Moravcová L, Pyšek P (2012) Effects of Soil Characteristics, Allelopathy and Frugivory on Establishment of the Invasive Plant Carpobrotus edulis and a Co-Occuring Native, Malcolmia littorea. Plos One 7: 1–11. doi:10.1371/journal.pone.0053166.
8. Novoa A, González L, Moravcová L, Pyšek P (2013) Constraints to native plant species establishment in coastal dune communities invaded by Carpobrotus edulis: Implications for restoration. Biological Conservation 164: 1–9. Available: http://linkinghub.elsevier.com/retrieve/pii/S0006320713001031. Accessed 20 June 2013.
9. Novoa A, Rodríguez R, Richardson D, González L (2014) Soil quality: a key factor in understanding plant invasion? The case of Carpobrotus edulis (L.) N.E.Br. Biological Invasions 16: 429–443. Available: http://link.springer.com/10.1007/s10530-013-0531-y. Accessed 30 June 2014.
10. Santoro R, Jucker T, Carranza ML, Acosta ATR (2011) Assessing the effects of Carpobrotus invasion on coastal dune soils. Does the nature of the invaded habitat matter? Community ecology 12: 234–240. doi:10.1556/ComEc.12.2011.2.12.
11. Brown CS, Anderson VJ, Claassen VP, Stannard ME, Wilson LM, et al. (2008) Restoration Ecology and Invasive Plants in the Semiarid West. Invasive plant science and management 1: 399–413.
12. Lorenzo P, Palomera-Pérez A, Reigosa MJ, González L (2011) Allelopathic interference of invasive Acacia dealbata Link on the physiological parameters of native understory species. Plant Ecology 212: 403–412. doi:10.1007/s11258-010-9831-9.
13. Mangla S, Sheley RL, James JJ, Radosevich SR (2011) Intra and interspecific competition among invasive and native species during early stages of plant growth. Plant Ecology 212: 531–542. Available: http://link.springer.com/10.1007/s11258-011-9909-z. Accessed 19 June 2014.
14. Lorenzo P, Rodríguez-Echeverría S, González L, Freitas H (2010) Effect of invasive Acacia dealbata Link on soil microorganisms as determined by PCR-

DGGE. Applied Soil Ecology 44: 245–251. Available: http://linkinghub.elsevier.com/retrieve/pii/S0929139310000065. Accessed 7 November 2012.
15. Thuiller W, Richardson DM, Rouget M, Thuiller W, Richardson DM, et al. (2006) Interactions between Environment, Species Traits, and Human Uses Describe Patterns of Plant Invasions. Ecology 87: 1755–1769.
16. D'Antonio C, Mahall BE (1991) Root profiles and competition between the invasive, exotic perennial, Carpobrotus edulis, and two native shrub species in California coastal scrub. American Journal of Botany 78: 885–894.
17. Combs JK, Reichard SH, Groom MJ, Wilderman DL, Camp PA (2011) Invasive competitor and native seed predators contribute to rarity of the narrow endemic Astragalus sinuatus Piper. Ecological Applications 21: 2498–2509.
18. De Wit CT, Van den Bergh JP (1965) Competition between herbage plants. Journal of Agricultural Science 13: 212–221.
19. Hussain MI, González-Rodríguez L, Reigosa MJ (2008) Germination and growth response of four plant species to different allelochemicals and herbicides. Allelopathy Journal 22: 101–110.
20. Maun MA (2009) The biology of coastal sand dunes. Oxford University Press.
21. Del Vecchio S, Giovib E, Izzia CF, Abbateb G, Acosta ATR (2012) Malcolmia littorea: The isolated Italian population in the European context. Journal for Nature Conservation 20: 357–363. doi:10.1016/j.jnc.2012.08.001.
22. Del Vecchio S, Marbà N, Acosta A, Vignolo C, Traveset A (2013) Effects of Posidonia oceanica beach-cast on germination, growth and nutrient uptake of coastal dune plants. PloS one 8: e70607. Available: http://www.pubmedcentral.nih.gov/articlerender.fcgi?artid=3720909&tool=pmcentrez&rendertype=abstract. Accessed 5 August 2014.
23. Lee JA, Caporn SJM (1998) Ecological effects of atmospheric reactive nitrogen deposition on semi-natural terrestrial ecosystems. New Phytologist 139: 127–134.
24. Tschirhart J (2002) Resource competition among plants: from maximizing individuals to community structure. Ecological Modelling 148: 191–212. Available: http://linkinghub.elsevier.com/retrieve/pii/S0304380001004392.
25. Huangfu CH, Zhang TR, Chen DQ, Wang N, Yang DL (2011) Residual effects of invasive weed yellowtop (Flaveria bidentis) on forage plants for ecological restoration. Allelopathy Journal 27: 55–64.
26. Ammondt S, Litton CM (2012) Competition between Native Hawaiian Plants and the Invasive Grass Megathyrsus maximus: Implications of Functional Diversity for Ecological Restoration. Restoration Ecology 20: 638–646. Available: http://doi.wiley.com/10.1111/j.1526-100X.2011.00806.x. Accessed 28 November 2012.
27. Holdredge C, Bertness MD (2010) Litter legacy increases the competitive advantage of invasive Phragmites australis in New England wetlands. Biological Invasions 13: 423–433. Available: http://www.springerlink.com/index/10.1007/s10530-010-9836-2. Accessed 3 November 2012.
28. Shadel WP, Molofsky J (2003) Habitat and population effects on the germination and early survival of the invasive weed, Lythrum salicaria L. (purple loosestrife). Biological Invasions 4: 413–423.
29. Ren M-X, Zhang Q-G (2009) The relative generality of plant invasion mechanisms and predicting future invasive plants. Weed Research 49: 449–460. Available: http://doi.wiley.com/10.1111/j.1365-3180.2009.00723.x. Accessed 6 November 2012.
30. Ens EJ, French K, Bremner JB (2009) Evidence for allelopathy as a mechanism of community composition change by an invasive exotic shrub, Chrysanthemoides monilifera spp. rotundata. Plant and Soil 316: 125–137. Available: http://www.springerlink.com/index/10.1007/s11104-008-9765-3. Accessed 7 January 2013.
31. Khurana E, Singh JS (2004) Germination and seedling growth of five tree species from tropical dry forest in relation to water stress: impact of seed size. Journal of Tropical Ecology 20: 385–396. Available: http://www.journals.cambridge.org/abstract_S026646740400135X. Accessed 26 November 2012.
32. Bubel N (1988) The new seed-starters handbook. Rodale Press, Inc.
33. Conser C, Connor EF (2008) Assessing the residual effects of Carpobrotus edulis invasion, implications for restoration. Biological Invasions 11: 349–358. Available: http://www.springerlink.com/index/10.1007/s10530-008-9252-z. Accessed 4 November 2012.

13

Strong Discrepancies between Local Temperature Mapping and Interpolated Climatic Grids in Tropical Mountainous Agricultural Landscapes

Emile Faye[1,2,3]*, Mario Herrera[3], Lucio Bellomo[4], Jean-François Silvain[1], Olivier Dangles[1,3,5]*

1 Institut de Recherche pour le Développement (IRD), UR 072, Laboratoire Evolution, Génomes et Spéciation, UPR 9034, Centre National de la Recherche Scientifique (CNRS), Gif sur Yvette, France et Université Paris-Sud 11, Orsay, France, **2** UPMC Univ Paris06, Sorbonne Universités, Paris, France, **3** Facultad de Ciencias Exactas y Naturales, Pontificia Universidad Católica del Ecuador, Quito, Ecuador, **4** Mediterranean Institute of Oceanography (MIO) CNRS/INSU, IRD, UM 110, Université de Toulon, La Garde, France, **5** Instituto de Ecología, Universidad Mayor San Andrés, Cotacota, La Paz, Bolivia

Abstract

Bridging the gap between the predictions of coarse-scale climate models and the fine-scale climatic reality of species is a key issue of climate change biology research. While it is now well known that most organisms do not experience the climatic conditions recorded at weather stations, there is little information on the discrepancies between microclimates and global interpolated temperatures used in species distribution models, and their consequences for organisms' performance. To address this issue, we examined the fine-scale spatiotemporal heterogeneity in air, crop canopy and soil temperatures of agricultural landscapes in the Ecuadorian Andes and compared them to predictions of global interpolated climatic grids. Temperature time-series were measured in air, canopy and soil for 108 localities at three altitudes and analysed using Fourier transform. Discrepancies between local temperatures vs. global interpolated grids and their implications for pest performance were then mapped and analysed using GIS statistical toolbox. Our results showed that global interpolated predictions over-estimate by 77.5±10% and under-estimate by 82.1±12% local minimum and maximum air temperatures recorded in the studied grid. Additional modifications of local air temperatures were due to the thermal buffering of plant canopies (from −2.7°K during daytime to 1.3°K during night-time) and soils (from −4.9°K during daytime to 6.7°K during night-time) with a significant effect of crop phenology on the buffer effect. This discrepancies between interpolated and local temperatures strongly affected predictions of the performance of an ectothermic crop pest as interpolated temperatures predicted pest growth rates 2.3–4.3 times lower than those predicted by local temperatures. This study provides quantitative information on the limitation of coarse-scale climate data to capture the reality of the climatic environment experienced by living organisms. In highly heterogeneous region such as tropical mountains, caution should therefore be taken when using global models to infer local-scale biological processes.

Editor: Michael Sears, Clemson University, United States of America

Funding: This work was partly conducted within the project "Adaptive management in insect pest control in thermally heterogeneous agricultural landscapes" (ANR-12-JSV7-0013-01) funded by the Agence Nationale pour la Recherche (ANR, http://www.agence-nationale-recherche.fr/). A financial support of the McKnight Foundation (http://www.mcknight.org/) to EF during the fieldwork of this study is greatly acknowledged. The funders had no role in study design, data collection and analysis, decision to publish, or preparation of the manuscript.

Competing Interests: The authors have declared that no competing interests exist.

* Email: ehfaye@gmail.com (EF); olivier.dangles@ird.fr (OD)

Introduction

Bridging the gap between the predictions of coarse-scale climate models and the fine-scale climatic reality of species is increasingly recognized as a key issue of climate change biology research [1,2,3,4]. Despite decades of study on microclimates [5,6,7,8] and evidence for habitat-related and topographical variations in local temperatures and their relevance for species ecology [2,9,10,11,12,13], most attempts to understand and model species distributions still do not integrate spatially-explicit fine-scale climatic data (e.g. [14,15,16]). Many work use global model of temperature interpolation to examine species vulnerability to climate change and, doing so, ignore the critical issue of habitat complexity in climate buffering [4,5,17]. Indeed, climate surfaces used in species distribution models (SDMs) are rarely generated or interpolated to a resolution finer than 1 km^2 (e.g. WorldClim database), a resolution that is still very coarse relative to the home ranges or body size of most species [13,18]. For instance, [8] showed that climate grid lengths used in SDMs are, on average, ~10,000-fold larger than studied animals, and ~1,000-fold larger than studied plants. Their meta-analysis showed that the WorldClim was the most widely used climatic dataset in global SDMs. As this commonly used coarse scale climatic data in SDMs overlook the spatiotemporal thermal heterogeneity experienced by organisms, there is an urgent need for a more sophisticated use of these datasets for making inferences about biological processes that are driven by hour to hour operative temperatures of organisms.

An important yet poorly studied issue in climate change biology is to quantify to what extent climatic conditions differ between widely used 1 km^2 interpolated grid cells of global climatic database and real-world landscapes of similar areas. While it is now well-known that most organisms, especially tiny ectotherms such as insects and other arthropods, do not experience the climatic conditions recorded at weather stations [9,12,18], there is little quantitative information on the spatial and temporal heterogeneity at the landscape scale of local climatic conditions (i.e. conditions at biologically relevant scales, e.g., from cm to km for insects) and their consequences for organisms' performance. A better quantification of the climatic conditions of ecologically-relevant habitats over relatively large landscape scales (e.g., 1 km^2) is therefore a necessary first step to better incorporate dynamical microclimate into global distribution models.

Here, we investigate the sources of variance between global interpolated and local temperatures by examining 1) how well WorldClim predicts local air temperatures in our study region (the tropical Andes), 2) to what extent temperatures in crop canopies and soils differ from local air temperatures, and 3) how relevant is to use WorldClim to infer the potential performance of an insect crop pest. Addressing these questions is not an easy task as the mosaic of climatic habitats relevant for small ectothermic species at a 1-km^2 scale in real-world landscapes may be outstandingly complex. In this study, we focused on highland agricultural landscapes of the tropical Andes as most prior similar data came from low elevation and temperate agroecosystems. In such systems, most crop pests experience, over their entire life cycle, climatic conditions in three well-defined environmental layers (air, air inside-canopy and soil) and these conditions are remarkably stable over the year [19]. In this context, we firstly decided to map over replicated 1-km^2 climatic grid cells the ecologically relevant local temperatures for ectothermic crop pests in agricultural landscapes, and to compare these maps to interpolated temperature grid cells of the widely used WorldClim database. We used Fourier analysis applied to local temperature time-series as a tool to fit daily variations of temperature and to feature microclimate discrepancies in space and in time (both in terms of amplitude and phase). We then explored the implication of our thermal landscape mapping for pest performance by comparing temperature frequencies in our grid cells with the temperature-dependent growth curve of the potato tuber moth (*Phthorimaea operculella*) a major crop pest species in the region and worldwide.

Materials and Methods

1. Study area

The Ecuadorian Andes are characterized by a low seasonality, with mean temperatures varying more within days (up to 30°K variation) than within months and years (less than 0.6°K and 0.2°K variations, respectively, see [19]). This region exhibits a marked altitudinal gradient in temperatures (between 2000 and 4000 m) with mean monthly air temperature roughly decreasing by 0.6°K every 100 m of elevation [20]. Agricultural landscapes dominate the altitudinal belt between 2600 and 3800 m, and are typically composed by small field crops (mainly potato *Solanum tuberosum* L., broad bean *Vicia faba* L., corn *Zea mays* L., alfalfa *Medicago sativa* L., and pasture), natural grasslands (páramos) and a few forest patches [21]. Under the climatic conditions of the region, crops can be planted and harvested all year round, thereby creating a landscape mosaic of a wide variety of crops at different phenological stages.

Our study area was located 115 km south from the equatorial line (01°01′36″S, 78°32′16″W) in the Cotopaxi province of Ecuador. It spread out on a 20-km^2 elevation transect (2.35×8.5 km), ranging from 2,600 to 3,800 m a.s.l. The gradient had a Southwest exposure and an average slope of 9.5° (±5.2) (based on a 30 m resolution digital elevation model). To investigate the elevation effect on local vs. global interpolated temperature variations, we divided our study area into three 400 m altitudinal belts which correspond to natural floors in the hillside (2,600–3,000 m, 3,000–3,400 m, and 3,400–3,800 m) with a mean monthly temperature of 13.2±0.4°C, 10.8±0.6°C, and 9.3±0.4°C, respectively. Beyond temperature, these belts also differed in terms of landscape composition (Appendix S1 in Supporting Information), with lower elevations dominated by small fields (0.3±0.1 Ha) of potato, corn, broad bean, and pasture while the higher band had larger fields (0.7±0.3 Ha) of mainly potato and pasture. Working in these agricultural landscapes no requires specific permissions expect the kind agreement of the field owner. The presented study did not involve endangered or protected species.

2. Temperature data collection

In each of the three-altitudinal belts, we measured temperature regimes in six habitats (five crops and natural grasslands) where insect pests can be found. In each habitat, we defined three layers: air, air inside-canopy (referred as "air canopy" in the text) and soil. These layers are all used by most insect pests over their life cycle: air layer by adults, air canopy layer by adults and leaf-eating larvae and pupae, soil layer by tuber feeding larvae and pupae. In each layer of each habitat, temperature was recorded with a 1 min time step using data loggers (Hobo U23-001-Pro-V2 internal temperature loggers, Onset Computer Corporation, Bourne, USA) with an accuracy of ±0.21°K over the 0–50°C range and a resolution of 0.02°K at 25°C. According to [4], 1) air loggers were fixed on a wooden stake at 1 m high to overstep most crop canopies and sheltered by a 20 cm^2 white plastic roof to minimize solar radiation heating; the roof was itself placed 5 cm above the logger to avoid warming by greenhouse effect, 2) air canopy loggers were placed 0.3 m high inside vegetation 5 cm bellow large leaves to minimize the effect of direct solar radiation and 3) soil loggers were buried 0.1 m into the ground where roots and tubers grow (see Appendix S2 for photographs). In each field, only one logger per layer measured the temperatures. Those triplets of loggers were located at the centre of the field to avoid edge effect (see Appendix S3 for an analysis of the spatial variability of temperatures within a field and [22]). As vegetation land cover influences microclimate beneath and around plants, see [5,6], we repeated these 54 measurements (3 elevations ×6 habitats ×3 layers) for three classes of leaf area index (LAI) [23] defined as follows: 0 (bare soil), 0.01–0.5 for and >0.5 of LAI. Minimum LAI was fixed to 0.01 to avoid confusion with bare soil and allowed enough leaf area to place the loggers underneath. At each measurement site, LAI values were visually estimated (twice) measuring the ratio of leaf area within a 1-m^2 quadrant sub-divided into 0.1 m^2 cells delimited by strings. This indirect method did not account for leaves that lie on each other however it relates to shaded areas that influence inside-canopy and soil microclimates [23].

Each of the 162 measurement combinations (3 altitudinal belts ×6 habitats ×3 layers × 3 LAI classes) was replicated 1–3 times depending on availability of habitats at a given elevation and phenology stage. In total 324 independent temperature time series were acquired over 15 days between September and December 2011 (data available in Appendices S9, S10 and S11). Importantly, under the climatic conditions of the study area, 15-days time series characteristics did not differ from those obtained over one year (see Appendix S4 for details). At each measurement site, we

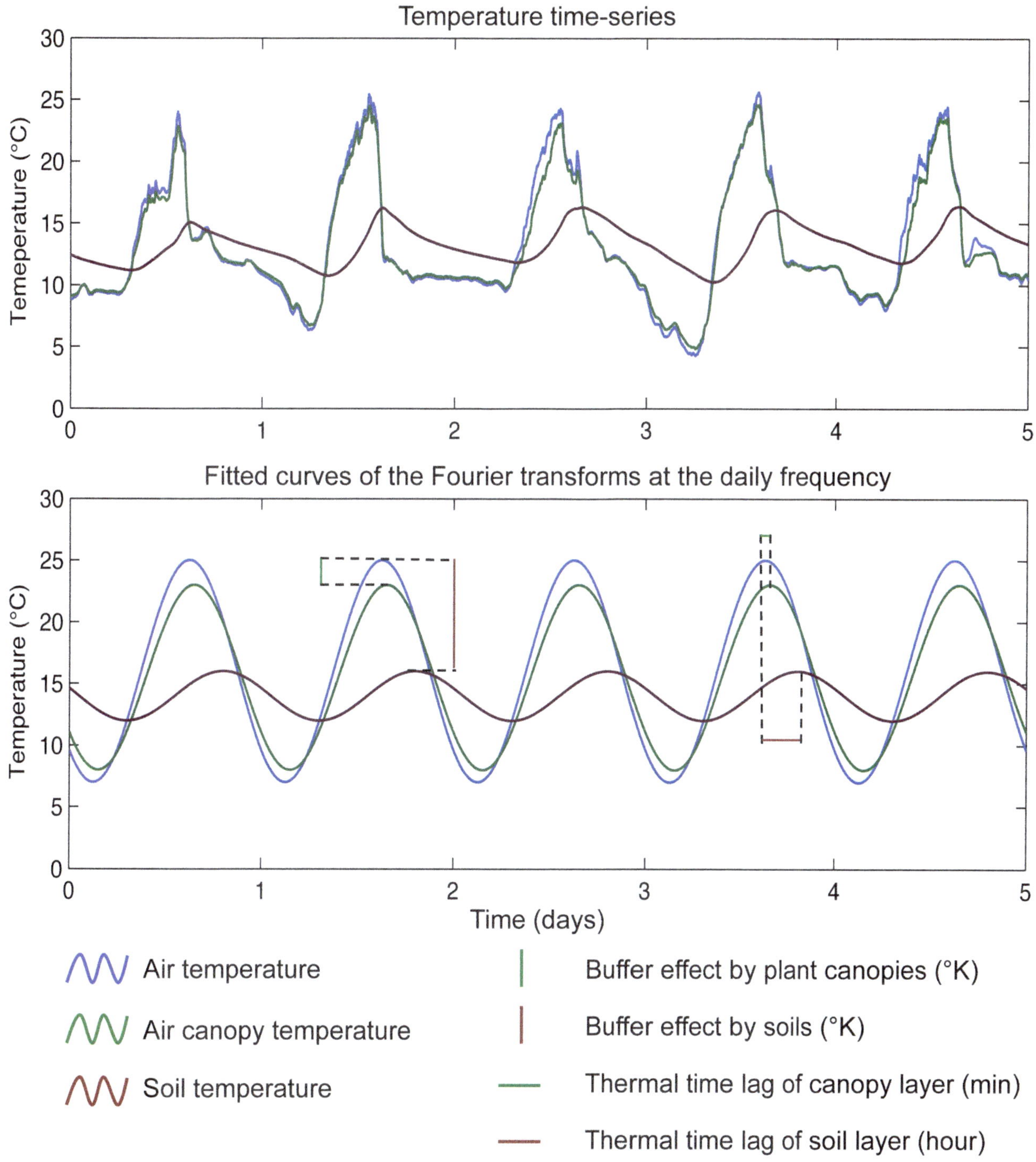

Figure 1. Fit of temperature time series with discrete Fourier transforms at the daily frequency K_d. Air temperatures are in blue, crop canopy temperatures are in green and soil temperatures are in brown.

recorded the UTM-WGS84 geographic coordinates with a handheld GPS Garmin Oregon 550 (Garmin, Olathe, USA).

3. Global solar radiations

Infrared and visible radiations (expressed in Watt/m^2) were monitored in each altitudinal belts using a LI-1400 LI-COR datalogger equipped with a LI-200 pyranometer sensor (LI-COR, Lincoln, USA) placed perpendicular to gravity. Between 9:00 AM and 4:00 PM, mean global solar radiations ranged from 500 to 1000 watts/m^2, with temporal variability mainly induced by short-term changes in cloud cover.

4. Data analyses

4.1. Times series analyses using Fourier transforms. Air and air canopy temperature time series showed extreme events during a few minutes that were certainly due to strong radiations experienced at the study sites – these affected loggers recording despite their plastic roofs. Therefore, we found

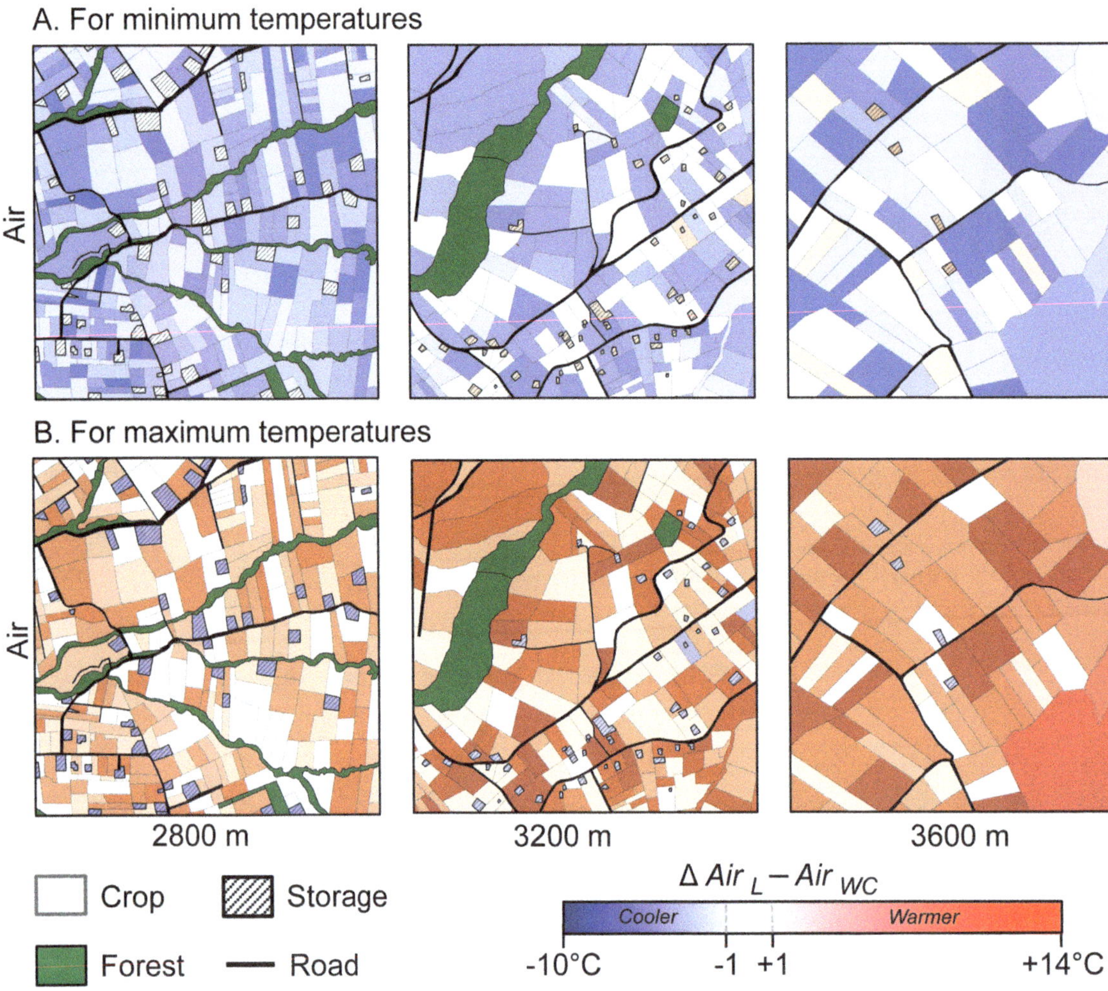

Figure 2. Maps showing the differences between local air temperatures and the WorldClim interpolated minimum (A) and maximum (B) (*Δ Air* $_L$ – *Air* $_{WC}$). Blue colours indicate *Δ Air* $_L$ – *Air* $_{WC}$ <0, i.e. area where local air temperatures are cooler than those gave by WorldClim. Red colours indicate *Δ Air* $_L$ – *Air* $_{WC}$ >0, i.e. area where air local temperatures are warmer than the ones gave by the WorldClim. White colours *Δ Air* $_L$ – *Air* $_{WC}$=0 indicate areas where air WorldClim temperatures equate air local temperatures (±1°C). The extent and position of each square is equal to the spatial resolution of the WorldClim database: 30-arc sec that is the equivalent of 0.86 km^2 for the study area. Temperatures in storages were obtained from [26].

relevant to fit our time series data with a discrete Fourier transform (DFT) at the daily frequency k_d (Fig. 1) as this allowed averaging daily minimum and maximum temperatures while limiting the effect of short extremes (mainly for maximum). Moreover fitting temperature time series with the DFT allowed us to circumvent (or partially resolve) the issue of comparing time series with different temporal resolution: a sinusoid built from a daily time step time series will be accurate enough to compare with another sinusoid built from a one minute time step time series (our operative temperatures vs. global climatic models).

DFT analyses allowed us estimating two important descriptors of the time series at the daily frequency k_d: the amplitude A_d and the phase ϕ_d of the DFT (see Appendix S5 for details). The thermal amplitude allowed us to measure the thermal buffer effect in Kelvin between air and canopy layers and air and soil layers (Fig. 1 and Appendix S5). The phase allowed us to measure the thermal time lag expressed in minute in inside-canopy and soil layers with respect to the air layer (Fig. 1 and Appendix S5). Thermal time lag therefore quantifies the time delay in time series to reach their maximum between air vs. canopy and air vs. soil layers. This is an important climatic parameter to test whether microclimate conditions below canopy (canopy and soil layers) would track air conditions with some time lag depending on habitat characteristics.

We also ran DFT analyses on a four-year monitoring (2008–2012) of air temperatures (recorded at one meter high with half an hour time step with the same shelter process described above) to measure the seasonality. Analyses were performed for the three-altitudinal belts of the study area (2800, 3200, 3600 m) by reading the amplitude at the seasonal frequencies (91, 182 and 364 days, see Appendix S6). On average the Fourier transform amplitudes at 91, 182 and 364 days were 0.14 (+/−0.01), 0.44 (+/−0.04), 0.97 (+/−0.03)°K indicating that the seasonality was negligible in the study area [24].

All Fourier analyses were performed in MATLAB R2011a (Mathworks, Natick, USA). The effects of habitat, elevation, LAI classes and the interaction "elevation × LAI classes" on daytime and nigh-time DFT amplitudes and on DFT thermal time lag were assessed using a two-way ANOVA with Bonferroni corrections. When habitat was found significant, we ran post-hoc

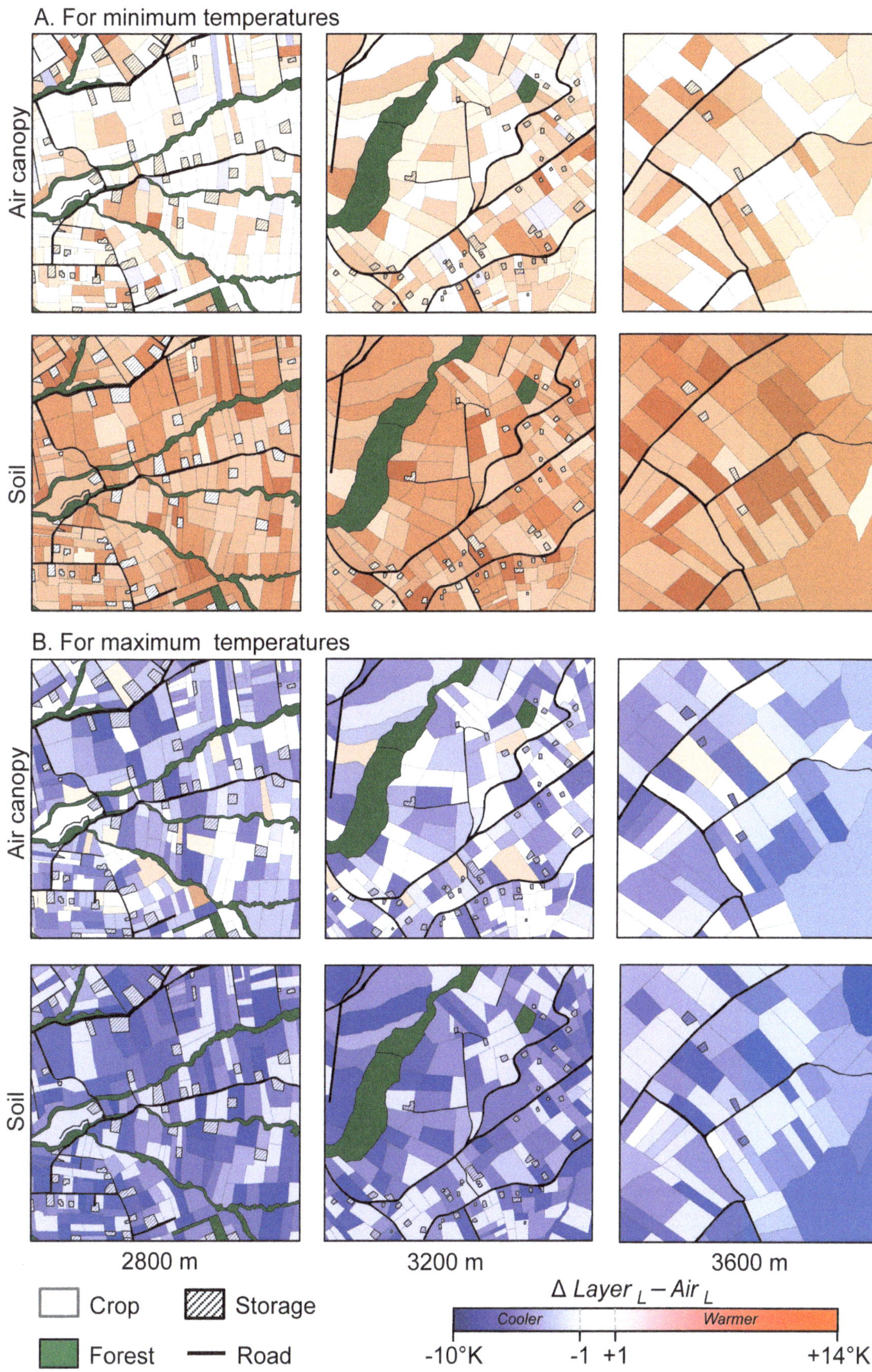

Figure 3. Maps showing the differences between local air canopy and soil temperatures with the air local for minimum (A) and maximum (B) *(Δ Layer* $_L$ *− Air* $_L$*)***.** Colour code is given in Figure 2.

multiple comparisons using a Tukey HSD test to identify differences among habitats. All statistical analyses were performed in R version 3.0.0 (R Development Core Team 2012).

4.2. Thermal landscape analyses. To compare local temperatures with global interpolated climate data employed in species distribution models, we considered one of the most widely used and readily available climate database, WorldClim [25]. The WorldClim database is a set of global climate layers (interpolated averages of monthly minimum, maximum and mean 1.5 m high air temperatures from weather stations spread out worldwide) with a spatial resolution of 30 arc seconds. Close to equator, this resolution is equivalent to squares of 0.86 km. In each altitudinal belt, we selected one WorldClim grid cell with homogenous slope (between 5.4° and 7.9°), micro-topography and exposition (south-west). Based on a digitized municipal cadastre (from the town council of Salcedo, Cotopaxi province) and a 5-m resolution digital orthophoto (Ecuadorian Military Geographical Institute, www.igm.gob.ec/site/index.php), we built the digital landscape of each grid cell in ArcGIS 10.01 (ESRI, Redlands, USA). In addition to the six studied habitats, crop storage infrastructures were also included into the digital maps as they significantly modify air temperature patterns, offering optimal conditions for crop pest development [26]. Outside air vs. inside air storage-temperature relationships for different elevations were derived from measurements made by [26] within the same area with

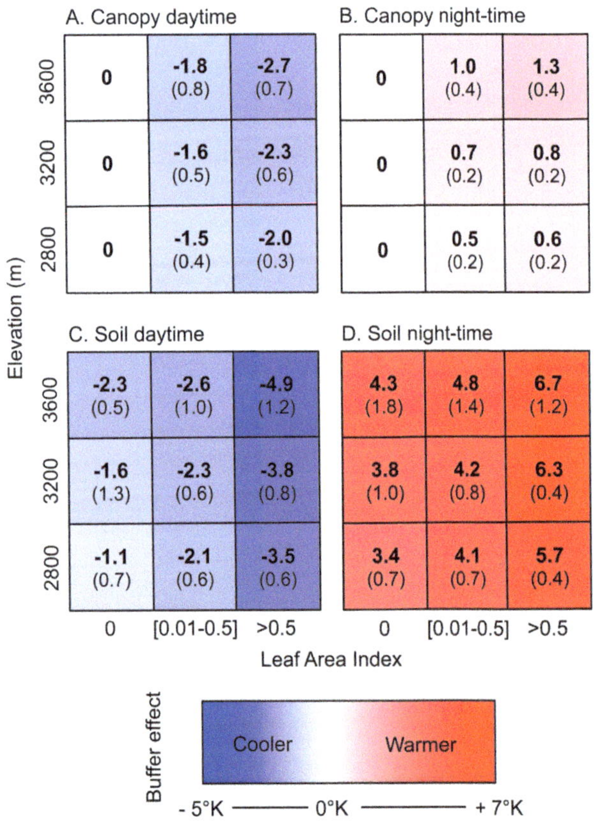

Figure 4. Mean thermal buffering from Fourier transforms at the daily frequency for canopy (A, B) and soil temperatures (C, D) as a function of elevation and leaf area index. (A, C) show the daytime temperature excursion with respect to air, whereas (B, D) are the equivalent results for night-time temperatures. The 95% interval of confidence is given between brackets. Blue colours show colder temperatures than air. Red colours show warmer temperatures than air.

similar temperature data design (see Fig. 1 in Appendix A2 of their paper). Roads and woodlots were also indicated on the maps even if they were not included in the temperature comparison analysis, as they do not constitute relevant habitats for crop pests.

In order to simulate landscape thermal heterogeneity, crop habitats were attributed with one crop type (potato, broad bean, corn, alfalfa or pasture) and one LAI classes (0, 0.01–0.5, >0.5) based on a survey of 85 sites in the region, in which we quantified landscape composition (% of each crop and LAI classes) in 100-m radius sampling circles (see Appendix S7). For each habitat, we assigned the corresponding air, air canopy and soil temperature values at each elevation. Finally, since we were particularly interested in minimal and maximal values, as they are the most biologically relevant for ectothermic crop pests [4], we focused on minimum and maximum temperatures obtained from the DFT analyses and the WorldClim database.

Afterwards, we decomposed the variance of temperatures between global interpolated grids and local temperatures measured in agricultural landscapes by mapping the differences in minimum and maximum temperatures between the air local temperatures (Air_L) and the WorldClim interpolated temperatures (Air_{WC}) for the three studied grid cells. Then, to illustrate the part of the variance due to microclimate effects, we mapped the differences in minimum and maximum temperatures between measured local air canopies, soil temperatures ($Layer_L$) and the air local temperatures (Air_L) for the three studied grid cells.

4.3. Pest performance in thermal landscape. As a final step of our analysis, we explored the implication of our thermal landscape mapping for pest performance by comparing temperature frequencies in our grid cells with the temperature-dependent growth curve of a major crop pest species in the region: *Phthorimaea operculella* (Lepidoptera: Gelechiidae). This pest is considered one of the most important potato pests worldwide, but also attacks a wide variety of other crops such as tomato (*Solanum lycopersicum* L.), eggplant (*Solanum melongena* L.) or tobacco (*Nicotiana tabacum* L.) (see [27] for a review). *P. operculella* feeds on different part of the plant (leaves, stems, and tubers) and also tubers in storage structures [26,28]. In agricultural landscapes, *P. operculella* is abundant in virtually all types of habitats (even far from its host plant) because 1) this pest is able to fly over large distances (100–250 m) to infest suitable host plants [29] and 2) a significant quantity of tubers are left in the field after harvest, and are rapidly colonized by the moth before the following crop is planted. It is therefore common to observe infested potato plants in corn or broad bean fields. These left-over potatoes are well know by farmers and agronomists as significant obstacle to the control of these pests [28].

The temperature-dependent growth rate curve of *P. operculella* larvae (in day-1) over a 0–40°C range was obtained using published temperature-response data of laboratory experiments performed in the Andean region (see [30] for details). PTM development rate data were then modeled with the [31] equation as modified by [32]:

$$D(T) = \frac{\frac{dT}{298.16}\exp\left[\frac{e}{R}\left(\frac{1}{298.16}-\frac{1}{T}\right)\right]}{1+\exp\left[\frac{f}{R}\left(\frac{1}{g}-\frac{1}{T}\right)\right]+\exp\left[\frac{h}{R}\left(\frac{1}{i}-\frac{1}{T}\right)\right]} \quad (1)$$

where T is temperature in Kelvin (°C+273.15), $R = 1.987$, and d, e, f, g, h, and i estimated parameters. This model has been widely used to describe the kinetics of insect development based on several assumptions about the underlying developmental control

Table 1. Results of the two-way ANOVA with a Bonferroni correction on the effects of habitat, elevation, LAI and elevation × LAI terms on daytime and nigh-time DFT amplitudes and thermal time lag on inside-canopy and soil temperature time series.

Effect	Canopy				Soil			
	Df	Mean sq	F value	P value	Df	Mean sq	F value	P value
Daytime amplitude								
Habitat	5	6.282	3.370	**0.007***	5	5.745	2.466	0.036
Elevation	2	12.491	6.701	**0.002***	2	5.722	2.456	0.089
LAI	1	40.171	21.551	**<0.001***	1	136.78	58.705	**<0.001***
Elevation × LAI	2	2.513	1.348	0.263	2	0.292	0.125	0.882
Residuals	132	1.864			127	2.330		
Night-time amplitude								
Habitat	5	0.936	3.895	**0.002***	5	1.390	0.839	0.525
Elevation	2	4.539	18.896	**<0.001***	2	2.143	1.293	0.278
LAI	1	1.083	4.509	0.035	1	157.52	95.041	**<0.001***
Elevation × LAI	2	1.754	7.302	0.010	2	0.097	0.059	0.943
Residuals	132	0.240			127	1.657		
Thermal Time Lag								
Habitat	5	0.001	1.297	0.269	5	0.009	3.881	**0.003***
Elevation	2	0.001	5.777	**0.004***	2	0.024	10.139	**<0.001***
LAI	1	0.001	29.322	**<0.001***	1	0.022	9.165	**0.003***
Elevation × LAI	2	0.001	2.374	0.097	2	0.005	2.334	0.101
Residuals	132	0.001			127	0.002		

Bold* indicates significant results ($P<0.05$).

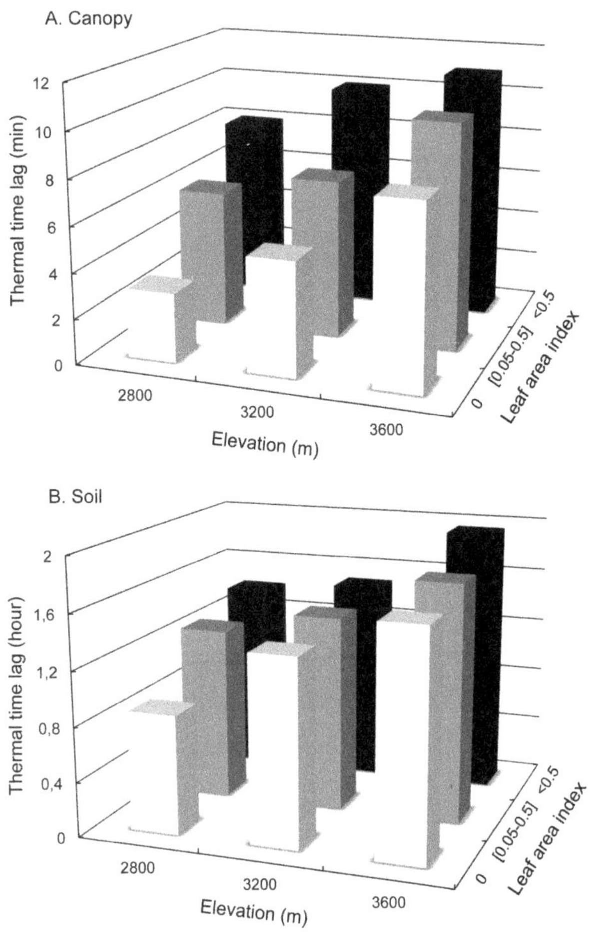

Figure 5. Thermal time lag from Fourier transforms at the daily frequency for canopy (A) and soil temperatures (B) as a function of elevation and leaf area index. The z-axis (log+1 transformed) is expressed in minutes (A) and in hours (B).

enzymes. For instance, it has been used to describe poikilotherms' temperature-dependent development [33].

We then compared the growth rate performance curve of *P. operculella* for local temperature distribution (canopy and soil layer temperatures) and for global interpolated ones (e.g., Fig. 3 in [3]). Distributions of canopy and soil minimum, maximum and mean temperatures were extracted from the three digitized landscapes using the geostatistical analyst extension of ArcGIS. Canopy and soil temperature frequencies were expressed as the percent of total grid cell area. The growth performance model of *P. operculella* given by Eqn. 1 was implemented with WorldClim minimum and maximum temperatures and the local minimum, maximum and mean temperature distribution. This allowed estimating insect growth rate within the range of WorldClim and measured field data.

Results

1. Local vs. global air temperature discrepancies in thermal landscapes

Differences in average minimum and maximum temperatures between local air temperatures and the global coarse grain interpolated air temperatures from the WorldClim (Δ Air $_L$ – Air $_{WC}$) were mapped for the three studied grid cells (Fig. 2). While minimum local air temperatures were cooler than those predicted by WorldClim in 77.5±10% of the studied areas (blue areas, average min Δ Air $_L$ – Air $_{WC}$ = −2.9°K) maximum local air temperatures were warmer than extrapolated temperatures in 82.1±12% of the studied areas (red areas, average max Δ Air $_L$ – Air $_{WC}$ = +5.6°K). This pattern was not influenced by elevation. Notably, for all elevations, local mean air temperatures were quite well predicted by the WorldClim (+/−1°K) as in average 55.3±3.4% of the studied areas felt in the range of Air $_L$ – Air $_{WC}$ ≤1°K (Appendix S8).

2. Temperature discrepancies due to microclimate in agricultural landscapes

Differences in average minimum and maximum temperatures between local canopy and soil temperatures and local air temperatures (Δ Layer $_L$ – Air $_L$) were mapped for the three studied grid cells (Fig. 3). Overall, canopy and soil areas were always cooler than maximum air temperature and were always warmer than air minimum temperatures resulting in a general buffer effect of minimum and maximum air temperatures by canopy and soil layers. The buffer effect on air temperatures was significantly stronger for soil than for canopy layer (see Fig. 4, Student's t-test, $t = -27.10$ and $t = 4.52$, $P<0.001$ for night-time and daytime, respectively). Interestingly, the buffer effect on air temperatures by soil was higher during night-time than daytime (Fig. 4D) while the opposite pattern was found in crop canopy (Fig. 4A).

Elevation had a significant effect on air temperature buffering in the canopy layer but not in the soil layer (Table 1). Contrastingly, LAI had a highly significant thermal buffering effect in both soils (night and daytime) and canopies (daytime, see Table 1). Buffer effect on air temperatures by bare soil (e.g. without plant cover, LAI = 0) ranged from −1.1°K to −2.3°K for daytime and from 3.4°K to 4.3°K for night-time. Crop type had no significant effect on buffering patterns except for potato in which higher buffer effects were recorded (Post-Hoc HSD test, $P<0.05$).

Overall, thermal time lag was much shorter in canopies (7.5±2.6 min) than in soils (1.5±0.3 hours, Fig. 5). LAI classes had a significant positive effect on thermal time lag for both canopy and soil layers (Table 1). On average, thermal time lag increased by 2 min. in canopies and 30 min. in soils between two LAI classes. Similarly, elevation had a significant positive effect on thermal time lag for both canopy and soil layers (Table 1) with an average increase of 2±0.3 min. in canopies and of 60±31 min. in soil between two altitudinal belts (Fig. 5).

3. Thermal performance curve using local vs. interpolated temperatures

To assess the implication of local vs. global interpolated temperature discrepancies for crop pest performances, we plotted the frequency distribution of the minimum (blue bars), maximum (red bars) and mean local (stripped bars) temperatures and those given by WorldClim (from minimum to maximum temperature, shaded region in the background) with the temperature-dependent growth rate curve of the potato moth *P. operculella* (Fig. 6). As a general pattern, global interpolated temperature ranges predicted lower growth rates of *P. operculella* than those predicted by local temperatures at all elevations, in both inside-canopy and soil layers (where the pest lives most of their time). While mean temperature distribution generally fell within the WorldClim min-max range, extreme temperatures (and especially maximum ones) largely exceeded this range.

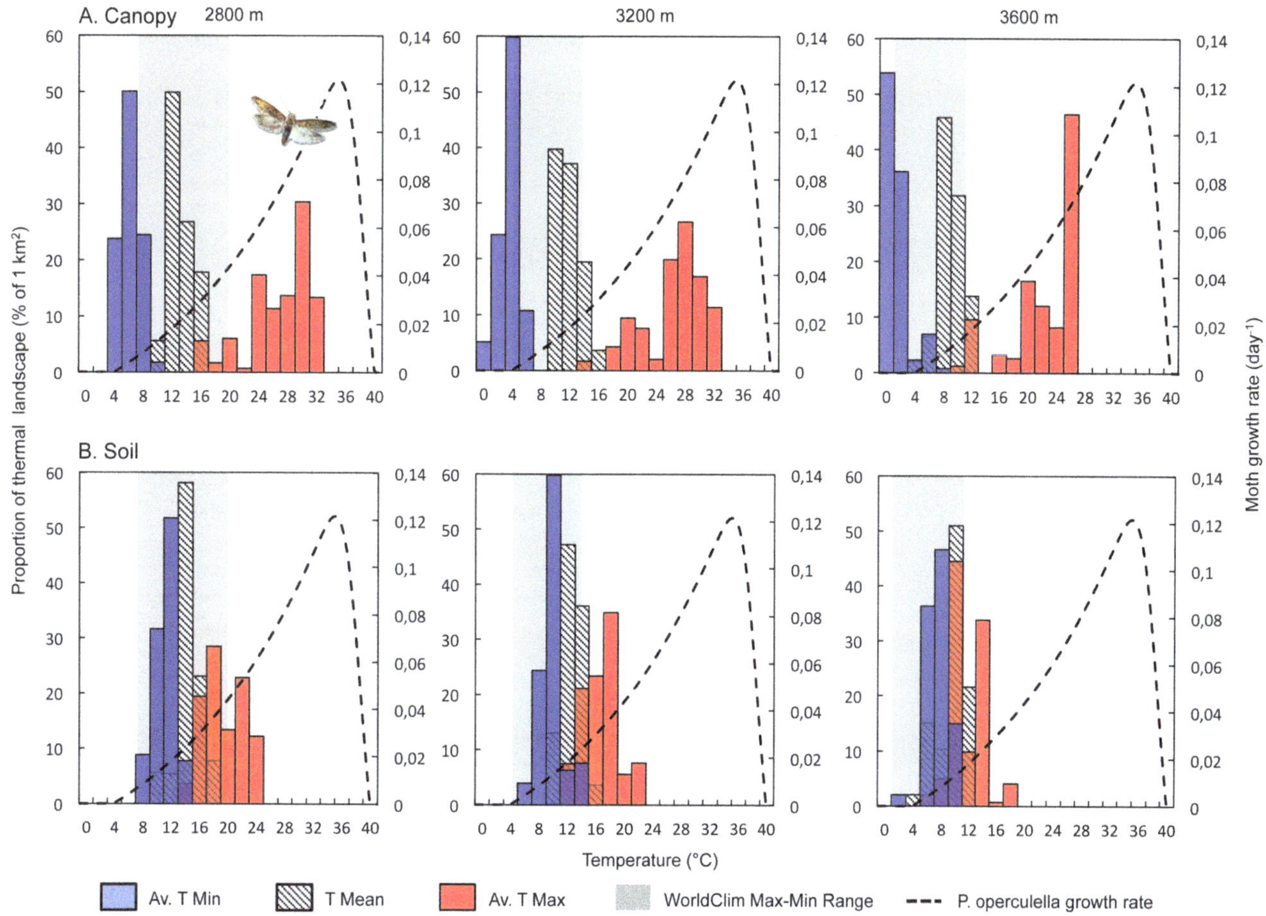

Figure 6. Superimposed plot of the temperature-dependent growth rate curve of the potato moth Phthorimaea operculella (dashed line) and the frequency distribution (% of area) of average minimum (blue), maximum (red) and mean (striped) temperatures for canopy and soil layers at the three studied elevations. Grey (shaded) bands in the background represent the WorldClim minimum and maximum temperature range.

The WorldClim estimations predicted *P. operculella* growth rates ranging between 0.007 and 0.045 day^{-1} at 2800 m, and between 0 and 0.018 day^{-1} at 3600 m, the maximum rates being slightly lower than those predicted by soil temperatures (0.068 day^{-1} at 2800 m and 0.037 day^{-1} at 3600 m). These differences were exacerbated in canopy layers where estimated maximum growth rates were 2.6–4.3 times higher than those predicted by WorldClim (0.118 day^{-1} at 2800 m and 0.079 day^{-1} at 3600 m). Discrepancies between WorldClim and local temperature-based growth rate estimations were not significantly affected by elevation (One-way ANOVA, $F = 7.79$, $P = 0.219$ and $F = 1.67$, $P = 0.419$ for canopies and soils, respectively).

Discussion

Accurate predictions of the responses of organisms to climate change using SDMs require knowledge of microclimates at spatial and temporal scales relevant for studied organisms [13,34,35]. To our knowledge, our study is the first to quantify the thermal heterogeneity among a set of agricultural habitats at fine spatial and temporal scales and to compare those thermal microhabitats to the most widely used global climatic dataset in SDMs. By documenting the mosaic of thermal habitats found in tropical agricultural landscapes, our study confirms previous evidence that microclimates strongly differ from nearby macroclimates due to the variability of air motion and solar radiation patterns created by complex topographies with heterogeneous elevation, slope angle, exposure or roughness [1,7,18,36]. Our study therefore supports the view that results from the long tradition of agrometeorological studies on microclimates (e.g. [6,17,22]) have to be revived in the new context of microhabitat modelling for predicting the response of organisms to climate change.

1. LAI-based and elevation-based climate heterogeneity

In contrast to many previous studies (see [7] for a review), our objective was not to examine the well-documented effect of topography on local temperatures but rather to examine the less-known effects of habitat types and vegetation land cover on thermal landscape features. We found significant thermal time lag and buffer effects on air temperatures by plant and soil layers below crop canopies during night-time and daytime. The top of canopies reflects and absorbs part of the solar radiation during the day, allowing less energy to reach the layers (plants and soils) below canopies. During the night, infrared heat released from both the ground and plants is partly held back by the canopy above [5]. As a consequence plants and soils limit night-time cooling and daytime warming [6], leading to a significant buffer effect of minimum and maximum temperatures [1,4,17]. That is also why we found a buffer effect on air temperatures by soil higher during night-time than daytime and the opposite pattern for crop canopies.

Our results indicate a strong effect of elevation on thermal buffering and thermal time lag by canopy and soil layers. This could result from the combination of a negative relationship between elevation and air temperature and a positive relationship between elevation and solar radiation exposure, part of which is absorbed by plants and soils [6]. As a result, the difference between air temperature and canopy and soil temperature increased with elevation. Interestingly, the modifications of local temperatures by habitats and LAI were of the same magnitude (from -2.70 to $4.82°C$ in average) than that generated by topography-related factors [7,36], supporting the need to better consider habitat effects on microclimates.

2. Fine scale variations in temperature vs. climatic units

Our findings show that the complex agricultural mosaic resulting from habitat types and LAI classes at the landscape scale was a major modifier of the thermal patterns in the studied tropical highlands. More importantly, our findings revealed that, at best, 55% of landscape habitats had real mean air temperatures that were well estimated by WorldClim predictions while in average less than 20% of these areas had minimum and maximum air temperatures well estimated. Additional thermal discrepancies between large and fine-scale temperatures resulted from heterogeneity in crop types and phenologies. This strongly supports the view that the common use of the WorldClim database arrayed into 1-km^2 grids may not adequately capture the reality of the climatic environment experienced by living organisms, in particular tiny ectothermic species [2,3,13,18]. It is important to note that to obtain the highest level of thermal heterogeneity we chose a complex mountainous agricultural study area that provided boundary conditions for climate modelling. Indeed, these mountainous areas provide strong climatic gradients and extreme habitat fragmentation which combined with un-seasonal agrosystem make up a mosaic of thermal patches that expanded the difficulties to faithfully assess climatic parameters for modelling [25]. In view of the urgent need of fine scale climate data with large extent [2,8,35] more research is necessary to develop accurate up- or down-scaling methods, in mountainous locations where thermal heterogeneity is large, and may be needed to properly describe the ecologically significant microclimates [7,37].

3. Microclimates and species distribution models

From tiny insects to mega-herbivores, it is well recognized that species ecology is strongly influenced by micro-climatic features of the landscape [2,10,11,12,13] yet quantitative information on how thermal landscape heterogeneity may affect species performance is scarce. Short-scale differences in temperatures may provide opportunities for individual organisms, even with limited dispersal capabilities, to escape unfavourable microclimates or to maximize physiological performances by selecting preferred microclimates [38,39]. Our analysis showed that predictions on *P. operculella* growth rates strongly differed between Wordclim-based and locally-measured temperatures, suggesting that global species distribution models using global coarse-scale climatic datasets without further microclimate modelling could be strongly limited to accurately predict species occurrence and performance, in particular that of ectotherms living in habitats such as mountain slopes. Such a spatial heterogeneity in thermal patches, where climatic conditions are strongly modified, provides a mosaic of favourable, sub-optimal or lethal thermal habitats that directly influences the performance of natural populations of ectotherms.

Coarse-extent modeling of microclimate is currently one of the major obstacles to predicting how organism will react to their experienced environments and forecast their distribution under climate change [8]. To date, two main types of models have been shown to provide relatively accurate, continent-wide calculations of microclimate: statistical model and mechanistic model [13]. The first one is statistical as the variables are not deterministically but stochastically related. These models perform statistical correlation of species occurrences with climatic data and have proven to be powerful interpolative tools for defining and projecting climatic envelopes [40,41]. A disadvantage of these statistical models is that they can only be applied to the conditions under which they are fitted. On the other hand, mechanistic models of the climatic responses of organisms [13,34] use fundamental knowledge of the interactions between process variables to define the model structure. Therefore they do not require much data for model development and validation. One of them is the Microclim model recently developed by [35,42] for all terrestrial landmasses – except Antarctica– which quantify key microclimatic parameters at macro-scales, with a relatively fine spatial (15 km^2) and temporal resolution (hours). The microclimatic parameters such as wind velocity, humidity, and solar radiation allow building energy and mass budgets of organisms, and therefore serve as key inputs for biophysical models of species distributions.

It is important to highlight that a better spatiotemporal resolution in temperature patterns should go in pair with the development of more accurate temperature-based population dynamics models to integrate it [2,13,34,43]. Existing predictions of models based on insect response measured in constant temperatures may yield different and less realistic results than those from predictions of models that include the effect of real temperature fluctuation on insect biology [33]. For example, to date, we still do not know the impact of a few hours of warm temperature for the performance of ectotherm species at longer time scales [33]. In this context, fine-scale spatiotemporal temperature mapping has revealed a key step for any studies aiming at understanding, predicting and managing the responses of species distributions to climate change.

Supporting Information

Appendix S1 Habitat and field size distribution in the three studied altitudinal belts.

Appendix S2 Photos of the temperature recording experiment.

Appendix S3 Spatial variability of temperatures within a field.

Appendix S4 Comparison of time series analysis outputs using 15 days vs. 1-year temperature data.

Appendix S5 Fourier analysis description.

Appendix S6 Seanonality measured on four year air temperature time series with Discrete Fourier Transform.

Appendix S7 Crop habitat composition survey used in the study area.

Appendix S8 Local and global air mean temperature discrepancies mapping.

Appendix S9 Microclimate temperature time-series used in this work #1.

Appendix S10 Microclimate temperature time-series used in this work #2.

Appendix S11 Microclimate temperature time-series used in this work #3.

Acknowledgments

We are grateful to the city council of Salcedo (Cotopaxi, Ecuador) for providing the digital shape files of the study area cadastre. We also thank all farmers who collaborated with us during fieldwork. And finally, we gratefully acknowledge Dr. Pincebourde S. and Dr. Duyck F. for their constructive scientific comments and suggestions.

Author Contributions

Conceived and designed the experiments: EF OD. Performed the experiments: EF MH. Analyzed the data: EF LB OD. Contributed reagents/materials/analysis tools: EF LB OD. Contributed to the writing of the manuscript: EF LB JFS OD.

References

1. Scherrer D, Korner C (2011) Topographically controlled thermal-habitat differentiation buffers alpine plant diversity against climate warming. J Biogeogr 38: 406–416.
2. Bennie J, Hodgson JA, Lawson CR, Holloway CTR, Roy DB, et al. (2013) Range expansion through fragmented landscapes under a variable climate. Ecol Lett 117: 285–229.
3. Logan ML, Huynh RK, Precious RA, Calsbeek RG (2013) The impact of climate change measured at relevant spatial scales: new hope for tropical lizards. Global Change Biol 19: 3093–3102.
4. Scheffers BR, Edwards DP, Diesmos A, Williams SE, Evans TA (2013) Microhabitats reduce animal's exposure to climate extremes. Global Change Biol 20: 495–503.
5. Geiger R (1965) The climate near the ground. Cambridge, USA.
6. Jones HG (1992) Plants and microclimate: a quantitative approach to environmental plant physiology. Cambridge University Press, Cambridge.
7. Dobrowski SZ (2011) A climatic basis for microrefugia: the influence of terrain on climate. Global Change Biol 17: 1022–1035.
8. Potter KA, Woods HA, Pincebourde S (2013) Microclimatic challenges in global change biology. Global Change Biol 19: 2932–2939.
9. Cloudsley-Thompson JL (1962) Microclimates and the distribution of terrestrial arthropods. Annu Rev Entomol 7: 199–222.
10. Tracy CR (1977) Minimum size of mammalian homeotherms: role of the thermal environment. Science 198: 1034–1035.
11. Willmer PG (1982) Microclimate and the environmental physiology of insects. Adv Insect Physiol 16: 1–57.
12. Unwin DM, Corbet SA (1991) Insects, plants and microclimate. Richmond Publishing Company Ltd.
13. Kearney M, Porter W (2009) Mechanistic niche modelling: combining physiological and spatial data to predict species' ranges. Ecol Lett 12: 334–350.
14. Beaumont LJ, Pitman AJ, Poulsen M, Hughes L (2007) Where will species go? Incorporating new advances in climate modelling into projections of species distributions. Global Change Biol 13: 1368–1385.
15. Deutsch CA, Tewksbury JJ, Huey RB, Sheldon KS, Ghalambor CK, et al. (2008) Impacts of climate warming on terrestrial ectotherms across latitude. P Natl Acad Sci USA 105: 6668–6672.
16. Warren RJ, Chick L (2013) Upward ant distribution shift corresponds with minimum, not maximum, temperature tolerance. Global Change Biol 19: 2082–2088.
17. Suggitt AJ, Gillingham PK, Hill JK, Huntley B, Kunin WE, et al. (2011) Habitat microclimates drive fine-scale variation in extreme temperatures. Oikos 120: 1–8.
18. Sears MW, Raskin E, Angilletta MJ (2011) The world is not flat: defining relevant thermal landscapes in the context of climate change. Integr Comp Biol 51: 666–675.
19. Dangles O, Carpio C, Barragan AR, Zeddam JL, Silvain JF (2008) Temperature as a key driver of ecological sorting among invasive pest species in the tropical Andes. Ecol Appl 18: 1795–1809.
20. McCain CM (2007) Could temperature and water availability drive elevational species richness patterns? A global case study for bats. Global Ecol Biogeogr 16: 1–13.
21. Dangles O, Carpio FC, Villares M, Yumisaca F, Liger B, et al. (2010) Community-based participatory research helps farmers and scientists to manage invasive pests in the Ecuadorian andes. Ambio 39: 325–335.
22. Baldocchi DD, Verma SB, Rosenberg NJ (1983) Microclimate in the soybean canopy. Agr Meteorol 28: 321–337.
23. Wilhelm WW, Ruwe K, Schlemmer MR (2000) Comparisons of three-leaf area index meters in a corn canopy. Crop Sci 40: 1179–1183.
24. Fitzpatrick EA (1964) Seasonal distribution of rainfall in Australia analysed by Fourier methods. Archiv Meteorologie, Geophysik und Bioklimatologie 13: 270–286.
25. Hijmans RJ, Cameron SE, Parra JL, Jones PG, Jarvis A (2005) Very high resolution interpolated climate surfaces for global land areas. Int J Climatol 25: 1965–1978.
26. Crespo-Perez V, Rebaudo F, Silvain JF, Dangles O (2011) Modeling invasive species spread in complex landscapes: the case of potato moth in Ecuador. Landscape Ecol 26: 1447–1461.
27. Rondon S (2010) The potato tuberworm: a literature review of its biology, ecology, and control. Am J Potato Res 87: 149–166.
28. Hanafi A (1999) Integrated pest management of potato tuber moth in field and storage. Potato Res 42: 373–380.
29. Cameron PJ, Walker GP, Penny GM, Wigley PJ (2002) Movement of potato tuberworm within and between crops, and some comparisons with diamondback moth. Environ Entomo 31: 65–75.
30. Crespo-Pérez V, Dangles O, Régnière J, Chuine I (2013) Modeling temperature-dependent survival with small datasets: insights from tropical mountain agricultural pests. Bul Entomol Res 103(03): 336–343.
31. Sharpe PJH, DeMichele DW (1977) Reaction-kinetics of poikilotherm development. J Theor Biol 64: 649–670.
32. Schoolfield RM, Sharpe PJH, Magnuson CE (1981) Non-linear regression of biological temperature-dependent rate models based on absolute reaction-rate theory. J Theor Biol 88: 719–731.
33. Gilbert E, Powell JA, Logan JA, Bentz BJ (2004) Comparison of three models predicting developmental milestones given environmental and individual variation. B Math Biol 66(6): 1821–1850.
34. Buckley LB, Urban MC, Angilletta MJ, Crozier LG, Rissler LJ, et al. (2010) Can mechanism inform species' distribution models? Ecol Lett 13: 1041–1054.
35. Kearney M, Shamakhy A, Tingley R, Karoly DJ, Hoffmann AA, et al. (2013) Microclimate modelling at macro scales: a test of a general microclimate model integrated with gridded continental-scale soil and weather data. Global Change Biol Doi : 10.1111/2041-210X.12148.
36. Scherrer D, Korner C (2010) Infra-red thermometry of alpine landscapes challenges climatic warming projections. Global Change Biol 16: 2602–2613.
37. Fridley JD (2009) Downscaling climate over complex terrain: high finescale (< 1000 m) spatial variation of near-ground temperatures in a montane forested landscape. J Appl Meteorol Clim 48: 1033–1049.
38. Kinahan AA, Pimm SL, van Aarde RJ (2007) Ambient temperature as a determinant of landscape use in the savanna elephant, Loxodonta africana. J Therm Biol 32: 47–58.
39. Dillon ME, Liu R, Wang G, Huey RB (2012) Disentangling thermal preference and the thermal dependence of movement in ectotherms. J Therm Biol 37: 631–639.
40. Guisan A, Thuiller W (2005) Predicting species distribution: offering more than simple habitat models. Ecol Lett 8: 993–1009.
41. Elith J, Leathwick JR (2009) Species distribution models: ecological explanation and prediction across space and time. Annu Rev Ecol Evol S 40: 677–697.
42. Kearney MR, Isaac AP, Porter WP (2014) microclim: Global estimates of hourly microclimate based on long-term monthly climate averages. Scientific Data 1.
43. Bakken GS, Angilletta MJ (2014) How to avoid errors when quantifying thermal environments. Func Ecol 8: 96–107.
44. Bloomfield P (2004) Fourier analysis of time series: an introduction. Wiley-Interscience, New York, USA.

14

Variation in Yield Gap Induced by Nitrogen, Phosphorus and Potassium Fertilizer in North China Plain

Xiaoqin Dai*, Zhu Ouyang, Yunsheng Li*, Huimin Wang

Key Laboratory of Ecosystem Network Observation and Modeling, Institute of Geographic Sciences and Natural Resources Research, Chinese Academy of Sciences, Beijing, China

Abstract

A field experiment was conducted under a wheat-maize rotation system from 1990 to 2006 in North China Plain (NCP) to determine the effects of N, P and K on yield and yield gap. There were five treatments: NPK, PK, NK, NP and a control. Average wheat and maize yields were the highest in the NPK treatment, followed by those in the NP plots among all treatments. For wheat and maize yield, a significant increasing trend over time was found in the NPK-treated plots and a decreasing trend in the NK-treated plots. In the absence of N or P, wheat and maize yields were significantly lower than those in the NPK treatment. For both crops, the increasing rate of the yield gap was the highest in the P omission plots, i.e., 189.1 kg ha^{-1} yr^{-1} for wheat and 560.6 kg ha^{-1} yr^{-1} for maize. The cumulative omission of P fertilizer induced a deficit in the soil available N and extractable P concentrations for maize. The P fertilizer was more pivotal in long-term wheat and maize growth and soil fertility conservation in NCP, although the N fertilizer input was important for both crops growth. The crop response to K fertilizers was much lower than that to N or P fertilizers, but for maize, the cumulative omission of K fertilizer decreased the yield by 26% and increased the yield gap at a rate of 322.7 kg ha^{-1} yr^{-1}. The soil indigenous K supply was not sufficiently high to meet maize K requirement over a long period. The proper application of K fertilizers is necessary for maize production in the region. Thus, the appropriate application of N and P fertilizers for the growth of both crops, while regularly combining K fertilizers for maize growth, is absolutely necessary for sustainable crop production in the NCP.

Editor: Carl J. Bernacchi, University of Illinois, United States of America

Funding: This study was financially supported by the Strategic Science Plan of Institute of Geographic Sciences and Natural Resources Research (2012ZD004)and National Science and Technology Support Projects of Ministry of Science and Technology of China (2013BAD05B03). The funders had no role in study design, data collection and analysis, decision to publish, or preparation of the manuscript.

Competing Interests: The authors have declared that no competing interests exist.

* E-mail: daixq@igsnrr.ac.cn (XD); liys@igsnrr.ac.cn (YL)

Introduction

The North China Plain (NCP) is the largest and the most important agricultural production region in China. It covers approximately 18.3% of total national farm lands (18 million hectares) and produces 21.6% of the total grain yield of edible crops in the country [1]. The main crops grown are wheat and maize, periodically rotated. The NCP now supplies more than 50% of the nation's wheat and 33% of its maize [2]. Grain production and the maintenance of soil fertility in the NCP are very important for the food security and agricultural sustainability of China.

Nutrient availability is the most yield-limiting factor. To produce higher yields, the over-application of chemical fertilizers has been a common practice in wheat-maize rotation systems and has led to severe environmental problems [3]. Improved nutrient management practices are urgently needed to maximize crop yields and maintain soil fertility while minimizing environmental impacts. The ability to better identify crop response to the application of fertilizers, soil indigenous nutrient supply capability, and the maintenance of soil fertility over time are crucial to the development of improved nutrient management practices. Various long-term experiments have been conducted to test the effects of fertilization on yield or soil fertility throughout the world [4–14], but over long time scales, crop response to the application of N, P and K fertilizers and soil indigenous nutrient supply capability has seldom been clearly understood.

The yield and soil fertility gap between a full NPK fertilizer plot and a fertilizer omission plot was used as a good diagnostic tool to assess the extent of macronutrient limitations. Mussgnug et al (2006) analyzed the yield gap resulting from nutrient limitation on a degraded soil in the Red River Delta (RRD) in Northern Vietnam. They found that in the absence of K application, the yield gap for rice and maize respectively averaged 1.7 Mg ha^{-1} and 3.4 Mg ha^{-1}, while when N or P was omitted the yield gap was less. Potassium was the most yield-limiting macronutrient in RRD of Vietnam [15]. The factors that primarily limit increasing crop yields varied by crop and region [16]. The objectives of the study were to examine the effects of the continuous use of inorganic fertilizers on wheat and maize yield and soil fertility gap to elucidate crop response to fertilizer inputs and the long-term N, P and K supply capability in the NCP.

Materials and Methods

Experimental Site

A long-term field experiment was conducted from 1990 to 2006 at the Yucheng Comprehensive Experimental Station (36°57′ N,

Table 1. Physical and chemical properties of soil at different layers at Yucheng, Shandong Province of China in 1990.

Soil parameters	0–20 cm	20–40 cm
Soil texture	Silt loam	Sandy loam
Clay (%)	17.3	13.7
Silt (%)	61.4	11.4
Sand (%)	21.3	74.9
Bulk density (g cm^{-3})	1.31	1.48
Field capacity (%)	23.3	22.7
Organic C (g kg^{-1})	5.4	4.8
Total N (g kg^{-1})	0.41	0.36
Total P (g kg^{-1})	1.11	1.13
Total K (g kg^{-1})	10.7	14.0
Available N (mg kg^{-1})	46.1	41.4
Available P (mg kg^{-1})	8.3	5.4
Available K (mg kg^{-1})	149.0	125.4

116°36′ E, 28 m a.m.s.l.) of the Chinese Academy of Sciences (CAS) located in Shandong Province, which lies in the NCP. This area represents the moderate- to high-yielding region of the NCP, where the dominant cropping system is double-cropped wheat-maize. The annual mean precipitation is 575 mm, with approximately 70% of the total precipitation falling between June and September. The soil is classified as fluvo-aquic loam soil. Before the initiation of the experiment, soil samples were collected from depths of 0–20 cm and 20–40 cm by stepwise soil augers in the experimental field in 1990. All samples from the same soil layers were mixed thoroughly to create one homogenous sample, and a representative sample was drawn to determine the soil texture, bulk density, organic carbon, total N, P and K, and available N, P and K. The physical and chemical properties of the soil in 1990 are listed in Table 1.

Experimental Design and Treatments

The study consisted of five treatments: no fertilization (control); N, P, and K applied (NPK); N and P applied (NP; K omission); N and K applied (NK; P omission); P and K applied (PK; N omission), with four replications performed for each condition in a randomized complete block design (Table 2). The application rates

Table 2. Treatment and fertilizer nutrient rates (kg ha^{-1}) applied to winter wheat and maize at Yucheng, Shandong Province of China.

Treatments	Winter wheat			Maize		
	N	P_2O_5	K_2O	N	P_2O_5	K_2O
NPK	253	90	171	262	52	298
PK	0	90	171	0	52	298
NK	253	0	171	262	0	298
NP	253	90	0	262	52	0
Control	0	0	0	0	0	0

of N, P and K, if they were applied, were the same in all of the treatments: 253 kg N ha^{-1} as urea (46% N), 90 kg P_2O_5 ha^{-1} as single superphosphate (16% P_2O_5), and 171 kg K_2O ha^{-1} as potassium sulfate (50% K_2O) for winter wheat and 262 kg N ha^{-1} as urea (46% N), 52 kg P_2O_5 ha^{-1} as single superphosphate (16% P_2O_5), and 298 kg K_2O ha^{-1} as potassium sulfate (50% K_2O) for maize (Table 2). Phosphate was applied before sowing and incorporated during land preparation. Winter wheat received 40% of the total N as a basal dressing before sowing, whereas the remainder was top-dressed in two splits (40% at green-up and 20% at flowering). Approximately 50% of the K was applied before sowing and 50% top-dressed at green-up (the first leaf growing in the early of spring growed to 1–2 cm from soil surface and the phenomena was seen on a half of wheat seedling in the field). During the maize season, all of the P and K fertilizers and 35% of the N fertilizer were applied as a basal fertilizer, and 65% of the N was applied at the elongation stage. Each plot area measured 6 m×5 m, and all plots were isolated from one another by a concrete wall to a soil depth of 1 m.

Winter wheat (*Triticum aestivum*) was double-cropped with maize (*Zea mays*), and the crops were grown along the border of the four blocks. Winter wheat was sown in mid-October and harvested in early June of the next year. Maize was sown in mid-June and harvested in early October of the same year. The winter wheat was seeded at a rate of 225 kg ha^{-1}, and the maize was planted at a rate of 75000 seeds ha^{-1}. Surface irrigations were conventionally conducted according to the conditions of crop growth and soil moisture. Wheat and maize were harvested to the level of the soil surface; thus, the stubble left in the field was negligible. However, the roots were left in the soil. All straws were removed from the field. All plots were kept free of weeds by the application of an herbicide. Pesticides were applied in accordance with good practices for crop protection.

Sampling and Chemical Analyses

At harvestable maturity, the grain yields were determined over the whole plot area. Soils samples were collected at a depth of 20 cm at six randomly selected points in every plot after wheat and maize were harvested each year. The soil samples were mixed thoroughly from all cores to obtain a representative soil sample for each plot. The soil samples were air-dried and sieved through a 1-mm mesh to perform available N, Olsen-P and available K measurements and through a 0.149-mm mesh to estimate the organic carbon and total N content. Soil available N was determined using the alkali-hydrolytic diffusion method [17]. Olsen-P was measured by ascorbic acid-molybdate blue colorimetry [18]. Available K was extracted with an ammonium acetate solution (NH_4OAc, 1 mol/L) and then determined with a flame photometer [17]. Soil organic carbon was determined using the wet oxidation method of Walkley and Black, and total N was measured following Kjeldahl digestion and distillation [17].

Data Analyses

The results of yield and soil fertility in the first three years of the experiment were not used because in the autumn of 1993 soybean was sown. Therefore, the results reported are from 1994 to 2006. Crop yield losses caused by N, P, K or NPK omission were calculated from the differences between an NPK treatment and the PK, NK, NP or control plot during the same years. The yield differences were called the "yield gap" induced by N (GYG_N), P (GYG_P), K (GYG_K) or NPK (GYG_{NPK}) fertilizer omission to assess the extent of macronutrient limitations in the study. The differences in soil available N, Olsen-P and available K between the NPK treatment and the PK, NK, NP or control plots were

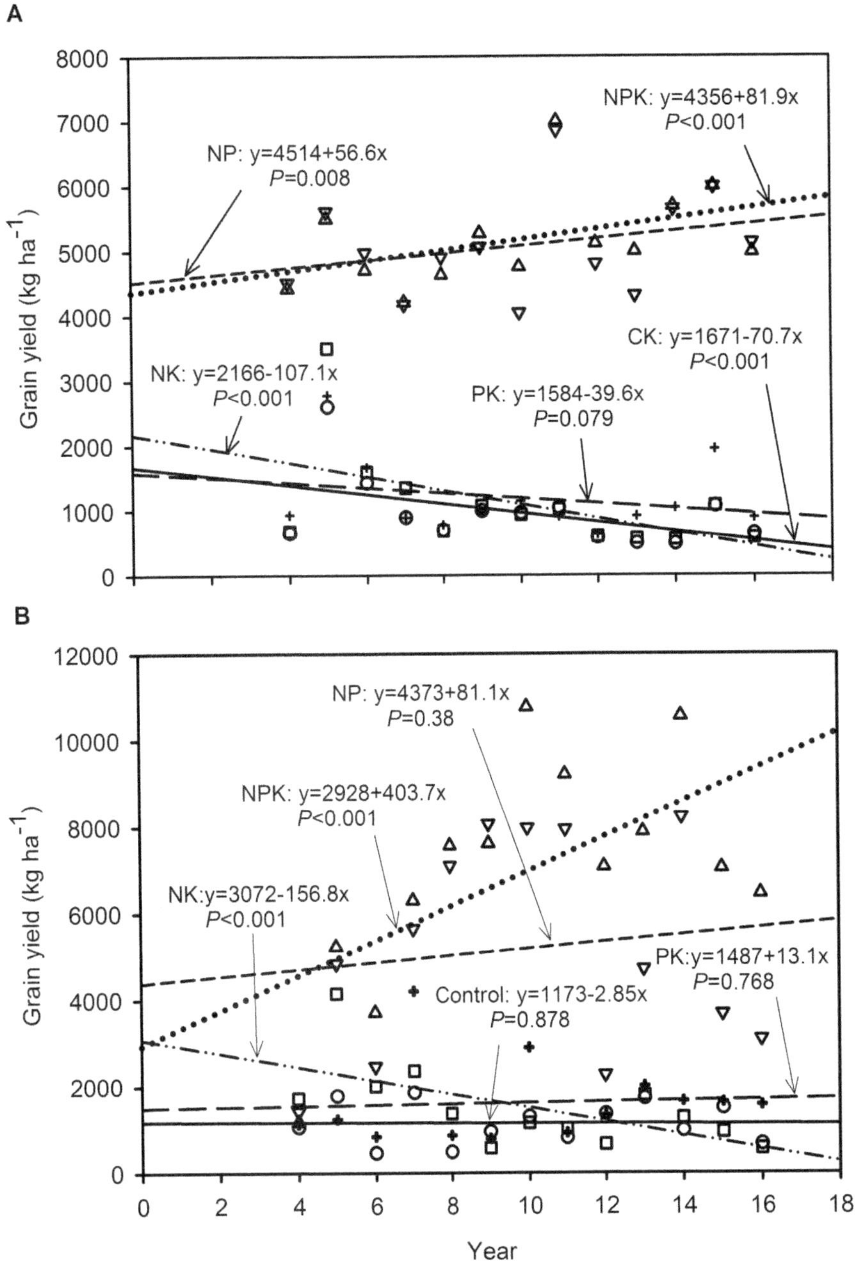

Figure 1. Grain yield of wheat (a) and maize (b) of different fertilizer treatments (Control, circle; NPK, triangle up; NP, triangle down; NK, square; PK, plus) under wheat-maize rotation system from 1994 to 2006 at Yucheng, Shandong Province of China.

calculated to evaluate the variation in soil fertility when nutrients were limited. The soil nutrient differences were called the "soil nutrient gap (SNG)" induced by N (SNG_N), P (SNG_P), K (SNG_P), or NPK (SNG_{NPK}) fertilizer omission.

All statistical analyses including the analysis of variance and regression were conducted using the SPSS package (version 16.0). Considering the data over time was lack of independence, the repeated measures ANOVA between differences in mean through years among treatments for various parameters (grain yield, soil organic C, total N, available N, Olsen-P, available K) were analyzed using a Fisher's protected least significant difference (LSD) test at $P=0.05$. A mixed models with random block and block×treatment were done to assess trends (slopes) of grain yield, yield gap and various soil nutrients gap over the years. The *P*-values of the slopes were used to test whether the observed changes were significantly different from 0.

Results

Grain Yield and Yield Gap

The fertilizer treatments significantly affected the grain yields (Table 3). The wheat and maize yields were highest in the NPK treatment, followed by those in the NP treatment. In the PK or NK treatments, the yields of wheat and maize were significantly

Table 3. Multi-year average grain yield of wheat and maize, as affected by various treatments with fertilizer in a wheat-maize rotation system at Yucheng, Shandong Province of China (mean±standard error).

Treatment	Wheat (kg ha^{-1})	Maize (kg ha^{-1})
Control	963±155 b	1145±133 c
NPK	5174±201 a	6965±724 a
NP	5080±216 a	5184±666 b
NK	1094±218 b	1504±270 c
PK	1188±165 b	1618±269 c

Different letters within a column indicate a significant difference at $P<0.05$ between fertilizer treatments.

lower than those in the NP and NPK treatments. Compared to those in the NPK treatment, the yields of wheat and maize decreased, respectively, by 77% with the N omission, 79% and 78% with the P omission, 2% and 26% with the K omission, and 81% and 84% with no fertilizer. The grain yield losses were much greater in either N or P omission than in plots fertilized combined N and P. Over time wheat yield significantly increased at a rate of 81.9 kg ha^{-1} yr^{-1} in NPK plots ($P=0.008$), and maize yield also significantly increased at a rate of 403.7 kg ha^{-1} yr^{-1} in NPK plots ($P<0.001$, Fig. 1). The yields of wheat and maize in NK treatment decreased significantly at a rate of 107.1 kg ha^{-1} yr^{-1} ($P<0.001$) and 156.8 kg ha^{-1} yr^{-1} ($P<0.001$) over time during the study period, respectively (Fig. 1). The long-term omission of K fertilizer also significantly reduced the maize yield compared to the maize yield of the NPK-treated plot (Table 3). But both crops yields had no significant upward or downward trends when K fertilizer was omitted (Fig. 1).

The variation in the yield gap of wheat and maize between the NPK and PK, NK, NP and control plots is illustrated in Figure 2. For wheat, the yield gap significantly increased at a rate of 152.6 kg ha^{-1} yr^{-1} for the control plots ($P<0.001$), a rate of 189.1 kg ha^{-1} yr^{-1} for the NK plots ($P<0.001$) and a rate of 121.5 kg ha^{-1} yr^{-1} for the PK plots ($P<0.001$). In the NP plots, the yield gap also gradually increased with the cumulative effect of K omission, but the relationship was not significant ($P=0.246$). However, for maize, when the N, P K or NPK fertilizer was omitted, the yield gap significantly increased with the continuous omission of nutrients. The significant increase in the yield gap was 390.6 kg ha^{-1} yr^{-1} for the PK treatments ($P<0.001$), 560.6 kg ha^{-1} yr^{-1} for the NK treatments ($P<0.001$), 322.7 kg ha^{-1} yr^{-1} for the NP treatments ($P<0.001$), and 406.6 kg ha^{-1} yr^{-1} for the control treatments ($P<0.001$). For both crops, the increasing rate of the yield gap was highest in the P omission plots, which indicated that the cumulative effect of P omission was the most significant for wheat and maize in the region. The slopes of the yield gap were the second highest in the N omission plots for both crops, which indicated that N fertilizer inputs were important for wheat and maize growth, especially the latter. It is worth noting that the slopes of the maize yield gap increased significantly with continuous K omission. This result indicates that the soil indigenous K supply was not sufficiently high to meet the maize K requirement over a long period.

Soil Nutrient Status

Long-term continuous treatment with different fertilizers significantly affected the soil organic carbon, available N, Olsen-P and available K concentrations (Table 4). The multi-year average soil organic carbon content was significantly lower in the control, NK and PK plots than in the NPK and NP plots for wheat and maize. The decrease was 6.7% to 15.6% for wheat and 7.4% to 15.8% for maize. The different treatments except control had no significant influences on soil total N. The available N, Olsen-P and available K concentrations were higher in the plots treated with the corresponding N, P or K fertilizer inputs (Table 4). For wheat, the available N concentration increased significantly in the acquired N fertilizer inputs plots compared to that in the PK plots. The Olsen-P concentration in plots with P fertilizer inputs increased significantly by 7.2 to 12.3 times the concentration in the control plots. Meanwhile, the Olsen-P concentration also increased by 9% and 62% in the NP and PK plots, respectively, compared with that in the NPK plots, indicating that the P uptake of the crops was lower in the NP and PK plots. Similar to the Olsen-P concentration, the available K concentration increased significantly in the plots with K fertilizer inputs, and the increase was as high as 1.2 to 2.3 times the K concentration in the control plots. In the NK and PK plots, the available K concentration increased by 48% and 55%, respectively, compared to that in the NPK plots.

The differences in soil nutrients between the NPK and PK, NK, NP or NPK omission plots are shown in Figure 3&4. During the wheat growth season, the continuous omission of nutrients had no significant effects on the difference in the soil available N concentration (Fig. 3a). The continuous omission of P fertilizer significantly enhanced the differences in the soil Olsen-P concentration in the NPK plots ($P<0.001$), indicating that the long-term absence of P fertilizer resulted in a great deficit in the soil available P concentration. Similarly to the P omission plots, the control plots also possessed a significantly enhanced soil Olsen-P gap ($P<0.001$). However, a surplus concentration of Olsen-P was observed when the continuous N was omitted ($P=0.006$, Fig. 3b). Also a surplus concentration of available K was observed when the continuous P and N inputs were omitted ($P<0.001$) due to a large amount of K fertilizer inputs and lower crop uptake. When the K fertilizer was omitted, the soil available K concentration gap significantly increased ($P=0.001$ for NP and $P=0.002$ for Control), indicating a deficit in soil available K in the NP and Control plots.

During the maize growth season, the soil available N concentration gradually became deficient when the nutrients were omitted, but only when P fertilizer was omitted, the effects were significant (Fig. 4a). Like wheat, P fertilizer omission induced a severe deficit in soil Olsen-P ($P<0.001$). Although the PK plots feature a larger amount of P fertilizer inputs and lower crop uptake, no significant surplus in the soil Olsen-P concentration was observed (Fig. 4b). In the plots fed K fertilizer inputs, a surplus in the soil available K concentration was observed, but the effects were not significant (Fig. 4c). It is worthy to note that the NP plots induced a deficit in soil available K (Table 4). Although the effects were not significant, maize growth was significantly affected due to the shortage of soil available K.

Discussion

The response of wheat and maize growth of combined N and P fertilization (i.e. NP) was much greater than that of K fertilizer application combined with N or P (i.e. NK and PK treatments) (Table 3). The continuous 16-yr omission of N or P fertilizer significantly reduced the wheat and maize yield (Table 3). The grain yield in the N omission plots exhibited no significant decreasing trends over time (Fig. 1a). On the one hand, the

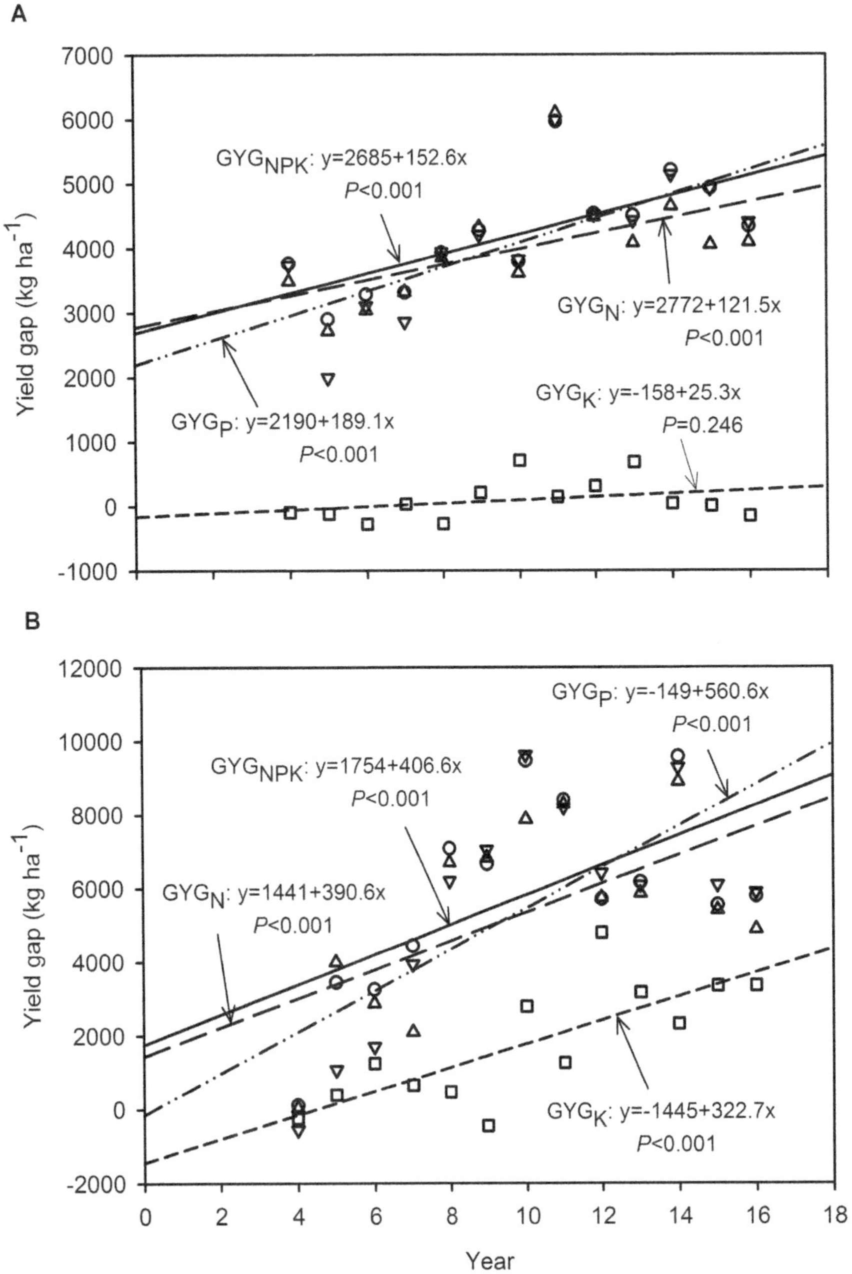

Figure 2. Yield gap variation of wheat (a) and maize (b) by N (GYG_N, the yield difference between NPK and PK plots in each year, triangle up), P (GYG_P, the difference between NPK and NK plots, triangle down), K (GYG_K, the difference between NPK and NP plots, square) or NPK (GYG_{NPK}, the difference between NPK and Control plots, circle) fertilizer omission from 1994 to 2006 at Yucheng, Shandong Province of China.

continuous N omission significantly reduced the soil available N concentration, but no significant effects were observed on the total N concentration (Table 4). The differences in soil available N between the NPK and PK plots also exhibited no significant increase over time (Fig. 3a&4a). It might be possible that atmospheric nitrogen deposition in the NCP soil, N mineralization and lower nitrogen depletion may have attributed to the trend in available N over the years in the nitrogen omitted plots. Liu et al. (2006) and Zhang et al. (2008) reported that the bulk deposition of inorganic N in the NCP is approximately 30 kg/ha per year, which has a significant effect on agricultural systems [19,20]. But yield gap significantly increased over time when the N was omitted (Fig. 2), indicating that the soil indigenous N supply combined with the deposition of N could not meet the N requirement for wheat and maize growth in the NCP.

Both of the wheat and maize yields had significant downward trends in P omission treatment (Fig. 1) and the yield gap of both crops significantly increased with the continuous omission of P fertilizer (Fig. 2), indicating that the cumulative absence of P fertilizers significantly inhibited wheat and maize growth. Tang

Table 4. Multi-year average soil nutrient status at wheat and maize harvest, as affected by various treatments with fertilizer in a wheat-maize rotation system at Yucheng, Shandong Province of China (mean±standard error).

Crops	Treatments	Organic C (g kg^{-1})	Total N (g kg^{-1})	Available N (mg kg^{-1})	Olsen-P (mg kg^{-1})	Available K (mg kg^{-1})
Wheat	Control	5.5±0.19 c	0.56±0.04b	52.5±5.15b	3.1±0.28c	104.5±5.50c
	NPK	6.5±0.21a	0.66±0.05a	63.3±4.35a	25.2±1.95b	226.1±16.0b
	NP	6.3±0.15ab	0.67±0.05a	62.7±4.07a	27.5±2.04b	91.6±8.72c
	NK	5.9±0.17abc	0.64±0.05a	62.5±4.17a	3.2±0.31c	333.4±40.3a
	PK	5.8±0.19bc	0.56±0.04ab	49.1±4.20b	40.8±3.98a	349.3±35.3a
Maize	Control	5.7±0.13c	0.66±0.02ab	56.9±6.11b	2.7±0.32c	106.5±5.22c
	NPK	6.7±0.15a	0.71±0.05a	65.8±6.45a	22.1±2.47b	283.7±24.6b
	NP	6.5±0.15ab	0.69±0.04ab	66.6±6.05a	21.7±1.78b	90.9±4.46c
	NK	6.0±0.13bc	0.66±0.04ab	68.4±6.76a	3.6±0.54c	423.8±43.7a
	PK	5.8±0.14bc	0.62±0.03b	52.1±5.90b	33.5±2.32a	438.8±39.8a

For each crop, different letters within a column indicate a significant difference at $P<0.05$ between fertilizer treatments.

et al. (2008) also reported that wheat and maize yields without P fertilization significantly decreased over time in several other sites in China [21]. The results are different from those reported under European conditions, most likely due to differences in the P supplying abilities of the different soils [22]. It is noted that the continuous omission of P fertilizer significantly reduced the soil Olsen-P concentration for both wheat and maize (Table 4). Meanwhile, the differences in the Olsen-P concentration between the NPK and NK plots significantly increased with continuous P omission (Fig. 3b&4b). The results indicate that the long-term absence of P fertilizer resulted in a great deficit in soil available P. It is interesting that the continuous omission of P fertilizer increased the available N concentration gap between the NPK and NK plots, and the effects were significant for maize (Fig. 4a), which indicates that a large amount of available N was depleted or lost, although the plots were fed N fertilizer inputs. The cumulative absence of P fertilizer intensified the shortage of N for crop growth. Because the treatments fertilized without P had a lower the number of cultivable microorganisms, microbial biomass and community functional diversity than in the treatments with P fertilization [23], which possibly induced the lower N and P mineralization in treatments fertilized without P.

The yield response to K fertilizer was much lower than that to N or P fertilizer (Table 3), most likely due to the high inherent soil K levels, which were in excess of the crop K demands. Most soils of the alluvial floodplain in Asia are high in K, and K is a rare limiting factor [24]. Shen et al. (2004) conducted a 14-yr field trial in Hebei Province of northern China and indicated that the application of N and P enhanced rice yields, while K had no yield-increasing effect due to the large resource of available soil K [25]. In our study, wheat and maize yield in the plots of combined NP and K fertilizer significantly improved, and the yield gap of maize significantly increased with continuous K omission; however, that of wheat yield did not (Fig. 1&2). Compared to that in the NPK plots, the soil available K concentration significantly decreased in the NP plots (Table 4). The differences in soil available K between NPK and NP also improved with the continuous omission of K, and the effects were significant during the wheat season (Fig. 3c). Although soil available K was gradually depleted, the soil indigenous K supply was sufficiently high to meet the requirement for normal wheat growth over the 16-yr period of the study. If the K fertilizer was continuously omitted for a longer period, the soil available K deficit would possibly inhibit wheat growth, which must be further verified. In addition, the soil available K gap between the NPK and NP plots during maize growth season showed an increasing trend with the omission of K, but the differences were not significant (Fig. 4c). This behavior is because the long-term omission of K fertilizer significantly reduced the maize yield, which resulted in lower K uptake. The results show that long-term maize production is much more sensitive to the absence of K than wheat production in the NCP. The results are supported by those of Tan et al. (2007), who concluded that the effect of K fertilizer on maize was higher than that on wheat under the wheat-maize rotation system of Hebei [26].

In our study, the combination of N and P fertilizers mostly sustained soil organic carbon, total N and available N, P and K levels over time (Table 4). This result is similar to that of Shen et al. (2004), who reported that the soil organic carbon and total N concentrations remained stable over time [25]. Manna et al. (2005) also showed that the recommended NPK plots are adequate for maintaining a constant SOC content under the sub-humid and semi-arid tropical conditions of India over a long period [27]. When N fertilizer was omitted, soil available P and K significantly improved in the plots with P and K fertilizer (PK). Meanwhile, when P fertilizer was omitted, soil available K was significantly accumulated in the plots with N and K fertilizer (NK, Table 4). Because the deficiency of other nutrients inhibited crop production, both crops exhibited lower P or K uptake. The application of inorganic fertilizer allowed the crops to exceed production needs and thus resulted in a substantial build-up of available P or available K, which inevitably induced resource waste and ecological pollution.

Conclusions

The P fertilizer was more pivotal in the long-term growth of wheat and maize and the conservation of soil fertility in the NCP, although the N fertilizer input was important for the growth of both crops as well. Although the crop yield response to K fertilizer was much lower than to N or P fertilizer, the proper application of K fertilizer is also necessary, especially for maize production in the

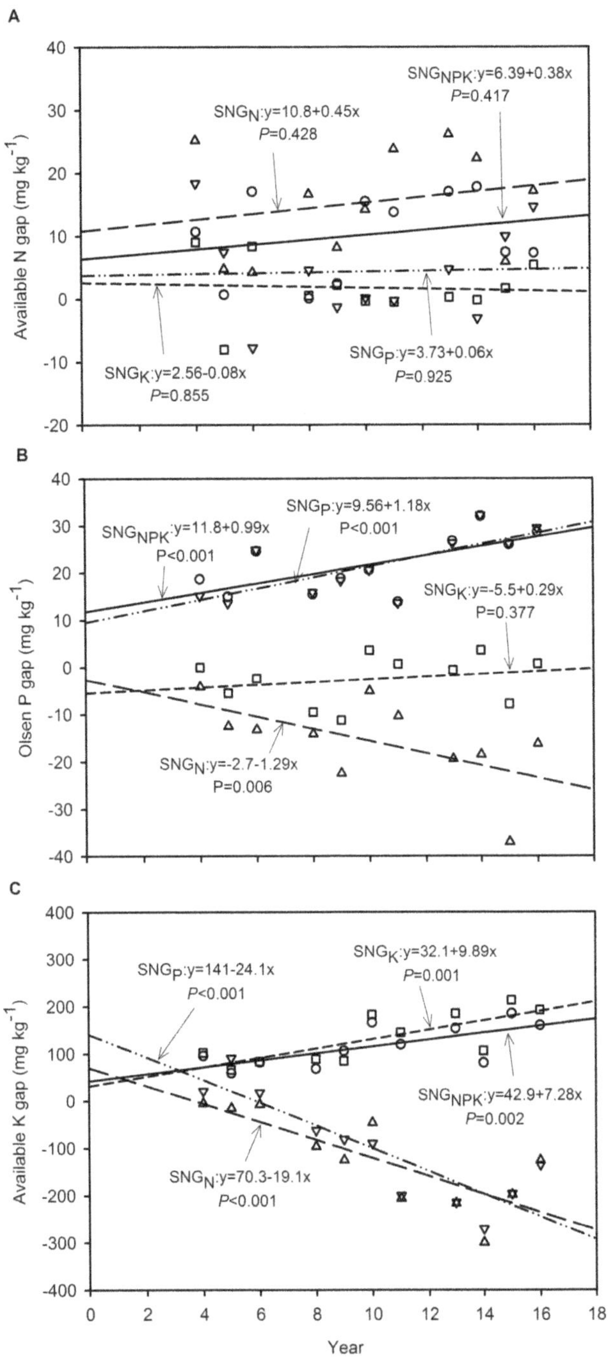

Figure 3. Soil available N (a), Olsen-P (b) and available K (c) gap variation by N (SNG_N, soil available nutrient difference between NPK and PK plots, triangle up), P (SNG_P, the difference between NPK and NK plots, triangle down), K (SNG_K, the difference between NPK and NP plots, square) or NPK (SNG_{NPK}, the difference between NPK and Control plots, circle) fertilizer omission from 1994 to 2006 during wheat season at Yucheng, Shandong Province of China.

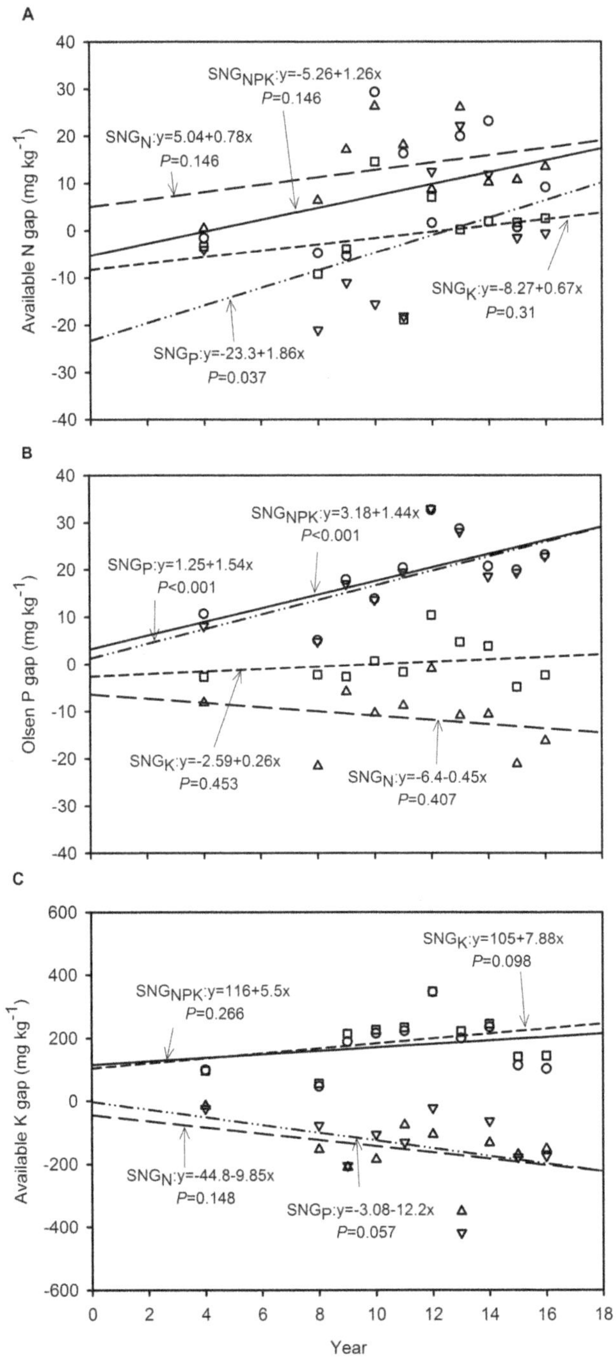

Figure 4. Soil available N (a), Olsen-P (b) and available K (c) gap variation by N (SNG_N, soil available nutrient difference between NPK and PK plots, triangle up), P (SNG_P, the difference between NPK and NK plots, triangle down), K (SNG_K, the difference between NPK and NP plots, square) or NPK (SNG_{NPK}, the difference between NPK and Control plots, circle) fertilizer omission from 1994 to 2006 during maize season at Yucheng, Shandong Province of China.

region. Thus, the appropriate application of N and P fertilizers for both crops, in combination with regular K fertilizers for maize, is absolutely necessary in terms of sustainable crop production in the NCP. However, a longer-range study is required to verify whether the soil indigenous K supply could continue to meet the requirement for wheat growth over many years.

Acknowledgments

The authors are grateful to Zhenrong Tian in the Yucheng Comprehensive Experimental Station of the Chinese Academy of Sciences (CAS) for many sampling work and Lynn M. Johnson in Cornell University for statistical analysis. Authors also thank the academic editor and anonymous reviewers for their constructive comments, which helped in improving the manuscript.

Author Contributions

Conceived and designed the experiments: ZO YL. Performed the experiments: XD YL. Analyzed the data: XD HW. Wrote the paper: XD.

References

1. Yu Q, Saseendran SA, Ma L, Flerchinger GN, Green TR, et al. (2006) Modeling a wheat-maize double cropping system in China using two plant growth modules in RZWQM. Agr Syst 89: 457–477.
2. China Statistics Bureau (2001) China Statistics Bureau, Annual Agricultural Statistics of Henan Province. Beijing, China Statistics Press. 112–115.
3. Vitousek PM, Naylor R, Crews T, David MB, Drinkwater LE, et al. (2009) Nutrient imbalances in agricultural development. Science 324: 1519–1520.
4. Berzsenyi Z, Győrffy B, Lap D (2000) Effect of crop rotation and fertilization on maize and wheat yields and yield stability in a long-term experiment. Eur J Agron 13: 225–244.
5. Bi LD, Zhang B, Liu GR, Li ZZ, Liu YR, et al. (2009) Long-term effects of organic amendments on the rice yields for double rice cropping systems in subtropical China. Agr Ecosyst Environ 129: 534–541.
6. Cai ZC, Qin SW (2006) Dynamics of crop yields and soil organic carbon in a long-term fertilization experiment in the Huang-Huai-Hai Plain of China. Geoderma 136: 708–715.
7. Fan T, Stewart BA, Wang Y, Luo J, Zhou G.(2005) Long-term fertilization effects on grain yield, water-use efficiency and soil fertility in the dryland of Loess Plateau in China. Agr Ecosyst Environ 106: 313–329.
8. Glendining MJ, Powlson DS, Poulton PR, Bradbury NJ, Palazzo D, et al. (1996) The effects of long-term applications of inorganic nitrogen fertilizer on soil nitrogen in the Broadbalk wheat experiment. J Agr Sci 127: 347–363.
9. Gong W, Yan XY, Wang JY, Hu TX, Gong YB (2009) Long-term manuring and fertilization effects on soil organic carbon pools under a wheat-maize cropping system in North China Plain. Plant Soil 314: 67–76.
10. Jiang D, Hengsdijk H, Dai T, de Boer W, Jing Q, et al. (2006) Long-term effects of manure and inorganic fertilizers on yield and soil fertility for a winter wheat-maize system in Jiangsu, China. Pedosphere16: 25–32.
11. Lu R, Shi Z (1998) Effect of long-term fertilization on soil properties. In: Lu R, Xie J, Cai G, Zhu Q (2010) editors. Soil-plant nutrients principles and fertilizer. Beijing: Chemical Industry Press. 102–110.
12. Wang YC, Wang E, Wang DL, Huang SM, Ma YB, et al. (2010) Crop productivity and nutrient use efficiency as affected by long-term fertilization in North China Plain. Nutr Cycl Agroecosys 86: 105–119.
13. Yan XY, Gong W (2010) The role of chemical and organic fertilizers on yield, yield variability and carbon sequestration – results of a 19-year experiment. Plant Soil 331: 471–480.
14. Zhang H, Xu M, Zhang F (2009) Long-term effects of manure application on grain yield under different cropping systems and ecological conditions in China. J Agr Sci 147: 31–42.
15. Mussgnug F, Becker M, Son TT, Buresh RJ, Vlek PLG (2006) Yield gaps and nutrient balances in intensive, rice-based cropping systems on degraded soils in the Red River Delta of Vietnam. Field Crop Res 98: 127–140.
16. Mueller ND, Gerber JS, Johnston M, Ray DK, Ramankutty N, et al. (2012) Closing yield gaps through nutrient and water management. Nature 490: 254–257.
17. Page AL, Miller RH, Keeney DR (1982) Methods of soil analysis. Part 2. Madison, WI: American Society of Agronomy. 539–871.
18. Olsen SR, Cole CV, Watanabe FS, Dean LA (1954) Estimation of available phosphorus in soils by extraction with sodium bicarbonate. USDA Circular 939.
19. Liu X, Ju X, Zhang Y, He C, Kopsch J, et al. (2006) Nitrogen deposition in agroecosystems in the Beijing area. Agr Ecosyst Environ 113: 370–377.
20. Zhang Y, Liu XJ, Fangmeier A, Goulding KTW, Zhang FS (2008) Nitrogen inputs and isotopes in precipitation in the North China Plain. Atmos Environ 42: 1436–1448.
21. Tang X, Li J, Ma Y, Hao X, Li X (2008) Phosphorus efficiency in long-term (15-years) wheat-maize cropping systems with various soil and climate conditions. Field Crop Res 108: 231–237.
22. Blake L, Mercik S, Koerschens M, Moskal S, Poulton PR, et al. (2000) Phosphorus content in soil, uptake by plants and balance in three European long-term field experiments. Nutr Cycl Agroecosys 56: 263–275.
23. Zhong W, Cai Z (2007) Long-term effects of inorganic fertilizers on microbial biomass and community functional diversity in a paddy soil derived from quaternary red clay. Appl Soil Ecol 36: 84–91.
24. Bajwa MI (1994) Soil K status, K fertilizer usage and recommendation in Pakistan. Potash Review No. 3/1994. Subject 1, 20th Suite. Basel: International Potash Institute. 67.
25. Shen J, Li R, Zhang F, Fan J, Tang C, et al. (2004) Crop yields, soil fertility and phosphorus fractions in response to long-term fertilization under the rice monoculture system on a calcareous soil. Field Crop Res 86: 225–238.
26. Tan D, Jin J, Huang S, Li S, He P (2007) Effect of long-term application of K fertilizer and wheat straw to soil on crop yield and soil K under different planting systems. Agr Sci China 6: 200–207.
27. Manna MC, Swarup A, Wanjari RH, Ravankar HN, Mishra B, et al. (2005) Long-term effect of fertilizer and manure application on soil organic carbon storage, soil quality and yield sustainability under sub-humid and semi-arid tropical India. Field Crop Res 93: 264–280.

15

Response of Vegetation and Soil Carbon and Nitrogen Storage to Grazing Intensity in Semi-Arid Grasslands in the Agro-Pastoral Zone of Northern China

Min-yun Xu[1,2]*, Fan Xie[3], Kun Wang[4]

1 Department of Ecology, College of Urban and Environmental Sciences, Peking University, Beijing, P.R. China, **2** College of Animal Science & Technology, Hebei Agricultural University, Baoding, Hebei province, P.R. China, **3** Department of Party Affairs, China National Light Industry Council, Beijing, P.R. China, **4** College of Animal Science & Technology, China Agricultural University, Beijing, P.R. China

Abstract

Overgrazing has been the primary cause of grassland degradation in the semi-arid grasslands of the agro-pastoral transition zone in northern China. However, there has been little evidence regarding grazing intensity impacts on vegetation change and soil C and N dynamics in this region. This paper reports the effects of four grazing intensities namely un-grazed (UG), lightly grazed (LG), moderately grazed (MG) and heavily grazed (HG) on vegetation characteristics and soil properties of grasslands in the Guyuan county in the agro-pastoral transition region, Hebei province, northern China. Our study showed that the vegetation height, canopy cover, plant species abundance and aboveground biomass decreased significantly with increased grazing intensity. Similarly, soil organic carbon (SOC) and total nitrogen (STN) in the 0–50 cm were highest under UG (13.3 kg C m^{-2} and 1.69 kg N m^{-2}) and lowest under HG (9.8 kg C m^{-2} and 1.22 kg N m^{-2}). Soil available nitrogen (SAN) was significantly lower under HG (644 kg N hm^{-2}) than under other treatments (725–731 kg N hm^{-2}) in the 0–50 cm. Our results indicate that the pasture management of "take half-leave half" has potential benefits for primary production and livestock grazing in this region. However, grazing exclusion was perhaps the most effective choice for restoring degraded grasslands in this region. Therefore, flexible rangeland management should be adopted in this region.

Editor: Xiujun Wang, University of Maryland, United States of America

Funding: This research was funded by Baoding Science and Technology Development Project (Grant No. 13zf063). The funder (URL: http://www.bdskjj.gov.cn/) had no role in study design, data collection and analysis, decision to publish, or preparation of the manuscript.

Competing Interests: The authors have declared that no competing interests exist.

* E-mail: xuminyun@pku.edu.cn

Introduction

Grazing, the most common use of grasslands, can influence plant community structure, soil properties and nutrient cycling within the plant-soil system[1]. For instance, selective grazing caused changes in plant species composition[2] and influenced herbage production[3]. Several previous studies proposed that light and moderate grazing favored grasses and stimulated grassland productivity[4][5]. However, increasing grazing intensity, to some extent, generally decreased grassland productivity, canopy height and sward cover[6], but increased unpalatable species proportion[4], and destroyed soil aggregation[7]. Milchunas and Lauenroth[2] analyzed 236 grazing studies worldwide and found different results with dynamics of species composition, root biomass, soil organic C and soil N of grasslands not closely associated with grazed or non-grazed measurements. These contradictory findings suggested that the effects of grazing on nutrient cycling and ecosystem functioning still needs further study [2].

Soil water content was the most limiting factor for primary productivity in semi-arid rangelands[8]. Grazing intensity affected soil water conditions of grasslands through animal trampling and intake behavior. Trampling compacted soil and increased soil bulk density[9], which led to the reduction of storage capacity and supply of soil water and nutrients, ultimately decreasing soil fertility and grassland productivity[10]. Naeth et al.[11] reported that soil water in grazed treatments was generally lower than in non-grazed grassland. However, grazing could have the opposite effect on soil water through removing vegetation, resulting in decreased evapo-transpiration [11].

With regard to soil C and N, previous investigations have produced very variable results. Grazing management usually has a negative effect [1][9][12] on soil C storage, with well-managed grazing was reported to enhance C sequestration [13][14]. However, Milchunas and Lauenroth[2] found that grazing management had no effect on soil C and N storage. In addition, Frank et al.[1] found heavy grazing did not reduce soil C but moderately grazed pasture contained less soil C compared to a non-grazed exclosure, which was attributed to changes in species composition. Bauer et al.[15] found greater total N content to a soil depth of 0.457 m in grazed than non-grazed grasslands. Schuman et al. [16] found 12 years of grazing under different stocking rates did not change the total C and N mass of in the top 60 cm of the plant-soil system, but changed the distribution of C and N among the system components, primarily via a significant increase in the of C and N mass in the root zone (0–30 cm) of the soil profile. Wu et al.[17] reported that the temperate grasslands of northern China could achieve significant C and N storage on decadal scales in the context of mitigating global climate change by grazing elimination. He et al.[18] found that C and N storage in both the 0–10 cm and 10–30 cm soil layers decreased linearly with increasing stocking rates in northern China. They also found there existed an underlying transformation from soil C sequestration under light grazing to C loss under heavy grazing, which implied that grasslands used for grazing have the capacity to sequester C in the soil under appropriate grazing pressure[18]. These varied findings were likely associated with the differences in climate, soil properties, study sites, plant community composition and grazing management measurements [19], indicating that more studies are needed to further clarify C and N dynamics under different grazing intensities and grazing regions.

Scientists have already conducted numerous studies on the effects of grazing pressure on plant and soil properties of grasslands in northern China, but most have been in the pastoral zone [12][20]. Relatively little research has focused on the productivity and sustainability of grasslands in agro-pastoral areas[21]. The agro-pastoral transition zone in semi-arid northern China [22], is one of the world's largest agro-pastoral ecotones, with an area of 69×10^4 km^2 and a population of approximately 6.7×10^7[23]. Grasslands in this area play an important role in providing ecological services, such as sequestering carbon, regulating nutrient cycling, affecting biodiversity and ecosystem functioning [24]. Overgrazing and land use changes that reduces the areas of forests and grasslands along with an increase in agricultural activity have intensified pressures on the regional grasslands, causing severe land degradation[25]. Therefore, it is very important to understand the effects of grazing intensity on plant characteristics and soil properties of grasslands in this region. The objectives of this study were to quantity the effects of grazing intensity on (i) biomass allocation patterns and vegetation characteristics, (ii) soil properties and C and N storage dynamics in soil, and (iii) the relationship between soil C and N storage and the grazing-induced variation in plant biomass allocation patterns.

Materials and Methods

Study sites

The research was conducted at the Guyuan Grassland Ecosystem Observation and Research Station of China Agricultural University (115°41′E, 41°46′N), in Hebei province, northern China (Figure 1). With elevation between 1300 to 1450 m a.s.l., this area belongs to a semi-arid agro-pastoral transitional region, with a typical continental climate. The mean annual precipitation and potential evaporation range from 350 to 450 mm and 1700 to 2300 mm, respectively (1990 to 2010), with 80% of precipitation occurring during the June through September growing season. The mean annual temperature ranges from 1 to 2°C (1990 to 2010). The region is characterized by open, flat land, alternating with undulating hilly land. Further details of physiographic characteristics of this region have been presented by Chen and Zhu[26]. Soils are classified as chestnut type soils (i.e. Calcic Kastanozems), which are equivalent to Calcic-orthic Aridisol in the US soil taxonomy classification system. Soil pH ranges from 7.79 to 8.49 [27]. The plant community is mainly dominated by *Leymus chinensis, Stipa grandis, Cleistogenes squarrosa, Agropyron michnoi,* and *Koeleria cristata* [28].

Experimental design

In this study, four contiguous sites were selected based on the availability of reliable information on the grassland management history under the same environmental conditions. Prior to the establishment of the Guyuan Grassland Ecosystem Observation and Research Station in 2000, the area had not been grazed or only lightly grazed by domestic livestock for about 20 years. These sites had been exposed to a 10-year controlled grazing experiment (2001 to 2011) using different grazing pressures by herds consisting of 70 to 80% goats and sheep and 20 to 30% Holstein and beef cattle. Stocking rate was the basis for the four grazing intensities used, un-grazed (UG), lightly grazed (LG), moderately grazed (MG) and heavily grazed (HG). Calculation of the stocking rate is defined in terms of number of sheep units/pasture area, where a sheep unit (in China) is defined as a 40 kg liveweight ewe nursing lamb(s) with daily forage consumption of 5 to 7.5 kg[29]. Based on their forage intake and this definition of a sheep unit, ten goats were equal to eight sheep, and five sheep were approximately equivalent to one Holstein or beef cow. Animals were kept in pens at night, so the input of faecal organic material was reduced. There was no application of additional fertilizers to the four experimental sites.

The UG grassland was a 10 ha area with no grazing or human disturbance for livestock production. The vegetation was dominated by *Leymus chinensis*, *Stipa grandis* and *Hordeum spontaneum*. The LG grassland carried 11.16 sheep units within a 12.0 ha area, and its stocking rate was 0.93 sheep units per ha, resulting in 20% grassland utilization. The vegetation was dominated by *Leymus chinensis*, *Potentilla lancinata*, and *Carex capricornis*. The MG grassland supported 74.52 sheep units within a 32.0 ha area, and its stocking rate was 2.33 sheep units per ha, resulting in 50% grassland utilization. The vegetation was dominated by *Leymus chinensis, Thermopsis lanceolata, Ixeris polycephala, Puccinellia altaica Tzvel* and *Iris lactea var. pall. chinensis (Fish) Koidz*. The HG grassland supported 114.50 sheep units within a 35.12 ha area, and its stocking rate was 3.26 sheep units per ha, resulting in 70% grassland utilization. The vegetation was dominated by *Leymus chinensis, Iris lactea var. pall. chinensis (Fish) Koidz* and *Stellera chamaejasme*.

Sampling and field investigation

In each of the four experimental sites, three sampling plots (50 m×50 m) were selected with the constraint that they were at least 2.0 m from the margin to avoid any edge effects. In each sampling plot, three sampling quadrats were randomly selected using a 0.5×0.5 m metal frame along the diagonal gradients. The experiment comprised 36 quadrats (3 plots x 3 quadrats x 4 grazing intensities). The vegetation composition survey was performed during the peak standing biomass in early September of 2012. It included measurements of plant canopy cover, mean vegetation height, aboveground biomass (comprising live biomass, standing dead material and litter on the soil surface), and plant species abundance across the four grazing intensities. Plant species composition was estimated using three sampling quadrats along a 10 m line intercept located at the centre of each plot. Plant canopy cover was determined using point frames (a 0.5 m×0.5 m metal frame with 50 grids) combined with a visual estimation method. The standing crop was harvested near ground-level using hand shears. Surface litter and standing dead plant biomass were hand brushed to bare the sampling area after herbage was clipped. The material was oven-dried at 65°C to a constant weight.

To estimate root biomass, five soil cores were collected at 10 cm intervals to 50 cm depth using a soil core sampler (6 cm diameter) from each quadrat, and placed in mesh bags, then rinsed in water to remove soil and debris. The root samples were dried at 80°C to a constant weight. Final biomass estimates were converted to a kilogram per hectare basis using the area of the sample quadrat, or the surface area in the case of the roots.

For soil C and N analysis, five soil samples were collected from three quadrats in each sampling plot after the aboveground material was harvested. Soil samples were collected using a soil

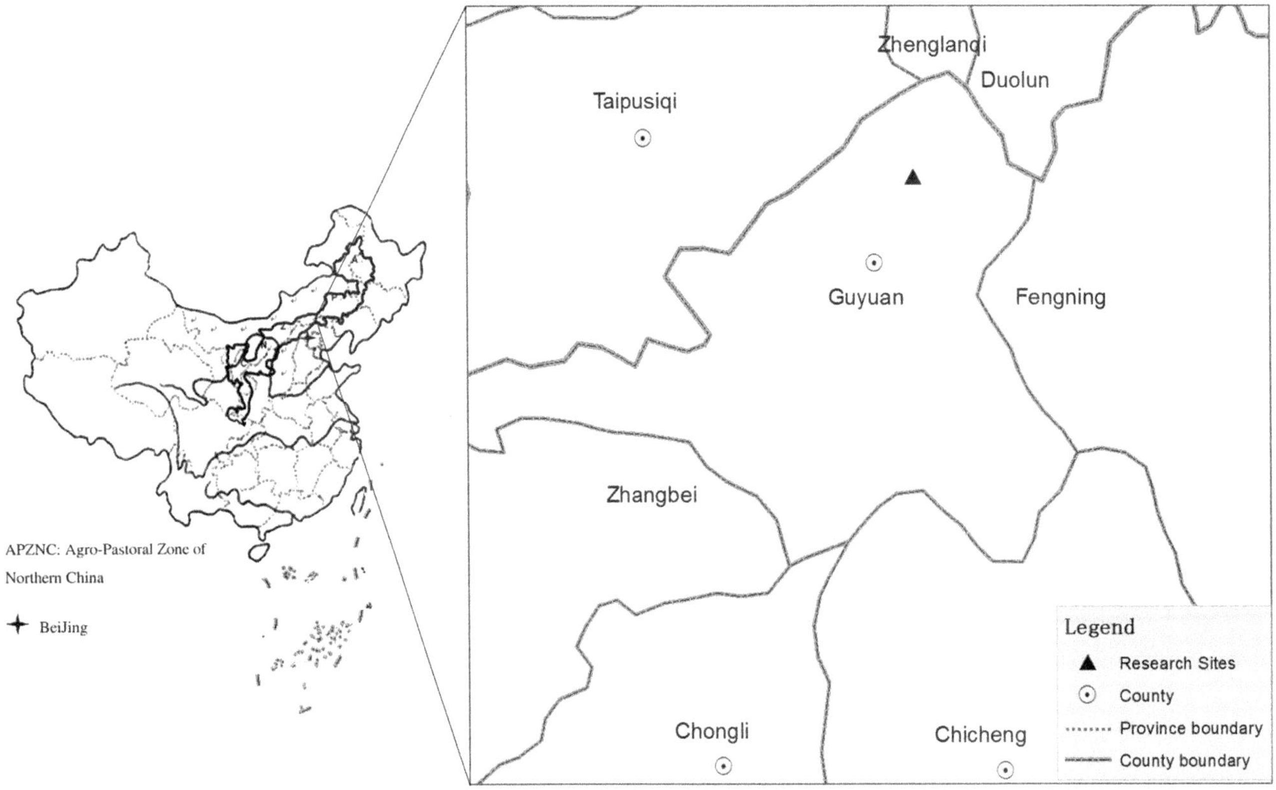

Figure 1. Location of the research sites in the agro-pastoral zone of northern China.

core sampler (6 cm diameter) and separated into 0–10, 10–20, 20–30, 30–40, and 40–50 cm increments. Five cores were taken from each layer and combined to provide an adequate sample for analyses. Samples were placed in sealed plastic bags in the field, then air-dried in a ventilated room, ground, then sieved to <2 mm to remove stones, root fragments and organic debris before chemical analyses. Soil samples were analyzed for organic C by the external heating method[30]. Soil total nitrogen (STN) content was determined using the Kjeldahl acid-digestion method with an auto-analyzer (Foss Inc., FIAstar5000, Sweden). Soil available nitrogen (SAN) was measured by the methods of Page et al.[31]. Soil bulk density was measured using samples taken with a volumetric (100 cm^3) steel ring and calculated as the mass of oven-dry soil (105°C) divided by the core volume [32] for each depth increment. The measurement of bulk density allows the estimation of the total soil C and N storage under different grazing intensities. Soil water content was determined by weighing samples before and after oven-drying at 105°C for 8 h.

Derived Variables Calculations

Soil organic C (SOC, kg C m^{-2}), STN (kg N m^{-2}), and SAN (kg N hm^{-2}) were calculated on an area basis to a soil depth of 50 cm as follows:

$$SOC = \frac{\sum_{i=1}^{n} D_i \times \rho_i \times OC_i}{100} \quad (1)$$

$$STN = \frac{\sum_{i=1}^{n} D_i \times \rho_i \times TN_i}{100} \quad (2)$$

$$SAN = 100 \sum_{i=1}^{n} D_i \times \rho_i \times AN_i \quad (3)$$

where i is the number of soil horizons, D_i is the depth interval (cm) of the horizon i from the top soil down, ρ_i is the soil bulk density (g cm^{-3}) in the horizon i, OC_i is the mean SOC content (g kg^{-1}) in the horizon i, TN_i and AN_i represent total N concentration (g kg^{-1}) and available N concentration (mg kg^{-1}) respectively.

Data analysis

All statistical analyses were conducted using SPSS 16.0 for Windows (SPSS Inc., Chicago, IL, USA). Graphical presentation was done with SigmaPlot 10 (SPSS Inc., Chicago, IL, USA) and Adobe Acrobat XI Pro (Adobe Systems Incorporated, California, USA). Statistical analysis of each measured variable was conducted with the GLM (general linear model) procedure. One-way analyses of variance (ANOVA) were conducted to assess treatment effects on grassland vegetation characteristics, soil C and N storage properties. Individual system components (litter, standing dead and live biomass, roots by depth and soil by depth) were tested with a separate analysis of variance with replicate pastures treated as blocks. Error variances for depths were heterogeneous, so only univariate analysis was reported for depth. Least-significant-differences (LSD) procedures were used for comparisons between means. Normal distribution and homogeneity of variance for each collective were tested using the Shapiro-Wilk's test and Levene's test. Significant differences for all of the statistical tests were evaluated at the level of $P = 0.05$.

Results

Vegetation characteristics

Our study demonstrated that the vegetation height, canopy cover, plant species abundance and aboveground biomass (live, standing dead and litter biomass) of grasslands in the agro-pastoral zone, decreased significantly with increased grazing intensity over 10-years (Table 1).

The UG plots had significantly higher plant species abundance than the other three plots with, the MG pasture displaying a similar species abundance to the LG and HG pastures. Our research also showed that increased grazing significantly decreased the canopy cover, with no significant difference between the UG and LG plots. The UG plots had significantly higher aboveground biomass, and HG plots had significantly lower value, while LG and MG plots showed intermediate values that were not significantly different. Belowground biomass of the MG plots was significantly higher than the other plots. Surprisingly, root biomass of the UG and HG pastures were similar and significantly lower than that of the LG pasture. This change was reflected in the higher root to shoot biomass ratios under the heavy grazing treatment compared to the light grazing treatment (Table 1). The total plant biomass in the LG and MG plots were similar, which were higher than those in the UG and HG plots. The HG plots had the lowest total biomass among the four grazing plots.

Soil moisture content

Increasing grazing intensity generally decreased soil moisture content (Table 2). The overall trend was that the mean soil moisture content was highest in the UG plots and lowest in the HG plots in the upper horizons (0–40 cm), and there was no significant difference in the 40–50 cm soil layer.

Soil bulk density

Soil bulk density increased with each soil depth increment for each grazing intensity (Table 2). The changes in soil bulk density due to grazing were significant in the 0–10 cm soil layer among the different stocking rates. However, grazing had no effect on the bulk density of the lower horizons (10–50 cm) among the four grazing intensities.

SOC, STN, and SAN concentrations

For SOC, the UG plots had significantly higher values, and the HG plots significantly lower values in the 0–10, 10–20 and 20–30 cm soil layers, while the LG and MG plots showed intermediate values that were not significantly different. In the 30–40 and 40–50 cm soil layers, the UG and LG plots had similar values that were significantly higher than the MG and HG plots, which did not differ. In all four plots, SOC decreased with successive soil depth increments (Figure 2A).

The STN values in the UG plots were significantly higher than the other three plots across all of the 0–50 cm soil layers. In the 0–10 and 10–20 cm soil layers, HG had the lowest STN (2.60 and 2.03 g kg^{-1}), and there was no difference between the LG (2.97 and 2.53 g kg^{-1}) and MG (2.93 and 2.46 g kg^{-1}) plots. In the 20–30, 30–40 and 40–50 cm soil profiles, STN in LG, MG, and HG did not differ (Figure 2B). SAN in the HG plots was significantly lower than in the UG, LG, and MG plots across all the 0–50 cm soil layers (Figure 2C). STN and SAN were all higher in the surface soil and decreased with soil depth among all four grazing intensities.

SOC, STN, and SAN storage

Trends in the storage of SOC, STN, and SAN (Figure 3) were mostly similar to the trends in their concentrations i.e. decreasing amounts of C and N with increasing grazing pressure. The

Table 1. Vegetation height, canopy cover, plant species abundance, aboveground biomass, belowground biomass, and R:S (root/shoot) ratio of grasslands under different grazing intensities in agro-pastoral zone.

System components	UG	LG	MG	HG
Vegetation height (cm)	20.23±1.10a	17.10±0.99b	10.61±0.94c	5.68±0.87d
Canopy cover (%)	82.78±1.12a	78.44±1.03a	63.89±2.34b	40.89±1.82c
Species abundance (species m^{-2})	41.67±1.94a	30.44±1.60b	33.89±1.52b	29.44±1.84b
Aboveground biomass (g m^{-2})	575.79±5.68a	462.30±6.20b	439.51±5.18b	161.88±4.95c
Live biomass	538.80±5.56a	436.13±6.26b	413.50±5.17b	149.36±4.93c
Standing dead biomass	21.92±0.28a	14.31±0.34b	13.39±0.23b	5.40±0.21c
Surface litter biomass	15.08±0.33a	11.96±0.17b	12.62±0.19b	7.11±0.17c
Belowground biomass (g m^{-2})	1297.38±4.14c	1599.42±4.34b	1694.09±4.62a	1277.04±11.05c
0–10 cm	940.90±6.17d	1162.92±8.64b	1277.66±7.15a	1026.85±8.69c
10–20 cm	162.17±1.99b	183.98±3.70a	186.13±2.42a	130.80±1.56c
20–30 cm	105.78±2.76b	128.04±2.19a	136.14±2.85a	57.08±0.64c
30–40 cm	62.61±2.36c	75.66±1.17a	68.16±2.11b	36.78±1.85d
40–50 cm	25.92±1.07b	48.81±1.03a	26.30±0.66b	25.53±2.23b
Total plant biomass (g m^{-2})	1873.18 ±4.79b	2061.72±8.12a	2133.61±7.70a	1438.92±10.80c
R:S	2.32±0.03c	3.57±0.05b	3.97±0.05b	8.32±0.39a

Note: UG, LG, MG, and HG are four grazing intensities: un-grazed; lightly grazed; moderately grazed; and heavily grazed. Values in the table are means ± SE (standard error). Different letters across the same component indicate significant difference for the different grazing intensities ($p<0.05$).

Table 2. Soil moisture content and soil bulk density in grasslands to 50 cm soil depth under different grazing intensities in the agro-pastoral zone.

Parameters	Depth intervals (cm)	Grazing intensities			
		UG	LG	MG	HG
Soil moisture content (%)	0–10	18.03±0.24a	16.58±0.35b	15.95±0.29b	10.66±0.23c
	10–20	20.09±0.27a	18.98±0.14b	19.23±0.22ab	14.60±0.42c
	20–30	20.72±0.43a	18.06±0.18b	17.56±0.32b	17.35±0.30b
	30–40	18.25±0.33a	17.47±0.47a	15.45±0.25b	15.66±0.30b
	40–50	18.85±0.23a	17.97±0.34a	18.26±0.30a	17.87±0.34a
Soil bulk density (g cm^{-3})	0–10	1.05±0.03c	1.20±0.03b	1.23±0.01b	1.47±0.01a
	10–20	1.38±0.01a	1.37±0.02a	1.40±0.03a	1.37±0.01a
	20–30	1.47±0.01a	1.45±0.01a	1.46±0.02a	1.44±0.02a
	30–40	1.46±0.01a	1.47±0.01a	1.47±0.02a	1.47±0.03a
	40–50	1.56±0.02a	1.55±0.02a	1.54±0.03a	1.61±0.02a

Note: UG, LG, MG, and HG are four grazing intensities: un-grazed; lightly grazed; moderately grazed; and heavily grazed. Values in the table are means±SE (standard error). Difference letters within each row indicate significant difference for the different grazing intensities ($p<0.05$).

exception was SAN, which increased with increasing grazing pressure in the 0–10 cm layer. The overall trend for SOC, STN, and SAN was a decline at each successive depth increment from 0 to 50 cm depth, except SAN, which was lower at 0–10 cm than 10–20 cm for the UG, LG, and MG. SOC differed significantly for the overall 0–50 cm profile decreasing from UG through LG and MG to HG (Figure 3A). The UG plots had significantly higher STN levels (1.69 kg m^{-2}), HG plots were generally the lowest (1.22 kg m^{-2}) and LG and MG were intermediate with no significant difference (Figure 3B). HG plots generally had the lowest SAN (644 kg hm^{-2}) while the other grazing intensities did not differ for the overall 0–50 cm profile (Figure 3C). The C:N ratio across the entire soil profile was significantly greater for the LG pasture (9.74) compared to UG (7.71), MG (8.22) and HG pasture (7.93) (Figure 3D).

Discussion

Vegetation characteristics

Our results indicate that vegetation height, canopy cover, and plant species abundance decreased significantly with increased grazing intensities. This change in plant species abundance in our study is consistent with a study conducted in Inner Mongolia [33], which showed reduced species richness of the *Aneurolepidium chinense* and *Stipa grandis* steppes with an increase in grazing intensity.

In our study, aboveground biomass decreased significantly with increased grazing intensity, which is not consistent with research from alpine meadow in the eastern Tibetan Plateau. Gao et al. [34] examined the seasonal dynamics of biomass and plant nitrogen content under light, moderate, and heavy grazing intensities and found that the highest live aboveground biomass occurred at sites with moderate grazing pressure. Possible explanations for heavy grazing reducing aboveground biomass include the decrease in abundance of the initially dominant and palatable graminoid species (e.g. *Leymus chinensis* and *Stipa grandis*) with increasing grazing intensity and the subsequent domination of unpalatable forbs and legumes became dominant. Likewise, the reduced canopy cover caused by selective grazing decreasing the area and resources available for plant production probably contributes to this change[3].

Our results showed that change of belowground biomass was different from that of aboveground biomass. These changes were not consistent with the findings of Su et al. [35], who observed that steppe vegetation grows better, produces more root exudates and develops a stronger root system in un-grazed areas compared to grazed areas in the semi-arid region of northern China. Other studies also showed variable findings for belowground biomass affected by grazing, such as an increase [34], decrease [36], or no effect[37]. In our grazing region, these results indicate that proper grazing intensity would stimulate vegetation growth in *Leymus chinensis* steppe, and overgrazing would decrease the steppe productivity.

Soil moisture content

Grazing has significant impacts on soil water through its influence on infiltration via treading and altering evapo-transpiration through defoliation[38][39], i.e. soil water content declines under poor infiltration and/or high evapo-transpiration [38]. Our study showed soil water content decreased with increasing grazing intensity, consistent with previous research conducted to quantify the water fluxes as affected by grazing intensity in Inner Mongolia Grassland of northern China [40]. This is likely associated with a reduction in vegetation height, canopy cover, and biomass as grazing intensity increased. The poor ground cover directly contributed to the increase of soil evapo-transpiration loss. Larger vegetation and litter biomass remaining in the low grazing intensity, compared to heavier grazing, would increase the holding capacity of surface soil water [38].

Soil bulk density

Increased bulk densities and lower soil moisture content, as a result of increased animal trampling, have been observed for different grazing animals in different grassland ecosystems [41]. Our results were consistent with previous research conducted on the Loess Plateau that showed grazing significantly increased soil bulk density and decreased soil moisture content[41]. He et al. [18] also found that bulk density of temperate grasslands in northern China increased linearly with increasing stocking rates in both the 0–10 cm and 10–30 cm soil layers.

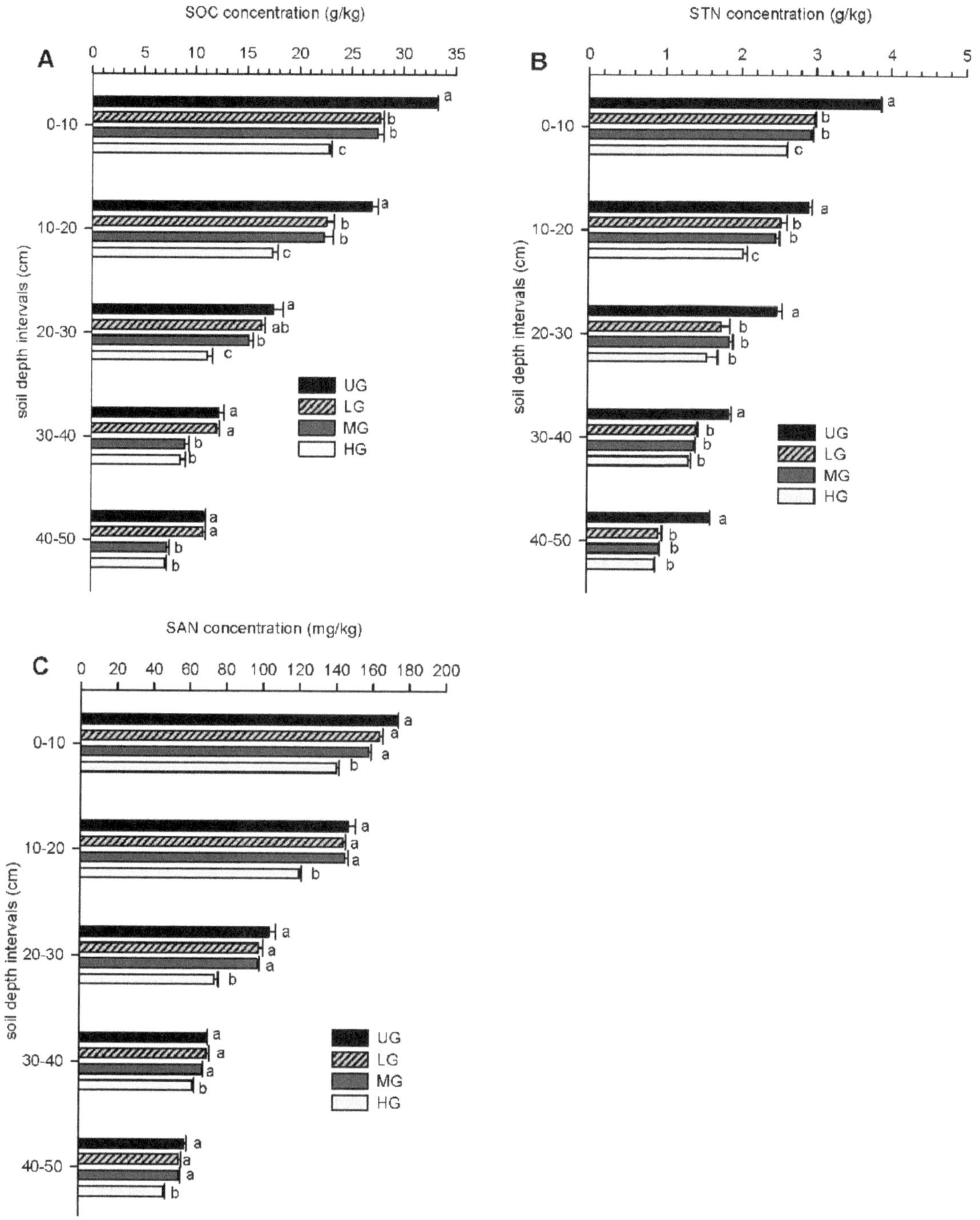

Figure 2. Concentration of SOC (A), STN (B), and SAN (C) under different grazing intensities. Soil organic carbon (SOC), soil total nitrogen (STN) and soil available nitrogen (SAN) concentrations in grasslands measured at 10 cm increments to depths of 50 cm in heavily grazed (HG), moderately grazed (MG), lightly grazed (LG), and un-grazed (UG) plots in the agro-pastoral zone. Within soil depth intervals, different letters indicate treatments are significantly different ($p<0.05$). Horizontal bars show S.E. $n=9$.

SOC, STN, and SAN

Grazing exclusion can enhance soil C and N storage in temperate grasslands in northern China[18]. Reduction in soil C and N storage under long-term heavy grazing was reported previously from research carried out in this region [9][18][20]. Our results were consistent with previous studies. The response of SOC, STN, and SAN storage to different grazing intensities in our study may be due, in part, to the effects of grazing on belowground biomass, litter and standing dead components of the aboveground biomass. Belowground biomass has been proven to be positively correlated with the size of the soil organic matter pool[42]. Repeated and frequent grazing resulted in decreased root elongation and biomass, and hence lower C inputs into the soil from the roots[43]. Livestock grazing altered plant composition, canopy cover, and community biomass. This may account for

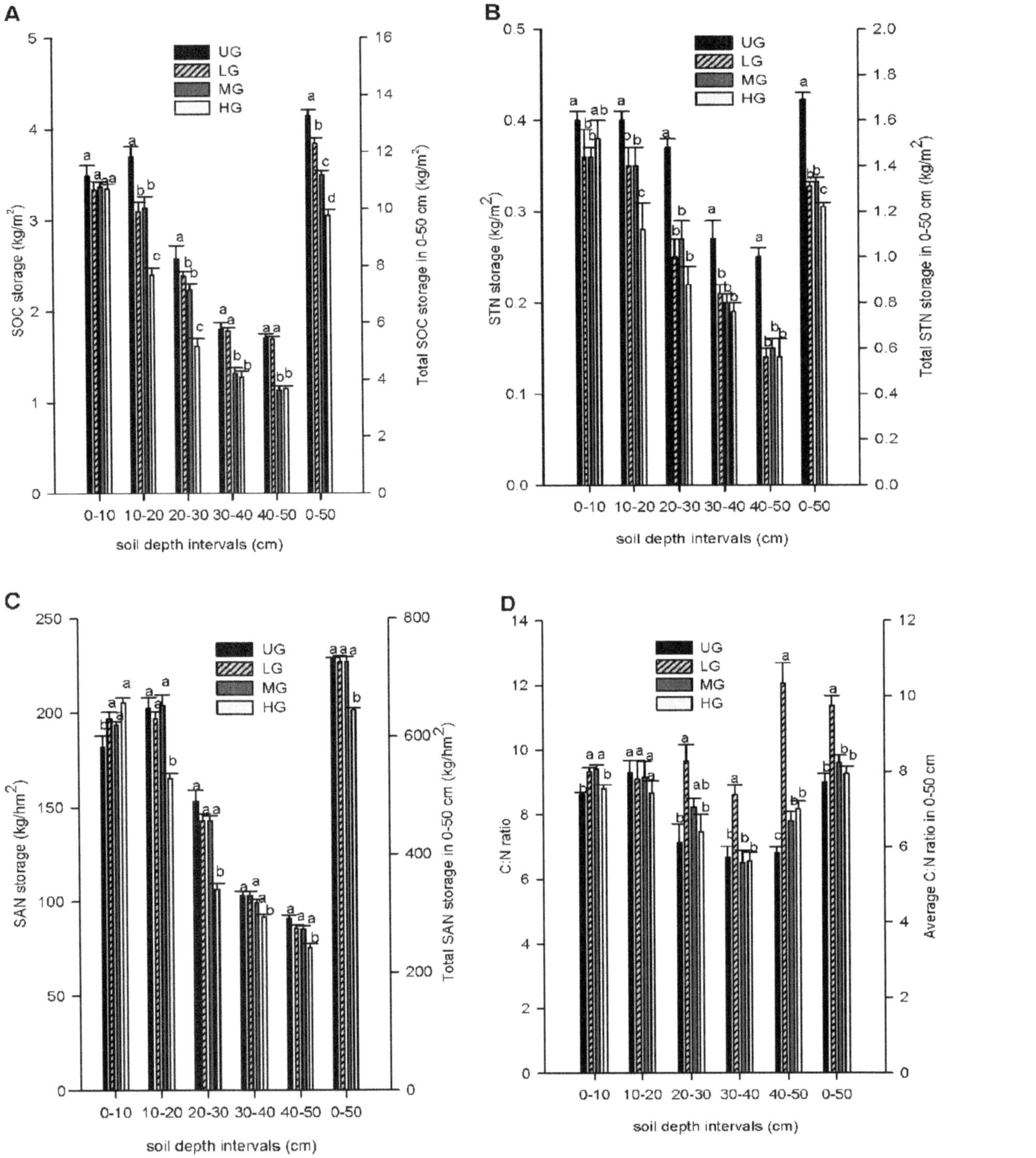

Figure 3. SOC (A), STN (B), SAN (C) storage and C:N ratio (D) under different grazing intensities. Soil organic carbon (SOC), soil total nitrogen (STN), soil available nitrogen (SAN) storage and C:N ratio in grasslands measured at 10 cm increments to depths of 50 cm in heavily grazed (HG), moderately grazed (MG), lightly grazed (LG) and un-grazed (UG) in the agro-pastoral zone. Within soil depth intervals, different letters indicate treatments are significantly different ($p<0.05$). Vertical bars show S.E. n = 9.

changes in the storage of SOC, STN, and SAN in the soil under different grazing intensities.

Possible explanations for overgrazing reducing SOC storage include vegetation production reduction, vegetation destruction, and change of environmental factors[1]. Grazing might also be expected to influence the organic C content of the soil by reducing vegetation growth, the amount of litter and exposing the soil surface to erosion, which leads to direct soil nutrient losses [39]. Disruption of soil aggregate structure and surface soil crust due to trampling increases the decomposition of soil organic matter and renders the soil susceptible to water and wind erosion[44]. In temperate grasslands, considerable loss of SOC and soil N is caused by wind and water erosion, particularly in areas with sandy soil and high wind speed [45]. Soils under HG in this region, with lower canopy cover and lower vegetation height, are vulnerable to wind erosion[21].

In the current study, a similar concentration of available N was found among UG, LG and MG plots, maybe due to organic matter and nutrient transfer via animal excreta[46], although there were significant differences in aboveground and belowground biomass among these plots. But defoliation in HG significantly decreased both STN and SAN by decreasing root and total production, resulting in reduced soil C and N [47]. The changes in plant community composition due to selective foraging can also indirectly influence soil N and available N by affecting plant litter decomposition rates and soil microbial activities [48].

The higher C:N ratio in LG suggests either a slower rate of decomposition or a greater portion of recent plant material in the residue[1]. The higher N content in the UG suggests that grazing reduced soil N [1]. A lower C:N ratio in UG perhaps means that higher soil moisture is stimulating litter decomposition, thereby releasing more N. Lower C:N ratios in MG and HG are perhaps because grazers return large amounts of N to the soil through urine and feces, increasing levels of available soil N [49]. However, grazing can also reduce N turnover and availability as grazers feed selectively on plants with high N content and thus increase the dominance of plant species with low N content[50].

Conclusions and Implications

Our study revealed that vegetation characteristics and soil C and N storage of semiarid grassland in the agro-pastoral zone were sensitive to grazing practice. With increasing grazing intensity, the vegetation height, canopy cover, plant species abundance and aboveground biomass decreased significantly. In addition, ten years of grazing not only changed the distributions of C and N through the soil profile (0–50 cm), but also the total storages of C and N in the soil profile. As a result, SOC and STN in the 0–50 cm were highest under UG (13.3 kg C m^{-2} and 1.69 kg N m^{-2}) and lowest under HG (9.8 kg C m^{-2} and 1.22 kg N m^{-2}). SAN was significantly lower under HG (644 kg N hm^{-2}) than under other treatments (725–731 kg N hm^{-2}) in the 0–50 cm.

Our results indicated that the pasture management of "take half-leave half" has potential benefits for primary production and livestock grazing, which would achieve a balance between protection of species diversity, livestock production and soil C and N management in this region. However, the "Reduce Livestock Return the Grasslands" program with grazing exclusion supported by the Ministry of Agriculture of the People's Republic of China in 2002 was perhaps the effective choice for restoring degraded grasslands in the agro-pastoral zone. Therefore, flexible rangeland management should be adopted that suits local circumstances by balancing the demand for grassland utilization and conservation in this region. These results would be useful to the decision making on rangeland management in terms of maximizing C and N sequestration while maintaining adequate productivity for servicing the regional socio-economic development of the agro-pastoral zone in northern China.

Acknowledgments

The authors thank Dr. I. A. White, Dr. X. Jing, and Dr. K.S. Liu for their careful language checking. We are grateful to P.G. Li, W. Yang, and X.G. Wang for their helpful assistance in the field and laboratory work. We greatly appreciate comments by the anonymous referees and academic editor on earlier versions of this paper. Our manuscript was substantially improved by the comments of Dr. I. A. White and Dr. Z.H. Shang.

Author Contributions

Conceived and designed the experiments: MYX KW. Performed the experiments: MYX FX. Analyzed the data: MYX FX. Contributed reagents/materials/analysis tools: MYX FX. Wrote the paper: MYX.

References

1. Frank AB, Tanaka DL, Hofmann L, Follett RF (1995) Soil carbon and nitrogen of northern Great-Plains grasslands as influenced by long-term grazing. J Range Manage 48:470–474.
2. Milchunas DG, Lauenroth WK (1993) Quantitative effects of grazing on vegetation and soils over a global range of environment. Ecol Monogr 63:327–366.
3. Willms WD, McGinn SM, Dormaar JF (1993) Influence of litter on herbage production in the mixed grass prairie. J Range Manage 46:320–324.
4. Hiernaux P (1998) Effects of grazing on plant species composition and spatial distribution in rangelands of the Sahel. Plant Ecol 138:191–202.
5. McNaughton SJ, Banyikwa FF, McNaughton MM (1997) Promotion of the cycling of diet-enhancing nutrients by African grazers. Science 278:1798–1800.
6. Milchunas DG, Lauenroth WK, Burke IC (1998) Livestock grazing: Animal and plant biodiversity of shortgrass steppe and the relationship to ecosystem function. Oikos 83:65–74.
7. Van Haveren BP (1983) Soil bulk density as influenced by grazing intensity and soil type on a shortgrass prairie site. J Range Manage 36:586–588.
8. Le Houerou HN, Bingham RL, Skerbek W (1988) Relationship between the variability of primary production and the variability of annual precipitation in world arid lands. J Arid Environ 15:1–18.
9. Steffens M, Kölbl A, Totsche KU, Kögel-Knabner I (2008) Grazing effects on soil chemical and physical properties in a semiarid steppe of Inner Mongolia (P.R.China). Geoderma 143:63–72.
10. Christensen L, Coughenour MB, Ellis JE, Chen ZZ (2004) Vulnerability of the Asian typical steppe to grazing and climate change. Climatic Change 63:351–368.
11. Naeth MA, Chanasyk DS, Rothwell RL, Bailey AW (1991) Grazing impacts on soil water in mixed prairie and fescue grassland ecosystems of Alberta. Can J Soil Sci 71:313–325.
12. He NP, Yu Q, Wu L, Wang YS, Han XG (2008) Carbon and nitrogen store and storage potential as affected by land-use in a *Leymus chinensis* grassland of northern China. Soil Biol Biochem 40:2952–2959.
13. Derner JD, Schuman GE (2007) Carbon sequestration and rangelands: A synthesis of land management and precipitation effects. J Soil Water Conserv 62:77–85.
14. Wen L, Dong SK, Li YY, Li XY, Shi JJ, et al. (2013) The impact of land degradation on the C pools in alpine grasslands of the Qinghai-Tibetan Plateau. Plant Soil 368:329–340.
15. Bauer A, Cole CV, Black AL (1987) Soil property comparisons in virgin grasslands between grazed and nongrazed management systems. Soil Sci Soc Am J 51:176–182.
16. Schuman GE, Reeder JD, Manley JT, Hart RH, Manley WA (1999) Impact of grazing management on the carbon and nitrogen balance of a mixed-grass rangeland. Ecol Appl 9:65–71.
17. Wu L, He NP, Wang YS, Han XG (2008) Storage and dynamics of carbon and nitrogen in soil after grazing exclusion in *Leymus chinensis* grasslands of northern China. J Environ Qual 37:663–668.
18. He NP, Zhang YH, Yu Q, Chen QS, Pan QM, et al. (2011) Grazing intensity impacts soil carbon and nitrogen storage of continental steppe. Ecosphere 2:1–10.
19. Reeder JD, Schuman GE (2002) Influence of livestock grazing on C sequestration in semi-arid mixed-grass and short-grass rangelands. Environ Poll 116:457–463.
20. Zhao Y, Peth S, Krümmelbeina J, Horn R, Wang Z, et al. (2007) Spatial variability of soil properties affected by grazing intensity in Inner Mongolia grassland. Ecol Modell 205:241–254.
21. Huang D, Wang K, Wu WL (2007) Dynamics of soil physical and chemical properties and vegetation succession characteristics during grassland desertification under sheep grazing in an agro-pastoral transition zone in northern China. J Arid Environ 70:120–136.
22. Gong DY, Shi PJ, Wang JA (2004) Daily precipitation changes in the semi-arid region over northern China. J Arid Environ 59:771–784.
23. Huang D, Wang K, Wu WL (2007) Problems and strategies for sustainable development of farming and animal husbandry in the agro-pastoral transition zone in northern China(APTZNC). Int J Sust Dev World 14:391–399.
24. Zhou ZY, Osbert JS, Huang JH, Li LH, Liu P, et al. (2007) Soil carbon and nitrogen stores and storage potential as affected by land-use in an agro-pastoral ecotone of northern China. Biogeochemistry 82:127–138.

25. Liu QY, Tong YP (2003) The effects of land use on the eco-environmental evolution of farming-pastoral region in northern China: With an emphasis on Duolun County in Inner Mongolia. Acta Ecologica Sinica 23:1025–1030.
26. Chen ZQ, Zhu ZD (2001) Development of land desertification in Bashang area in the past 20 years. J Geogr Sci 11:433–437.
27. Xu MY, Wang K, Xie F (2011) Effects of grassland management on soil organic carbon density in agro-pastoral zone of northern China. Afr J Biotechnol 10:4844–4850.
28. Chen ZZ, Wang SP (2000) Typical steppe ecosystem of China. Beijing: Science Press.
29. Ren JZ (1985) Grassland survey and planning. Beijing: Chinese Agriculture Press.
30. Bao SD (2000) Soil and agricultural chemistry analysis (third edition). Beijing: China Agricultural Press.
31. Page AL, Miller RH, Keeney DR (1983) Methods of soil analysis, Part 2: Chemical and microbiological properties (2nd edition). Madison, Wisconsin, USA: American Society of Agronomy.
32. Institute of soil science, Chinese Academy of Sciences (1978) Soil physical and chemical analysis. Shanghai: Shanghai Science & Technology Press.
33. Li YH (1993) Grazing dynamics of the species diversity in *Aneurolepidium chinense* steppe and *Stipa grandis* steppe. Acta Bot Sin 35:877–884.
34. Gao YH, Luo P, Wu N, Yi SL, Chen H (2007) Biomass and nitrogen responses to grazing intensity in an alpine meadow on the eastern Tibetan Plateau. Pol J Ecol 55:469–479.
35. Su YZ, Zhao HL, Zhang TH, Zhao XY (2004) Soil properties following cultivation and non-grazing of a semi-arid sandy grassland in northern China. Soil Till Res 75:27–36.
36. Gao YZ, Giese M, Lin S, Sattelmacher B, Zhao Y, et al. (2008) Belowground net primary productivity and biomass allocation of a grassland in Inner Mongolia is affected by grazing intensity. Plant Soil 307:41–50.
37. McNaughton SJ, Banyikwa FF, McNaughton MM (1998) Root biomass and productivity in a grazing ecosystem: The Serengeti. Ecology 79:587–592.
38. Naeth MA, Chanasyk DS (1995) Grazing effects on soil water in Alberta foothills fescue grasslands. J Range Manage 48:528–534.
39. Stavi I, Eugene DU, Hanoch L, Pariente S (2008) Grazing induced spatial variability of soil bulk density and content of water moisture, organic carbon and calcium carbonate in a semiarid rangeland. Catena 75:288–296.
40. Gan L, Peng XH, Peter S, Horn R (2012) Effects of grazing intensity on soil water regime and flux in Inner Mongolia grassland, China. Pedosphere 22:165–177.
41. Zhou ZC, Gan ZT, Shangguan ZP, Dong ZB (2010) Effects of grazing on soil physical properties and soil erodibility in semiarid grassland of the northern Loess Plateau (China). Catena 82:87–91.
42. Gill RA, Burke IC, Milchunas DG, Lauenroth WK (1999) Relationship between root biomass and soil organic matter pools in the shortgrass steppe of eastern Colorado. Ecosystems 2:226–236.
43. Holland EA, Detling JK (1990) Plant response to herbivory and belowground nitrogen cycling. Ecology 71:1040–1049.
44. Neff JC, Reynolds RL, Belnap J, Lamothe P (2005) Multi-decadal impacts of grazing on soil physical and biogeochemical properties in southeast Utah. Ecol Appl 15:87–95.
45. Liu YR, Yang C, Zhu ZM, Liu ML (2004) Soil C and N dynamics during desertification of grassland in northern China. Ying Yong Sheng Tai Xue Bao 15:1604–1606.
46. Liu TZ, Nan ZB, Hou FJ (2011) Grazing intensity effects on soil nitrogen mineralization in semi-arid grassland on the Loess Plateau of northern China. Nutr Cycl Agroecosys 91:67–75.
47. Holland EA, Parton WJ, Detling JK, Coppock DL (1992) Physiological responses of plant populations to herbivory and their consequences for ecosystem nutrient flow. Am Nat 140:685–706.
48. Tracy BF, Frank DA (1998) Herbivore influence on soil microbial biomass and nitrogen mineralization in a northern grassland ecosystem: Yellowstone national park. Oecologia 114:556–562.
49. Haynes RJ, Williams PH (1999) Influence of stock camping behaviour on the soil microbiological and biochemical properties of grazed pastoral soils. Biol Fertil Soils 28:253–258.
50. Leibold MA (1989) Resource edibility and the effects of predators and productivity on the outcome of trophic interactions. Am Nat 134:922–949.

Soil Microbial Substrate Properties and Microbial Community Responses under Irrigated Organic and Reduced-Tillage Crop and Forage Production Systems

Rajan Ghimire[1], Jay B. Norton[1]*, Peter D. Stahl[1], Urszula Norton[2]

1 Department of Ecosystem Science and Management, University of Wyoming, Laramie, Wyoming, United States of America, **2** Department of Plant Sciences, University of Wyoming, Laramie, Wyoming, United States of America

Abstract

Changes in soil microbiotic properties such as microbial biomass and community structure in response to alternative management systems are driven by microbial substrate quality and substrate utilization. We evaluated irrigated crop and forage production in two separate four-year experiments for differences in microbial substrate quality, microbial biomass and community structure, and microbial substrate utilization under conventional, organic, and reduced-tillage management systems. The six different management systems were imposed on fields previously under long-term, intensively tilled maize production. Soils under crop and forage production responded to conversion from monocropping to crop rotation, as well as to the three different management systems, but in different ways. Under crop production, four years of organic management resulted in the highest soil organic C (SOC) and microbial biomass concentrations, while under forage production, reduced-tillage management most effectively increased SOC and microbial biomass. There were significant increases in relative abundance of bacteria, fungi, and protozoa, with two- to 36-fold increases in biomarker phospholipid fatty acids (PLFAs). Under crop production, dissolved organic C (DOC) content was higher under organic management than under reduced-tillage and conventional management. Perennial legume crops and organic soil amendments in the organic crop rotation system apparently favored greater soil microbial substrate availability, as well as more microbial biomass compared with other management systems that had fewer legume crops in rotation and synthetic fertilizer applications. Among the forage production management systems with equivalent crop rotations, reduced-tillage management had higher microbial substrate availability and greater microbial biomass than other management systems. Combined crop rotation, tillage management, soil amendments, and legume crops in rotations considerably influenced soil microbiotic properties. More research will expand our understanding of combined effects of these alternatives on feedbacks between soil microbiotic properties and SOC accrual.

Editor: Jose Luis Balcazar, Catalan Institute for Water Research (ICRA), Spain

Funding: Agricultural Prosperity for Small and Medium-Sized Farms Competitive Grant no. 2009-55618-05097 from the USDA National Institute of Food and Agriculture. The funders had no role in study design, data collection and analysis, decision to publish, or preparation of the manuscript.

Competing Interests: The authors have declared that no competing interests exist.

* Email: jnorton4@uwyo.edu

Introduction

Many changes in soil properties after conversion from one agricultural management system to another result from changes in soil microbiotic properties, defined here as the quality of microbial substrate and its effects on soil microbial communities [1,2]. It is well known that management practices such as reduced-tillage, cover crops, and crop diversification increase soil microbial activity in general, and microbial biomass and diversity in particular [2]. Similarly, practices used in certified-organic food and feed production, including amendments and legume crops in rotations, support increased microbial biomass [1,4], arbuscular mycorrhizal fungi (AMF) [5], and soil fauna [4]. It is not as clear how beneficial these management practices are under marginally productive conditions of cold, semiarid agroecosystems. In the study reported here, we evaluated whole-system effects on soil microbiotic properties after conversion from irrigated maize monoculture to conventional, reduced-tillage, and organic crop rotation systems in the central High Plains region of North America. Each management system combines a different suite of practices, including cultivation methods, crop rotations, and soil amendments.

Soil microbiotic properties are influenced by soil amendments, crop rotations, and tillage practices by different mechanisms. Organic amendments contribute diverse microbial substrates as heterogeneous organic materials in different states of decomposition [1,4], while crop rotations diversify the supply of plant residues, including fine roots, root exudates, sloughed off tissues, and rhizodeposited materials, which drive diversification of soil microbial communities [2,6,7]. Intensive tillage drives pulses of microbial activity that mineralize labile soil organic matter (SOM) and shift microbiotic properties toward C-limited conditions that favor bacteria and reduce SOM concentrations, while reduced-tillage conserves labile substrates and creates a more consistent soil environment for microbial activity [2,3,8,9]. In reduced-tillage systems, plant- and root-derived residues provide nucleation sites for fungal and bacterial growth, which further colonize soil particles to form aggregates and increase aggregate-protected, labile SOM and efficiency of substrate utilization (less C respired per unit of microbial biomass) [10,11]. Perennial legume and non-legume forage crops in rotations further reduce soil disturbance compared with annual crops, and stimulate SOM accrual and microbial activity through increases in root biomass and residues [11–13]. Combinations of organic amendments and perennial legumes in rotations, which are common practices in certified organic crop and forage production, support more efficient soil N utilization than conventional, synthetic fertilizer-based management [4,14] and can shift microbiotic properties toward N-limited conditions that favor fungi and accrue SOM.

Such management systems may be especially important in the central High Plains agroecosystem, where the semiarid environment, with inherently low SOM, cold winters, hot, dry summers, and irrigation-driven wetting-drying cycles, exacerbate mineralizing microbiotic conditions that drive losses of SOM [15–18]. Improved understanding of how reduced-tillage and organic crop and forage production systems affect soil microbial substrate quality, microbial biomass, community structure, substrate utilization, and soil organic C (SOC) sequestration in this cold and dry agroecosystem will help to design more sustainable systems during a time of uncertainty due to the changing climate, increasing operation costs, and changing markets [24–26].

The aim of this study was to evaluate SOC, DOC, C:N ratios of microbial substrates, soil microbial biomass and community structure, and substrate utilization after transition from monocropped corn to crop rotations under conventional, organic, and reduced-tillage crop and forage production. The experiments were set up on inherently low fertility, irrigated soils in the dry and cold central High Plains agroecosystem. We hypothesized that crop rotations developed in the previously monocropped field would increase microbial biomass and microbial community diversity by increasing the quantity and changing the quality of microbial substrates. In addition, organic and reduced-tillage management systems would favor greater increases in soil microbial biomass and more diverse microbial communities with higher substrate utilization efficiency compared with conventional management.

Materials and Methods

Experimental Site

The four-year study was established in 2009 at the University of Wyoming Sustainable Agriculture Research and Extension Center (SAREC) near Lingle, Wyoming (42°7′15.03″N; 104°23′13.46″W). The study area has cool temperatures and a short growing season with an average frost-free period of about 125 days and 60-year average maximum and minimum temperature of 17.8°C and 0.06°C, respectively, and precipitation of 332 mm [19]. In addition, maximum and minimum air temperature and precipitation were monitored at the SAREC weather station within 1 km of the research plots during the study period. Monthly average maximum and minimum air temperature and monthly total precipitation throughout the study period are presented in Figure S1. Soil at the study site is mapped as Mitchell loam (loamy, mixed, active, mesic Ustic Torriorthent) with low SOM content (<1%), and slightly alkaline soil pH [20]. Soil texture of the study site was loamy with sand, silt and clay content of 41.0 (13.5)%, 41.4 (10.5)% and 17.6 (4.0)%, respectively (standard deviation in parentheses; n = 24).

Experimental Design and Treatments

The study was designed as two independent randomized complete block experiments (row-crop production and forage production) laid out on a 15-ha half-circle under an irrigation pivot (305-m radius) that was divided into four wedge-shaped blocks (replications) (Figure S2). Each block was further separated into six plots consisting of three 0.405-ha crop production plots (outer three circles) and three 0.81-ha forage production plots (inner three circles). The three management-system treatments (conventional, certified organic, and reduced-tillage) were then randomly assigned to the crop and forage production plots. Before establishment of the experiment the entire area was under conventionally managed corn for at least six years.

All treatments were managed under four-year rotations starting in 2009. Table 1 shows the specific rotations, which were determined by a project advisory committee consisting of local producers and the SAREC management team. Under the conventional system inputs are applied as needed to maximize production, namely commercial synthetic fertilizer based on soil-test recommendations to supply nutrients, and chemical pesticides to control weeds, insects, and diseases. Specific management details are provided in Table S1. Conventional plots were moldboard ploughed, disked, and harrowed, which typically incorporates crop residues into soils leaving <15% of the soil surface covered by residues. The reduced-tillage system used conservation tillage that does not invert surface soil and leaves > 15% residue cover on the soil surface. In the organic system, tillage was done as in conventional plots, and pest control and nutrient management were based on practices allowed by the USDA National Organic Program standards (http://www.ams.usda.gov/AMSv1.0/nop). Conventional and reduced-tillage systems had chemical weed and pest control.

For soil fertility management, conventional and reduced-tillage plots received chemical fertilizer based on soil-test recommendations (Table S1). Organic management received composted cattle manure (dry matter 78% and C:N:P:S = 24.6:0.88:0.22:0.25%) in both crop and forage system in 2010. Because of the limited availability of composted cattle manure in 2011 and 2012, the organic crop system received raw manure (dry matter 29.2% and C:N:P:S = 21.3:1.42:0.35:0.40%) and the organic forage system received composted manure.

In the crop production experiment, management systems had different crops in rotation (Table 1). In forage system plots, a legume-grasses mixture was planted at 22 kg ha^{-1} in all plots in 2009, and included 50% alfalfa (*Medicago sativa* L.), 30% orchard grass (*Dactylis glomerata* L.), 10% meadow brome (*Bromus riparius* Rehmann), and 10% oat (*Avena sativa* L.) by weight. The forage production system plots were winter grazed for three months during 2011–2012 at stocking density of 1.6 fall-weaned calves ha^{-1}.

Soil Sampling

Soil samples were collected during spring, early summer, late summer, and fall seasons of the first (Year 1; 2009) and the fourth year (Year 4; 2012) from each of the 24 plots. During each sampling event, soil cores (3.2-cm diameter) were collected from 0–15 cm at 16 sampling points along a 50-m transect set in each plot, composited, thoroughly homogenized, subsampled (~500 g), and placed on ice for transport to the laboratory. The 0–15 cm depth was considered to be sufficient because the focus was on near-surface microbial properties. Sampling transects were mapped using GPS (Trimble GeoXT, Sunnyvale, CA) to locate transects for subsequent sampling. In the laboratory, soil samples were stored at −20°C for PLFA analysis and at 4°C for DOC, TDN, and potential soil respiration. Phospholipid fatty acid contents in soil were analyzed within two weeks of soil sample collection. Soil bulk density was measured in a separate set of 2.1×15 cm cores collected from 8 sampling points along the 50-m

Table 1. Crop rotations and management practices under different conventional (CV), organic (OR), and reduced-tillage (RT) management systems for crop and forage production (see Table S2 for detailed dates and management activities).

System		year	Crop in rotation	Management practices
Crop	CV	2009	Pinto bean	Tillage with moldboard plow and disk (5–7 passes each year), use of chemical fertilizers based on soil test recommendation for each crop, pesticides application as needed, and no livestock grazing.
		2010	Corn	
		2011	Sugar beet	
		2012	Corn	
	OR	2009	Alfalfa	Tillage with moldboard plow and disk (5–7 passes each year) and use of USDA-NOP certified practices for soil fertility (organic manure application) and pest management (e.g., cultivation), and no livestock grazing.
		2010	Alfalfa	
		2011	Corn	
		2012	Pinto bean	
	RT	2009	Pinto bean	Reduced-tillage (1–2 tillage passes each year that leave >15% crop residue on surface), use of chemical fertilizers based on soil test recommendation, pesticides application as needed, and no livestock grazing.
		2010	Corn	
		2011	Sugar beet	
		2012	Corn	
Forage	CV	2009	Alfalfa/grasses	Conventional tillage (5–7 passes in year 1 and 4), use of chemical fertilizers based on soil test recommendation, pesticides application as needed, and grazing with fall weaned calves during winter 2011/12.
		2010	Alfalfa/grasses	
		2011	Alfalfa/grasses	
		2012	Corn	
	OR	2009	Alfalfa/grasses	Conventional tillage and use of USDA certified practices for soil fertility (compost application) and pest management (no pesticides), and grazing with fall weaned calves during winter 2011/12.
		2010	Alfalfa/grasses	
		2011	Alfalfa/grasses	
		2012	Corn	
	RT	2009	Alfalfa/grasses	Reduced-tillage in the first year and no-tillage after, use of chemical fertilizers based on soil test recommendation, pesticides application as needed, and grazing with fall weaned calves during winter 2011/12.
		2010	Alfalfa/grasses	
		2011	Alfalfa/grasses	
		2012	Corn	

transects. Soil samples from the first and the last sampling dates were analyzed for other soil properties described below.

Laboratory Analysis

Total soil C and N were analyzed by dry combustion (EA1100 Soil C/N analyzer, Carlo Erba Instruments, Milan, Italy), inorganic C by modified pressure-calcimeter [21], and soil moisture by the gravimetric method [22]. Soil organic C was determined by subtracting inorganic C from total soil C. Soil pH was measured in a 1:1 soil:water mixture using an electrode [23]. Soil texture was determined by the hydrometer method [24]. Microbial substrate quality was determined as the ratio of DOC to total dissolved N (TDN) present in soils expressed as the C:N ratio of microbial substrate. For this, 10 g of field-moist soil was extracted with 50 ml of 0.5 M K_2SO_4 and amounts of DOC and TDN were determined by 720°C combustion catalytic oxidation/chemiluminescence with a Schimadzu TOC Analyzer (TOC-VCPH with TNM-1, Schimadzu Scientific Instruments, Inc.) coupled with TOC-Control V Ver.2 analysis software. Dissolved inorganic C was removed by automatic acidification and sparging within the instrument. Potential soil respiration was determined as the amount of CO_2-C mineralized during a two-week incubation period [25]. Soil bulk density was determined by the core method [26] and water filled soil pore space was calculated from bulk density and gravimetric moisture content [27].

Microbial biomass and community structure was analyzed by the Blight and Dyre [28] method of fatty acid methyl ester (FAME) analysis as modified by Frostegård et al. [29] and Buyer et al. [30]. Fatty acids were directly extracted from soil samples using a 1:2:0.8 chloroform:methanol:phosphate buffer mixture (0.15 *M*, pH 4.0), and PLFAs were separated from neutral and glycolipid fatty acids in a solid-phase extraction column (Agilent Technologies Inc.). The PLFAs were methylated using a mild methanoic KOH, and the FAMEs were analyzed using an Agilent 6890 gas chromatograph with autosampler, split-splitless injector (7683B series), and flame ionization detector (Agilent Technologies Inc.). The system was controlled with Agilent Chemstation and MIDI Sherlock software, and the fatty acid peaks were identified using the MIDI peak identification software (MIDI, Inc., Newark, DE, USA). All solvents and chemicals used were of analytical grade, and all glassware used was rinsed 10 times with deionized water, and sterilized overnight in 450°C in a Blue M lab heat box type muffle furnace (Blue M Electric, Richardson, TX). The PLFA signatures of 16 different fatty acids, which were quantified in almost all the field plots, were used to study soil microbial community structure and these fatty acids were grouped into gram positive, gram negative and other bacteria, AMF and other fungi, and protozoa (Table S3). In addition, the Shannon diversity index [31] was calculated as an index of soil microbial diversity as influenced by management systems in crop and forage production. The ratio of potential soil respiration to total PLFA microbial biomass was also calculated as an index of microbial substrate utilization.

Statistical Analysis

Crop and forage production experiments were each analyzed as separate randomized complete block designs (RCBD) with three management-system treatments (conventional, organic, reduced-tillage) and four replicates. The analysis of soil properties that were measured at the beginning and end of the study, such as SOC, STN, pH and EC, were analyzed as split plot in time analysis of variance set in an RCBD for each system ($p = 0.05$). This analysis considered year as a repeated observation and replication as a random term in the model. Soil properties measured four times each year, such as soil microbial PLFA contents, DOC, C:N ratio of microbial substrate, potential soil respiration, water filled pore space, and soil bulk density, were analyzed as a split plot in time analysis of variance that considered season and year as repeated observation terms in the model. Statistical computations for both designs were facilitated by the mixed model (Proc Mixed) procedure of the Statistical Analysis System (SAS, ver. 9.3, SAS Institute, Cary, NC). Means were separated using the PDIFF test in the LSMEANS procedure ($p = 0.05$) unless otherwise stated. There were no significant season×management system interactions for either system in the three way split plot in time analysis of variance, therefore, results are reported as average of all four seasons within a year. In addition, PLFA data for individual microbial groups were normalized to the total microbial PLFAs and the data (mole percent of total PLFAs) were reanalyzed through a multivariate method (principal component analysis) to compare shifts in microbial community structure. Relationships between soil microbial substrate properties, microbial biomass and community structure, and substrate utilization were analyzed using Pearson correlation. Principle component and Pearson correlation analyses were performed using a Minitab V.16.0 (Minitab Inc., State College, PA, USA) and the first two principal components are graphed to summarize the results.

Results

Monthly average maximum and minimum temperatures during growing seasons (May to September) of 2009–2012 varied from year to year (Figure S1). The average minimum temperature was lowest in December 2009 and February 2011. Average precipitation was the lowest in 2012, followed by 2009 and 2010, compared to that in 2011. The amount of irrigation water depended on crop demand and the amount of precipitation received, and more water was applied to meet the crop water requirement in 2012 than in 2009–2011. All plots were irrigated to 60% of field capacity. Water filled porosity was consistent across management systems, seasons, and study years in both production systems (data not presented).

Soil pH was consistent across management systems and study years (range 7.3–7.8) under both crop and forage production, as was SOC concentration (Table 2). Soil organic C concentrations were, however, significantly influenced by a management system × year interaction. Soils under reduced-tillage ($p = 0.034$) and organic ($p = 0.004$) crop production had significantly more SOC than soils under conventional crop production. In addition, soils under organic crop production in the fourth year had significantly more SOC than in the first year ($p = 0.02$). Soils under reduced-tillage forage production had significantly more SOC than those under conventional forage production ($p < 0.01$). Soil total N concentrations were not significantly influenced by management systems, years, or management system×year interactions under either crop or forage production. Soil bulk density was not significantly influenced by management system, season, or year, but was significantly influenced by a management system×year interaction under crop production. Specifically, soil bulk density was significantly higher under reduced-tillage than organic crop production in the fourth year ($p = 0.007$).

Soil microbial biomass concentrations were significantly influenced by a management system×year interaction, but not by season. Specifically, in the fourth year under crop production there was significantly more soil microbial biomass under organic than conventional and ($p < 0.001$) and reduced-tillage management ($p = 0.01$) (Fig. 1a). In the fourth year under forage production there was significantly more microbial biomass in soils under reduced-tillage than conventional ($p = 0.002$) and organic management ($p = 0.047$) (Fig. 1b). Under crop production, the greatest increase in total microbial biomass over the four-year period was observed in soils under organic management (353%) followed by conventional (262%) and reduced-tillage (202%) (based on year-one values). Under forage production, the increase in total microbial biomass concentrations were statistically similar at 396, 378 and 361% higher in the fourth year than in the first year in soils under conventional, organic, and reduced-tillage management systems, respectively.

We also observed increases in soil bacterial PLFAs, fungal biomarker PLFAs, DOC, and TDN across all treatments (only DOC data presented in Table 2), but the increases differed in magnitude. Dissolved organic C per unit SOC was 0.28–0.56% in the first year and 1.49–2.04% in the fourth year, with highest amount of DOC per unit SOC under conventional management. These changes corresponded with significantly higher fungal to bacterial ratios (F:B ratios) (Figure 2) and C:N ratios of microbial substrates (Figure 3) in the fourth year than in the first year. In addition, C:N ratios of microbial substrates were greater in soils under organic forage production than under conventional and reduced-tillage forage production. Similarly, microbial substrate utilization (potential soil respiration per unit PLFA) was consistent across management systems (Figure 4), but was 85–90% lower under crop production and 61–77% lower under forage production in the fourth year than in the first year.

Principal component analysis of microbial community structure revealed that the first two principle components explained 64.9% and 25.4% of the total sample variance under the crop production, and 74.3% and 21.0% of total sample variance under the forage production (Figure 5). The soil samples collected in the first year clustered on the left side of the figures 5.a1 and b1, and those collected in the fourth year clustered on the right side, corresponding to the increase in microbial substrate quality and decrease in potential soil respiration. There was greater variance in microbial community data collected in the first year than in the fourth year under both crop and forage production. In addition, loading scores for management systems separated more clearly along PC2 in the fourth year than in the first year. Among microbial groups, protozoa, other bacteria, and other fungi had positive loadings, while AMF and gram-negative bacteria had

Table 2. Soil properties as influenced by conventional (CV), organic (OR), and reduced-tillage (RT) management systems for crop and forage production one and four years after transition from continuous conventional corn.

System		SOC[†‡]			STN			DOC			Db		
		Y1	Y4	Δ%	Y1	Y4	Δ%	Y1	Y4	Δ%	Y1	Y4	Δ%
		g kg^{-1} soil			g kg^{-1} soil			mg kg^{-1} soil			g cm^{-3}		
Crop	CV	6.70aA(1.31)	5.83bA(0.98)	−13.0	0.65(0.08)	0.73(0.12)	+12.3	19.5aB(2.08)	119cA(4.46)	+510	1.46aA(0.02)	1.43abA(0.02)	−2.05
	OR	6.52aB(0.16)	8.95aA(0.37)	+37.3	0.79(0.05)	0.90(0.08)	+13.3	36.4aB(5.45)	154aA(4.43)	+323	1.48aA(0.02)	1.40bA(0.02)	−5.41
	RT	7.31aA(0.86)	8.00aA(0.52)	+9.44	0.72(0.04)	0.80(0.05)	+11.8	20.6aB(3.78)	137bA(8.49)	+565	1.42aA(0.02)	1.48aA(0.03)	+4.23
Forage	CV	8.96aA(1.36)	7.23bA(0.46)	−19.3	0.77(0.09)	0.81(0.08)	+5.19	38.6aB(3.85)	139aA(13.3)	+360	1.47aA(0.03)	1.44aA(0.02)	−2.04
	OR	9.26aA(0.76)	8.50abA(0.35)	−8.31	0.77(0.02)	0.89(0.07)	+15.5	26.5aB(5.73)	147aA(7.36)	+455	1.47aA(0.02)	1.42aA(0.03)	−3.40
	RT	8.13aA(0.70)	10.2aA(1.04)	+25.5	0.86(0.14)	1.04(0.09)	+20.9	27.7aB(3.31)	152aA(2.67)	+449	1.39bA(0.02)	1.42aA(0.02)	+2.16

[†]Number in parenthesis indicates standard error.
[‡]SOC = soil organic carbon, STN = soil total nitrogen, DOC = dissolved organic carbon (g kg^{-1} soil) and Db = soil bulk density (g cm^{-3}). Different lowercase letters within a column indicate significant difference among management systems within a year and different uppercase letters indicate significant difference among years within a management system in crop as well as forage production (p = 0.05). No letter within a column or row indicates no significant management system × year difference.

negative loadings along the PC1 axis (Figure 5.a2 and b2). Along the PC2 axis, gram-positive bacteria under crop production, and both gram-positive bacteria and AMF under forage production, had positive loadings. Across management systems and crop and forage production, gram-positive bacteria, gram-negative bacteria, and AMF together constituted of 93% of total soil microbial biomass in the first year and 76–78% in the fourth year (Table 3). After four years under alternative management systems, biomarker PLFAs for these three microbial groups had increased 2–4 fold, while biomarker PLFAs for other bacteria, fungi, and protozoa had increased up to 26, 9, and 36 fold, respectively, corresponding with significant shifts in microbial community structure. Changes in microbial community structure over the four-year study period are also indicated by Shannon's diversity index in Table 3, which was significantly higher in the fourth year than in the first year across both crop and forage production and all three management systems.

Increases in microbial biomass and F:B ratios, along with other changes in microbial community structure, along PC1 were strongly positively correlated with substrate availability (DOC concentrations) and quality (C:N ratio of microbial substrate) (Table 4) under both crop and forage production. Microbial substrate utilization decreased significantly with increasing substrate availability, increasing C:N ratios of microbial substrates, and increasing microbial biomass. Similarly, microbial community changes along PC2 were not related with substrate properties and substrate utilization.

Discussion

Our results support our hypotheses and indicate that conversion from continuous corn to crop rotations positively impacted soil microbiotic properties across all three management systems, with higher substrate availability, substrate C:N ratios, and soil microbial biomass contents, but lower substrate utilization in the fourth year than in the first year following transition (Tables 2 and 4; Figures 1a, 3a, and 4a). Both reduced-tillage and organic management systems added to these effects.

Under organic crop production, combined effects of perennial legumes in the rotations, which eliminated tillage for two of the four years, with additions of manure and compost, apparently offset losses of microbial substrates due to heavy tillage during the annual crop phases and supported the highest year-four microbial biomass concentrations (Table 2). Inclusion of legumes in rotations and organic amendments typically favor microbial growth, SOC and N accumulation, and diversification of microbial substrates [6,7,33]. Under reduced-tillage crop production, a more consistent soil environment apparently facilitated higher soil microbial biomass concentrations and diversity, as well as higher fungal productivity than under conventional management (Table 3; Figure 1a). Similar effects of reduced disturbance have been noted [2,9,34] in which fungal hyphae improve soil aggregation, which protects labile SOM components and regulates microbial substrate utilization [3,12,13,33].

Under forage production, reduced-tillage management had the highest year-four microbial biomass concentrations of the three systems, probably due to lack of plowing with conversion from perennial forage to corn. Organic management, with applications of composted manure, created significantly higher C:N ratios of microbial substrates in the fourth year than under conventional and reduced-tillage with chemical fertilizer application (Figure 3b). The fact that this difference in substrate quality did not occur under the crop production experiment (Figure 3a) suggests that it resulted from a combined effect of the alfalfa-grasses mixture and

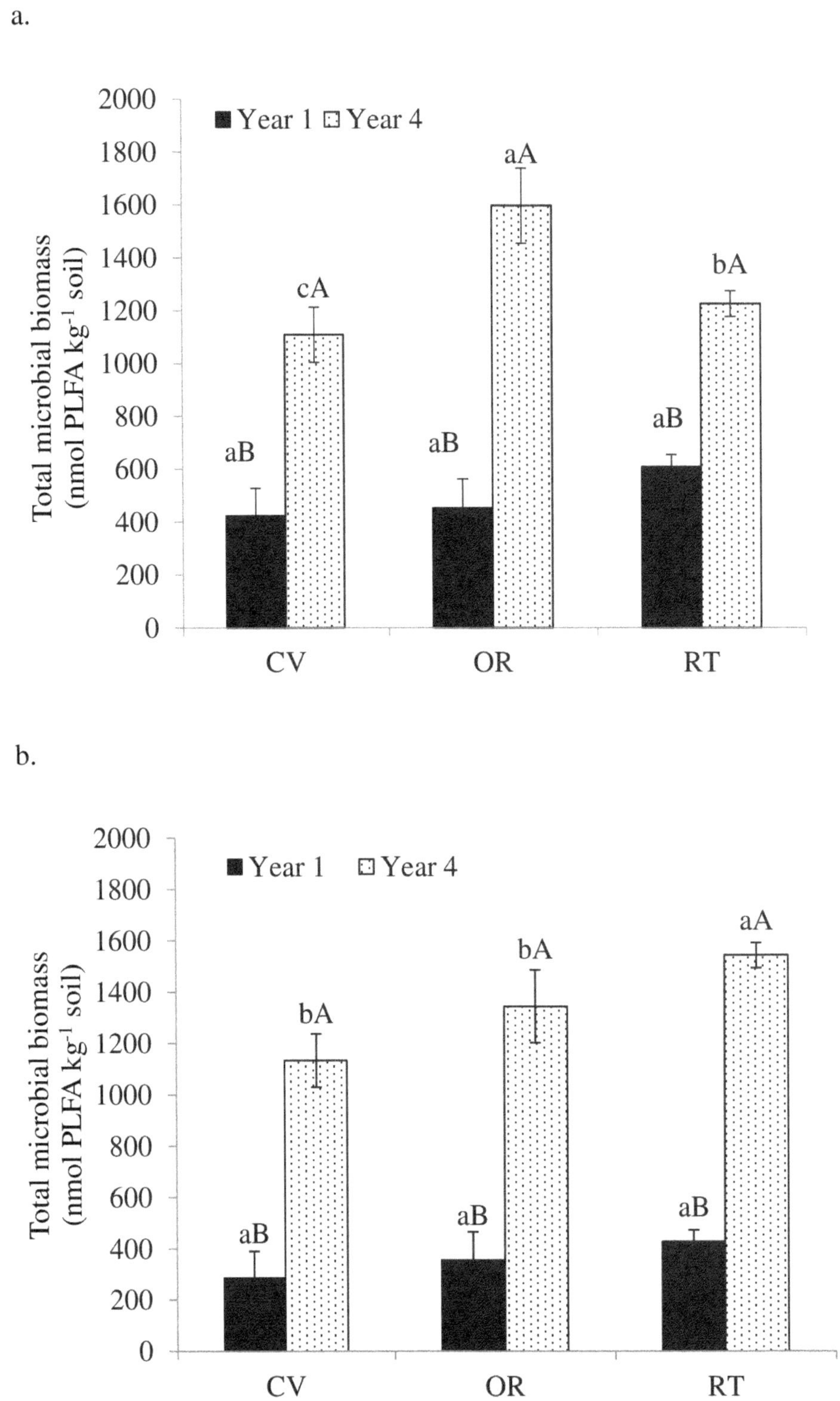

Figure 1. Total soil microbial biomass as influenced by conventional (CV), organic (OR), and reduced-tillage (RT) management systems for crop (a) and forage (b) production in the first and fourth years. Different lowercase letters indicate significant differences among management systems within a year and different uppercase letters indicate significant difference among years within a management system ($p = 0.05$).

compost applications. In the conventional and reduced-input forage systems, N from chemical fertilizer and alfalfa may have contributed to lower substrate C:N ratios than under the organic forage system.

Changes in soil microbiotic properties we observed in response to alternative management systems are consistent with results of previous studies, but greater in magnitude. Previous studies have reported two- to three-fold increases in microbial biomass with

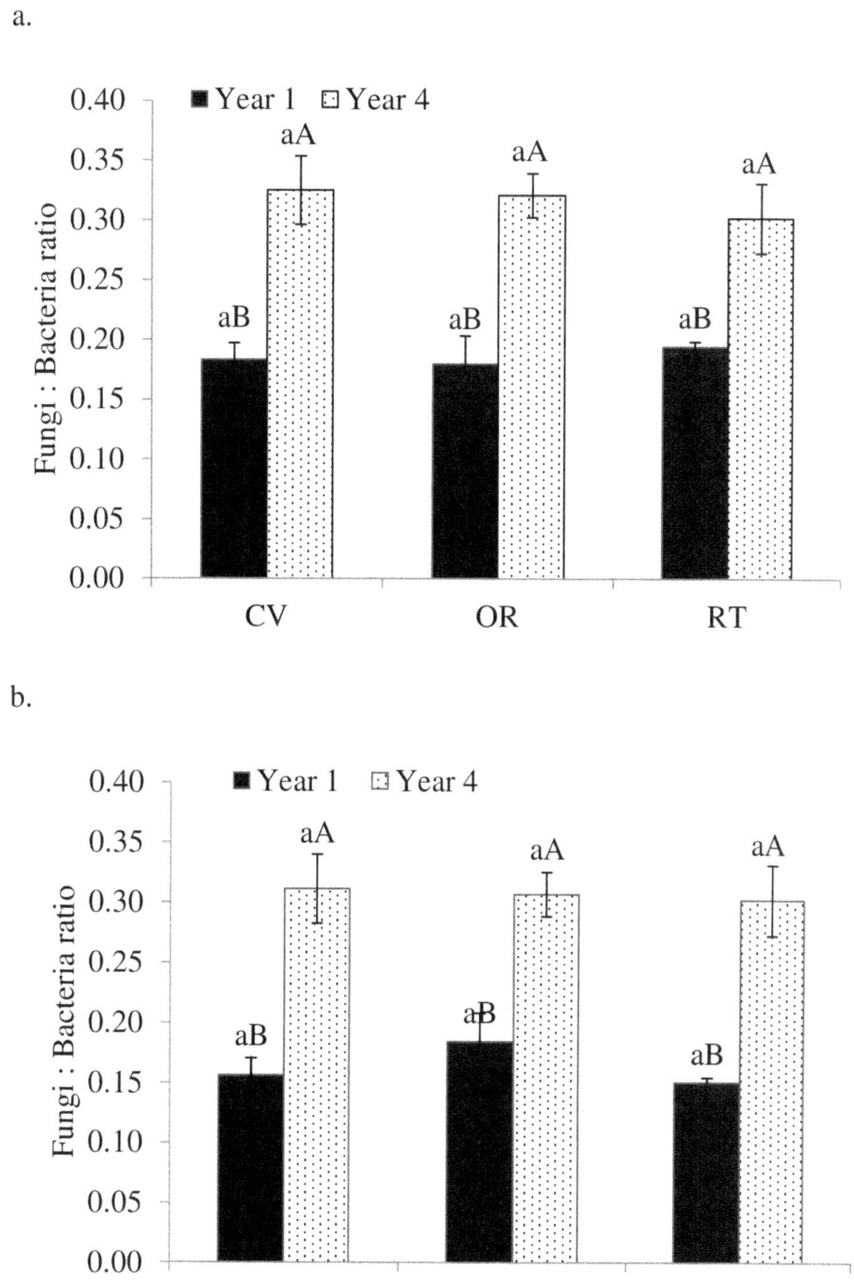

Figure 2. Fungal-to-Bacterial ratio as influenced by conventional (CV), organic (OR), and reduced-tillage (RT) management systems for crop (a) and forage (b) production in the first and fourth years. Different lowercase letters indicate significant differences among management systems within a year and different uppercase letters indicate significant difference among years within a management system (p = 0.05).

diversified rotations or reduced disturbance in place for several years [2,8,32,35]. The greater magnitudes we observed probably resulted from combined effects of transition from continuous corn to rotations with legumes, manure applications, and reduced-tillage on depleted, inherently-low-fertility soils. While variable precipitation during the seasons of the study may have influenced comparisons between years 1 and 4, air temperatures (Figure S1) and soil water filled pore space were consistent among management systems across years under both crop and forage production. Therefore, we believe that the changes in management, rather than annual climatic variability, drove the observed changes in soil microbiotic properties.

The changes in quantity and quality of microbial substrates during the study period drove notable shifts in microbial community structure, including greater increases in fungal relative to bacterial PLFAs (Table 3; Figure 2), in gram-positive relative to gram-negative bacterial PLFAs, and in saprophytic fungi and protozoa relative to other groups. Higher DOC and C:N ratios of microbial substrates at the end of the study drove greater increases in saprophytic fungi, which rely on carbonaceous substrates, than AMF, which are often associated with more mineral-rich, low C:N a.

b.

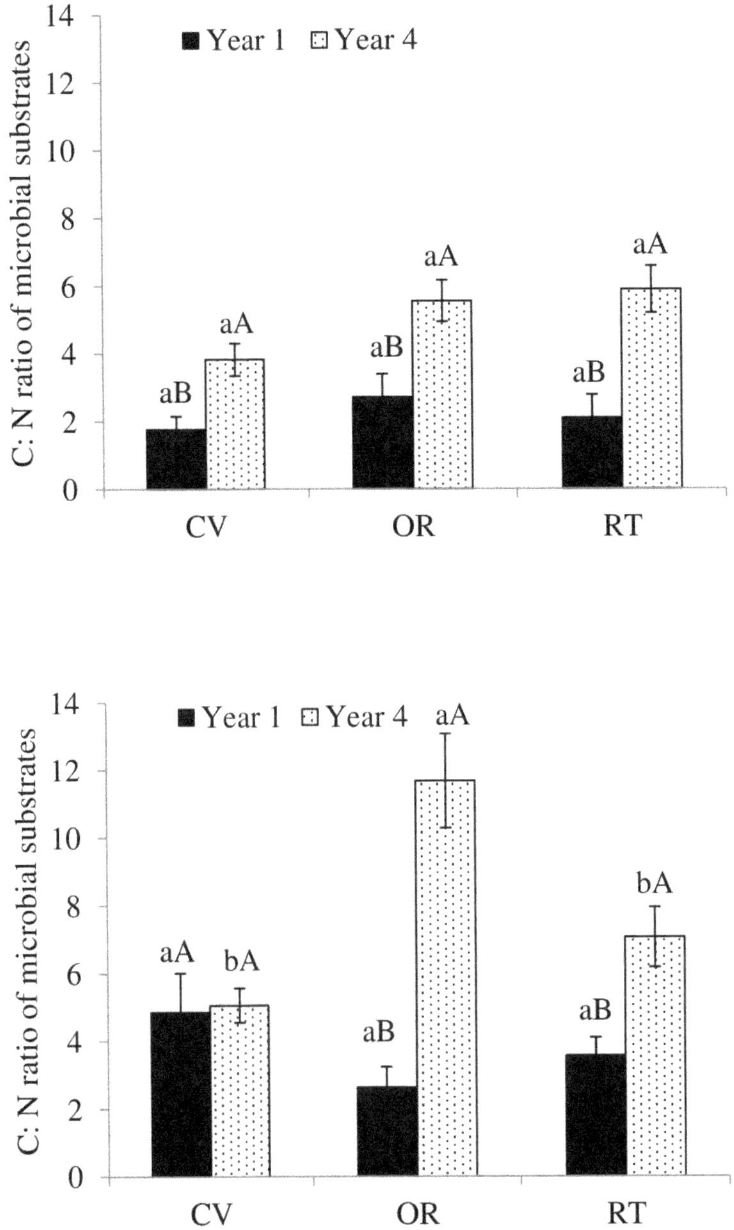

Figure 3. Carbon-to-nitrogen ratio of microbial substrate as influenced by conventional (CV), organic (OR), and reduced-tillage (RT) management systems for crop (a) and forage (b) production in the first and fourth years. Different lowercase letters indicate significant differences among management systems within a year and different uppercase letters indicate significant difference among years within a management system ($p = 0.05$).

substrates [36–38]. The observed increases in the non-mycorrhizal fungi strongly correlated with increased substrate availability, indicating changes in C:N ratios favored more fungal growth (saprotrophic fungi) than bacterial growth.

Increases in gram-positive relative to gram-negative bacteria are also often associated with increases in diversity of C sources in soils and decreases in mechanical soil disturbance, [36–39]. Gram-positive bacteria are associated with low substrate availability (high C:N) environments, while gram-negative bacteria dominate soils with more easily decomposable substrates [38]. Decreases in relative abundance of gram-negative bacteria may be beneficial because many plant pathogenic microorganisms such as *Pseudomonas* and *Xanthomonas* species are gram negative [40,41]. Large increases in protozoa parallel the increases soil bacteria, which are their food source [42].

Greater amounts of higher C:N-ratio microbial substrates, more diverse communities of microorganisms with higher F:B ratios, and reduced potential soil respiration in the fourth year across all three management systems in general, and under reduced-tillage forage and organic crop systems in particular, suggests that minimum soil disturbance, application of organic amendments, and more legume crops in rotation increase soil microbial biomass, a.

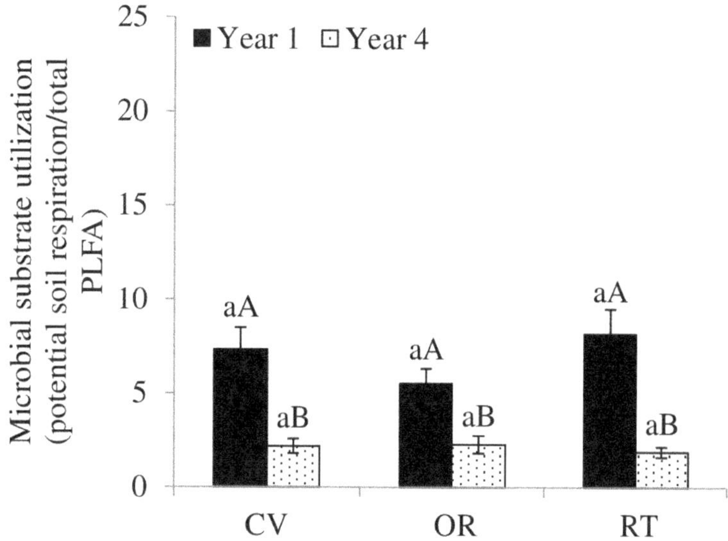

b.

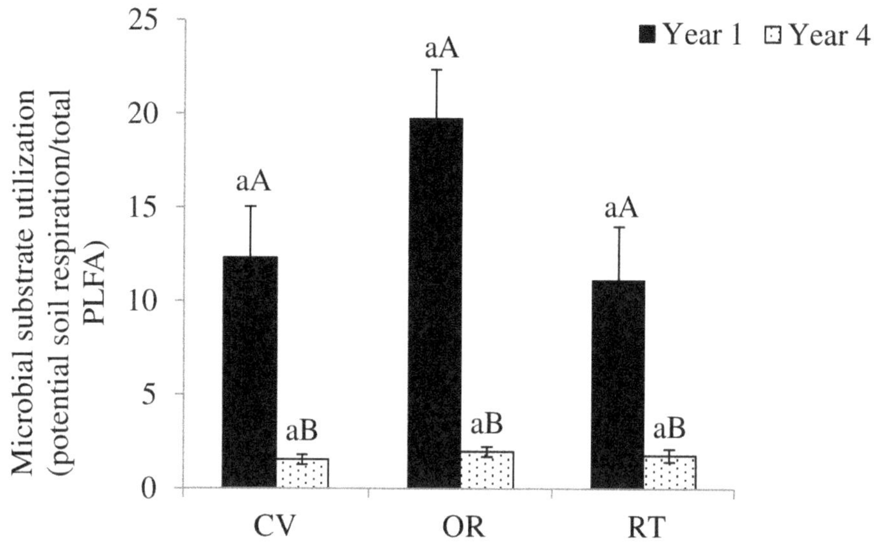

Figure 4. Microbial substrate utilization as influenced by conventional (CV), organic (OR), and reduced-tillage (RT) management systems for crop (a) and forage (b) production in the first and fourth years. Different lowercase letters indicate significant differences among management systems within a year and different uppercase letters indicate significant difference among years within a management system ($p = 0.05$).

alter community structure, and thereby influence SOC accrual (Figure 6). Although mechanisms of SOC regulation by specific groups of microorganisms are not well defined, given similar site characteristics, higher SOC sequestration potential is typically observed in soils with higher F:B ratios [43,44]. Higher SOC in fungal-dominated systems is mainly attributed to higher biomass C production per unit of C metabolized by fungus than by soil bacteria [44]. The changes in microbiotic properties we observed indicate that substrate C was mainly transformed into microbial biomass or less labile SOM components, which may be reflected in year-four SOC contents.

Under crop production, both our organic system, with two years of alfalfa, and reduced-tillage system involved considerably less soil disturbance than our conventional system, and both had more year-4 SOC than the conventional system (Table 1, S1). Under forage production, with the same rotation across the three

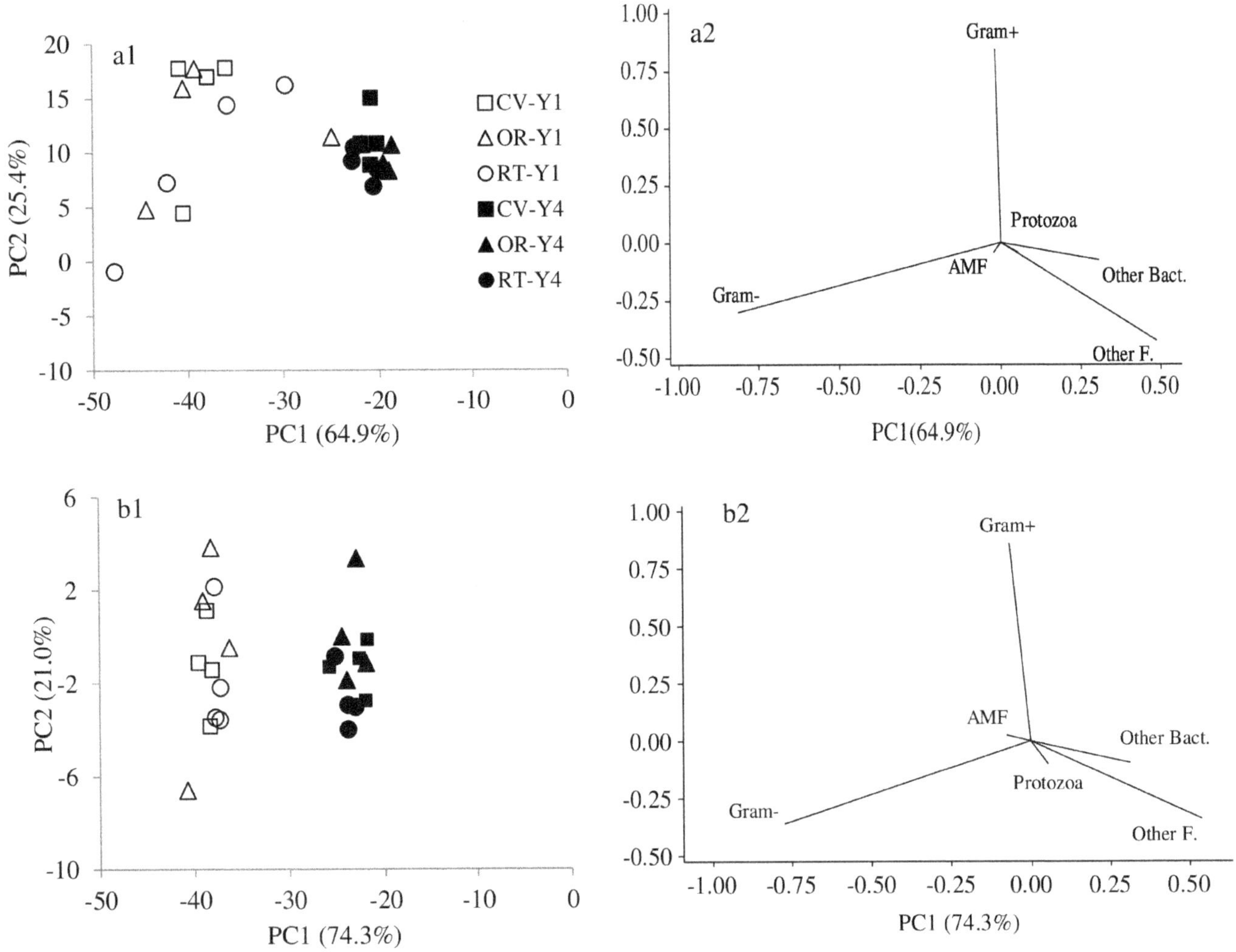

Figure 5. Score plots of the first two principle components (1) and loading of different microbial groups (2) as influenced by conventional (CV), organic (OR), and reduced-tillage (RT) management systems for crop (a) and forage (b) production. Gram+ = gram-positive bacteria, Gram− = gram-negative bacteria, AMF = arbuscular mycorrhizal Fungus, Other Bact. = other bacteria and Other F. = other fungus.

management systems, reduced tillage had significantly more year-4 SOC than the other two system, indicating that organic amendments combined with intensive tillage did not increase SOC. Taken together, these results suggest that reduced-tillage combined with legumes in rotation had the largest impacts on SOC accrual. Repeated tillage to plow down the grasses and alfalfa and establish corn in year 4 might have caused significant loss of SOC accrued during three years under forages in both organic and conventional systems.

Overall, increases in microbial substrate availability and microbial biomass over the 4-year study represent a small fraction of SOC reservoirs, even in this low-SOM environment. Therefore, longer-term evaluation of the effects of tillage, crop rotations, soil amendments, and legume integration in crop and forage production may further our understanding of the influence of microbiotic properties on SOC sequestration. Results of this cropping systems study bundle effects of reduced tillage, crop rotation, legumes in rotation, and soil fertility options into three management systems for crop and forage production. While overall effects are crucial to understanding how management alternatives affect system sustainability, evaluating individual components will complement these results and contribute to design of best management practices for irrigated agriculture in cold, semiarid agroecosystems like the central High Plains.

Conclusions

In this study, management systems that included reduced tillage, perennial legumes, and organic amendments improved soil microbiotic properties that support SOC accrual. The greatest changes occurred with transition from continuous corn to crop rotation. The changes were enhanced under organic management in cash-crop production and reduced-tillage management under forage production. Under the different rotations of our crop production systems, more legume crops in rotation had greater influence on soil microbiotic properties than fewer legume crops. Under the same rotations of our forage production systems, reduced-tillage management had the greatest influences on soil microbiotic properties. These effects were driven by interactions between soil microbial community structure and microbial substrate quantity and quality that resulted in increases in fungal biomass that support SOC accrual. The results indicate that

Table 3. Soil microbial communities as influenced by conventional (CV), organic (OR), and reduced-tillage (RT) management systems for crop and forage production one and four years after transition from continuous conventional corn.

System		Gram positive bacteria†			Gram negative bacteria			Other Bacteria			AMF			Other Fungi			Protozoa			Shannon's diversity index		
		Y1	Y4	Δ%	Y1	Y4	Δ%	Y1	Y4	Δ%	Y1	Y4	Δ%	Y1	Y4	Δ%	Y1	Y4	Δ%	Y1	Y4	Δ%
		nmol kg^{-1} soil			nmol kg^{-1} soil			nmol kg^{-1} soil			nmol kg^{-1} soil			nmol kg^{-1} soil			nmol kg^{-1} soil					
Crop	CV‡	200aB (53.3)	414bA (43.7)	+107	160bB (33.8)	391cA (60.4)	+144	12.2aB (5.64)	60.3bA (11.7)	+394	28.7aB (6.21)	76.5bA (7.24)	+167	20.0bB (5.01)	152bA (15.8)	+665	3.30 (1.81)	14.4bA (2.70)	+336	3.50aB (0.25)	5.53aA (0.23)	+58.2
	OR	213aB (51.6)	555aA (48.8)	+161	166bB (34.4)	559aA (84.4)	+237	12.0aB (4.26)	105aA (18.9)	+775	28.8aB (7.00)	118aA (13.9)	+310	29.1aB (6.70)	230aA (16.1)	+201	3.02 (1.63)	28.7aA (2.17)	+850	3.55aB (0.29)	5.89aA (0.20)	+66.0
	RT	270aB (60.7)	467bA (54.1)	+72.9	236aB (45.9)	479bA (78.2)	+106	22.4aB (6.97)	70.9bA (13.5)	+217	44.3aB (8.88)	89.6bA (10.5)	+116	31.7aB (7.17)	173bA (23.0)	+147	4.38 (2.36)	21.4abA (3.28)	+754	3.80aB (0.27)	5.71aA (0.27)	+50.2
Forage	CV	124aB (21.5)	391bA (23.7)	+215	117aB (17.5)	408bA (55.1)	+249	2.30aB (1.61)	51.5aA (10.6)	+2139	24.9aB (3.50)	86.7bA (7.99)	+248	18.2aB (3.59)	176aB (15.8)	+867	0.00	21.1 (2.31)	-	3.85aB (0.22)	5.69aA (0.20)	+47.9
	OR	157aB (22.3)	474abA (55.3)	+197	142aB (18.8)	490bA (93.0)	+245	3.69aB (2.26)	70.7aA (11.5)	+1575	29.5aB (3.64)	111aA (22.7)	+276	23.0aB (3.77)	177aB (16.1)	+670	0.00	22.8 (3.87)		4.00aB (0.16)	5.64aA (0.22)	+41.0
	RT	180aB (24.2)	526aA (41.6)	+192	177aB (19.5)	566aA (72.1)	+219	3.11aB (2.19)	80.4aA (19.4)	+2485	36.9aB (4.22)	116aA (12.3)	+214	29.1aB (3.30)	218aB (23.0)	+649	1.13 (1.13)	37.1 (2.94)	+3597	4.19aB (0.04)	5.71aA (0.22)	+36.1

†Number in parenthesis indicates standard error.
‡AMF = Arbuscular Mycorrhizal Fungi. Different lowercase letters within a column indicate significant difference among management systems within a year and different uppercase letters indicate significant difference among years within a management system in crop as well as forage production ($p = 0.05$). No letter within a column or row indicates no significant management system × year difference.

Table 4. Significant correlation coefficients (r) between soil microbial substrate properties, microbial community and substrate utilization in crop and forage production systems.

	DOC[†‡]	C:N MAS	Microbial biomass	F:B ratio	PC1[§]
Crop system					
Substrate C:N	0.69(<0.001)	-			
Microbial biomass	0.83(<0.001)	0.86(<0.001)	-		
F:B ratio	0.83(<0.001)	0.81(<0.001)	0.91(<0.001)	-	
PC1	0.78(<0.001)	0.63(0.001)	0.83(<0.001)	0.91(<0.001)	-
Soil Resp.	−0.64(0.001)	−0.47(0.02)	−0.70(<0.001)	−0.74(<0.001)	−0.84(<0.001)
Forage system					
Substrate C:N	0.65(0.001)	-			
Microbial biomass	0.85(<0.001)	0.56(0.006)	-		
F:B ratio	0.91(<0.001)	0.64(0.001)	0.83(<0.001)	-	
PC1	0.93(<0.001)	0.67(0.001)	0.91(<0.001)	0.97(<0.001)	-
Soil Resp.	−0.79(<0.001)	−0.40(<0.001)	−0.87(<0.001)	−0.73(<0.001)	−0.84(<0.001)

[†]Number in parenthesis indicates Pearson correlation p values.
[‡]DOC = dissolved organic carbon, F:B ratio = fungi to bacteria ratio, PC1 = First principal component and Soil Resp. = potential soil respiration.
[§]PC1 explains shift in soil microbial community structure from the first to the fourth year.

reducing disturbance, including legumes, and applying organic amendments positively impact soil processes in ways that enhance sustainable productivity of inherently low-fertility soils in a cold, semiarid environment, even over a relatively short time period. Further research may confirm the combined effects of crop rotations and alternative management systems on SOC accrual

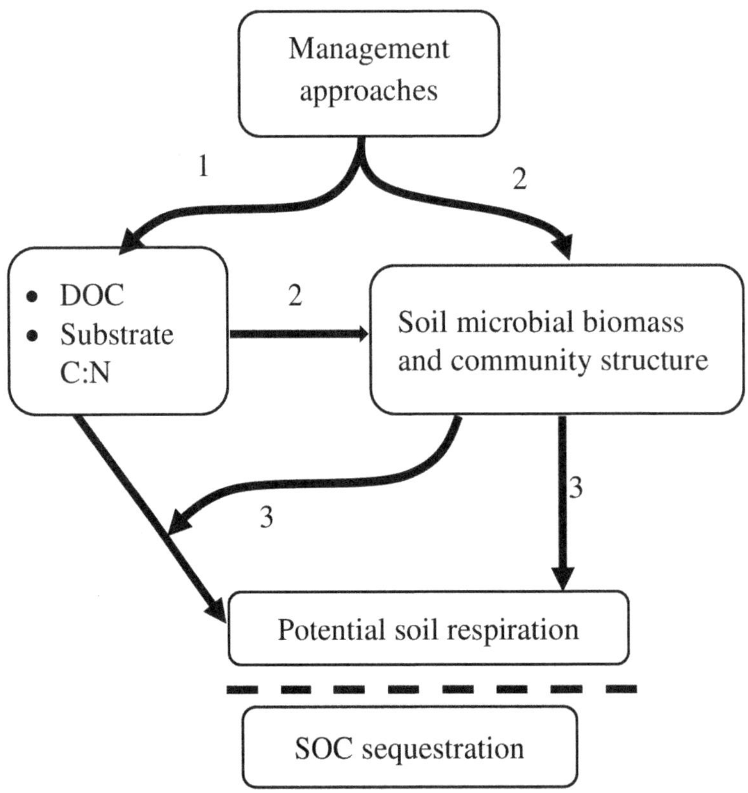

Figure 6. Conceptual framework illustrating influence of management system on microbial substrate properties, microbial communities, and SOC sequestration. Alternative management systems influence (1) microbial substrate availability and quality, (2) soil microbial biomass and community structure, and (3) soil respiration and SOC sequestration. DOC = dissolved organic carbon (microbial substrate).

and ecosystem services as influenced by soil microbial biomass and fungal productivity.

Supporting Information

Figure S1 Monthly average maximum and minimum temperature, and total precipitation at the SAREC weather station, Lingle, Wyoming (Apr. 2009–Sept. 2012).

Figure S2 Experimental design of crop and forage production plots under alternative management systems (C, conventional; O, organic; R, reduced tillage). Tiers A, B, and C are 1-acre cash-crop plots; tiers D, E, and F are 2-acre forage plots.

Table S1 Crop rotations under conventional (CV), organic (OR) and reduced-tillage (RT) management systems in crop and forage production.

Table S2 Timing of management practices by under conventional (CV), organic (OR), and reduce-tillage (RT) management systems on crop and forage production (2009–2012).

Table S3 Biomarker phospholipid fatty acids used for identifying taxonomic microbial groups.

Acknowledgments

We thank Jenna Meeks and Caley Kristian Gasch for field and laboratory assistance, Professor David Legg for assistance with statistical analyses, and Prakriti Bista and Professor Elise Pendall for their helpful comments to improve earlier drafts of this manuscript.

Author Contributions

Conceived and designed the experiments: JBN RG UN PDS. Performed the experiments: RG JBN. Analyzed the data: RG JBN. Contributed reagents/materials/analysis tools: JBN PDS UN. Wrote the paper: RG JBN. Reviewed the manuscript: UN PDS.

References

1. Berthrong ST, Buckley DH, Drinkwater LE (2013) Agricultural management and labile carbon additions affect soil microbial community structure and interact with carbon and nitrogen cycling. Microbial Ecol 66: 158–170.
2. Acosta-Martinez V, Mikha MM, Vigil MF (2007) Microbial communities and enzyme activities in soils under alternative crop rotations compared to wheat-fallow for the Central Great Plains. Appl Soil Ecol 37: 41–52.
3. Stahl PD, Parkin TB, Christensen M (1999) Fungal presence in paired cultivated and undisturbed soils in central Iowa. Biol Fertil Soils 29: 92–97.
4. Drinkwater LE, Letourneau DK, Workneh F, Vanbruggen AHC, Shennan C (1995) Fundamental differences between conventional and organic tomato agroecosystems in California. Ecol Appl 5: 1098–1112.
5. Oehl F, Sieverding E, Ineichen K, Mader P, Boller T, et al. (2003) Impact of land use intensity on the species diversity of arbuscular mycorrhizal fungi in agroecosystems of Central Europe. Appl Environ Microbiol 69: 2816–2824.
6. Kandeler E, Tscherko D, Spiegel H (1999) Long-term monitoring of microbial biomass, N mineralisation and enzyme activities of a Chernozem under different tillage management. Biol Fertil Soils 28: 343–351.
7. Shi Y, Lalande R, Ziadi N, Sheng M, Hu Z (2012) An assessment of the soil microbial status after 17 years of tillage and mineral P fertilization management. Appl Soil Ecol 62: 14–23.
8. Acosta-Martínez V, Dowd SE, Bell CW, Lascano R, Booker JD, et al. (2010) Microbial community structure as affected by dryland cropping systems and tillage in a semiarid sandy soil. Diversity 2: 910–931.
9. Halvorson AD, Wienhold BJ, Black AL (2002) Tillage, nitrogen, and cropping system effects on soil carbon sequestration. Soil Sci Soc Am J 66: 906–912.
10. Blanco-Canqui H, Lal R (2007) Regional assessment of soil compaction and structural properties under no-tillage farming. Soil Sci Soc Am J 71: 1770–1778.
11. Ghimire R, Norton J, Pendall E (2014) Alfalfa-grass biomass, soil organic carbon, and total nitrogen under different management systems in an irrigated agroecosystem. Plant Soil: 374: 173–184.
12. Six J, Feller C, Denef K, Ogle SM, Sa MJC, et al. (2002) Soil organic matter, biota and aggregation in temperate and tropical soils: Effects of no-tillage. Agronomie 22: 755–775.
13. Liebig M, Carpenter-Boggs L, Johnson JMF, Wright S, Barbour N (2006) Cropping system effects on soil biological characteristics in the Great Plains. Renew Agric Food Syst 21: 36–48.
14. Jenkinson DS, Parry LC (1989) The nitrogen-cycle in the broadbalk wheat experiment - a model for the turnover of nitrogen through the soil microbial biomass. Soil Biol Biochem 21: 535–541.
15. Ghimire R, Norton JB, Norton U, Ritten JP, Stahl PD, et al. (2013) Long-term farming systems research in the central High Plains. Renew Agric Food Syst 28: 183–193.
16. Krall JM, Delaney RH, Taylor DT (1991) Survey of nonirrigated crop production practices and attitudes of Wyoming producers. J Agron Educ 20: 120–122.
17. Norton JB, Mukhwana EJ, Norton U (2012) Loss and recovery of soil organic carbon and nitrogen in a semiarid agroecosystem. Soil Sci Soc Am J 76: 505–514.
18. Frey SD, Elliott ET, Paustian K (1999) Bacterial and fungal abundance and biomass in conventional and no-tillage agroecosystems along two climatic gradients. Soil Biol Biochem 31: 573–585.
19. Western Regional Climate Center (2013) Historical climate information. Reno, NV: Desert Research Institute. http://www.wrcc.dri.edu/. Accessed: 06-15, 2013.
20. Soil Survey Staff (2013) Web Soil Survey. Natural Resources Conservation Service, United States Department of Agriculture. http://websoilsurvey.sc.egov.usda.gov/ Accessed: 06-15, 2013.
21. Sherrod LA, Dunn G, Peterson GA, Kolberg RL (2002) Inorganic carbon analysis by modified pressure-calcimeter method. Soil Sci Soc Am J 66: 299–305.
22. Gardner WH (1986) Water content. In: Klute A, editor. Methods of Soil Analysis Part 1: Physical and Mineralogical Methods. 2nd ed. Madison, WI: Agronomy Monograph 9. ASA, SSSA pp. 493–541.
23. Thomas GW (1996) Soil pH and soil acidity. In: Sparks DL, editor. Methods of Soil Analysis, part 3: Chemical Methods. Madison, WI: Agronomy Monograph 9. ASA, SSSA pp. 475–490.
24. Gee GW, Bauder JW (1986) Particle-size analysis. In: Klute A, editor. Methods of Soil Analysis Part 1: Physical and Mineralogical Methods. 2nd ed. Madison, WI: Agronomy Monograph 9. ASA, SSSA pp. 383–411.
25. Nie M, Pendall E, Bell C, Gasch CK, Raut S, et al. (2012) Positive climate feedbacks of soil microbial communities in a semi-arid grassland. Ecol Letters 16: 234–241.
26. Blake GR, Hartge KH (1986) Bulk density. In: Klute A, editor. Methods of Soil Analysis, part 1: Physical and Mineralogical Methods. 2nd ed Madison, WI. ASA, SSSA 363–375.
27. Linn DM, Doran JW (1984) Effect of water-filled pore-space on carbon-dioxide and nitrous-oxide production in tilled and nontilled soils. Soil Sci Soc Am J 48: 1267–1272.
28. Blight EG, Dyre WJ (1959) A rapid method of total lipid extraction and purification. Can J Biochem Physiol 37: 911–917.
29. Frostergard A, Tunlid A, Baath E (1991) Microbial biomass measured as total lipid phosphate in soils of different organic content. J Microbiol Methods 14: 151–163.
30. Buyer JS, Roberts DP, Russek-Cohen E (2002) Soil and plant effects on microbial community structure. Can J Microbiol 48: 955–964.
31. Shannon CE, Weaver W (1949) The Mathematical Theory of Communication. Urbana, IL: Univ. of Illinois Press. 132 p.
32. Minoshima H, Jackson LE, Cavagnaro TR, Sanchez Moreno S, Ferris H, et al. (2007) Soil food webs and carbon dynamics in response to conservation tillage in California. Soil Sci Soc Am J 71: 952–963.
33. Delate K, Cambardella CA (2004) Agroecosystem performance during transition to certified organic grain production. Agron J 96: 1288–1298.
34. Ngosong C, Jarosch M, Raupp J, Neumann E, Ruess L (2010) The impact of farming practice on soil microorganisms and arbuscular mycorrhizal fungi: Crop type versus long-term mineral and organic fertilization. Appl Soil Ecol 46: 134–142.
35. Reganold JP, Andrews PK, Reeve JR, Carpenter-Boggs L, Schadt CW, et al. (2010) Fruit and soil quality of organic and conventional strawberry agroecosystems. Plos One 5: e12346.
36. Fliessbach A, Oberholzer HR, Gunst L, Maeder P (2007) Soil organic matter and biological soil quality indicators after 21 years of organic and conventional farming. Agric Ecosyst Environ 118: 273–284.
37. Carpenter-Boggs L, Stahl PD, Lindstrom MJ, Schumacher TE, Barbour NW (2003) Soil microbial properties under permanent grass, conventional tillage, and no-till management in South Dakota. Soil Till Res 71: 15–23.
38. Fierer N, Schimel JP, Holden PA (2003) Variations in microbial community composition through two soil depth profiles. Soil Biol Biochem 167–176.
39. Schaad NW, Jones JB, Chun W, editors (2001) Laboratory guide for identification of plant pathogenic bacteria: APS press, Minnesota. 398 p.

40. De Vos P, Goor M, Gills M, De Ley J (1985) Ribosomal ribonucleic acid Cistron similarities of phytopathogenic Pseudomonas species. Int J Syst Evol Microbiol 35: 169–184.
41. Stout JD (1980) The role of Protozoa in nutrient cycling and energy flow. Adv Microbial Ecol 4: 1–50.
42. Jastrow JD, Amonette JE, Bailey VL (2007) Mechanisms controlling soil carbon turnover and their potential application for enhancing carbon sequestration. Clim Change 80: 5–23.
43. Bailey VL, Smith JL, Bolton Jr H (2002) Fungal:bacterial ratios in soils investigated for enhanced C sequestration. Soil Biol Biochem 34: 997–1007.
44. Strickland MS, Rousk J (2010) Considering fungal: bacterial dominance in soils - Methods, controls, and ecosystem implications. Soil Biol and Biochem 42: 1385–1395.

17

An Invasive Plant Promotes Its Arbuscular Mycorrhizal Symbioses and Competitiveness through Its Secondary Metabolites: Indirect Evidence from Activated Carbon

Yongge Yuan[1], Jianjun Tang[1], Dong Leng[1], Shuijin Hu[2], Jean W. H. Yong[3], Xin Chen[1]*

1 College of Life Sciences, Zhejiang University, Hangzhou, China, **2** Department of Plant Pathology, North Carolina State University, Raleigh, North Carolina, United States of America, **3** Singapore University of Technology and Design, Singapore, Singapore

Abstract

Secondary metabolites released by invasive plants can increase their competitive ability by affecting native plants, herbivores, and pathogens at the invaded land. Whether these secondary metabolites affect the invasive plant itself, directly or indirectly through microorganisms, however, has not been well documented. Here we tested whether activated carbon (AC), a well-known absorbent for secondary metabolites, affect arbuscular mycorrhizal (AM) symbioses and competitive ability in an invasive plant. We conducted three experiments (experiments 1–3) with the invasive forb *Solidago canadensis* and the native *Kummerowia striata*. Experiment 1 determined whether AC altered soil properties, levels of the main secondary metabolites in the soil, plant growth, and AMF communities associated with *S. canadensis* and *K. striata*. Experiment 2 determined whether AC affected colonization of *S. canadensis* by five AMF, which were added to sterilized soil. Experiment 3 determined the competitive ability of *S. canadensis* in the presence and absence of AMF and AC. In experiment 1, AC greatly decreased the concentrations of the main secondary metabolites in soil, and the changes in concentrations were closely related with the changes of AMF in *S. canadensis* roots. In experiment 2, AC inhibited the AMF *Glomus versiforme* and *G. geosporum* but promoted *G. mosseae* and *G. diaphanum* in the soil and also in *S. canadensis* roots. In experiment 3, AC reduced *S. canadensis* competitive ability in the presence but not in the absence of AMF. Our results provided indirect evidence that the secondary metabolites (which can be absorbed by AC) of the invasive plant *S. canadensis* may promote *S. canadensis* competitiveness by enhancing its own AMF symbionts.

Editor: Mari Moora, University of Tartu, Estonia

Funding: This study was supported by the Zhejiang Provincial Natural Science Foundation of China (No. Z5090089) and the Research Fund for the Doctoral Program of Higher Education of China (RFDP, No. 20110101110077). The funders had no role in study design, data collection and analysis, decision to publish, or preparation of the manuscript.

* E-mail: chen-tang@zju.edu

Introduction

Why invasive plants have increased competitive ability at their introduced range is the central question in understanding the mechanism of successful plant invasion [1–3]. Published studies have shown that invasive plants enhanced their competitive ability via the production of secondary metabolites [4,5]. Secondary metabolites can mediate invasive plants' competition via strong effects on nutrient cycling [6], native plants [7], and soil microbes [8]. These influences could be mediated through directly and indirect ways.

The direct mediation of secondary metabolites from invasive plants on plant-plant competition can be best illustrated by observing native plants inhibiting their neighboring plants [9,10]. For example, roots exudes phytotoxin (−)-catechin from the invasive *Centaurea maculosa* directly inhibited the growth of native plants [7]. Secondary metabolites released from other invasive plants also directly inhibited seed germination of native plants [9,10]. The inhibition resulting from such direct effects can obviously increase the competitive ability of the invasive plant [11].

Secondary metabolites released by invasive plants also mediate competition indirectly. For example, secondary metabolites can indirectly promote the growth of invaders via influencing the nutrient cycling, like the pools and fluxes of inorganic and organic soil nutrient [6], or indirectly mediate competition via native soil microorganisms. Plant pathogens and symbionts [12–17], and the soil community [18–20] can be altered by secondary metabolites in general. Such effects on soil microorganisms may alter the competition between invasive and native plants [21].

Arbuscular mycorrhizal fungi (AMF) are important and ubiquitous soil microorganisms that form symbioses with many plant species in terrestrial ecosystems [22,23]. Interestingly, AMF also influence invasive plants' competition [24,25]. For example, invasive Asian knapweed (*Centaurea maculosa*) in North America is able to tap into and benefit from the mycorrhizal network linking native plant roots [26,27]. Invasive *Solidago canadensis* in China can alter the AMF composition in the introduced range in ways that promote its competition against native plants [20]. In some cases, the secondary compounds released by invasive plants disrupt the symbioses between the AMF and native plants. For example, secondary chemicals released by the invasive plant *Alliaria petiolata*, which cannot form symbioses with the AMF, disrupted the symbioses between native canopy tree seedlings and AMF [28]. A specific flavonoid fraction released by *A. petiolata* has a far stronger inhibitory effect on AMF in the invaded soil than in its original soils [29]. Additionally, the secondary compounds from *A. petiolata* not only inhibited AMF hyphal growth and spore germination [30] but also altered the AMF community associated with native sugar maple seedlings [31].

Although substantial evidence demonstrates that invasive plants can affect AMF symbioses of local hosts via secondary metabolites [30,32–36], whether those secondary metabolites can affect symbioses between AMF and the invasive plants themselves has not been documented. We address this question by using activated carbon (AC) to absorb secondary metabolites. Generally the effect of AC addition on plant growth and the formation of mutualistic relationships can reflect the putative role of secondary metabolites [37].

We used the perennial herb *Solidago canadensis* L. (goldenrod) as a model plant in our study. *S. canadensis* is native to North America but forms a near monoculture in areas of China where it has invaded [38]. *S. canadensis* has a strong allelopathic effect on native plants and is always associated with AMF, making it suitable for testing our hypothesis [39]. In a previous study, we found that the concentrations of three main secondary metabolites in *S. canadensis* plants were greater in the invaded range than in its native range and that these compounds substantially increased the competitive ability of *S. canadensis* in the invaded range [11]. Experiments have also shown that invasive *S. canadensis* is always associated with AMF and can alter the AMF community in ways that enhance its competitive ability [20,40]. In the current study, we separated the indirect effects (i.e., those mediated by AMF) and direct effects (i.e., those not mediated by AMF) of secondary metabolites from *S. canadensis* on the competition between *S. canadensis* and a native plant.

Materials and Methods

Activated Carbon

Activated Carbon (AC) is often used as an experimental tool to absorb secondary metabolites in soil [41–43]. Researchers have noted, however, that even in the absence of secondary metabolites, AC can affect plant growth by altering nutrient availability and other soil properties [44,45]. In a preliminary experiment, we tested whether AC affected soil properties with and without plants. There were five replicates for each treatment. The experiment lasted for six months. Soil samples were air-dried and sieved carefully, prior to testing. The concentration of soil organic matter (SOM) was measured by the K_2CrO_4–H_2SO_4 oxidation method of Walkey and Black [46]. Total nitrogen (TN) and total phosphorus (TP) in soil samples (1.0 g per sampling treatment) was determined with a San++ Continuous Flow Analyzer (Skalar, Netherlands). The pH of the soil from each sampling plot was measured using a soil–water slurry (1:5 soil:water). Soil available phosphorus (AP) was analyzed by the Olsen method [47]. AC at a rate of 20 ml l^{-1} did not affect soil properties (Table 1), and we therefore used AC at 20 ml l^{-1} in this study. The P and N concentration in the AC was 0 and 0.18%, respectively, and the pH was 6.0. The effect of AC on soil properties was also evaluated in experiment 1 of the current study, as described later.

Plant Species

In a previous study with six *S. canadensis* populations (three invasive populations from China and three native populations from America) and one *K. striata* population (from the area invaded by *S. canadensis* in China), we determined that the enhanced competitive ability of *S. canadensis* relative to *K. striata* in China resulted from increased root exudates [11]. Because neither biomass nor height of *S. canadensis* in our previous study was significantly affected by the source of the *S. canadensis* population (either from the original or invaded lands), the current study used one invasive *S. canadensis* population originating from the invaded area in China. One native *K. striata* population from China was also used. Seeds of *S. canadensis* and *K. striata* were collected in the same locations (30°16′N, 120°11′E) in Hangzhou, Zhejiang province (China), where the area was invaded by *S. canadensis*. Seeds of *S. canadensis* and *K. striata* were air-dried and stored at 4°C.

Experiment 1

A greenhouse microcosm experiment was performed to determine whether AC affects soil properties, the levels of three groups of secondary metabolites (total flavones, total phenolics, and total saponins) in soil, plant growth, and the AMF community associated with invasive *S. canadensis* and native *K. striata*.

The experiment was a two factorial design, with two AC treatments (with and without AC), two host plants (invasive *S. canadensis* and native *K. striata*), and five replicates (five blocks). Each microcosm measured 20×15×20 cm (length×width×height) and contained 6 kg of soil. The soil used in the experiment was collected from Cixi City, Zhejiang Province, China (30°18′N, 121°10′E), where *S. canadensis* had invaded the locality.

To ensure the initial consistency of plants in the experiment, we germinated and pre-cultured the plants in a plastic mesh plate with vermiculite and peat in the greenhouse with natural light and temperature [11]. When seedlings were 4 cm tall, two seedlings were transplanted into each microcosm so that the microcosm contained two *S. canadensis* seedlings or two *K. striata* seedlings. For the AC treatment, finely ground AC was mixed into the soil at 20 ml l^{-1} [11].

Microcosms were randomly arranged in each block in a greenhouse with natural light and with average daily temperatures ranging from 18–30°C during the experiment, which began in April and ended in October. Each microcosm was watered daily to maintain soil moisture at 70–90% of water-holding capacity. No additional nutrients were added during the experiment.

Experiment 1 was terminated 6 months after transplanting, which coincided with floral initiation. Root systems were separated from shoots. Half of each root sample was frozen at −80°C for molecular analysis. The remaining half of each root sample was used for quantification of AMF colonization. AMF colonization of roots was quantified using a microscope (×20 magnification) and utilizing the gridline intersection method [48], and 200 transects were examined per replicate.

Soil samples in each microscope were air-dried and sieved for the test of soil properties. Soil properties were measured with the methods described in the preliminary experiment. Soil organic matter (SOM), total nitrogen (TN), total phosphorus (TP), available phosphorus (AP) and pH were tested. Three main secondary metabolites (total flavones, total phenolics, and total saponins) in the soil were also quantified using the methods described by Zhang et al. [49]. First, the crude soil extracts were obtained by placing the 10-g sample in 100 ml of 70% ethanol and later, filtering the sample through a Whatman no. 44 filter paper. Following which, total flavones in the crude extracts were determined by employing the $NaNO_2$–$Al(NO_3)_3$–$NaOH$ colorimetric assay with rutin as the reference substance. Total phenolic acids were measured by the Folin-Ciocalteu assay with gallic acid as the reference substance. The total saponins were detected using vanillin–$HClO_4$ as the chromogenic reagent.

Table 1. Effects of AC treatments on soil properties with and without plants.

	Without plants			With plants		
Soil property	**No AC**	**AC**	**significance**	**No AC**	**AC**	**significance**
pH	8.18	8.19	ns	7.96	8.03	ns
TN (g/kg)	0.55	0.56	ns	0.58	0.61	ns
TP (g/kg)	0.62	0.63	ns	0.49	0.50	ns
SOM (g/kg)	12.23	13.25	ns	12.34	11.80	ns
AP (mg/kg)	21.87	21.08	ns	18.06	18.69	ns

TN refers to total nitrogen; TP refers to total phosphorous; SOM refers to soil organic matter; AP refers to available phosphorous. No effect was significant at $P<0.05$. ns, not significant.

To examine the effects of AC on AMF in roots of *S. canadensis* and *K. striata*, we used a nested PCR-denaturing gradient gel electrophoresis (DGGE)-cloning-sequencing method. Each plant species was represented by three replicates. The details of the methods are described in Methods S1. The DGGE banding pattern obtained using this method was analyzed with Quantity-One software (Bio-Rad).

Each DNA sequence obtained from the AMF was subjected to similarity comparison using an online program (BLAST, http://www.ncbi.nlm.gov/BLAST). Sequences of possible chimeric origin were detected using the online CHIMERA DETECTION program (http://rdp8.cme.msu.edu/html/analyses.html). Sequences were subsequently registered in the GenBank database under accession numbers KC507871–KC507891. For the phylogenetic analysis, sequences with gaps were treated as missing data, and then a neighbor-joining tree (1,000 replicates) including the obtained sequences and their closest relatives from GenBank was constructed using MEGA version 4.0 with the Kimura 2-Parameter model. Sequences that were 2% different from each other were subsequently assigned to separate phylotypes. It is noteworthy that the tree topology and bootstrap value were always included in the analysis.

To analyze the potential influence of secondary metabolites and soil properties on the AMF community composition, we

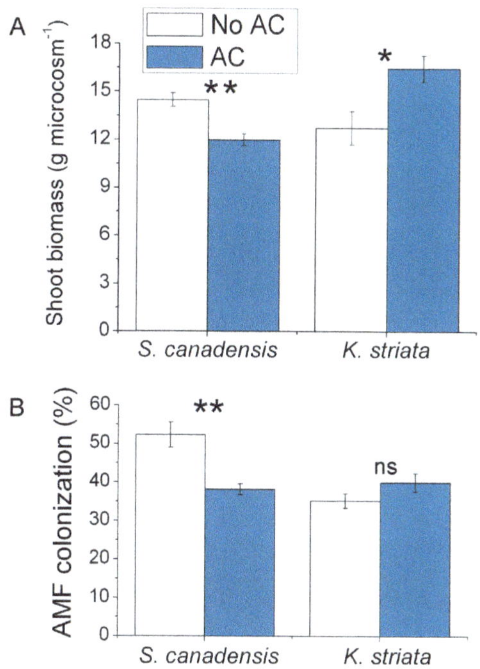

C

Phylotypes	*S. canadensis*		*K. striata*	
	No AC	AC	No AC	AC
GA1	■	■	■	■
GA2	■			
GA3	■	■	■	■
GA4		■		
GA5	■			
GA6	■	■	■	■

Figure 1. Effects of activated carbon (AC) on shoot biomass and AMF community of plants (experiment 1). Effects of AC on shoot biomass of *S. canadensis* and *K. striata* (A), AMF colonization rate (B) and on the phylotypes of AMF communities in the roots of *S. canadensis* and *K. striata* (C) growing in field soil in experiment 1. GA1–GA6 are phylotypes (Fig. S2). Values are means ± SE. *P*-value: *<0.05; **<0.01; ns, not significant. Means with different letters are significantly different.

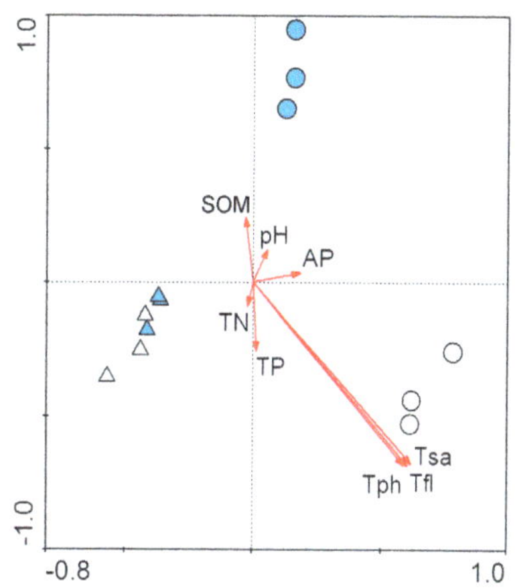

Figure 2. Factors affecting the changes of AMF communities of host plants (experiment 1). CCA-biplot depicting the relationship between AMF communities and soil properties. Circles represent *S. canadensis*, and triangles represent *K. striata*; open symbols indicate No AC treatment while filled symbols indicate AC treatment. TN refers to total nitrogen; TP refers to total phosphorous; SOM refers to soil organic matter; AP refers to available P; Tph refers to total phenolics; Tfl refers to total flavones; Tsa refers to total saponins. The length of the arrows indicates the relative influence of the factor on community structure, while the angle between the arrows indicates the degree to which factors are correlated.

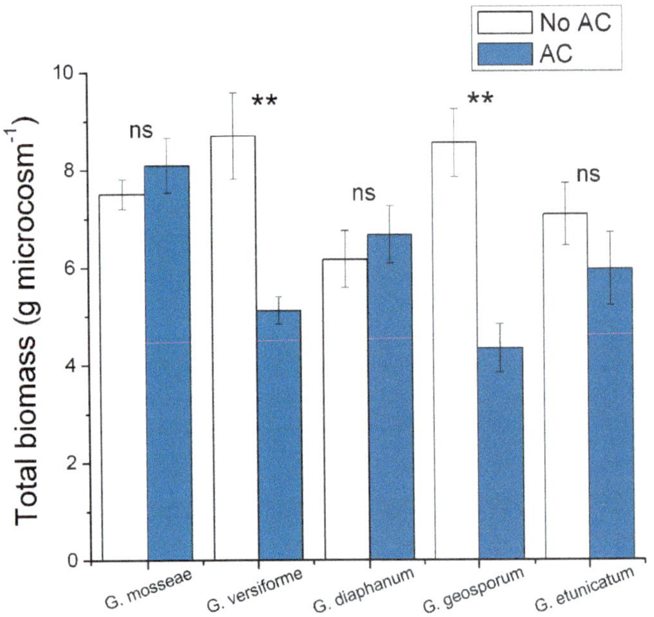

Figure 3. Effects of AC on total biomass of *S. canadensis* (experiment 2). Total biomass of *S. canadensis* inoculated with five AMF species under different activated carbon (AC) treatments in experiment 2. Values are means ± SE. *P*-value: **<0.01; ns, not significant.

performed a canonical correspondence analysis (CCA) with Monte-Carlo permutation tests (n = 499) using the Canoco for Windows version 4.52. In the CCA plot, the length of the arrows illustrated the relative importance of factors affecting community structure, while the angle between the arrows indicated the degree to which factors are correlated.

The presence (or absence) and intensity of the bands in a DGGE profile were used to construct a two-dimensional matrix. Two-way analysis of similarity (ANOSIM) was performed using the Past, version 1.91 program [50] to analyze AMF communities differences subjected to different AC treatments.

After the data for soil properties, shoot biomass, and AMF colonization were tested for normality and homogeneity of variances, we used a one-way ANOVA in the general linear model to determine the effects of AC on soil properties, shoot biomass, and AMF colonization of *S. canadensis* and *K. striata*.

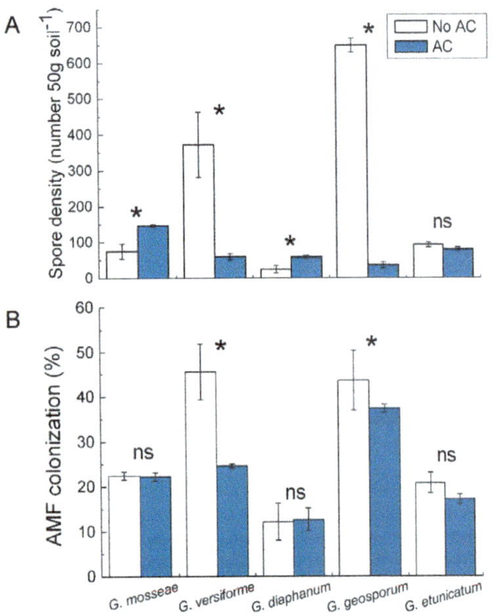

Figure 4. Effects of AC on AMF spore density and colonization of *S. canadensis* (experiment 2). Spore density in soil (A) and colonization of *S. canadensis* roots (B) for five AMF species as affected by addition of activated carbon (AC) in experiment 2. Values are means ± SE. *P*-value: *<0.05; ns, not significant.

Experiment 2

This experiment was conducted to study the effects of AC on the symbioses between *S. canadensis* and five AMF species respectively. In contrast to experiment 1, experiment 2 used sterilized soil that was subsequently inoculated with selected AMF. Based on the results from experiment 1 and the experiments of Zhang et al. [20], which indicated that *Glomus* was the dominant AMF genus in *S. canadensis* roots, we used the following five *Glomus* species in experiment 2: *G. mosseae* (BGC501, XJ-01), *G. versiforme* (BGC504, BJ08), *G. diaphanum* (BGC506, SC05), *G. geosporum* (BGC507, GZ01), and *G. etunicatum* (BGC505, TW01). The AMF species were provided by the Glomales Germplasm Bank in China. These five *Glomus* species existed naturally in abandoned fields where the seeds of *S. canadensis* and *K. striata* were collected [20,51]. Soil with spores was used as the inoculum for the experiment.

Experiment 2 employed the following experimental design: two factors (AC and AMF) with five replicates (five blocks). The soil was the same as in experiment 1. AC was added to half of the microcosms at 20 ml l^{-1} to absorb secondary compounds. After each microcosm was filled with 3 kg soil that had been sterilized by γ-radiation, an equal number of spores of each AMF species was mixed into the soil (one species per microcosm).

Seeds of *S. canadensis* were germinated and pre-cultured as described in experiment 1. Two seedlings were transplanted into each microcosm. The microcosms were randomly arranged in each block in the greenhouse. All growth conditions were the same as in experiment 1. Six months after transplanting, the plants were

Table 2. Effects of AC treatments and plants (invasive *S. canadensis* or native *K. striata*) on soil properties in experiment 1.

	S. canadensis			*K. striata*		
Soil property	**No AC**	**AC**	**significance**	**No AC**	**AC**	**significance**
pH	7.77	7.78	ns	7.77	7.77	ns
TN (g/kg)	1.03	1.01	ns	1.02	1.05	ns
TP (g/kg)	0.50	0.48	ns	0.49	0.50	ns
SOM (g/kg)	20.54	23.04	*	20.34	21.20	ns
AP (mg/kg)	49.02	48.57	ns	48.06	48.69	ns
Tph (ppm)	204.40	5.61	**	107.81	2.24	**
Tfl (ppm)	47.67	0.48	**	23.50	0.17	**
Tsa (ppm)	467.67	23.80	**	231.00	19.9	**

TN refers to total nitrogen; TP refers to total phosphorous; SOM refers to soil organic matter; AP refers to available phosphorous; Tph refers to total phenolics; Tfl refers to total flavones; Tsa refers to total saponins. For each plant species, a significant difference between No AC and AC is indicated as follows:
*$P<0.05$;
**$P<0.01$;
ns $P>0.05$ (not significant).

harvested, and roots were separated from shoots. Shoot and root samples were oven-dried (65°C for 48 h) for the measurement of dry total biomass. AMF colonization of roots was quantified as described for experiment 1. Spores were separated from the soil by the wet-sieving method and counted using a microscope [52].

The effects of AC on total biomass of *S. canadensis*, AMF colonization and spore numbers in soil were determined with a two-way ANOVA in the general linear model after the data had been checked for normality and homogeneity of variances. When ANOVAs were significant, mean values were compared by least significant difference (LSD) at the 5% significance level.

Experiment 3

Experiment 3 examined how AC altered the competitive ability of *S. canadensis* in the presence and absence of AMF. The AMF inoculum was a mixture of three *Glomus* species: *G. intraradices*, *G. mosseae*, and *G. geosporum*. The three AMF species were also provided by the Glomales Germplasm Bank in China.

The experiment was a three factorial design, with two AC treatments (No AC and AC), two AMF treatments (AMF and No AMF), and three plants culture types (monoculture of invasive, monoculture of native, and a mixture of invasive and native), and five replicates (five blocks). A total of $2\times2\times3\times5=60$ microcosms were produced. The microcosms and soil were the same as in experiment 1. Each microcosm was filled with 6 kg of soil that had been sterilized by γ-radiation and subsequently inoculated or not inoculated with AMF. Control microcosms without AMF received equal amounts of AMF inoculum sterilized by γ-radiation plus filtrate from the AMF inoculum that contained no AMF propagules; this was done to provide experimental control to account for potential mineral and non-mycorrhizal microbial components in the inoculum. AC was added to half the microcosms as indicated in experiment 1.

Two *S. canadensis* seedlings, two *K. striata* seedlings, or one *S. canadensis* seedling and one *K. striata* seedling were transplanted into each microcosm. The microcosms were randomly placed in each block in the greenhouse. All growth conditions were the same as in experiment 1.

The plants were harvested after 6 months, just before they began to flower. Roots were separated from shoots. Shoot samples were oven-dried (65°C for 48 h) for measurement of dry shoot biomass. AMF colonization of roots was quantified as described earlier for experiment 1. The competitive ability of *S. canadensis* was determined using an aggressivity index (AI), which was based on shoot biomass data. The AI was calculated as described previously [11,53]. $AI=(Y_{ij}/Y_{ii})-(Y_{ji}/Y_{jj})$, where Y_{ij} and Y_{ii} are the shoot biomasses of *K. striata* growing in mixture and monoculture, and Y_{ji} and Y_{jj} are the shoot biomasses of *S. canadensis* growing in mixture and monoculture. The competitive ability of *S. canadensis* relative to *K. striata* is inversely related to the AI value.

After the data for shoot biomass and AMF colonization were checked for normality and homogeneity of variances, we used a two-way ANOVA in the general linear model to test the effect of AC and AMF on shoot biomass and AMF colonization of *S. canadensis* and *K. striata* in monoculture or in mixture.

After checking the AI data for homogeneity, we used a two-way ANOVA in the general linear model to examine the effects of AC and AMF on AI. Mean values were compared by least significant difference (LSD) at the 5% significance level.

Results

Effects of AC on Soil Properties, Plant Growth, and AM Symbiosis (Experiment 1)

AC did not affect the key attributes of soil properties (soil organic matter, total nitrogen, total phosphorus, available phosphorus and pH) ($P>0.05$) but significantly decreased the levels of the main groups of secondary compounds (total flavones, total phenolics, and total saponins) ($P<0.01$) in soil planted with *S. canadensis* or *K. striata* (Table 2).

The addition of AC to soil significantly reduced the shoot biomass of *S. canadensis* ($F_{1,8}=58.169$, $P=0.007$) but increased the shoot biomass of *K. striata* ($F_{1,7}=9.561$, $P=0.018$) (Fig. 1A). AC reduced AMF colonization of *S. canadensis* ($F_{1,8}=47.287$, $P<0.001$) but did not affect AMF colonization of *K. striata* ($F_{1,7}=2.523$, $P=0.131$) (Fig. 1B).

DGGE profiles showed that the AMF composition differed in invasive *S. canadensis* versus native *K. striata*. AC affected the pattern and signal intensity of bands of the 18S rDNA fragments of AMF in roots of *S. canadensis* and *K. striata* (Fig. S1). The phylogenetic analysis indicated that sequences retrieved from all root samples could be separated into six AMF groups (GA1–GA6) (Fig. S2). AC

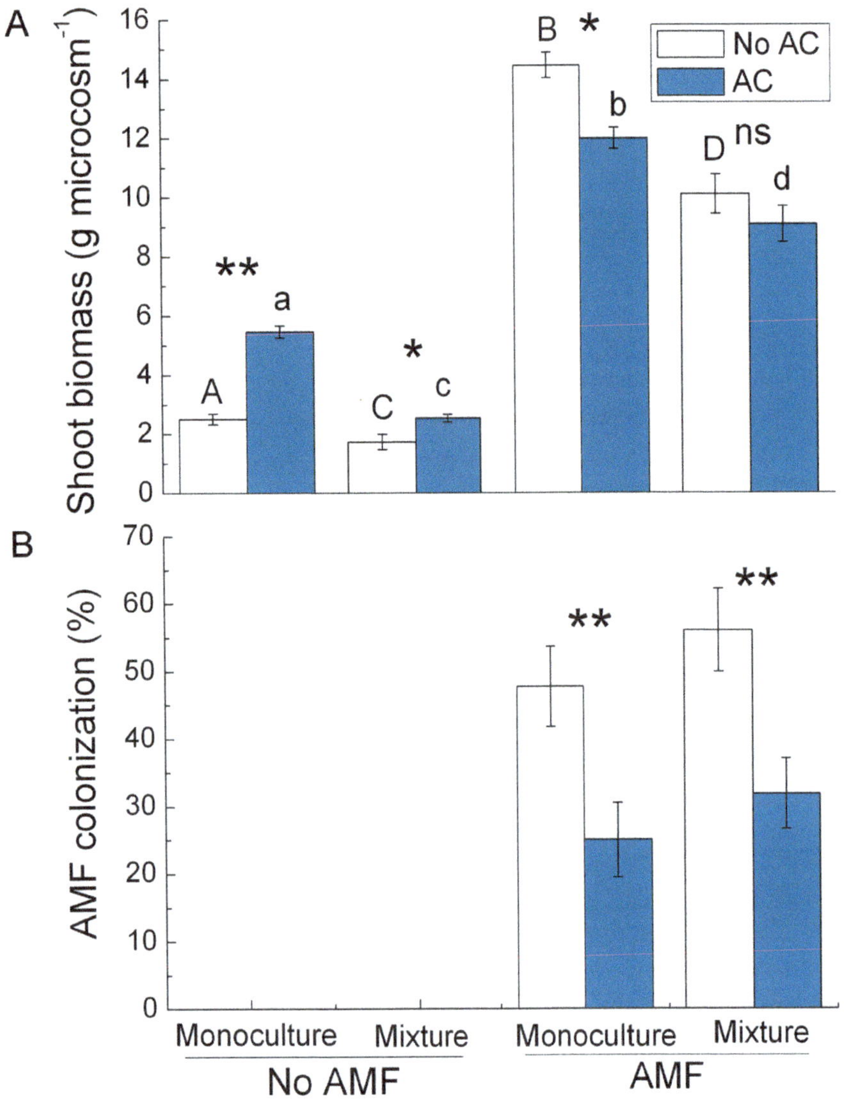

Figure 5. Effects of AMF and AC on growth and AMF colonization of *S. canadensis* (experiment 3). Shoot biomass (A) and AMF colonization (B) of *S. canadensis* under monoculture or mixture (with *K. striata*) as affected by AMF treatments (AMF or No AMF) and activated carbon treatments (AC or No AC) in experiment 3. Values are means ± SE. Means for No AC with different uppercase letters and means for AC with different lowercase letters are significantly different at $P<0.05$. For comparison of paired bars (No AC vs. AC), asterisks indicate significant differences (*<0.05; **<0.01), and ns indicates not significant.

treatments affected AMF phylotypes in *S. canadensis* roots but not in *K. striata* roots (Fig. 1C). Roots of *S. canadensis* were colonized by five AMF phylotypes when AC was not added (group GA4 was missing) and by four phylotypes when AC was added (GA2 and GA5 were missing) (Fig. 1C). AMF phylotypes GA1, GA3, and GA6 in roots of *S. canadensis* or *K. striata* were not affected by the addition of AC (Fig. 1C).

The ANOSIM output showed that the AMF communities was significantly different with different AC treatments ($P=0.011$) and host plants ($P=0.010$).

The CCA-biplot showed that AC treatment significantly altered the AMF community structure of *S. canadensis* but not of *K. striata* (Fig. 2). The CCA-biplot also demonstrated that the differences in AMF community structures of *S. canadensis* between the No AC and AC treatments were more closely related to the three groups of secondary compounds (total flavones, total phenolics, and total saponins) than to soil properties.

Effects of AC on Growth of *S. canadensis* and AMF, when Five AMF Species were Added to Sterilized Soil (Experiment 2)

AC treatment reduced the total biomass of *S. canadensis* when inoculated with *G. geosporum* and *G. versiforme* ($P<0.05$), but did not affect the total biomass of *S. canadensis* when inoculated with *G. mosseae*, *G. diaphanum* and *G. etunicatum* ($P>0.05$). In the absence of AC, *S. canadensis* attained higher total biomass with *G. geosporum* and *G. versiforme* than with other three AMF species (*G. mosseae*, *G. diaphanum* and *G. etunicatum*) (Fig. 3).

Spore density in soil and colonization of *S. canadensis* roots were high for *G. geosporum* and *G. versiforme* in the absence of AC but were reduced ($P<0.05$) when AC was added (Fig. 4). Addition of AC increased ($P<0.05$) the spore density for *G. mosseae* and *G. diaphanum* (Fig. 4A). By contrast, the addition of AC did not significantly affect ($P>0.05$) spore density or colonization for *G. etunicatum* (Fig. 4).

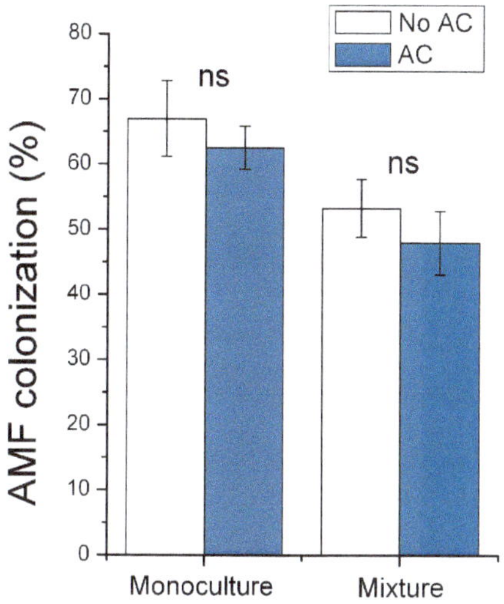

Figure 6. Effects of AC treatments on AMF colonization of *K. striata* (experiment 3). AMF colonization of *K. striata* under monoculture or mixture as affected by AC treatments in experiment 3. Values are means ± SE. For comparison of paired bars (No AC vs. AC), ns indicates not significant.

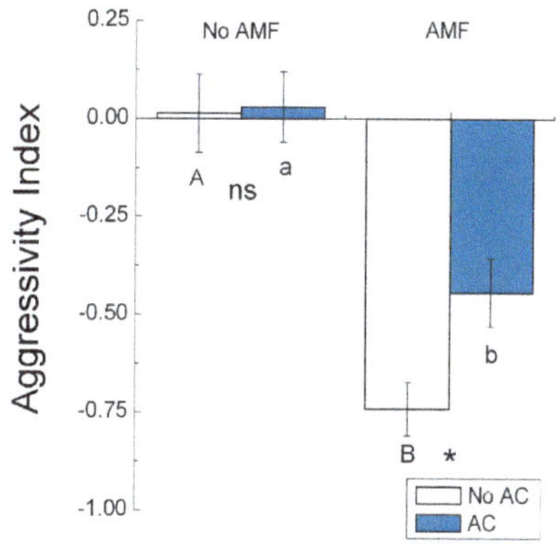

Figure 7. Effects of AMF and AC treatments on the competition of *S. canadensis* (experiment 3). Competitive ability (as indicated by the aggressivity index) of *S. canadensis* as affected by AMF (AMF or No AMF) and activated carbon (AC or No AC) (experiment 3). A lower value for the index indicates greater competitive ability. Values are means ± SE. Means for No AC with different uppercase letters and means for AC with different lowercase letters are significantly different at $P<0.05$. For comparison of paired bars (No AC vs. AC), asterisks indicate significant differences (*<0.05), and ns indicates not significant.

Effects of AC on the Competition between *S. canadensis* and *K. striata* in the Presence and Absence of AMF in Soil that had been Previously Sterilized (Experiment 3)

In the absence of AMF, the addition of AC increased *S. canadensis* shoot biomass in monoculture ($F_{1,8}=39.526$, $P<0.001$) (Fig. 5A) and in mixture ($F_{1,8}=8.914$, $P=0.013$) (Fig. 5A). In the presence of AMF, addition of AC decreased *S. canadensis* shoot biomass in monoculture ($F_{1,8}=9.429$, $P=0.017$) (Fig. 5A) but did not cause a reduction in biomass when planted in a mixture ($F_{1,8}=0.022$, $P=0.883$) (Fig. 5A).

No AMF spores or any other evidence of AMF colonization were detected in microcosms that were not inoculated with AMF. In soil inoculated with AMF, adding AC reduced AMF colonization of *S. canadensis* in monoculture ($F_{1,8}=19.253$, $P<0.001$) (Fig. 5B) and in mixture ($F_{1,8}=21.452$, $P<0.001$) (Fig. 5B) but did not affect AMF colonization of *K. striata* in monoculture ($F_{1,8}=0.088$, $P=0.775$) (Fig. 6) and in mixture ($F_{1,8}=0.782$, $P=0.080$) (Fig. 6).

When AMF were not added, AC did not affect the competitive ability (as indicated by AI values) of *S. canadensis* ($F_{1,8}=0.023$, $P=0.745$), i.e., the AI values were near zero with or without AC (Fig. 7). When AMF were added, *S. canadensis* AI values were lower (indicating higher competitive ability) without AC than with AC ($F_{1,8}=13.776$, $P=0.012$) (Fig. 7). Regardless of AC addition, AI values were much lower with AMF than without AMF (without AC, $F_{1,8}=17.851$, $P<0.001$; with AC, $F_{1,8}=10.415$, $P<0.001$) (Fig. 7).

Discussion

Because activated carbon (AC) absorbs secondary metabolites in soil, it is often used to test the effects of secondary metabolites *in situ* [3,41–43]. AC, however, may also affect nutrient availability [54] and soil physio-chemical properties [45,55,56]. Thus, when AC is used in research concerning plant secondary compounds, its putative effects should be checked carefully. In our experiment 1, addition of AC did not change the general soil properties (soil organic matter, total nitrogen, total phosphorus, available phosphorus and pH) but greatly reduced the concentrations of total flavones, phenolics, and saponins. CCA also indicated that variation in the AMF community in *S. canadensis* roots was more closely related to these secondary metabolites than to other soil properties and that these secondary metabolites enhanced the colonization of *S. canadensis* roots by certain AMF species (experiment 2). We therefore deduced that the effects of AC in our study resulted predominantly from the absorption of secondary metabolites by AC.

Our community similarity analysis demonstrated that through the absorption of secondary metabolites, AC affected the AMF community in the roots of *S. canadensis*. Although it was documented that secondary metabolites from invasive plants can affect AM symbiosis of neighboring plants [28,30], in our study, AC effects suggest that secondary metabolites from an invasive plant may promote its own AM symbiosis and thereby enhancing its own growth.

It remains unclear how plant secondary metabolites affect the process of AMF symbiosis [29,30,32,33]. One plausible avenue could be that some compounds (e.g., flavones, sesquiterpenes, and strigolactones) could act as signals that induce AMF spore germination [57], hyphal branching [32], and/or AM symbiosis formation [36]. Flavones, phenolics, and saponins are commonly produced by many plants [58–60]. These three types of compounds had been reported to accumulate in the soil where *S. canadensis* had invaded [49]. Phenolics have been shown to reduce AMF colonization [35], while flavones like chrysin and luteolin can promote AM symbiosis [36]. In this study, AC greatly reduced the levels of three types of secondary metabolites, indicating that the effects of AC on AMF symbiosis in *S. canadensis* may be due to the decrease in the concentrations of secondary metabolites.

Interestingly, not all AMF species were affected in the same way by the secondary metabolites from *S. canadensis*. In experiment 2, for example, without the addition of AC, *S. canadensis* secondary compounds greatly enhanced the growth (as indicated by spore density and root colonization) of *G. geosporum* and *G. versiforme* but did not affect the growth of *G. etunicatum*. The DGGE data and the phylogenetic tree also indicated some AMF phylotypes were present in *S. canadensis* roots only when secondary metabolites had been absorbed by AC. These differential effects on AMF growth confirm that different AMF species can respond differently to the same chemicals. For example, spore germination of *Glomus intraradices* and *G. claroideum* can respond differently to strigolactones [57]. These results suggested that the development of symbioses between specific AMF and *S. canadensis* was mediated by *S. canadensis* secondary metabolites; this selective effect resulted in changes in the AMF community in the roots and also in the soil. In experiment 2, the beneficial AMF species *G. geosporum* and *G. versiforme* were able to promote greater growth of *S. canadensis* than the AMF species that were considered to be less beneficial (Fig. 3). These observations implied that *S. canadensis*' secondary metabolites (which can be absorbed by AC in the soil) may specifically favor certain compatible AMF partners.

Results from experiment 3 showed that the competitive ability of invasive *S. canadensis* was influenced by both AC and AMF, i.e., release of chemicals by *S. canadensis* enhanced AMF colonization of *S. canadensis* roots. When AMF were absent, secondary metabolites did not increase the competitive ability of *S. canadensis* and autotoxicity might have been generated instead (Fig. 5A). As a consequence, *S. canadensis* grew better with AC treatment than without AC treatment in the absence of AMF. Autotoxicity is also exhibited by other exotic plants, including *Centaurea maculosa* [61], and this phenomenon is a way to reduce intraspecific competition. When AMF inoculum was added to the soil, however, the autotoxic effect was not evident (Fig. 5A), and secondary metabolites enhanced the competitive ability of *S. canadensis* (Fig. 7). These results suggested that AMF can offset the negative effects of secondary metabolites on *S. canadensis*. In other words, secondary metabolites indirectly promoted the competitive ability of *S. canadensis* via AMF. Given that secondary metabolites affected the degree of AMF colonization and the composition of the AMF community in the roots of the invasive *S. canadensis* but had little effect on AMF in the roots of the native *K. striata* in experiments 1 and 3, we concluded that the enhanced competitive ability of *S. canadensis* might have been the consequence of its own secondary metabolites affecting its own AM symbioses.

Secondary metabolites released by an invasive plant may enhance invasiveness by both direct and indirect effects. The direct effect involves the inhibition of native competitors through secondary metabolites without mediation by microorganisms. The indirect effect involves microorganisms that inhibit the native competitors or enhancing the invader [21,62]. Although both field and common garden experiments have demonstrated that plant secondary metabolites can promote an invader's competitive ability, past research was unable to separate the direct effects from indirect, microbially mediated effects of invasive non-indigenous plants [14]. It follows that the reported impact of secondary metabolites on competition may in many or all cases have resulted from combined direct and indirect effects. In this study, we separated the direct and indirect effects of secondary metabolites on plant-plant competition using the intrinsic properties of activated carbon. Although our study was performed in microcosms and not in the field, it provided authentic evidence that plant secondary metabolites are able to enhance the competitive ability of an invader by increasing AMF colonization of the invader itself.

In conclusion, the effects of secondary metabolites produced by invasive plants on local plants have been well documented [29], but few studies had focused on how secondary chemicals released by an invasive plant affect that invasive plant itself. Our study demonstrated an interesting phenomenon of "self-promotion" by a plant invader resulting from the positive effects of its own secondary metabolites on the AMF that colonize its roots. The results presented here further expanded our understanding of the interesting role of secondary metabolites during plant invasion.

Supporting Information

Figure S1 Effect of AC on DGGE pattern of AMF communities in experiment 1. DGGE pattern of 18S rDNA fragments of AMF in roots of *S. canadensis* and *K. striata* as affected by addition of activated carbon (AC and No AC) in experiment 1.

Figure S2 Phylogenetic tree of AMF communities in experiment 1. Neighbor-joining phylogenetic tree based on partial SSU rRNA gene sequences of all identified AMF and referenced sequences in the root samples. Numbers above branches indicate bootstrap values from 1,000 replicates. The internal identification number represents sequences retrieved from specific DGGE profile bands. A total of 21 partial SSU rRNA sequences were thus obtained in the present study. They are shown in bold and labeled with the GeneBank database accession numbers (i.e., KC507871–KC507891). Sequence groups (GA1, GA2, etc.) identify distinct clusters of sequences with similarity ≥ 99%.

Acknowledgments

We thank Liangliang Hu for the design of the figures.

Author Contributions

Conceived and designed the experiments: XC SH. Performed the experiments: YY JT DL. Analyzed the data: YY JWHY. Contributed reagents/materials/analysis tools: YY JT DL. Wrote the paper: YY XC SH.

References

1. Blossey B, Nötzold R (1995) Evolution of increased competitive ability in invasive nonindigenous plants: a hypothese. Journal of Ecology 83: 887–889.
2. Siemann E, Rogers WE (2001) Genetic differences in growth of an invasive tree species. Ecology Letters 4: 514–518.
3. Siemann E, Rogers WE (2003) Increased competitive ability of an invasive tree limited by an invasive beetle. Ecological Applications 13: 1503–1507.
4. Yang Q, Ye W, Liao F, Yin X (2005) Effects of allelochemicals on seed germination. Chinese Journal of Ecology 24: 1459–1465.
5. Alford ÉR, Perry LG, Qin B, Vivanco JM, Paschke MW (2007) A putative allelopathic agent of Russian knapweed occurs in invaded soils. Soil Biology and Biochemistry 39: 1812–1815.
6. Hättenschwiler S, Vitousek PM (2000) The role of polyphenols in terrestrial ecosystem nutrient cycling. Trends in Ecology and Evolution 15: 238–243.
7. Thorpe AS, Thelen GC, Diaconu A, Callaway RM (2009) Root exudate is allelopathic in invaded community but not in native community: field evidence for the novel weapons hypothesis. Journal of Ecology 97: 641–645.
8. Yu X (2005) A new mechanism of invader success: Exotic plant inhibits natural vegetation restoration by changing soil microbe com-munity. Chinese Science Bulletin 50: 1105.
9. Ens EJ, Bremner JB, French K, Korth J (2008) Identification of volatile compounds released by roots of an invasive plant, bitou bush (*Chrysanthemoides monilifera* spp. *rotundata*), and their inhibition of native seedling growth. Biological Invasions 11: 275–287.
10. Cipollini D, Stevenson R, Enright S, Eyles A, Bonello P (2008) Phenolic metabolites in leaves of the invasive shrub, *Lonicera maackii*, and their potential phytotoxic and anti-herbivore effects. Journal of Chemical Ecology 34: 144–152.

11. Yuan Y, Wang B, Zhang S, Tang J, Tu C, et al. (2012) Enhanced allelopathy and competitive ability of invasive plant *Solidago canadensis* in its introduced range. Journal of Plant Ecology. DOI:10.1093/jpe/rts033.
12. Liu SQ, Wu FZ, Wen XY (2013) Allelopathic effects of root exudates of Chinese onion on tomato growth and the pathogen *Fusarium oxysporum* (Sch1) f.sp lycopersici. Allelopathy Journal 31: 387–403.
13. Xu N, Wei M, Wang C, Shi W, Tian FM, et al. (2013) Composition of Welsh onion (*Allium fistulosum* L.) root exudates and their allelopathy on cucumber sprouts and *Fusarium oxysporum* f.sp cucumerinum. Allelopathy Journal 32: 243–256.
14. Mangla S, Inderjit, Callaway RM (2008) Exotic invasive plant accumulates native soil pathogens which inhibit native plants. Journal of Ecology 96: 58–67.
15. Malmstrom CM, McCullough AJ, Johnson HA, Newton LA, Borer ET (2005) Invasive annual grasses indirectly increase virus incidence in California native perennial bunchgrasses. Oecologia 145: 153–164.
16. Eppinga MB, Rietkerk M, Dekker SC, De Ruiter PC (2006) Accumulation of local pathogens: a new hypothesis to explain exotic plant invasions. OIKOS 114: 168–176.
17. Lodhi MAK, Killingbeck KT (1980) Allelopathic inhibition of nitrification and nitrifying bacteria in a ponderosa pine (*Pinus ponderosa* Dougl.) community. American Journal of Botany 67: 1423–1429.
18. Lankau RA (2011) Intraspecific variation in allelochemistry determines an invasive species' impact on soil microbial communities. Oecologia 165: 453–463.
19. Kourtev PS, Ehrenfeld JG, Häggblom M (2002) Exotic plant species alter the microbial community structure and function in the soil. Ecology 83: 3152–3166.
20. Zhang Q, Yang R, Tang J, Yang H, Hu S, et al. (2010) Positive feedback between mycorrhizal fungi and plants influences plant invasion success and resistance to invasion. PLoS One 5: e12380.
21. Reinhart KO, Callaway RM (2006) Soil biota and invasive plants. New Phytologist 170: 445–457.
22. Smith SE, Read DJ (1997) Mycorrhizal symbiosis. London, UK: Academic Press.
23. Rillig MC (2004) Arbuscular mycorrhizae and terrestrial ecosystem processes. Ecology Letters 7: 740–754.
24. Richardson D, Allsopp N, Antonio C, Milton S, Rejmanek M (2000) Plant invasions – the role of mutualisms. Biological Reviews 75: 65–93.
25. Pringle A, Bever JD, Gardes M, Parrent JL, Rillig MC, et al. (2009) Mycorrhizal symbioses and plant invasions. Annual Review of Ecology, Evolution, and Systematics 40: 699–715.
26. Marler MJ, Zabinski CA, Callaway RM (1999) Mycorrhizae indirectly enhance competitive effects of an invasive forb on a native bunchgrass. Ecology 80: 1180–1186.
27. Carey EV, Marler MJ, Callaway RM (2004) Mycorrhizae transfer carbon from a native grass to an invasive weed,evidence from stable isotopes and physiology. Plant Ecology 172: 133–141.
28. Stinson KA (2006) Invasive plant suppresses the growth of native tree seedlings by disrupting belowground mutualisms. PLoS BIOLOGY 4: e140.
29. Callaway RM (2008) Novel weapons: invasive plant suppresses fungal mutualists in America but not in its native Europe. Ecology 89: 1043–1055.
30. Cantor A, Hale A, Aaron J, Traw MB, Kalisz S (2011) Low allelochemical concentrations detected in garlic mustard-invaded forest soils inhibit fungal growth and AMF spore germination. Biological Invasions 13: 3015–3025.
31. Barto EK, Antunes PM, Stinson K, Koch AM, Klironomos JN, et al. (2011) Differences in arbuscular mycorrhizal fungal communities associated with sugar maple seedlings in and outside of invaded garlic mustard forest patches. Biological Invasions 13: 2755–2762.
32. Akiyama K, Matsuzaki K-I, Hayashi H (2005) Plant sesquiterpenes induce hyphal branching in arbuscular mycorrhizal fungi. Nature 435: 824–827.
33. Buee M, Rossignol M, Jauneau A, Ranjeva R, Bécard G (2000) The pre-symbiotic growth of arbuscular mycorrhizal fungi is induced by a branching factor partially purified from plant root exudates. The American Phytopathological Society 13: 693–698.
34. Nagahashi G, Douds DD (2000) Partial separation of root exudate components and their effects upon the growth of germinated spores of AM fungi. Mycological Research 104: 1453–1464.
35. Piotrowski JS, Morford SL, Rillig MC (2008) Inhibition of colonization by a native arbuscular mycorrhizal fungal community via *Populus trichocarpa* litter, litter extract, and soluble phenolic compounds. Soil Biology and Biochemistry 40: 709–717.
36. Scervino JM, Ponce MA, Erra-Bassells R, Bompadre J, Vierheilig H, et al. (2007) The effect of flavones and flavonols on colonization of tomato plants by arbuscular mycorrhizal fungi of the genera *Gigaspora* and *Glomus*. Canadian Journal of Microbiology 53: 702–709.
37. Callaway RM, Aschehoug ET (2000) Invasive plants versus their new and old neighbors: a mechanism for exotic invasion. Science 290: 521–523.
38. Weber E (1998) The dynamics of plant invasions: a case study of three exotic goldenrod species (*Solidago* L.) in Europe. Journal of Biogeography 25: 147–154.
39. Jin L, Gu Y, Xiao M, Chen J, Li B (2004) The history of *Solidago canadensis* invasion and the development of its mycorrhizal associations in newly-reclaimed land. Functional Plant Biology 31: 979–986.
40. Zhang Q, Yao L, Yang R, Yang X, Tang J, et al. (2007) Potential allelopathic effects of an invasive species *Solidago canadensis* on the mycorrhizae of native plant species. Allelopathy Journal 20: 71–78.
41. Mahall BE, Callaway RM (1992) Root communication mechanisms and intracommunity distributions of two Mojave desert shrubs. Ecology 73: 2145–2151.
42. Ridenour WM, Callaway RM (2001) The relative importance of allelopathy in interference: the effects of an invasive weed on a native bunchgrass. Oecologia 126: 444–450.
43. Inderjit, Callaway RM (2003) Experimental designs for the study of allelopathy. Plant and Soil 256: 1–11.
44. Lau JA, Puliafico KP, Kopshever JA, Steltzer H, Jarvis EP, et al. (2008) Inference of allelopathy is complicated by effects of activated carbon on plant growth. New Phytologist 178: 412–423.
45. Gundale MJ, DeLuca TH (2006) Charcoal effects on soil solution chemistry and growth of *Koeleria macrantha* in the ponderosa pine/Douglas-fir ecosystem. Biology and Fertility of Soils 43: 303–311.
46. Nelson DW, Sommers LE (1982) Total carbon, organic carbon, and organic matter, In: Rh, M., Dr, K. (Eds.), Methods of soil analysis, agronomy, part 2, 2nd. WI: ASA and SSSA, Madison, 539–577.
47. Olsen SR, Cole CV, Watanabe FS, Dean LA (1954) Estimation of available phosphorus in soils by extraction with sodium bicarbonate. USDA circular, 939.
48. Giovannetti M, Mosse B (1980) An evaluation of techniques for measuring vesicular arbuscular mycorrhizal infection in roots. New Phytologist 84: 489–500.
49. Zhang S, Zhu W, Wang B, Tang J, Chen X (2011) Secondary metabolites from the invasive *Solidago canadensis* L. accumulation in soil and contribution to inhibition of soil pathogen *Pythium ultimum*. Applied Soil Ecology 48: 280–286.
50. Hammer Ø, Harper DAT, Ryan PD (2001) PAST: paleontological statistics software package for education and data analysis. Palaeontologia Electronica 4: 1–9.
51. Tang J, Zhang Q, Yang R, Chen X (2009) Effects of exotic plant *Solidago canadensis* L. on local arbuscular mycorrhizal fungi. Bulletin of Science and Technology 25: 233–237.
52. Gerdemann J, Nicolson TH (1963) Spores of mycorrhizal *Endogone* species extracted from soil by wet sieving and decanting. Transactions of the British Mycological Society 46: 235–244.
53. Scheublin TR, Van Logtestijn RSP, Van Der Heijden MGA (2007) Presence and identity of arbuscular mycorrhizal fungi influence competitive interactions between plant species. Journal of Ecology 95: 631–638.
54. Lehmann J Jr. JPdS, Steiner C, Nehls T, Zech W, et al. (2003) Nutrient availability and leaching in an archaeological anthrosol and a ferralsol of the central amazon basin: fertilizer, manure and charcoal amendments. Plant and Soil 249: 343–357.
55. DeLuca TH, MacKenzie MD, Gundale MJ, Holben WE (2006) Wildfire-produced charcoal directly influences nitrogen cycling in ponderosa pine forests. Soil Science Society of America Journal 70: 448–453.
56. Matsubara Y, Hasegawa N, Fukui H (2002) Incidence of Fusarium root rot in asparagus seedlings infected with arbuscular mycorrhizal fungus as affected by several soil amendments. Journal of the Japanese Society for Horticultural Science 71: 370–374.
57. Besserer A, Puech-Pagès V, Kiefer P, Gomez-Roldan V, Jauneau A, et al. (2006) Strigolactones stimulate arbuscular mycorrhizal fungi by activating mitochondria. PLoS BIOLOGY 4: e226.
58. Harborne JrB, Williams CA (2000) Advances in flavonoid research since 1992. Phytochemistry 55: 481–504.
59. Mert-Türk F (2006) Saponins versus plant fungal pathogens. Journal of Cell and Molecular Biology 5: 13–17.
60. Kim YO, Lee EJ (2010) Comparison of phenolic compounds and the effects of invasive and native species in East Asia: support for the novel weapons hypothesis. Ecological Research 26: 87–94.
61. Perry LG, Thelen GC, Ridenour WM, Weir TL, Callaway RM, et al. (2005) Dual role for an allelochemical: (±)-catechin from *Centaurea maculosa* root exudates regulates conspecific seedling establishment. Journal of Ecology 93: 1126–1135.
62. Cappuccino N, Arnason JT (2006) Novel chemistry of invasive exotic plants. Biology Letters 2: 189–193.

18

Soil Chemical Property Changes in Eggplant/Garlic Relay Intercropping Systems under Continuous Cropping

Mengyi Wang, Cuinan Wu, Zhihui Cheng*, Huanwen Meng, Mengru Zhang, Hongjing Zhang

College of Horticulture, Northwest A&F University, Yangling, Shaanxi, China

Abstract

Soil sickness is a critical problem for eggplant (*Solanum melongena* L.) under continuous cropping that affects sustainable eggplant production. Relay intercropping is a significant technique on promoting soil quality, improving eco-environment, and raising output. Field experiments were conducted from September 2010 to November 2012 in northwest China to determine the effects of relay intercropping eggplant with garlic (*Allium sativum* L.) on soil enzyme activities, available nutrient contents, and pH value under a plastic tunnel. Three treatments were in triplicate using randomized block design: eggplant monoculture (CK), eggplant relay intercropping with normal garlic (NG) and eggplant relay intercropping with green garlic (GG). The major results are as follows: (1) the activities of soil invertase, urease, and alkaline phosphatase were generally enhanced in NG and GG treatments; (2) relay intercropping significantly increased the soil available nutrient contents, and they were mostly higher in GG than NG. On April 11, 2011, the eggplant/garlic co-growth stage, the available nitrogen content in GG was 76.30 $mg \cdot kg^{-1}$, significantly higher than 61.95 $mg \cdot kg^{-1}$ in NG. For available potassium on April 17, 2012, they were 398.48 and 387.97 $mg \cdot kg^{-1}$ in NG and GG, both were significantly higher than 314.84 $mg \cdot kg^{-1}$ in CK; (3) the soil pH showed a significantly higher level in NG treatment, but lower in GG treatment compared with CK. For the last samples in 2012, soil pH in NG and GG were 7.70 and 7.46, the highest and lowest one among them; (4) the alkaline phosphatase activity and pH displayed a similar decreasing trend with continuous cropping. These findings indicate that relay intercropping eggplant with garlic could be an ideal farming system to effectively improve soil nutrient content, increase soil fertility, and alleviate soil sickness to some extent. These findings are important in helping to develop sustainable eggplant production.

Editor: Wen-Xiong Lin, Agroecological Institute, China

Funding: This research was supported by a project of the National Natural Science Foundation of China (No. 31171949), the Special Fund for Agro-scientific Research in the Public Interest (No. 200903018), and the University Undergraduates Innovating Experimentation Project (No. 2201110712036). The funders had no role in study design, data collection and analysis, decision to publish, or preparation of the manuscript.

* Email: chengzh2004@163.com

Introduction

Eggplant (*Solanum melongena* L.), because of its rich nutritional value, good taste and easily cultivation, is grown in most parts of the world, among which China is the greatest producer [1]. However, in recent years, particularly under continuous cropping in facilities, many problems have affected the sustainable production of eggplant, such as aggravating plant disease and pests, degraded soil physical and chemical characteristics, and declining production and stress resistance of plants [2,3].

As a simple repetitive agronomic practice, continuous cropping is the practice of growing the same crop year after year in the same field [4]. It makes the soil susceptible to erosion hazard and weed invasion, soilborne pathogens increase, and the survival of certain pathogens enhancing that a certain degree of damage has to be accepted [5]. One of the root causes is that long-term monoculture with a single plant eliminates crop and biological diversity [6]. Consequently, the diversification of crop systems by increasing the number of cultivated species in the same or nearby areas has been proposed to overcome those continuous cropping obstacles. In modern agriculture, crop rotation is the most common for a vast range of crop species worldwide [7]. If properly designed, crop rotation is the most efficient practice to reduce the incidence and severity of soilborne diseases [8]. However, crop rotation is not always practiced because of the difficulty in design of a proper rotation and relatively high risk of losing the lower-value crop. In addition, as the cultivated land is limited and Chinese farmers are accustomed to plant crops of the same species or the same families, it is difficult to carry out crop rotation under protected cultivation in China [9]. Another significant cropping technique - relay intercropping, which is defined as after-crop planting between the rows or plants in later periods of the preceding crops' growth with a shorter symbiotic time [7], is claimed to promote biodiversity and diversify agricultural outcome compared with monocropping in sustainable agriculture. Intercropping, being looked as an efficient and most economical production system, is drawing more and more attention of small growers [10]. The most common advantage of intercropping is to produce a greater yield and diversified production per unit area and time by achieving higher efficient use of the available growth resources that would not be utilized by each crop grown alone. From the view of diversity restoration, intercropping provides high insurance against crop failure and overall provides greater financial stability for farmers, making the system particularly suitable for labor-intensive small farms or greenhouses [11]. Besides, intercropping offers effective weed suppression [12,13] and pest and disease control. Crops grown simultaneously enhance the predators and parasites, which in turn prevent the pest build-up, thus minimizing the use of expensive and dangerous chemical insecticides. Mixed crop species can also delay the diseases onset by reducing the spread of disease carrying spores and by modifying environmental conditions so that they are less favorable to the spread of certain pathogens. Moreover, intercropping is an excellent practice for controlling soil erosion and improving soil fertility [11] and quality [14].

Soil sickness, which means serious decline in soil quality, determines the sustainability and productivity of agroecosystems [15]. The changes of the physical, chemical, and biochemical properties of the soil must be taken into account in assessing changes in soil quality [16]. Soil enzyme activities, as mediators and catalysts of most soil transformation processes, have been proposed as appropriate indicators of the health and sustainability of soil quality and ecosystems [17] due to their sensitivity to ecosystem stress [18], intimate relationship with soil biology, and rapid response to changes in soil management [19]. Invertase, widely exists in the soil, plays an important role in increasing the soluble nutrients in the soil [20]. Urease is of great importance in the soil nitrogen cycle and utilization because it can hydrolyze urea to ammonia, one of the sources of plant nitrogen. Another hydrolase, alkaline phosphatase, can mineralize organic phosphorus (P) to inorganic P [21] for plant absorption. Thus, soil enzymes can provide indications of changes in metabolic capacity and nutrient cycling in management practices [22]. Soil nutrients are important factors affecting the growth and development of plants. Nitrogen (N) is the most important element for plant development; it stimulates shoot growth and produces the rich green color that is characteristic of a healthy plant. P is the second most frequently limiting macronutrient for plant growth [23] and plays a major role in the processes requiring a transfer of energy in plants. Another essential nutrient, potassium (K), is a key factor in plant tolerance to stresses such as cold/hot temperatures, drought, and pests. Soil nutrient contents relate to the soil productivity, and soil enzymatic characteristics can reflect the status of key biochemical reactions that participate in the transformation of soil nutrients [24]. However, the reaction rates of soil enzymes are markedly dependent on pH and the presence or absence of inhibitors [25]. In addition, the availability of mineral elements to plants may also be affected by soil pH. Soil pH may affect plant root growth directly or indirectly by impairment of nutrient relations [26]. In turn, growing roots affect the pH of the rhizosphere during plant growth processes and nutrient uptake [27,28].

In relay intercropping systems, although the increased crop species are expected to overcome the continuous cropping obstacles, the selection of companion crops is still critical. Garlic (*Allium sativum* L.), belonging to the Liliaceae family, is a common vegetable and food spice that is used widely throughout the world. Especially, it is an important economic crop and a good cover crop in vegetable production in China [29]. Garlic products-green garlic, garlic bulbs, and garlic sprouts are all important vegetables favored especially for Asian. It is also commonly used as natural broad-spectrum antibiotic. Khan and Cheng [30] found that garlic root exudates is an effective and environment-friendly management measure against *Phytophthora* blight of pepper and may be used in the organic vegetables production. In addition, it has been reported that the exudates secreted by the rooting system of garlic can produce noteworthy effects on soil structure and ecology [9,31]. Thus, garlic is expected to be an ideal companion crop for relay intercropping with eggplant.

There is an increasing population and decline in arable land in China, so sustainable agriculture has gained more attention owing to its efficient use of resources, balance with the environment, and the ultimate goal of providing human benefit [32]. In recent years, an increasingly number of studies have focused on intercropping of different grain crops [33–35], in addition to intercropping cucumbers [9], peppers [31] or Chinese cabbages [36] with garlic, but eggplant/garlic relay intercropping systems are rarely studied, and intensive study of the soil properties change has been considerably less. For this reason, we have concentrated on comparing the activity of enzymes, content of available nutrients, and pH value in the soil of eggplant/normal garlic or green garlic relay intercropping systems with those in an eggplant monoculture system under continuous cropping to assess if relay intercropping eggplant with normal or green garlic is effective on improving the soil fertility level and maintaining the soil quality, which will help ensure the sustainable long-term development of eggplant cropping.

Materials and Methods

Experimental site

The experiment was conducted from September 2010 to November 2012 under a plastic tunnel at the Horticultural Experimental Station (34°17' N, 108°04' E) of Northwest A&F University, Yangling, Shaanxi Province, China, where it is hot in summer and cold in winter, and the annual mean temperature is 12.9°C, with a frost-free period over 200 days. Under plastic tunnel, the highest temperature can achieve around 50°C, and the lowest is around −10°C (Fig. 1).

The soil used for this experiment was brown loamy, alkaline Orthic Anthrosol (Table 1). The soil pH was 7.8 (1:1 water), and it contained 27.02 g of organic matter, 1.38 g of total N, 0.96 g of total P, and 14.31 g of total K per kilogram dry soil. The ammonium N, available P, and exchangeable K concentrations were 57.17 $mg{\cdot}kg^{-1}$, 57.65 $mg{\cdot}kg^{-1}$, and 224.01 $mg{\cdot}kg^{-1}$ and invertase, urease, and alkaline phosphatase activities were 18.12 glucose $mg{\cdot}g^{-1}$, 1.99 NH_3-N $mg{\cdot}g^{-1}$ and 0.94 P_2O_5 $mg{\cdot}g^{-1}$ respectively in the 0–20 cm soil layer before crop transplanting in March 2010. Cucumber had been planted for approximately ten years before the planting of eggplant in the spring of 2010.

Experimental design

Eggplant as the main crop was relay intercropped with garlic (Fig. 2). A completely randomized design was used consisting of three treatments with three replications: eggplant monoculture (CK) (Fig. 2 A and D), eggplant relay intercropping with normal garlic cv. G110 (NG) (Fig. 2 B and E), and eggplant relay intercropping with green garlic cv. G064 (GG) (Fig. 2 C and F). In NG treatment, normal garlic means garlic bulb, where garlic cloves of uniform size were manually planted into the soil among the eggplant plants in autumn (Fig. 2 B), and expanded garlic bulbs were harvested one by one using shovel in the next spring (Fig. 2 E). In GG treatment, green garlic means green garlic sprouts, where garlic bulbs of uniform size were planted directly by hand in autumn (Fig. 2 C), and the green sprouts were harvested three or four times within the next three months once they were about 20 cm high (Fig. 2 F).

There were two beds per plot of the three treatments. Each bed was 3.5 m long and 1.2 m wide (Fig. 3). There were two rows of eggplants per bed and seven plants per row, and it was 50 cm for plant spacing and 80 cm for row spacing in both the monoculture and relay intercropping treatments. In the relay intercropping treatments, three rows of garlic cloves (20 cm for row spacing and

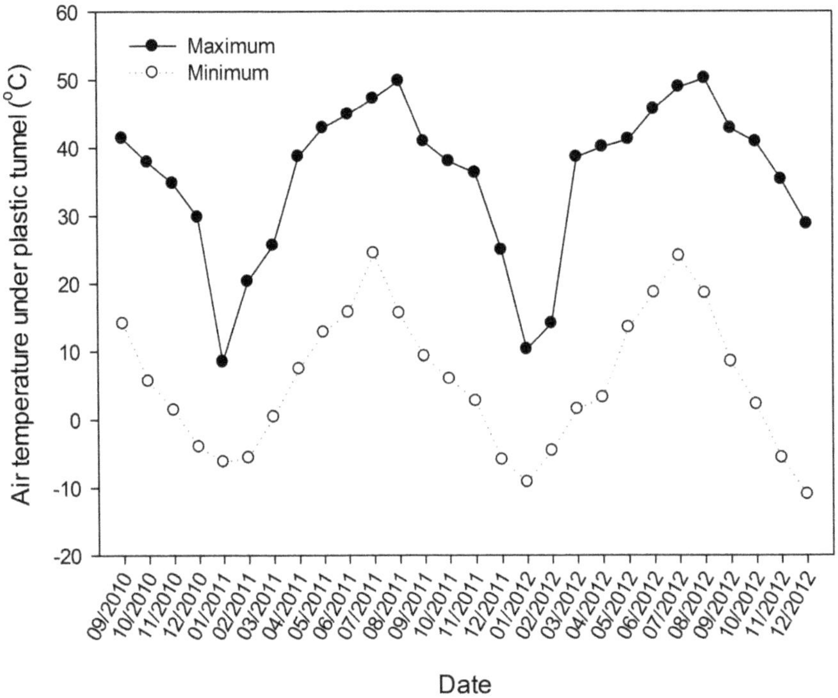

Figure 1. Monthly maximum and minimum air temperature under plastic tunnel in the three experimental years from September 2010 to December 2012. Fig. 1 was drawn using the software program Sigmaplot 12 (Systat Software, Inc.).

6 cm for plant spacing, with 141 cloves for each bed) for NG treatment and four rows of garlic bulbs (10 cm for row spacing and adjacent in each row, with 8.48 kg of bulbs for each bed) for GG treatment were planted in the middle of the bed within the eggplant rows.

Under plastic tunnel, all field operations were performed manually because of the limited land area. Eggplants were manually transplanted on March 19, 2010, March 22, 2011 and March 24, 2012, and uprooted one by one using spade around November 25 in the three years. In 2010, seed cloves and bulbs were planted on September 15, and the three treatments were applied; whereas in 2011 and 2012, the bulbs and seed cloves were planted on August 1/July 20 and September 15, respectively. Every year before eggplant transplantation, the experimental areas were plowed into two shallow furrows and fertilized with 1.5 kg of "PengDiXin" (organic fertilizer, manure substitute; made in Henan Province, China, containing organic matter≥30%, N+ P_2O+K_2O≥4%, humic acid≥20%, trace element≥2%, and organic sylvite≥5%), 0.15 kg of double superphosphate (chemical fertilizer which can be used to improve alkali soil and supply phosphorus and calcium plant needs; total P≥46%, available P≥ 44%) and 0.15 kg of "SaKeFu" fertilizer (NPK compound chemical fertilizer; made in SACF, Hebei Province, China, containing total primary nutrient≥40%) per bed as base fertilizer following local farming convention. In the eggplant-only period and eggplant/garlic co-growth period, the same amount of "JinBa" fertilizer (compound chemical fertilizer which is most often used in local vegetable production; made in Beijing, China, containing humic acid ≥3%, trace element ≥6%, N+K_2O ≥18%, and phosphate and K-solubilizing agent ≥5%) was top dressed on each bed according to the instructions. In the garlic-only stage for NG and GG treatments, only water was administered as required. For eggplant, ving tying, pruning, and other farm management were administered following local convention. No other farm management techniques were needed on garlic (Table 2).

Measurements

Soil samples were collected from the plow layer (0–20 cm) in the plots of each treatment (Table 3). Eight soil cores (40 mm in diameter) were removed in a serpentine pattern from the center of two eggplant rows of each bed resulting in 16 soil cores per plot. Subsequently, all sub-samples taken from a single plot were pooled. The first sampling dates per year were nine days before garlic planting in 2010 (September 6) and five days before eggplant transplantation in 2011 (March 17) and 2012 (March 19). In 2010, the other two soil sampling dates were October 16 (eggplant/garlic co-growth period) and November 26 (before eggplant uprooting). Then, in 2011 and 2012, soil samples were taken on April 11/ April 17 (eggplant/garlic co-growth period), June 20/June 17 (eggplant full bearing period after all garlic harvested), July 25/ July 15 (five days before planting green garlic), August 30/ September 10 (fifteen/five days before planting normal garlic), October 10/October 20 (eggplant/garlic co-growth period), and November 20/November 23 (five days before eggplant uprooted) separately.

Determination of soil enzyme activity, available nutrient content, and pH value

The soil collected from each treatment was put in a well-ventilated place to air-dry then sieved (1 mm) to analyze the activity of enzymes, content of available nutrients, and pH value in the soil. Determinations of all parameters were performed in triplicate, with values reported as means of each treatment.

The activities of invertase, urease, and alkaline phosphatase in soil were assayed on the basis of the release and quantitative determination of the products of glucose, NH_3-N and P_2O_5; soil

Table 1. Basic characteristics of original soil in the experiment.

Soil type	pH value (1soil: 1water)	Organic matter ($g \cdot kg^{-1}$)	Total nitrogen ($g \cdot kg^{-1}$)	Total phosphorus ($g \cdot kg^{-1}$)	Total potassium ($g \cdot kg^{-1}$)	Ammonium nitrogen ($mg \cdot kg^{-1}$)	Available phosphorus ($mg \cdot kg^{-1}$)	Exchangeable potassium ($mg \cdot kg^{-1}$)	Invertase activity (glucose $mg \cdot g^{-1}$)	Urease activity (NH_3-N $mg \cdot g^{-1}$)	Alkaline phosphatase activity (P_2O_5 $mg \cdot g^{-1}$)	Cultivation history
Brown loamy, alkaline Orthic Anthrosol	7.8	27.02	1.38	0.96	14.31	57.17	57.65	224.01	18.12	1.99	0.94	Cucumber planted for ten years

samples were incubated with 8% sucrose solution, 10% urea solution or 0.5% disodium phenyl phosphate solution, respectively, in suitable buffer solutions for 24 hours at 37°C, and then, spectrophotometric measurements were performed at 508 nm, 578 nm, and 660 nm, respectively [37].

Alkaline hydrolysis diffusion was used to determine the available N content in soil according to the method of Bermner and Shaw [38] with some modifications. The available N in soil was hydrolyzed by 1.0 $mol \cdot L^{-1}$ NaOH solution to NH_3, which was absorbed by H_3BO_3 indicator solution, and then the NH_3 absorbed by H_3BO_3 was titrated with 0.005 $mol \cdot L^{-1}$ (1/2 H_2SO_4) standard acid. Available P was extracted with 0.5 $mol \cdot L^{-1}$ $NaHCO_3$ solution, and then the level using molybdenum-antimony-D-isoascorbic-acid-colorimetry (MADAC) at 880 nm by the modified method of Olsen and Dean [39]. Available K was extracted with 1 $mol \cdot L^{-1}$ ammonium acetate neutral solution, and then the level using atomic absorption spectrometry according to the method of Pratt [40] with some modifications.

The soil pH value was determined in a soil:water suspension (1:1 ratios) with glass electrodes [41].

Data analyses

Data obtained for each year in this study were analyzed by analysis of variance (ANOVA) using PASW Statistics 18.0 software (IBM, Armonk, New York, USA). The significant differences between the means of the soil enzyme activities, available nutrient contents, and pH values among the monoculture and relay intercropping systems were examined according to Duncan's multiple range test at a $P<0.05$ level.

Results

Effects of relay intercropping eggplant with normal or green garlic on soil enzyme activities

Soil invertase activity. The activities of soil invertase from 2010 to 2012 are shown in Fig. 4 A. In 2010, the invertase activities of all treatments declined as the weather became cold. Then, in 2011 and 2012, the overall trend of invertase activity for all treatments first rose and then fell over time, although there were slight fluctuations in different periods. In 2011, the second experimental year, the soil invertase activity in NG and GG treatments was higher than that in CK treatment during co-growth periods both on April 11 and October 10. On June 20, during the eggplant full bearing period, the invertase activity in NG was 26.87 glucose $mg \cdot g^{-1}$, which was significantly higher than 23.84 glucose $mg \cdot g^{-1}$ in CK. At the same time, it was 29.78 glucose $mg \cdot g^{-1}$ in GG which was significantly higher than in NG. However on August 30, when the eggplant was relay intercropped with green garlic, but the normal garlic had not been planted, the invertase activity in GG treatment was significantly lower than that in NG treatments. The peak activity appeared on June 20 for the GG treatment and on July 25 for the CK and NG treatments. However, the maximum value of invertase activity of all three treatments appeared earlier in 2012 (on April 17) than that in 2011. On March 19, 2012, the garlic-only stage for NG and GG treatments, the invertase activity in NG and GG treatments showed a significantly lower level than in CK treatment. Then on April 17, during the co-growth period, the activity level in GG treatment was 47.54 glucose $mg \cdot g^{-1}$, significantly higher than that of the CK and NG treatments (33.66 and 33.60 glucose $mg \cdot g^{-1}$). From September 10, 2012, the invertase activity of both NG and GG treatments was higher than that of CK treatment, and the difference even reached a significant level on November 23. In

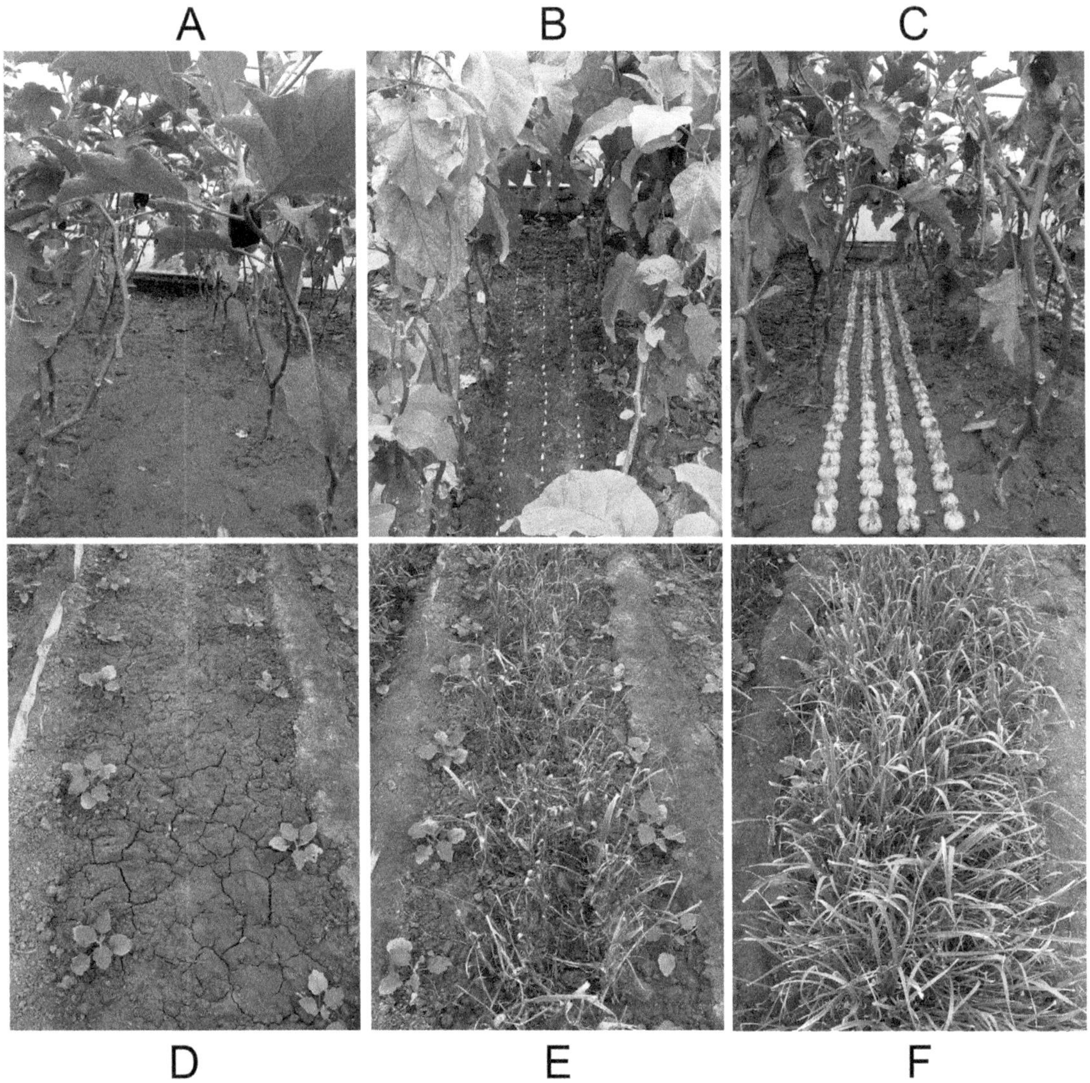

Figure 2. Three experimental treatments in autumn (A–C) and spring (D–F). Fig. 2 was made by graphics software Adobe Photoshop CS6 (Adobe Systems, Inc.). Eggplant monocropping (CK, A and D), eggplant relay intercropping with normal garlic (NG, B and E), and eggplant relay intercropping with green garlic (GG, C and F).

summary, soil invertase activity in relay intercropping treatments was higher than that of the monoculture treatment during the eggplant/garlic co-growth periods, but on other sampling dates, there was no regular routine.

Soil urease activity. The activity of soil urease fluctuated in different periods (Fig. 4 B). In the eggplant/garlic co-growth period on October 16, 2010, the urease activity of all three treatments increased over that on September 6 before the garlic was planted. On November 26, though a slight decline, the values of both the NG and GG treatments were still higher than that of the CK treatment, and the difference between NG and CK was significant.

In 2011, the second year of continuous cropping, the overall trend of urease activity continued increasing until it reached the maximum on August 30. On March 17, when there was only garlic in the field, the urease activity of the NG and GG treatments was 3.30 and 3.33 NH_3-N mg·g^{-1}, significantly higher than 2.74 NH_3-N mg·g^{-1} in the CK treatment. However, the urease activity of GG was significantly lower than that of CK and NG before the green garlic were planted on July 25. In the subsequent eggplant/green garlic co-growth period on August 30, the urease activity of GG treatment was no longer lower than CK despite the fact that it was still significantly lower than NG, and the activity of the CK treatment was significantly lower than that of NG as well. Later, the urease activity dropped markedly with the decrease of temperature; yet, it presented an abnormal phenomenon for the CK and GG treatment on November 20 that the activity increased again.

Then in 2012, in the third continuous cropping year, the urease activity of the NG and GG treatments was significantly higher than that of the CK treatment during the eggplant/garlic co-growth period, and this positive effect lasted until the green garlic were planted on July 15. However, there was a sharp decrease for all the three treatments on June 17, which may be the result of the application of adequate fertilizer in time during the eggplant vigorous growth stage. On September 10 and October 20, there was no marked difference among the treatments; but on November 23, the urease activity of the

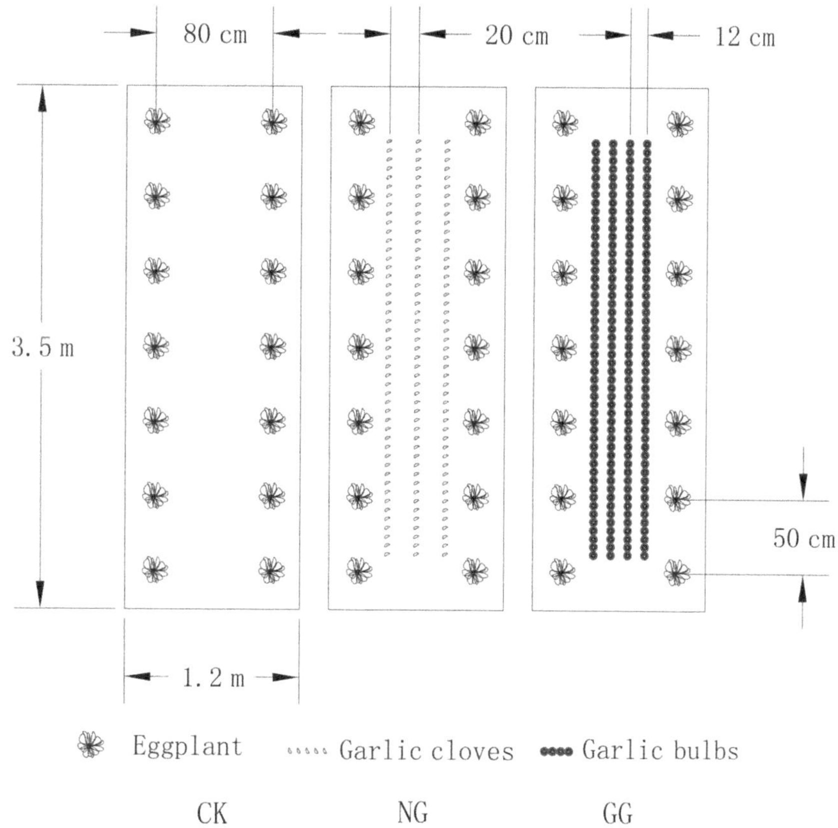

Figure 3. Diagram of three cropping systems in the experiment. Fig. 3 was drawn using the software program AutoCAD 2012 (Autodesk Inc.). CK: eggplant monoculture; NG: eggplant relay intercropping with normal garlic cv. G110; GG: eggplant relay intercropping with green garlic cv. G064.

relay intercropping treatments was again significantly higher than that of CK.

Soil alkaline phosphatase activity. As shown in Fig. 4 C, the overall change trend of the soil alkaline phosphatase activity was similar to urease activity, but varied among treatments and sampling dates. In addition, the alkaline phosphatase activity in 2012 displayed a general decline compared to that in 2011. For the two relay intercropping treatments, the activity was generally higher than that of the CK treatment during the three years of continuous cropping.

In 2010, the alkaline phosphatase activities decreased slightly in the eggplant/garlic co-growth stage on October 16 compared with that on September 6 before the garlic was planted but rebounded before eggplant uprooting. Then in the second year of eggplant continuous cropping, on March 17, 2011 when there was only garlic in the field, the enzyme activity of the GG treatment was 1.64 P_2O_5 mg·g^{-1}, significantly higher than that of CK (1.15 P_2O_5 mg·g^{-1}). For the rest of 2011, the three treatments had no significant difference on the alkaline phosphatase activity. In the third year of continuous cropping in 2012, the alkaline phosphatase activity of NG treatment was significantly higher than that of CK treatment on March 19. And for the GG treatment, the activity was always higher than CK, even on March 19 (garlic-only stage), June 17 (eggplant-only stage), and October 20 (eggplant/garlic co-growth stage), the difference reached a significant level.

Effects of relay intercropping eggplant with normal or green garlic on soil available nutrients

Soil available N content. Relay intercropping eggplant with garlic affected the content of the main available nutrients in soil. Fig. 5 A shows that, in 2010, the available N content kept rising during the three sampling dates, but there was no significant difference among the three treatments. Then in 2011, the soil available N content of the NG and GG treatments was almost higher than that of the CK treatment, and most reached significant levels. However on November 20, soil available N content in CK was 110.31 mg·kg^{-1}, significantly higher than 80.79 mg·kg^{-1} in NG and 75.31 mg·kg^{-1} in GG treatments. In the third continuous cropping year of 2012, the soil available N content of the NG or GG treatments was always significant higher than that of the CK treatment in different periods. The results also indicate that the available N content of the GG treatment was higher than that of the NG treatment in many cases. As a general view, the available N content in the soil of the NG and GG treatments was higher than that in the CK treatment, highlighting a positive effect of eggplant/garlic relay intercropping patterns on increasing the soil available N content.

Soil available P content. As shown in Fig. 5 B, the soil available P content of the NG and GG treatments kept increasing over time in 2010 and increased most rapidly for the GG treatment, which was significantly higher than that of the CK treatment on November 26; but for the CK treatment, the available P content first increased and then slightly decreased. In 2011, for the NG treatment, it was always higher than that of the CK treatment, and the difference was significant on most of the

Table 2. Field operations in the three years.

Year	Date	CK		NG		GG	
		growth stage	field management	growth stage	field management	growth stage	field management
	03/19		eggplant transplant		eggplant transplant		eggplant transplant
	03/20–09/14			eggplant only	eggplant management		
2010	09/15				plant garlic cloves		plant garlic bulbs
	09/16–11/27	eggplant only	eggplant management	eggplant/garlic co-growth	eggplant management	eggplant/garlic co-growth	eggplant management
	10/01 10/16 11/02 12/07			eggplant/garlic co-growth	eggplant management	eggplant/garlic co-growth	green garlic harvest
	11/28		eggplant uprooted		eggplant uprooted		eggplant uprooted
	11/29–12/31	blank	without any operation	garlic only	garlic watering	garlic only	garlic watering
	01/01–03/21						
	03/22		eggplant transplant		eggplant transplant		eggplant transplant
	03/23–04/14			eggplant/garlic co-growth	eggplant transplant	eggplant/garlic co-growth	eggplant transplant
	04/15				normal garlic (garlic bulbs) harvest		green garlic uprooted
	04/16–07/31			eggplant only	eggplant management	eggplant only	eggplant management
2011	08/01	eggplant only	eggplant management	eggplant only	eggplant management		plant garlic bulbs
	08/02–09/14			eggplant only	eggplant management	eggplant/garlic co-growth	eggplant management
	09/15				plant garlic cloves	eggplant/garlic co-growth	eggplant management
	09/16–11/24			eggplant/garlic co-growth	eggplant management	eggplant/garlic co-growth	eggplant management
	09/15 09/30 10/19			eggplant/garlic co-growth	eggplant management	eggplant/garlic co-growth	green garlic harvest
	11/25		eggplant uprooted		eggplant uprooted		eggplant uprooted
	11/26–12/31	blank	without any operation	garlic only	garlic watering	garlic only	garlic watering
	01/01–03/23						
	03/24		eggplant transplant		eggplant transplant		eggplant transplant
	03/25–04/17			eggplant/garlic co-growth	eggplant management	eggplant/garlic co-growth	eggplant management
	04/18				normal garlic (garlic bulbs) harvest		green garlic uprooted
2012	04/19–07/19			eggplant only	eggplant management	eggplant only	eggplant management
	07/20						plant garlic bulbs
	07/21–09/14	eggplant only	eggplant management				
	09/15				plant garlic cloves	eggplant/garlic co-growth	eggplant management
	09/16–11/27			eggplant/garlic co-growth	eggplant management	eggplant/garlic co-growth	eggplant management
	09/11 09/27 10/21					eggplant/garlic co-growth	harvest green garlic
	11/28		eggplant uprooted		eggplant uprooted		eggplant uprooted

CK: eggplant monoculture; NG: eggplant relay intercropping with normal garlic cv. G110; GG: eggplant relay intercropping with green garlic cv. G064.
Eggplant managements include irrigation, fertilization, pruning, and ving tying. When there was eggplant in the field, water and fertilizer were given only when eggplant needed. During the growth stages, eggplant was pruned to dichotomous branching. The two branches were ving tying when they grew 1 m high between June and July every year. When there was only garlic in the field, garlic managements include only watering.

Table 3. Soil sampling dates and corresponding eggplant/garlic growth stages in the three experimental years.

Sampling dates	2010			2011						
	09/06	10/16	11/26	03/17	04/11	06/20	07/25	08/30	10/10	11/20
				2012						
				03/19	04/17	06/17	07/15	09/10	10/20	11/23
Stage	nine days before planting normal garlic and green garlic	eggplant/garlic co-growth stage	eggplant/garlic co-growth and before eggplant uproot	five days before eggplant transplant	eggplant/garlic co-growth stage	eggplant full bearing period after all garlic harvested	five days before planting green garlic	fifteen days before planting normal garlic	eggplant/garlic co-growth stage	eggplant/garlic co-growth and five days before eggplant uprooted

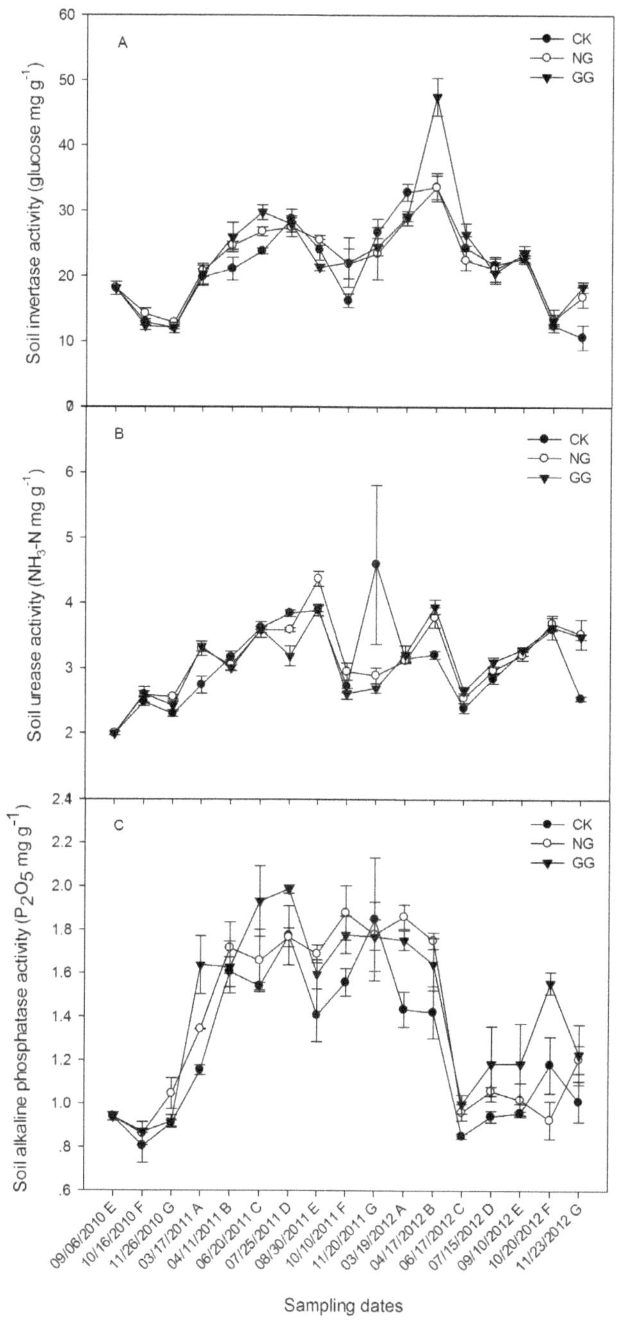

Figure 4. Effects of relay intercropping eggplant with garlic on the activities of invertase (A), urease (B), and alkaline phosphatase (C) in soil from September 2010 to November 2012. Fig. 4 was drawn using the software program Sigmaplot 12 (Systat Software, Inc.). CK: eggplant monoculture; NG: eggplant relay intercropping with normal garlic cv. G110; GG: eggplant relay intercropping with green garlic cv. G064 Error bars represent the standard error of the mean. The capital letters from A to G behind dates in the X-axis represent different periods of crop cycles in the experiment: A represents five days before eggplant transplanted in spring (garlic-only); B represents eggplant/garlic co-growth stage in spring; C represents eggplant-only stage; D represents five days before green garlic planted (eggplant-only); E represents several days before normal garlic planted; F represents eggplant/garlic co-growth stage in autumn; G represents several days before eggplant uprooted (co-growth).

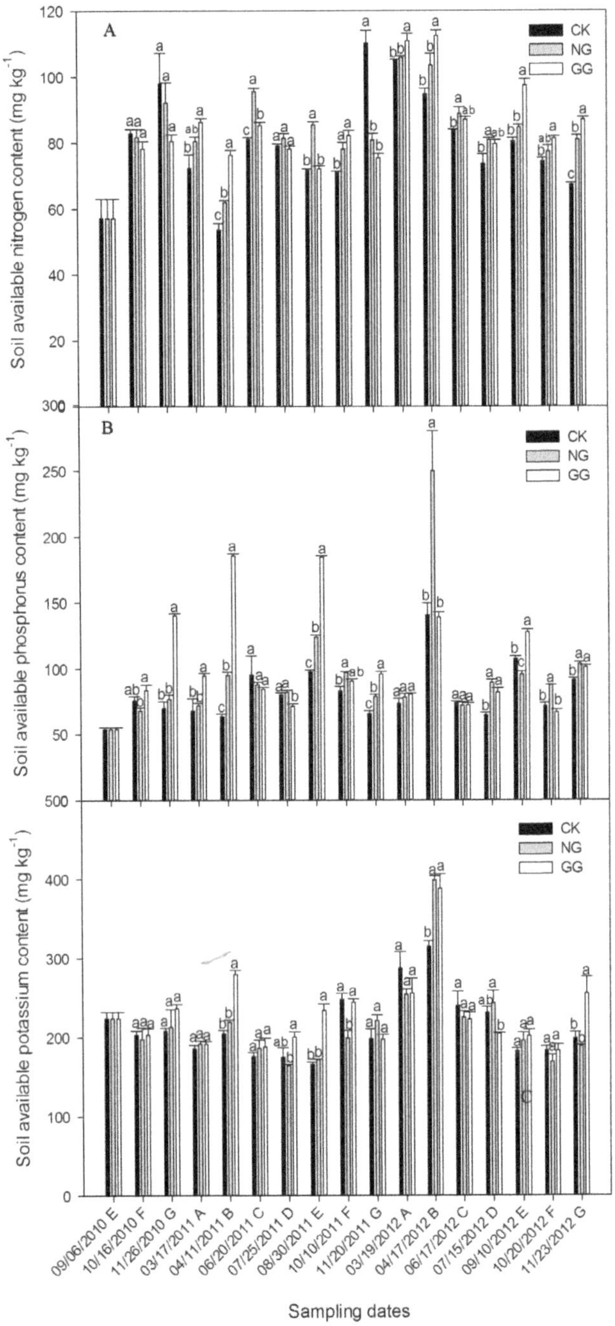

Figure 5. Effects of relay intercropping eggplant with garlic on the contents of available nitrogen (A), available phosphorus (B), and available potassium (C) in soil from September 2010 to November 2012. Fig. 5 was drawn using the software program Sigmaplot 12 (Systat Software, Inc.). CK: eggplant monoculture; NG: eggplant relay intercropping with normal garlic cv. G110; GG: eggplant relay intercropping with green garlic cv. G064 Error bars represent the standard error of the mean. Different letters above the bars indicate significant differences at a $P<0.05$ level (ANOVA and Duncan's multiple range test), n = 3. The capital letters from A to G behind dates in the X-axis represent different periods of crop cycles in the experiment: A represents five days before eggplant transplanted in spring (garlic-only); B represents eggplant/garlic co-growth stage in spring; C represents eggplant-only stage; D represents five days before green garlic planted (eggplant-only); E represents several days before normal garlic planted; F represents eggplant/garlic co-growth stage in autumn; G represents several days before eggplant uprooted (co-growth).

sampling dates, but for the GG treatment, the P content was significantly lower than that of the CK before the green garlic were planted on July 25. In addition to that, the available P content in GG treatment was significantly higher than that in CK treatment on many stages, including garlic-only stage (March 17, 2011) and eggplant/garlic co-growth stages (April 11, August 30, and November 20, 2011). Besides, the available P content of the GG treatment was significantly higher than that of the NG treatment on most sampling dates. In the third continuous cropping year in 2012, no fixed change rule was observed among the three treatments. Before the eggplant was transplanted on March 19, the available P content of the three treatments was about the same. Then on April 17, the eggplant/garlic co-growth stage, the P content of the NG treatment was 249.73 mg·kg^{-1}, significantly higher than 140.45 mg·kg^{-1} in the CK treatment, but for GG, there was no difference with CK. In the subsequent eggplant-only stage on June 17, when the eggplant grew vigorously, there were no significant differences among the three treatments. On September 10, 2012, when the green garlic had started rooting but the normal garlic had not planted, the available P content of the GG treatment was 127.56 mg·kg−1, significantly higher than 107.24 mg·kg-1 in the CK treatment, but in the NG treatment was 95.73 mg·kg-1, significantly lower than CK. Then on October 20, when the normal garlic grew together with eggplant, the available P content was also significantly higher in the NG treatment than that in the CK treatment. For the last samples in 2012, the available P content in the soil of the NG and GG treatments was significantly higher than that of the CK treatment.

Soil available K content. The soil available K content of the relay intercropping treatments on most sampling dates was higher than that of the CK treatment from September 2010 to November 2012 (Fig. 5 C). In 2010 from September to November, the available K content firstly decreased and then increased for all the three treatments. Then in the following spring, decline was seen on the soil available K content on March 17, 2011. In the eggplant/garlic co-growth period on April 11, the available K content of all three treatments increased again compared with that on March 17, and it was 279.52 mg·kg^{-1} in the GG treatment which was significantly higher than 204.49 mg·kg^{-1} in the CK treatment. On July 25 before the garlic was planted, while during the eggplant vigorous growth period, there was no significant difference between the CK and NG or GG treatment. After the green garlic grew up again on August 30, the available K content of the GG treatment was significantly higher than that of the CK treatment. In contrast, the K content was significantly lower than the CK treatment level in the NG treatment after the normal garlic rooting on October 10.

In 2012, the third year of eggplant continuous cropping, the soil available K content in the NG treatment was significantly higher than the CK treatment only during the eggplant/garlic co-growth stage on April 17. For the GG treatment, the K content was significantly higher than that of the CK treatment at two eggplant/garlic co-growth stages on April 17 and November 23.

Effects of relay intercropping eggplant with normal or green garlic on soil pH

The soil pH value varied from 7.36 to 8.00 during the three experimental years (Fig. 6). In 2010 from the pre-planting garlic on September 6 to pre-uprooting eggplant on November 26, the soil pH first increased and then decreased in CK and NG treatments, but kept decreasing in GG treatment. Then in both 2011 and 2012, the soil pH of all the three treatments was an initial decrease followed by an increase and a decrease, which

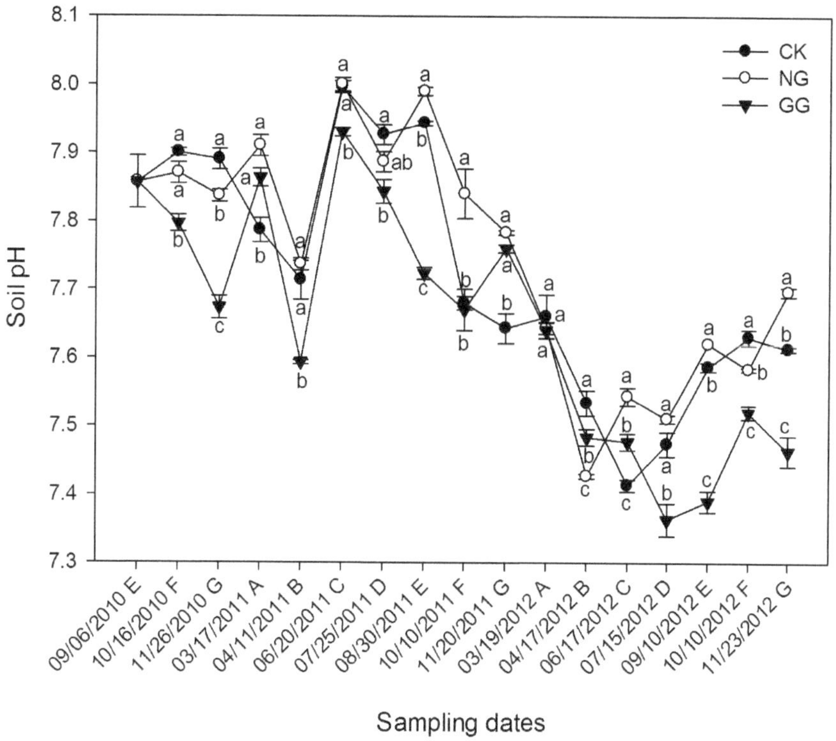

Figure 6. Effects of relay intercropping eggplant with garlic on the soil pH from September 2010 to November 2012. Fig. 6 was drawn using the software program Sigmaplot 12 (Systat Software, Inc.). CK: eggplant monoculture; NG: eggplant relay intercropping with normal garlic cv. G110; GG: eggplant relay intercropping with green garlic cv. G064 Error bars represent the standard error of the mean. Different letters above the bars indicate significant differences at a $P<0.05$ level (ANOVA and Duncan's multiple range test), n = 3. The capital letters from A to G behind dates in the X-axis represent different periods of crop cycles in the experiment: A represents five days before eggplant transplanted in spring (garlic-only); B represents eggplant/garlic co-growth stage in spring; C represents eggplant-only stage; D represents five days before green garlic planted (eggplant-only); E represents several days before normal garlic planted; F represents eggplant/garlic co-growth stage in autumn; G represents several days before eggplant uprooted (co-growth).

might be largely associated with the temperature under plastic tunnel and the water and fertilizer application situation.

On September 6, 2010, the initial value of soil pH was 7.86. Then on October 16, the preliminary growth stage of the normal garlic and green garlic, the soil pH in the GG treatment dropped to 7.80, which was significantly lower than that of CK, might because of a large amount of root exudates from green garlic or the interaction of eggplant and green garlic. However, for the NG treatment, the soil pH decreased unremarkably might because of the lower quantity of garlic. However, by the late eggplant growth period when the garlic thrived, the soil pH of both the NG and GG treatments was significantly below the soil pH value of the CK treatment. Contrary to the results of 2010, in the spring of 2011, when there was only garlic in the fields on March 17, the soil pH in the NG and GG treatments was significantly higher than that in the CK treatment. However, after the eggplant transplanted, the situation changed again. Except on July 25, the soil pH in the NG treatment was higher than CK, and some of the differences were significant. Especially on June 20, soil pH in NG reached up to 8.00. However, contradictory results were found in the GG treatment except on November 20, and the minimum value even dropped to 7.59 on April 11. These results indicate that the soil pH increased in NG with fewer root exudates but decreased in GG with more root exudates. Then in 2012, the results were similar to those in 2011 that soil pH of NG was always higher and in GG was lower in most cases than CK, and the differences were significant on some sampling dates. In addition, the soil pH levels were generally lower in 2012 than that in 2011 in the corresponding period under continuous cropping.

Discussion

Conventional continuous monocropping may degrade soil quality and negatively affect soil physical processes, and even crop growth potential and yield. Relay intercropping is believed to reduce these negative aspects by maintaining soil quality and it continues to be an important farming practice in developing countries. Higher species richness may be associated with nutrient cycling characteristics that often can regulate soil fertility [42] and limit nutrient losses [43]. Enzyme assays can indicate the situation in terms of soil quality improvement, functional diversity of critical soil processes, rapid responses to changes in management, and sensitivity to environmental stresses [44–46]. In turn, soil enzymes are also mainly influenced by vegetation species [21] and land management practices [47–49].

Invertase and urease are the most important enzymes in the transformation of carbon and nitrogen in soils [50]. The activity of phosphatase is also positively correlated with the content of soil carbon and nitrogen, and it is also related to the soil pH and organic phosphorus content. Therefore, higher enzyme activities are expected to hold in soil. According to many studies, soil enzyme activities and soil nutrient contents are higher under intercropping systems than under monoculture system [34,51]. Dai [52] found that intercropping of peanut with *A. lancea* effectively increased soil urease and invertase activities. Li [53] also found that urease activities of intercropping sugarcane and soybean were promoted by 89% and 81%, respectively, compared with that of the monoculture models. In addition, Ahmad et al. [31] found that pepper intercropping with green garlic significantly increased the activities of invertase and alkaline phosphatase in soil. Our results of soil enzymes are in agreement with their conclusions. In our work, the invertase activity in relay intercropping systems was always higher than in eggplant monocropping system during the eggplant/garlic co-growth periods in the three experimental years. For urease and alkaline phosphatase activities, the relay intercropping treatments were higher than those of the CK treatment for most sampling dates. These indicate that the garlic relay intercropped with eggplant stimulated the soil invertase, urease, and alkaline phosphatase activities. In the eggplant/garlic relay intercropping systems, the eggplant and garlic secrete different root exudates, and the root exudates of the two crops interact with each other, affecting the microorganism in the soil, thus increase the soil enzyme activities. It is also possible that the garlic root exudates stimulate the soil enzyme activity by directly acting on them. Exceptional cases appear may result from the influence of many complex factors beyond cropping patterns, such as temperature, fertilizer and water management, or the plant growth situation. In 2011 and 2012, the overall trend of invertase activity of all treatments first increased then decreased over time. This trend could be related to the temperature in the plastic tunnel, which was increasing from March to July then decreasing from August to the end of the year (Fig. 1). In addition, continuous monoculture is detrimental to soil enzyme activities. This was obviously demonstrated by the alkaline phosphatase activity, which displayed a general decrease in 2012, the third year of eggplant continuous cropping, compared with that in 2011. The higher activity of soil alkaline phosphatase in relay intercropping treatments compared with the eggplant monoculture treatment alleviated this decline caused by continuous cropping.

Soil enzyme activity can reflect the level of soil fertility [21,54]. Increased enzyme activities promote the transformation of soil nutrients and improve the soil fertility. Conversely, soil enzyme activity is also affected by the soil nutrient contents. N, P, and K are the main nutrients in the soil, and plants absorb nutrients in available nutrient form. Previous research has demonstrated that the utilization efficiency of N, P, and K in a maize/mung bean intercropping system were significantly higher than that of a monoculture system [55]. It was also reported by Li [53] that the effective N and P contents of rhizospheric soil of intercrops sugarcane and soybean were increased by 66% and 311.7%, respectively, compared with those in monoculture systems. These results were similar to our work that the available N, P, and K content in relay intercropping treatments were always higher than that in the CK treatment. Urease activity is strongly indicative of enhanced nitrogen transformation in soil [56,57]. At the initial stage of relay intercropping in 2010, the available N content kept rising, which might be related to the increased urease activity because the urease can promote the replacement of organic nitrogen with available nitrogen. However, the available N content in NG and GG treatments was slightly lower than that in CK, which might because the superiority of relay intercropping was not so obvious at the early stage of relay intercropping, and two crops grown together could absorb more available N than there was only one crop. However, the absorption by garlic did not cause a significant drop in soil available N content and the bad effect is negligible. The results of 2011 and 2012 indicated that the available N content in NG and GG treatments was significantly higher than CK in most stages, but on November 20, 2011, it was significantly lower in NG and GG treatments than CK. This exactly reversed result might because that, the available N was hardly needed for eggplants growing weakly or even dying before being uprooted, but the normal garlic and green garlic were still thriving at that time and needed more available N.

The higher available N, P, and K content in relay intercropping systems compared with that in monoculture systems demonstrates that the root exudates of normal garlic or green garlic stimulated the nutrition availability in soil. In addition, this may be the result of higher enzyme activity stimulated by garlic root exudates increasing the soil available nutrients [58,59]. Soil enzyme enrichment clearly occurs in response to soil nutrients and vegetation types [21]. This implies that increased enzyme activity is proportionally linked to the improved nutrient cycling and availability. Our study demonstrates that soil enzyme activities and nutrient contents had a similar variation trend in general and relay intercropping eggplant with garlic is better to improve soil fertility. As a result, the external input of N, P, and K chemical fertilizers can be reduced. Furthermore, increased soil fertility leads to good results on crop growth, yield, and land use efficiency. In our study, we also found that the eggplant grew stronger in relay intercropping systems than that in monoculture one, and the eggplant yield and combined output value of per unit area were also slightly higher. Although the eggplant yield declined with the continuous cropping year, relay intercropping could retard the production decrease to ensure the eggplant sustainable production (Data not shown). All these positive results on crop growth and yield could well be related to the higher soil nutrition in relay intercropping systems.

Soil pH is another important property related to soil characteristics and crop growth. Soil pH affects the activity of enzymes and the availability of nutrients [60]. As Acosta-Martínez [61] reported, phosphatase was significantly affected by soil pH, which controlled P availability by the transformation between organic and inorganic P. In other words, the availability of phosphorous in soil depends on the pH. Apparently in our work, the changing patterns of the soil phosphatase activity and pH displayed a similar downward trend in the three years of continuous cropping, which verified that the phosphatase activity was not only harmed by the continuous monoculture but also affected by the decreased soil pH. Results in this study also demonstrated the soil pH in the NG treatment was higher than that in the CK treatment. Increased soil pH led to large increases in nutrient availability [62]. The changes of available N, P, and K content in the CK and NG treatments were the same with soil pH. These results can be explained that the soil urease hydrolyzes urea to form ammonium carbonate, resulting in increased pH [63]. However, for the GG treatment, the soil pH was lower than CK, but the available nutrients contents were still higher than CK. This result was consistent with a study reported that in a wheat/faba bean intercropping system, the rhizosphere pH decreased, but the rhizosphere P availability increased compared with monocropped faba beans and wheat [64]. As is well-known, cropping systems have significant but different effects on soil with time. In relay intercropping systems, the roots of different crops can come into direct or indirect contact, change nutrient conditions and increase interactions, such as competition or mutualism of the two plants [65]. Soil properties based on biological and biochemical activities, especially those involved in energy flow and nutrient cycling, have often been demonstrated to respond to small changes in soil, thus providing sensitive information regarding subtle alterations in soil quality [66]. In eggplant/normal garlic or green garlic relay intercropping systems, both the soil enzyme activities and soil available nutrition content were promoted. However, some results revealed a discrepant change between the two parameters on the same sampling dates. These changes may arise from many intricate aspects of the system that are not yet clear.

Relay intercropping has been shown to motivate higher soil quality and produce more stable yields in a wide range of crop combinations. However, nothing is perfect. We can infer from the results that, the relay intercropping, although being better than monocropping, still exist some problem for improving the sustainability of vegetable production worldwide. Relay intercropping stimulated nutrition more available in soil than monocropping, making crops capture more nutrition, so in the long term, yield advantages of intercropping would have to pay for the higher fertilizer inputs. Besides, mechanization is another problem in intercropping, especially under plastic tunnel with limited land area, intercropping is very labor intensive. However in the developing countries, where manual labor is plentiful and cheap and the work is mainly done by hand in vegetable production, intercropping is still a better cultivation mode.

Conclusion

Conclusions are drawn from the study that the patterns of eggplant relay intercropping with normal garlic or green garlic can increase soil enzyme activities and available nutrient content and change soil pH, thus improve soil quality and ecological environment. They are expected to help overcome soil sickness and continuous cropping obstacles. It certainly suggests that relay intercropping eggplant with garlic represents a potentially important contribution to meet challenge to sustainable increase the supply of vegetables in China.

Furthermore, it is a reasonable hypothesis that enhanced soil fertility is related to microbial community functions, thus contributing to increased crop growth and yield of the two relay intercropping crops. Clearly, further work is needed to test the microbial community in soil and to elucidate relationships among the soil microorganism, enzyme activity, nutrition, and crop growth and yield. Besides, soil sickness is a result of long term continuous cropping. Longer study periods and larger study plots will help get more convincing results. However, in order to approach local actual vegetable production practice, natural greenhouse environment and field technique used by local farmers need to be kept.

Author Contributions

Conceived and designed the experiments: ZHC MYW. Performed the experiments: MYW CNW MRZ. Analyzed the data: HWM HJZ. Contributed reagents/materials/analysis tools: HWM MRZ. Wrote the paper: MYW.

References

1. FAOSTAT website (2014) Available: http://faostat3.fao.org/faostat-gateway/go/to/compare/Q/QC/E.
2. Kibunja CN (2007) Impact of long-term application of organic and inorganic nutrient sources in a maize-bean rotation to soil nitrogen dynamics and soil microbial populations and activity: University of Nairobi.
3. Kibunja CN, Mwaura FB, Mugendi DN, Gicheru PT, Wamuongo JW, et al. (2012) Strategies for maintenance and improvement of soil productivity under continuous maize and beans cropping system in the sub-humid highlands of kenya: case study of the Long-term trial at Kabete. In: Bationo A, editor. Lessons learned from Long-term Soil Fertility Management Experiments in Africa: Springer. 59–84.
4. Nafziger E (2009) Cropping systems. Illinois Agronomy Handbook.24 ed. University of Illinois at Urbana-Champaign: Cooperative Extension Service. 49–63.
5. Shipton PJ (1975) Take-all decline during cereal monoculture. In: Bruchl GW, editor. Biology and Control of Soil-Borne Plant Pathogens. St. Paul, Minnesota, USA: The American Phytopathological Society. 137–144.
6. Blanco-Canqui H, Lal R (2008) Principles of Soil Conservation and Management. Heidelberg, Germany: Springer. 617 p.
7. Hiddink GA, Termorshuizen AJ, van Bruggen AHC (2010) Mixed Cropping and Suppression of Soilborne Diseases. In: Lichtfouse E, editor. Genetic Engineering, Biofertilisation, Soil Quality and Organic Farming. France: Springer. 119–146.
8. Cook RJ, Veseth RJ (1991) Wheat Health Management. St. Paul, Minnesota, USA: The American Phytopathological Society Press.
9. Xiao X, Cheng Z, Meng H (2012) Intercropping with garlic alleviated continuous cropping obstacle of cucumber in plastic tunnel. Acta Agr Scand B-S P 62: 696–705.
10. Bhatti IH, Ahmad R, Jabbar A, Nazir MS, Mahmood T (2006) Competitive behaviour of component crops in different sesame-legume intercropping systems. Int J Agr Biol 8: 165–167.
11. Lithourgidis AS, Dordas CA, Damalas CA, Vlachostergios DN (2011) Annual intercrops: an alternative pathway for sustainable agriculture. Aust J Crop Sci 5: 396–410.
12. Famaye AO, Iremiren GO, Olubamiwa O, Aigbekaen AE, Fademi OA (2011) Intercropping cocoa with rice and plantain influencing cocoa morphological parameters and weed biomass. J Agric Sci Technol: 745–750.
13. Gomes JKO, Silva PSL, Silva KMB, Rodrigues Filho FF, Santos VG (2007) Effects of weed control through cowpea intercropping on mayze morphology and yield. Planta Daninha 25: 433–441.
14. Latif MA, Mehuys GR, Mackenzie AF, Alli I, Faris MA (1992) Effects of legumes on soil physical quality in a maize crop. Plant Soil 140: 15–23.
15. Sebastiana M, Engracia M, Carlos RJ, Francisco HJ (2007) Chemical and biochemical properties of a clay soil under dryland agriculture system as affected by organic fertilization. Eur J Agron 26: 327–334.
16. Yakovchenko V, Sikora LJ, Kaufman DD (1996) A biologically based indicator of soil quality. Biol Fertil Soils 21: 245–251.
17. Dick RP (1994) Soil enzyme activities as indicators of soil quality. In: Doran JW, editor. Defining Soil Quality for a Sustainable Environment. Madison, WI: Soil Science Society of America Journal. 107–124.
18. Zhang W, Zhang M, An S, Xiong B, Li H, et al. (2012) Ecotoxicological effects of decabromodiphenyl ether and cadmium contamination on soil microbes and enzymes. Ecotoxicol Environ Saf 82: 71–79.
19. Ndiaye EL, Sandeno JM, McGrath D, Dick RP (2000) Integrative biological indicators for detecting change in soil quality. Am J Alternative Agr 15: 26–36.
20. Li H, Shao H, Li W, Bi R, Bai Z (2012) Improving soil enzyme activities and related quality properties of reclaimed soil by applying weathered coal in opencast-mining areas of the Chinese Loess Plateau. Clean - Soil, Air, Water 40: 233–238.
21. Wang B, Xue S, Liu GB, Zhang G, Li G, et al. (2012) Changes in soil nutrient and enzyme activities under different vegetations in the Loess Plateau area, Northwest China. Catena: 186–195.
22. Saha S, Prakash V, Kundu SS, Kumar N, Mina BL (2008) Soil enzymatic activity as affected by long term application of farm yard manure and mineral fertilizer under a rainfed soybean-wheat system in N-W Himalaya. Eur J Soil Biol 44: 309–315.
23. Schachtman DP, Reid RJ, Ayling SM (1998) Phosphorus uptake by plants: from soil to cell. Plant Physiol 116: 447–453.
24. Zhang Y, Chen L, Chen Z, Sun C, Wu Z, et al. (2010) Soil nutrient contents and enzymatic characteristics as affected by 7-year no tillage under maize cropping in a meadow brown soil. R C Suelo Nutr Veg 10: 150–157.
25. Burns RG (1978) Enzyme activity in soil: Some theoretical and practical considerations. In: Burns RG, editor. Soil Enzymes. New York: Academic Press. 295–340.
26. Marschner H (1995) Mineral nutrition of higher plants. Harcourt, Brace, London: Academic Press.
27. Jaillard B, Ruiz L, Arvieu JC (1996) pH mapping in transparent gel using color indicator videodensitometry. Plant Soil 183: 85–93.
28. Ruiz I, Arvieu JC (1990) Measurement of pH gradients in the rhizosphere. Symbiosis 9: 71–75.
29. Han X, Cheng Z, Meng H (2012) Soil properties, nutrient dynamics, and soil enzyme activities associated with garlic stalk decomposition under various conditions. PLoS ONE 7: e50868.
30. Khan MA, Cheng Z (2010) Influence of garlic root exudates on cyto-Morphological alteration of the hyphae of Phytophthora capsici, the cause of phytophthora blight in pepper. Pak J Bot 42: 4353–4361.
31. Ahmad I, Cheng Z, Meng H, Liu T, Wang M, et al. (2013) Effect of pepper-garlic intercropping system on soil microbial and bio-chemical properties. Pak J Bot 45: 695–702.
32. Mousavi SR, Eskandari H (2011) A general overview on intercropping and its advantages in sustainable agriculture. J Appl Environ Biol Sci 1: 482–486.
33. Dahmardeh M, Rigi K (2013) The influence of intercropping maize (*Zea mays* L.) green gram (*Vigna Radiata* L.) on the changes of soil temperature, moisture and nitrogen. Int J Ecosyst 3: 13–17.

34. Inal A, Gunes A, Zhang F, Cakmak I (2007) Peanut/maize intercropping induced changes in rhizosphere and nutrient concentrations in shoots. Plant Physiol Biochem 45: 7.
35. Zhang F, Li L (2003) Using competitive and facilitative interactions in intercropping systems enhances crop productivity and nutrient-use efficiency. Plant Soil 248: 305–312.
36. Unlu H, Sari N, Solmaz I (2010) Intercropping effect of different vegetables on yield and some agronomic properties. J Food Agric Environ 8: 723–727.
37. Guan S (1986) Soil enzyme and research method (in Chinese). Beijing, China: Agriculture Press 376 p.
38. Bermner JM, Shaw K (1955) Determination of ammonia and nitrate in soil. J Agr Sci 46: 320–328.
39. Olsen SR, Cole CV, Watanabe FS, Dean LA (1954) Estimation of available phosphorus in soils by extraction with sodium bicarbonate. U S Dept Agr Circular 939: 1–19.
40. Pratt PF (1965) Potassium. In: Black CA, Evans DD, Ensminger LE, White JL, Clark FE et al., editors. Methods of Soil Analysis. Madison, Wisconsin, USA: American Society of Agronomy, Inc. 1025–1027.
41. Barber SA, Chen JH (1990) Using a mechanistic model to evaluate the effect of soil pH on phosphorus uptake. Plant Soil 124: 183–186.
42. Russell AD (2002) Antibiotic and biocide resistance in bacteria: Introduction. J Appl Microbiol 92: 1S–3S.
43. Hauggaard-Nielsen H, Ambus P, Jensen ES (2003) The comparison of nitrogen use and leaching in sole cropped versus intercropped pea and barley. Nutr Cycl Agroecosys 65: 289–300.
44. Dick RP (1997) Soil enzyme activities as integrative indicators of soil health. In: Pankhurst CE DB, Gupta VVSR, editor. Biological Indicators of Soil Health. Wallingford: CAB. 121–156.
45. Nannipieri P, Kandeler E, Ruggiero P (2002) Enzyme activities and microbiological and biochemical processes in soil. In: Burns RG, Dick RP, editors. Enzymes in the Environment. New York: Marcel Dekker, Inc. 1–33.
46. Caldwell BA (2005) Enzyme activities as a component of soil biodiversity: a review. Pedobiologia 49: 637–644.
47. Acosta-Martínez V, Burow G, Zobeck TM, Allen VG (2010) Soil microbial communities and function in alternative systems to continuous cotton. Soil Sci Soc Am J 74: 1181–1192.
48. Carney KM, Matson PA, Bohannan BJM (2004) Diversity and composition of tropical soil nitrifiers across a plant diversity gradient and among land-use types. Ecol lett 7: 684–694.
49. Yao HY, Bowman D, Wei S (2006) Soil microbial community structure and diversity in a turfgrass chronosequence: land-use change versus turfgrass management. Appl Soil Ecol 34: 209–218.
50. Eivazi F, Bayan MR (1996) Effects of long-term prescribed burning on the activity of select soil enzymes in an oak hickory forest. Can J For Res 26: 1799–1804.
51. Ghosh PK, Mohanty M, Bandyopadhyay KK, Painuli DK, Misra AK (2006) Growth, competition, yields advantage and economics in soybean/pigeonpea intercropping system in semi-arid tropics of India: II. Effect of nutrient management. Field Crops Res 96: 90–97.
52. Dai C, Chen Y, Wang X, Li P (2013) Effects of intercropping of peanut with the medicinal plant *Atractylodes lancea* on soil microecology and peanut yield in subtropical China. Agroforest Syst 87: 417–426.
53. Li X, Mu Y, Cheng Y, Liu X, Nian H (2013) Effects of intercropping sugarcane and soybean on growth, rhizosphere soil microbes, nitrogen and phosphorus availability. Acta Physiol Plant 35: 1113–1119.
54. Burke DJ, Weintraub MN, Hewins CR, Kalisz S (2011) Relationship between soil enzyme activities, nutrient cycling and soil fungal communities in a northern hardwood forest. Soil Biol Biochem 43: 795–803.
55. Chowdhury MK, Rosario EL (1994) Comparison of nitrogen, phosphorus and potassium utilization efficiency in maize/mungbean intercropping. J Agr Sci 122: 193–199.
56. Edwards IP, Bürqmann H, Miniaci C, Zeyer J (2006) Variation in microbial community composition and culturability in the rhizosphere of *Leucanthemopsis alpina* (L.) heywood and adjacent bare soil along an alpine chronosequence. Microb Ecol 52: 679–692.
57. Xing S, Chen C, Zhou B, Zhang H, Nang Z, et al. (2010) Soil soluble organic nitrogen and active microbial characteristics under adjacent coniferous and broadleaf plantation forests. J Soils Sediments 10: 748–757.
58. Kroehler CJ, Linkins AE (1988) The root surface phosphatases of *Eriophorum vaginatum*: Effects of temperature, pH, substrate concentration and inorganic phosphorus. Plant Soil 105: 3–10.
59. Johnson D, Leake JR, Read DJ (2005) Liming and nitrogen fertilization affects phosphatase activities, microbial biomass and mycorrhizal colonisation in upland grassland. Plant Soil 271: 157–164.
60. Dick WA, Cheng L, Wang P (2000) Soil acid and alkaline phosphatase activity as pH adjustment indicators. Soil Biol Biochem 32: 1915–1919.
61. Acosta-Martínez V, Upchurch DR, Schubert AM, Porter D, Wheeler T (2004) Early impacts of cotton and peanut cropping systems on selected soil chemical, physical, microbiological and biochemical properties. Biol Fertil Soils 40: 44–54.
62. Bagayoko M, Alvey S, Neumann G, Buerkert A (2000) Root-induced increases in soil pH and nutrient availability to field-grown cereals and legumes on acid sandy soils of Sudano-Sahelian West Africa. Plant Soil 225: 117–127.
63. Zahir ZA, Malik MAR, Arshad M (2001) Soil enzymes research: a review. Online Journal of Biological Sciences: Asian Network for Scientific Information. 299–307.
64. Song YN, Zhang FS, Marschner P, Fan FL, Gao HM, et al. (2007) Effect of intercropping on crop yield and chemical and microbiological properties in rhizosphere of wheat (*Triticum aestivum* L.), maize (*Zea mays* L.), and faba bean (*Vicia faba* L.). Biol Fertil Soils 43: 565–574.
65. Zhang NN (2010) Effets of intercropping and Rhizobium inoculation on yeild and rhizosphere bacterial community of faba bean (*Vicia faba* L.). Biol Fertil Soils 46: 625–639.
66. Pascual JA, Garcia C, Hernandez T, Moreno JL, Ros M (2000) Soil microbial activity as a biomarker of degradation and remediation processes. Soil Biol Biochem 32: 1877–1883.

19

Soil Type Dependent Rhizosphere Competence and Biocontrol of Two Bacterial Inoculant Strains and Their Effects on the Rhizosphere Microbial Community of Field-Grown Lettuce

Susanne Schreiter[1,2], Martin Sandmann[2], Kornelia Smalla[1], Rita Grosch[2]*

1 Julius Kühn-Institut – Federal Research Centre for Cultivated Plants (JKI), Institute for Epidemiology and Pathogen Diagnostics, Braunschweig, Germany, **2** Leibniz Institute of Vegetable and Ornamental Crops Großbeeren/Erfurt e.V., Department Plant Health, Großbeeren, Germany

Abstract

Rhizosphere competence of bacterial inoculants is assumed to be important for successful biocontrol. Knowledge of factors influencing rhizosphere competence under field conditions is largely lacking. The present study is aimed to unravel the effects of soil types on the rhizosphere competence and biocontrol activity of the two inoculant strains *Pseudomonas jessenii* RU47 and *Serratia plymuthica* 3Re4-18 in field-grown lettuce in soils inoculated with *Rhizoctonia solani* AG1-IB or not. Two independent experiments were carried out in 2011 on an experimental plot system with three soil types sharing the same cropping history and weather conditions for more than 10 years. Rifampicin resistant mutants of the inoculants were used to evaluate their colonization in the rhizosphere of lettuce. The rhizosphere bacterial community structure was analyzed by denaturing gradient gel electrophoresis of 16S rRNA gene fragments amplified from total community DNA to get insights into the effects of the inoculants and *R. solani* on the indigenous rhizosphere bacterial communities. Both inoculants showed a good colonization ability of the rhizosphere of lettuce with more than 10^6 colony forming units per g root dry mass two weeks after planting. An effect of the soil type on rhizosphere competence was observed for 3Re4-18 but not for RU47. In both experiments a comparable rhizosphere competence was observed and in the presence of the inoculants disease symptoms were either significantly reduced, or at least a non-significant trend was shown. Disease severity was highest in diluvial sand followed by alluvial loam and loess loam suggesting that the soil types differed in their conduciveness for bottom rot disease. Compared to effect of the soil type of the rhizosphere bacterial communities, the effects of the pathogen and the inoculants were less pronounced. The soil types had a surprisingly low influence on rhizosphere competence and biocontrol activity while they significantly affected the bottom rot disease severity.

Editor: Martha E. Trujillo, Universidad de Salamanca, Spain

Funding: The study was funded by 'German Research Foundation' (DFG), SM 59/11-1; GR 1729/8-1: KS RG. The funders had no role in study design, data collection and analysis, decision to publish, or preparation of the manuscript.

Competing Interests: The authors have declared that no competing interests exist.

* Email: grosch@igzev.de

Introduction

Plant pathogens are a limiting factor in crop productivity worldwide and responsible for yield losses [1]. Crop rotation, use of resistant cultivars and application of chemicals are strategies to minimize disease incidence and severity in Integrated Pest Management (IPM). However, resistant cultivars and effective fungicides for the control of diseases caused by soil-borne pathogens such as *Rhizoctonia solani* (Kühn) are often not available. Moreover, adverse eco-toxicological effects of chemical fungicides urge the development of alternative strategies to combat fungal diseases on crops [2–4]. In terms of disease control, it is well-documented that microbial inoculants as part of IPM can contribute to the reduction of adverse environmental effects caused by the exclusive reliance on fungicides [5-7] and thus represent a promising strategy for more sustainable agriculture [8]. Currently, the worldwide bio-pesticide market offers products including 60 bacterial and 60 fungal species [9]. Nonetheless, the exploitation of microbial inoculants as biocontrol agents in agriculture has been hampered by inconsistent results at the field scale [10,11]. The inconsistency observed in biocontrol effects limits the attraction of microbial inoculants for growers but the reasons for this variability remain largely unexplored. Variation in the colonization ability of bacterial inoculants in the rhizosphere (rhizosphere competence) is assumed to be one of the factors contributing to this inconsistency. Several studies showed that the expression of genes, responsible for the capability of a biocontrol strain to suppress a disease, is often regulated in a cell density dependent manner [12,13]. Therefore, the ability of inoculants to colonize the rhizosphere at sufficiently high numbers for an extended period was identified as a prerequisite for their beneficial effect on plants [11,14,15]. Thanks to the application of advanced genomics and microscopy methods the understanding of factors contributing to the biocontrol activity of bacterial strains has clearly improved [11,15–18], especially for *Pseudomonas* strains such as *P. fluorescens* CHA0 or Pf-5. The complex regulation of genes involved in biocontrol and plant-microbe interaction has been studied in more detail [19–23]. Although the mode of action differs from strain to strain, numerous studies supported the assumption that biocontrol activity most likely results from multifactorial processes such as antibiosis, production of cell wall degrading enzymes, surfactants, volatile substances or siderophores, competition for nutrients and space and/or the enhancement of plant innate defense responses [24,25]. Several properties of bacterial inoculants such as motility [26], attachment [27], growth [16], production of antifungal metabolites or siderophores [24] and uptake and catabolism of root exudates [17,28] have been shown to be linked to rhizosphere competence.

However, knowledge of factors influencing rhizosphere competence of bacterial inoculants under field conditions is largely lacking. Only in a few studies efforts were made to quantify the inoculant densities in the rhizosphere of field-grown crops or to evaluate the influence of inoculants and/or pathogens on the indigenous rhizosphere microbial community [7,29,30]. In agriculture, crops are cultivated under various ecological conditions. Therefore, a better understanding of the complex relationships among inoculant, pathogen, plant, and ecological factors such as the soil type is a prerequisite for improved and reliable biocontrol effects. So the goal of the present study was to investigate the influence of soil types on the rhizosphere competence and biocontrol activity of bacterial inoculants and their effects on the indigenous soil bacterial community.

The strains *Pseudomonas jessenii* RU47 [31] and *Serratia plymuthica* 3Re4-18 [32], which revealed remarkably good control effects against bottom rot in previous experiments [7,32], were selected for this study. The causal agent of bottom rot on lettuce, the soil-borne fungus *R. solani* AG1-IB whose genome was recently sequenced [33] was used as model pathogen and lettuce as model host plant. The experimental plot system with three soil types under the same cropping history at the same field site enabled us to study the effects of different soil types on the rhizosphere competence and the biocontrol activity of the bacterial inoculants for the first time. We hypothesized that the soil types influence the rhizosphere competence and the biocontrol activity of the inoculant strains applied. Furthermore, we hypothesized that both the inoculants and the presence of the pathogen (*R. solani* AG1-1B) also influences the structural diversity of microbial communities in the rhizosphere of lettuce, and that the extent of this effect depends on the soil type.

Materials and Methods

Bacterial inoculants

The bacterial inoculant *P. jessenii* RU47 was isolated from a disease-suppressive soil [31,34] and the strain *S. plymuthica* 3Re4-18 originated from the endorhiza of potato [35]. To monitor the survival of inoculants in the rhizosphere spontaneous rifampicin resistant mutants were used [34]. Both strains were stored at −80°C in Luria-Bertani broth (Carl Roth GmbH & Co. KG, Karlsruhe, Germany) with 20% glycerol.

Design of field experiments

To evaluate the effect of soil types on the rhizosphere competence and disease suppression of the bacterial inoculants *P. jessenii* RU47 and *S. plymuthica* 3Re4-18 without and with *R. solani* inoculation, two independent field experiments were carried out in a unique experimental plot system at the Leibniz Institute of Vegetable and Ornamental Crops (Großbeeren, Germany, 52° 33′ N, 13° 22′ E). The first experiment was performed in unit 5, with planting on 8 June, harvest on 18 July 2011, the second experiment in unit 6, with planting on 27 July and harvest on 5 September 2011. Each unit was comprised out of three blocks (one block per soil type). The three soil types were characterized as Arenic-Luvisol (diluvial sand, DS), Gleyic-Fluvisol (alluvial loam, AL) and Luvic-Phaeozem (loess loam, LL) [36,37]. Each block consisted of 24 plots of 2 m×2 m in size and a depth of 75 cm. In unit 5, the following crops were cultivated from 2000 to 2010: pumpkin, nasturtium, nasturtium, phacelia, amaranth, wheat, pumpkin, nasturtium, wheat, broccoli, wheat, Teltow turnip and lettuce, and in unit 6 pumpkin, nasturtium, pumpkin, amaranth, wheat, wheat, pumpkin, nasturtium, wheat, wheat and lettuce.

Lettuce seeds (cv. Tizian, Syngenta, Bad Salzuflen, Germany) were sown in seedling trays filled with the respective soil type and incubated at 12°C for 48 h and then grown in the greenhouse at approximately 20/15°C (day/night). To maintain the substrate moisture all seedling trays were watered daily and fertilized weekly (0.2% Wuxal TOP N, Wilhelm Haug GmbH & Co. KG, Düsseldorf, Germany). Lettuce seedlings were transplanted at the 3–4-leaf stage (BBCH 14) in the experimental plot system. Plants were placed in a within-row and intra-row distance of 30 cm (36 plants per plot), and lettuce plantlets were overhead irrigated based on the computer program 'BEREST' [38]. The daily soil water content in the rooted soil layer using the water holding capacity of the soil under consideration of the plant growth stage and the potential evapotranspiration were the input variables for the irrigation program. Irrigation decisions were made on the basis of the calculated soil water content and the expected evapotranspiration and precipitation of the next five days. The soil temperature (reflectometer PT100b1/3 DIN, Messtechnik Geraberg GmbH, Martinroda, Germany) and the matric potential (CS616-L water content reflectometer, Campbell Scientific, North Logan, Utah, USA) were recorded by data logger (4MbSRAm data logger, Campbell Scientific). Both reflectometers determine an average value of a 20 cm top soil layer. The fertilizer was added to each plot based on a chemical analysis of soils before planting, done according to the certified protocols of Agricultural Tests and Research Institutions Association (VdLUFA, Germany). Each soil type was adjusted to the same amount of nitrogen (162 mg/100 g) by fertilization with Kalkamon (27% N, TDG mbh Lommatzsch, Germany). Lettuce was harvested six weeks after planting (6WAP, BBCH 49) in both experiments. The lettuce shoot dry mass (SDM) of each plant and the disease severity of bottom rot were scored at harvest. For assessment of SDM each lettuce head was cut in four portions and dried at 80°C until a constant dry mass was achieved. The disease severity was rated in four categories: 1–without bottom rot symptoms; 3–symptoms only on first lower leaves and small brown spots on the underside of leaf midribs; 5–brown spots on leaf midribs on lower and next upper leaf layer and 7–severe disease symptoms on upper leaf layers and beginning of head rot to total head rot according to Grosch et al. [39].

The following treatments of lettuce were investigated for each soil type: no treatment with inoculants without (control) and with *R. solani* (*Rs*) inoculation (control+*Rs*), plants treated with inoculants without (RU47; 3Re4-18) and with *R. solani* inoculation (RU47+*Rs*; 3Re4-18+*Rs*). Each treatment included four replicates with 36 plants per replicate.

Preparation of pathogen inoculum and inoculation

The *R. solani* AG1-IB isolate 7/3 from the strain collection of the Leibniz Institute of Vegetable and Ornamental Crops (Großbeeren) was used in the present study. The inoculum was

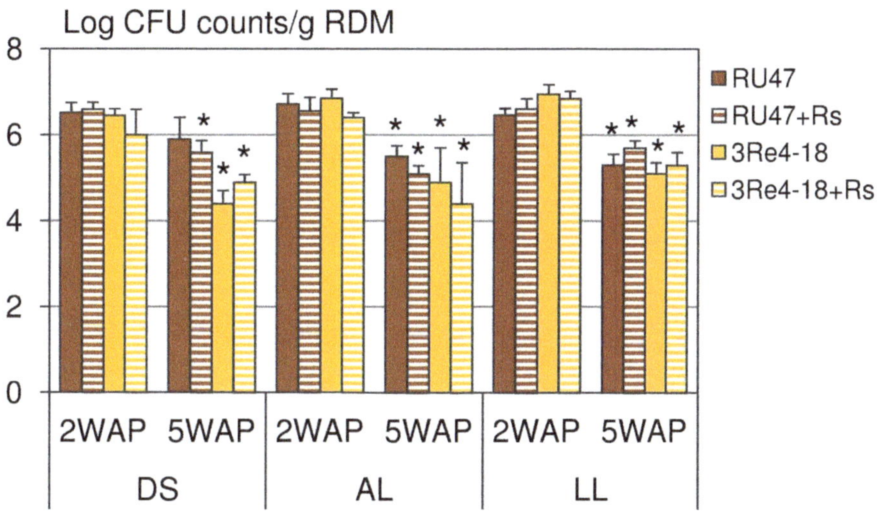

A)

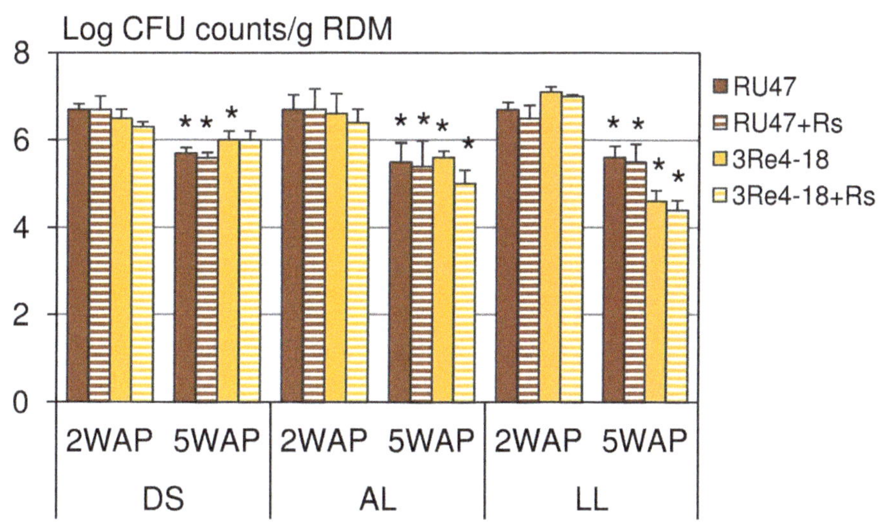

B)

Figure 1. CFU counts of *Pseudomonas jessenii* RU47 and *Serratia plymuthica* 3Re4-18 per gram of root dry mass (RDM) without (RU47, 3Re4-18) and with *Rhizoctonia solani* inoculation (RU47+*Rs;* 3Re4-18+*Rs*) in two experiments, A) and B), two and five weeks after planting (2WAP, 5WAP) in the 2011-season. Plants were grown in three soil types (DS, AL, LL) at the same field site. An asterisk indicates significant differences in CFU counts of RU47 or 3Re4-18 between 2WAP and 5WAP for each soil type (Tukey post-hoc test, $P<0.05$). The bars show the standard deviation.

Table 1. ANOVA results: Factor [soil type, plant growth development stage (PGDS), pathogen] dependent *P*-values for CFU counts of *Pseudomonas jessenii* RU47 and *Serratia plymuthica* 3Re4-18 ($P<0.05$).

	Experiment 1		**Experiment 2**	
Factor	**RU47**	**3Re4-18**	**RU47**	**3Re4-18**
Soil type	0.079	0.0007	0.544	<0.0001
PGDS	<0.0001	<0.0001	<0.0001	<0.0001
Pathogen	0.623	0.2609	0.3872	0.006

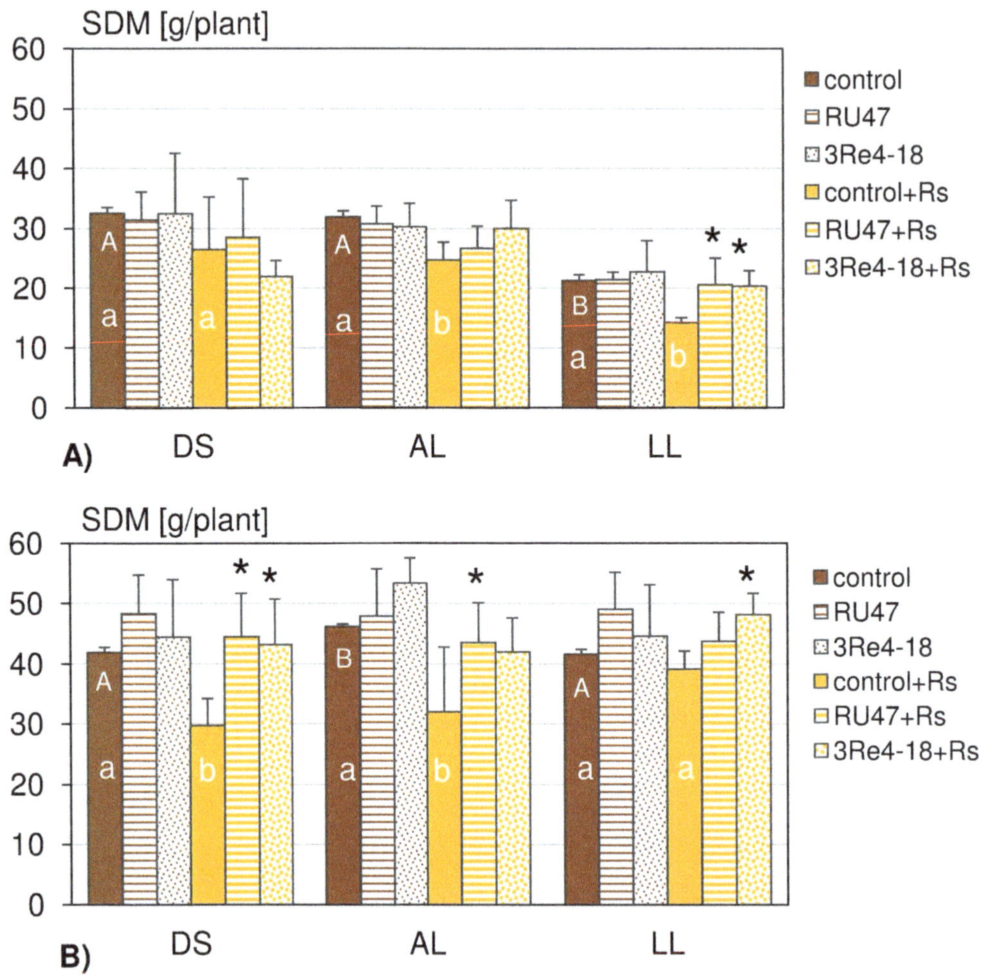

Figure 2. Shoot dry mass (SDM) of lettuce (cv. Tizian) determined for the following treatments: control, RU47, 3Re4-18, control+*Rs*, RU47+*Rs*, 3Re4-18+*Rs*) of two experiments, A) and B), in the 2011-season. Plants were grown in three soil types (DS, AL, LL) for six weeks each, at the same field site. Different capital letters denote significant differences in SDM of lettuce in controls or control+*Rs* between soil types (ANOVA; $P<0.1$). Different lower-case letters indicate significant differences in SDM between control and control+*Rs* within each soil type (Tukey post-hoc test, $P<0.1$). An asterisk denotes significant effects of the inoculants RU47 and 3Re4-18 on SDM to the corresponding control or control+*Rs* in each soil type. The bars show the standard deviation.

prepared as described by Schneider et al. [40] on barley kernels. To ensure a higher pathogen pressure the following inoculation procedure was applied: 36 lettuce plants were planted in each of the 24 plots as described above; after a cultivation time of three weeks they were evenly incorporated into the top soil (10 cm) by means of a rotary hoe, together with 40 g of barley kernels without or with *R. solani* infestation. The experiment started two weeks later assuming a decomposition of incorporated infested or non-infested lettuce plant material.

Preparation of bacterial inocula and application mode

For seed treatment King's B agar (Merck KGaA, Darmstadt, Germany) supplemented with rifampicin (75 µg/ml) were inoculated with the *P. jessenii* RU47 of *S. plymuthica* 3Re4-18 and incubated overnight at 29°C. The bacterial cells were harvested from the Petri dishes by resuspension in 15 ml sterile 0.3% NaCl and the concentration was adjusted in a spectrophotometer to a density of 10^8 colony forming units (CFU)/ml. A total of 200 lettuce seeds (cv. Tizian) were coated with 500 µl of a bacterial cell suspension dripping on the seed during vortexing in a 50 ml Falcon tube.

To prepare the inoculum for the treatment of young plants the inoculant strains were grown in nutrient broth (NB II, SIFIN GmbH, Berlin, Germany) amended with rifampicin (75 µg/ml) on a rotary shaker (90 rpm) at 29°C. After a cultivation time of 16 h the overnight culture was centrifuged at 13,000 *g* for 5 min, the supernatant discarded and the pellet was resuspended in sterile 0.3% NaCl solution. The cell density was adjusted to 10^7 CFU/ml or 10^8 CFU/ml for the drenching before and after planting, respectively. Lettuce plants were treated by drenching with 20 ml bacterial cell suspension per plant at the 3-leaf stage one week before planting in the field. A second treatment of young plants with 30 ml bacterial cell suspension 10^8 CFU/ml per plant was carried out at the 4-leaf stage two days after planting. The control plants were drenched with 20 ml or 30 ml of 0.3% NaCl solution, respectively, instead of bacterial suspension.

Sampling and sample processing

Rhizosphere samples were collected two and five weeks after planting (2WAP and 5WAP; BBCH 19 and BBCH 49) the lettuce in the experimental plot system. For each treatment and sampling time the roots of three plants per replicate (plot) were combined as

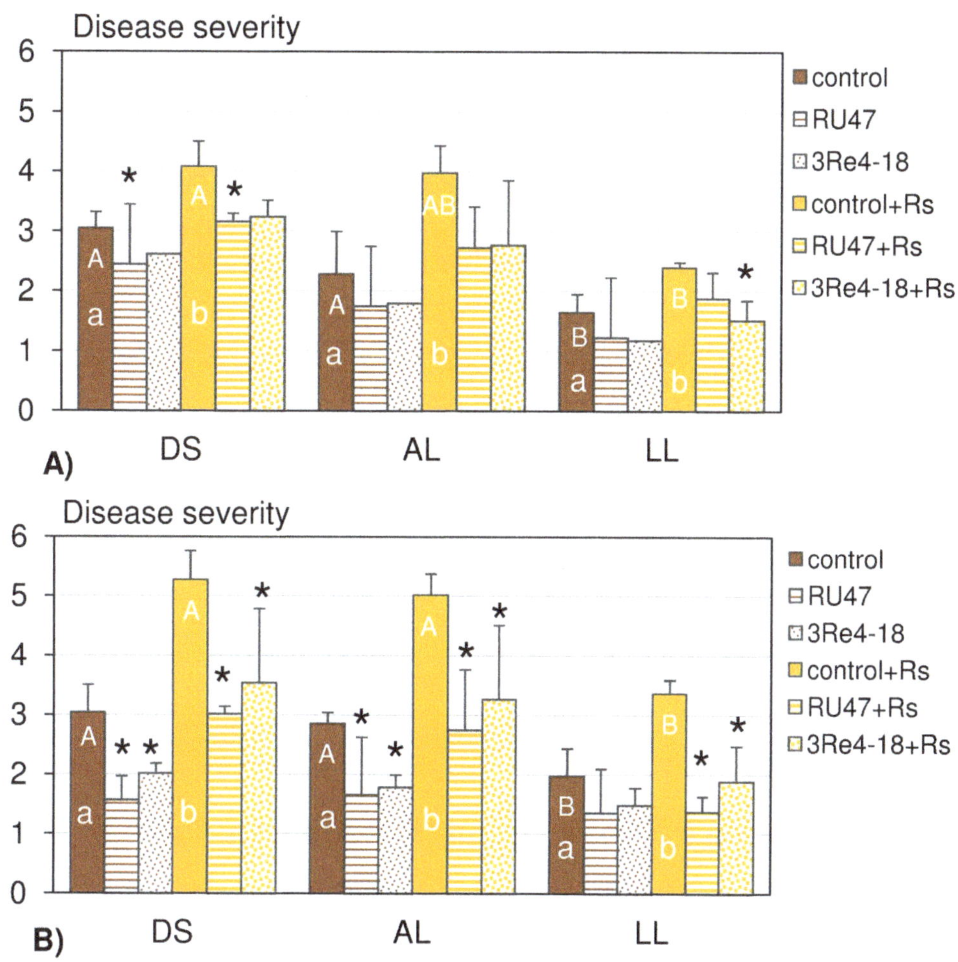

Figure 3. Disease severity of bottom rot on lettuce (cv. Tizian) determined for the following treatments: control, RU47, 3Re4-18, control+*Rs,* RU47+*Rs* and 3Re4-18+*Rs* the in two experiments, A) and B) in the 2011-season. Plants were grown in three soil types (DS, AL, LL) for six weeks each, at the same field site. Different capital letters indicate significant differences in disease severity of bottom rot in controls or control+*Rs* between soil types and different lower-case letters indicate significant differences between control and control+*Rs* within each soil type (Mann-Whitney U-test; $P<0.1$). An asterisk denotes significant differences in disease severity of the RU47 or 3Re4-18 treatment compared to the control and of RU47+*Rs* or 3Re4-18+*Rs* treatments compared to the control+*Rs* in each soil type. The bars show the standard deviation.

a composite sample and considered as one replicate; four replicates were used per treatment. Adhering soil was removed by washing the roots with sterile tap water before microbial cells were extracted as follows: the roots were cut into pieces of approximately 1 cm length and carefully mixed. Five gram of roots were placed in sterile Stomacher bags and treated by a Stomacher 400 Circulator (Seward Ltd, Worthing, UK) for 30 s at high speed after adding 15 ml of sterile 0.3% NaCl. The Stomacher blending step was repeated three times and followed by centrifugation steps as described by Schreiter et al. [37].

Analysis of rhizosphere competence of the bacterial inoculants

The ability of *P. jessenii* RU47 and *S. plymuthica* 3Re4-18 to colonize the rhizosphere of lettuce grown in the three soil types was determined 2WAP and 5WAP. Aliquots of the rhizosphere microbial cell suspension resulting from the combined supernatants of three Stomacher blending steps were immediately processed to determine the inoculant CFU counts by plating serial dilutions onto King's B agar supplemented with rifampicin (75 μg/ml) and cycloheximide (100 μg/ml) and incubated at 29°C for 48 h. The CFU counts were calculated per gram of root dry mass (RDM). For all soil types Stomacher supernatants obtained from the control plots were plated as well to determine the background of the rifampicin resistant indigenous bacteria.

Data analysis

Data of SDM, disease severity and inoculant plate counts were analyzed with the STATISTICA program (StatSoft Inc., Tulsa, OK, USA). The impact of the soil type, the pathogen and the inoculants on SDM was determined using three-way ANOVA ($P<0.1$) combined with Tukey post-hoc test ($P<0.1$). The data of disease severity was evaluated using the nonparametric Kruskal Wallis test followed by Mann-Whitney U-test ($P<0.1$). The determined inoculants density (CFU counts/g RDM) was calculated and logarithmically (Log_{10}) converted before the impact of the soil type, plant growth development stage, and the presence of the pathogen *R. solani* on the plate counts of each inoculant strain was analyzed using three-way ANOVA ($P<0.05$) combined with Tukey post-hoc test ($P<0.05$). Average values for soil temperature of each soil type were analyzed by Tukey post-hoc test using

Table 2. Treatment-dependent differences (d-values) of bacterial communities in the rhizosphere of lettuce, grown in three soil types (DS, AL, LL), in the 2011-season.

			Differences between control and				
Soil type	**Experiment**	**Figure**	**control+*Rs***	**RU47**	**3Re4-18**	**RU47+*Rs***	**3Re4-18+*Rs***
DS	1	S1	6.1*	2.6*	6.2*	3.6*	10.3*
AL	1	S2	1.9*	13.6*	20.6*	12.3*	13.2*
LL	1	S3	1.7*	6.8*	10.1*	7.8*	10.8*
DS	2	S4	4.9*	4.4*	8.8*	5.7*	7.6*
AL	2	4a	2.9*	2.8	10.0*	9.0*	17.3*
LL	2	S5	3.5	14.6*	16.1*	16.2*	21.8*

The asterisks indicate the significant differences ($P<0.05$) between the untreated control and the respective treatment.

standard errors of difference values and calculation of variance ($P<0.05$).

Analysis of 16S rRNA gene fragments PCR amplified from total community DNA by denaturing gradient gel electrophoresis (DGGE)

Total community DNA (TC-DNA) was extracted from the microbial pellets using the FastDNA SPIN Kit (MP Biomedicals, Heidelberg, Germany) as described by the manufacturer after a harsh lysis step with the FastPrep-24 Instrument (MP Biomedicals, Heidelberg, Germany). The TC-DNA was purified with GENE-CLEAN SPIN Kit (MP Biomedicals, Heidelberg, Germany) according to the manufacturer. The purified TC-DNA was diluted 1:10 with 10 mM Tris HCl before use.

For amplification of 16S rRNA gene fragments, PCR reactions were performed with TC-DNA obtained from rhizosphere samples with the primers F984-GC and R1378 as described by Heuer [41] using Taq DNA polymerase (Stoffel fragment, ABI, Darmstadt, Germany). The PCR products were analyzed by DGGE approach as described by Weinert et al. [42].

Bacterial fingerprints were evaluated with GELCOMPAR II version 6.5 (Applied Maths, Sint-Martens-Latern, Belgium) as described by Schreiter et al. [37]. The obtained Pearson similarity matrices were used for construction of a dendrogram by an Unweighted Pair-Group Method with Arithmetic mean (UPGMA) as well as of statistical analysis by the permutation test, calculating the d-value from the average overall correlation coefficients within the groups minus the average overall correlation coefficients between samples from treatments compared as suggested by Kropf et al. [43].

Results

Soil parameters of both field experiments

The concentration of N, P and K measured for each soil type before planting revealed only minor variations between both experiments (Table S1). In the second experiment, the temperature measured for all three soils in the 20 cm top layer was approximately 0.6°C below that of the first experiment (Table S2).

In contrast to the temperature the volumetric soil water content (VWC) in the top soils varied significantly among the soil types and experiments (except for AL and LL soil in the first experiment; Table S3). The lowest percentage of VWC was recorded in DS, and the highest in AL soil in both experiments. The calculated percentages of VWC in the first experiment were 18.9%, 39.6% and 39.1% in DS, AL and LL soil, and 17.0%, 35.7% and 27.4% in DS, AL and LL soil, in the second experiment. In the first experiment the VWC of AL and LL soil did not differ significantly. In both experiments VWC values for DS soil were significantly lower than those for AL and LL (Tukey post-hoc tests and confidence limits in Table S3).

Effect of the soil types on rhizosphere competence of the bacterial inoculants

In both experiments the inoculants were able to colonize the rhizosphere of lettuce grown in the different soil types at comparable CFU counts (Figure 1). Three-way ANOVA (soil type, plant growth development stage, pathogen) revealed that in both experiments the soil type had no significant effect on the CFU of RU47 in the rhizosphere of lettuce ($P>0.05$) but on the CFU of 3Re4-18 ($P<0.05$) (Table 1). In the first and second experiment 2WAP the CFU counts of 3Re4-18 were significantly higher in LL soil than in DS and AL soil. With increasing plant

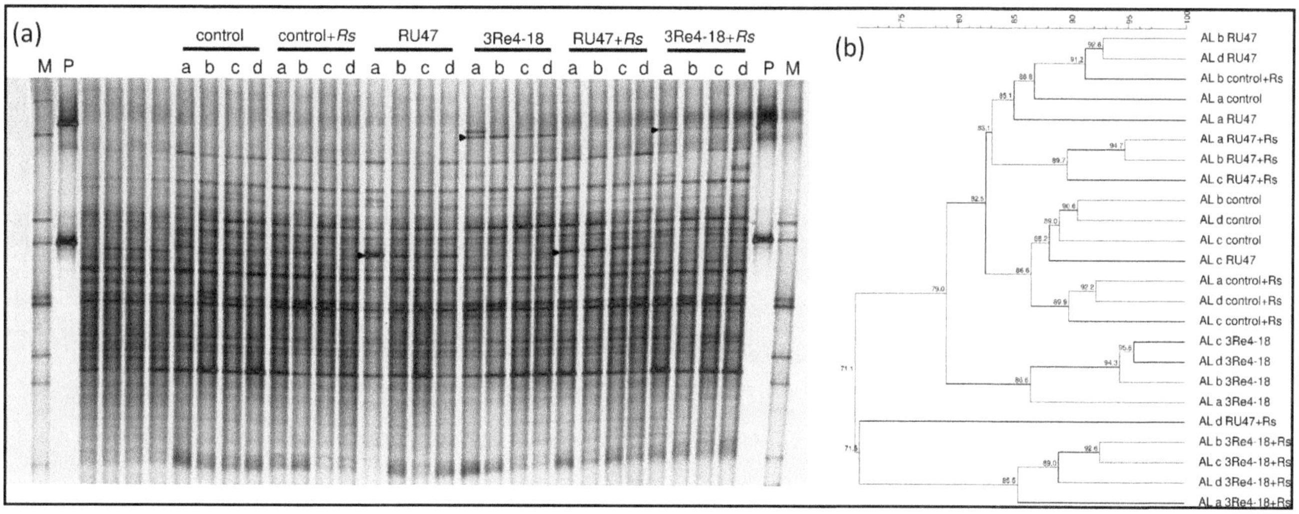

Figure 4. DGGE fingerprints (a) of bacterial 16Sr-RNA gene fragments from community DNA extracts were obtained from lettuce rhizosphere 2WAP of lettuce into the AL soil (second experiment). The corresponding UPGMA dendrogram (b) of these DGGE fingerprints based on the Pearson similarity matrix. The four replicates of the treatments: control, control+*Rs*, RU47, 3Re4-18, RU47+*Rs*, 3Re4-18+*Rs* were indicated by a–d. M: marker [57].

age the CFU counts of both inoculants decreased in the rhizosphere of lettuce grown in all three soils and in both experiments ($P<0.05$) except for the treatment with RU47 in DS soil in the first experiment and the treatment with 3Re4-18+*Rs* in DS soil in the second experiment (Figure 1). No significant effect of *R. solani* on CFU counts of both inoculants was revealed ($P>0.05$) in both experiments. The natural background of the rifampicin mutation was very low in all soil types.

Effects of the soil type, the bacterial inoculants and the pathogen on the lettuce growth

The effects of the soil type, the inoculants and *R. solani* on lettuce growth assessed by comparing the SDM of lettuce plants harvested 6WAP revealed lower lettuce growth in the first experiment compared with the second experiment in particular for LL soil (Figure 2). A significant effect of the soil type on lettuce growth was observed in both experiments. However, the results differed between the experiments. In the first experiment plants grown in LL soil showed the lowest SDM (21.3 g/plant) compared to SDM of plants from DS (32.5 g/plant) and AL soil (31.9 g/plant). In the second experiment the SDM was highest for plants grown in AL soil (46.2 g/plant) while the SDM of plants from DS (41.9 g/plant) and LL soil (41.6 g/plant) were comparable. In contrast to the first experiment an improved lettuce growth was observed in the treatments with the inoculants RU47 and 3Re4-18 in all three soils in the second experiment. However, the inoculant effects on lettuce growth were not significant (Figure 2).

In general the pathogen *R. solani* had a negative effect on lettuce growth in all three soils and in both experiments. The effect was significant in AL soil in both experiments, in LL soil only in the first experiment, and in DS soil in the second experiment (Figure 2). The treatment of lettuce with the inoculants resulted in an improved lettuce growth in RU47+*Rs* and 3Re4-18+*Rs* in all three soils and in both experiments but significant effects were only recorded in the first experiment for the treatments RU47+*Rs* in DS soil and 3Re4-18+*Rs* in LL soil, and in the second experiment for RU47+*Rs* and 3Re4-18+*Rs* in DS soil, for RU47+*Rs* in AL soil and for 3Re4-18+*Rs* in LL soil (Figure 2).

Effect of the soil type on the biocontrol activity of the bacterial inoculants

Bottom rot symptoms were also recorded in the untreated controls in each soil and in both experiments (Figure 3). The inoculation of *R. solani* (control+*Rs*) resulted in a significantly increased disease severity in all three soils and in both experiments. More severe bottom rot symptoms were scored in the second compared to the first experiment based on the disease severity in the pathogen controls in all three soils. The soil type influenced the disease severity of bottom rot in both experiments (Figure 3). The lowest disease severity was always recorded on lettuce grown in LL soil (control and control+*Rs*) compared to disease severity observed for lettuce grown in DS and AL soil.

RU47 and 3Re4-18 were able to reduce severity of bottom rot on lettuce in the treatments without and with *R. solani* inoculation in all three soils and in both experiments (Figure 3). Significant biocontrol effects were observed by the inoculant RU47 in DS soil (RU47; RU47+*Rs*) in both experiments and in AL (RU47; RU47+*Rs*) and LL (RU47+*Rs*) soil in the second experiment only. The inoculant 3Re4-18 showed a significant control effect in LL (3Re4-18+*Rs*) in the first experiment and in all soils in the second experiment (3Re4-18; 3Re4-18+*Rs*) except for 3Re4-18 in LL soil (Figure 3).

Effects of the pathogen and the bacterial inoculants on the bacterial community composition in the rhizosphere of lettuce

The effect of *R. solani* on the bacterial community composition were assessed by comparing the bacterial DGGE fingerprints of the control treatments (control) with those of the pathogen control (control+*Rs*) for each of the soils. In both experiments, the inoculation of *R. solani* had low but an significant effect in all soils except LL in the second experiment, the highest d-values were observed in DS soil (Table 2).

The effects of the inoculants on the bacterial community composition in the rhizosphere of lettuce analyzed by DGGE varied between the two experiments. The bacterial 16S rRNA gene fragments, amplified from TC-DNA of all treatments, were analyzed for each soil type separately and the DGGE gel for AL soil of the second experiment that had high d-values (Figure 4a). The DGGE fingerprint (Figure 4a) revealed that 2WAP the inoculant strain 3Re4-18 belonged to the dominant members of the bacterial community in the rhizosphere as bands with the electrophoretic mobility of 3Re4-18 were detected in the corresponding treatments. However, bands with electrophoretic mobility of the inoculant RU47 could not be detected as it co-migrated with bands of the indigenous bacterial community. Thus RU47 could only be identified in the second experiment in DS and AL soil, while 3Re4-18 was detected in AL and LL soil in the first, and in DS and AL soils in the second experiment (Figures S1–S5). Furthermore, the analysis of DGGE fingerprints by UPGMA revealed that the bacterial community composition of the different treatments shared approximately 70% similarity for all soils (Figure 4b). Although treatment dependent clusters were not always observed, the treatment effects were significant based on the permutation test ($P<0.05$) for all treatments with the inoculants RU47 or 3Re4-18 except for RU47 in AL soil in the second field experiment. However, the differences between the fingerprints of the controls and the treatments given by the calculated d-values varied for both experiments (Table 2). In the first experiment highest d-values were observed for AL soil for both inoculants while in the second experiment the highest d-values were observed for LL soil (Table 2). In both experiments and in all soils higher d-values for the treatments with 3Re4-18 indicated a stronger effect of this inoculant on the indigenous rhizosphere bacterial community compared to RU47 (Table 2). The effects of both inoculants on the bacterial community composition were typically increased for the treatments with *R. solani* inoculation in both experiments (except for RU47 and 3Re4-18 in AL soil in the first experiment).

Discussion

The availability of three soil types differing in their properties but sharing the same cropping history and weather conditions at the same field site enabled us to investigate the effect of the soil type on the rhizosphere competence of the two bacterial inoculants *P. jessenii* RU47 and *S. plymuthica* 3Re4-18 without and with *R. solani* inoculation for the first time. We hypothesized that the soil types with their distinct microbial community composition and physico-chemical properties but also the soil type dependent root exudates of the model plant lettuce [44] might affect the rhizosphere competence of the inoculants and in consequence their biocontrol efficacy. The composition of root exudates collected from lettuce grown in DS, AL and LL soil was recently shown to differ only quantitatively as similar compounds were detected as independent of the soil type [44]. Several studies showed that root exudates are important drivers of the rhizosphere microbial community composition [45,46] and the metabolic activity of the bacterial community including inoculant strains [47]. Recently, we could show by means of DGGE and pyrosequence analysis of 16S rRNA gene fragments, amplified from TC-DNA of bulk soil samples taken from the experimental unit 6 in 2010, that the three soil types indeed harbored a distinct bacterial community structure indicating the importance of the mineral composition and the soil organic matter for shaping the bacterial community composition [37]. Interestingly, in the rhizosphere of lettuce grown in the different soils numerous similar genera were increased in relative abundance, and a high proportion of dominant operational taxonomic units were shared among the rhizosphere samples from different soils [37]. Among the genera, significantly increased in the rhizosphere, was the genus *Pseudomonas* which might explain the negligible effect of soil types on the ability of RU47 to colonize the lettuce rhizosphere in both experiments performed in 2011. Remarkably, similar CFU counts of RU47 were observed 2WAP in all soil types and in both experiments. The decrease of RU47 CFU counts in the rhizosphere observed 5WAP in both experiments might well be a decrease in its relative abundance per gram of RDM as the rhizosphere pellet was obtained from the complete root system due to changes in the rhizoplane surface/RDM ratio or/and possibly newly emerging roots were not colonized by RU47 as previously reported for *gfp*-tagged inoculants [48]. A decrease in inoculant CFU densities with increasing plant age of lettuce confirmed previous observations for the inoculants used in this study [34,49] and other biocontrol strains [24]. Nonetheless, our results showed an influence of the soil types on the rhizosphere competence of 3Re4-18. While higher CFU counts were observed in LL soil in both experiments 2WAP in comparison to DS and AL soil, a pronounced decline of 3Re4-18 counts was recorded in LL soil in the second experiment 5WAP. Indeed, in the second experiment the SDM was almost doubled for lettuce grown in LL soil compared to the first experiment which might have resulted in a decreased relative abundance of 3Re4-18. Neumann et al. [44] reported that the lettuce growth and the root morphology were influenced by soil type which might have contributed to the soil type dependent differences in rhizosphere competence of 3Re4-18. The differences in lettuce growth between the first and the second experiment in particular in LL soil were not unusual for agricultural practice. Factors that might have contributed to these differences range from slightly lower VWC, temperatures in the field directly after planting to differences in sun light intensity. In addition, the slightly different cropping history in both experimental units could have affected the soil microbiome with potential effects on plant growth. Plant growth in turn is assumed to influence the quantity of root exudates [50]. Several studies have underlined a clear relationship between the level of disease suppression and inoculant densities in the rhizosphere of an introduced biocontrol strain [17,51,52]. Inconsistent results in biological control were often assumed to be associated with inefficient root colonization at the field scale [11,53]. However, despite the good colonization of the lettuce rhizosphere by both inoculants in the present study, the biocontrol effects varied in both experiments. Although in the presence of the inoculant strains the disease severity was reduced. Compared to the control and the pathogen control (control+*Rs*), the biocontrol effects were not always significant. In fact, in the first experiment significant biocontrol effects were only observed for RU47 and RU47+*Rs* in DS soil and 3Re4-18+*Rs* in LL soil while biocontrol effects were found to be much more pronounced and significant in almost all treatments in the second experiment. The improved biocontrol effect seems to be correlated with an increased lettuce growth in the second experiment. It is tempting to speculate that an improved plant growth might be linked to an increased amount of photosynthates released via the roots. The increased root exudation in turn might have not only affected the metabolic activity of the inoculants but also facilitated the interaction of *R. solani* with the lettuce plant. Therefore, more severe bottom rot symptoms in the control and the control+*Rs* treatments were recorded in the second experiment. Furthermore, the disease severity in the control (natural background of *R. solani*) and in the pathogen control (control+*Rs*) was highest in DS soil and lowest in LL soil in both experiments. The higher conduciveness of the DS soil may be caused by higher oxygen availability because of the bigger pore sizes in sandy soils but might also be due to differences in the microbial community structure. Also, Steinberg et al. [54] reported that the capacity of a soil to antagonize the action of a plant pathogen is related to the structure and function of the microbial communities in soil.

Molecular fingerprinting techniques confirmed that the inoculants were indeed dominant members of the bacterial communities in the rhizosphere of lettuce. The additional DGGE bands of the inoculants strains 3Re4-18 and the increased intensity of a band co-migrating with RU47 which did not occur in the control treatments likely contributed to the effects of the inoculants on the rhizosphere bacterial community composition determined by the permutation test (Table 2). An approach to prevent the amplification of the inoculant strains is the use of taxon-specific primers which would exclude the taxonomic group to which the inoculant belongs, as proposed by Gomes et al. [55]. The actinobacterial fingerprint, done for the samples from 2WAP of the second experiment still revealed a significant effect except for RU47 in DS soil (Schreiter et al. unpublished data). The low effects of *R. solani* on the bacterial community composition can most likely be explained by the pathogenesis of *R. solani* AG1-IB which is primarily not a root pathogen but attacks the lower leaves of lettuce [56]. In conclusion, in the present study which was based on two independent field experiments performed in the 2011-season we could not confirm our hypothesis that the soil type influences the rhizosphere competence of biocontrol strains RU47 and 3Re4-18. We are aware that this finding cannot be generalized and can be different for other plants–inoculant combinations. The soil type independent enrichment of *Gammaproteobacteria* such as *Pseudomonas*, as observed by Schreiter et al. [37], in the rhizosphere of lettuce may be an explanation for the fact that the rhizosphere competence of RU47 was not influenced by soil type. Only minor changes in the bacterial rhizosphere composition were observed due to the inoculation of RU47, 3Re4-18 and *R. solani*, and these did not depend on the soil type. However, the plant growth and the disease severity of bottom rot were influenced by the soil type which in turn also influenced the biocontrol effects. The present study is unique as the rhizosphere competence, biocontrol effects, plant growth and treatment effects on the indigenous rhizosphere bacterial community were assessed in three soil types at the same site in two independent field experiments. The present study showed that the often reported inconsistency of biocontrol is likely more due to plants which in turn are influenced by weather conditions. Thus we conclude that the multitrophic interaction between plant, inoculant, pathogen and indigenous microbial community deserves far more attention in the future.

Supporting Information

Figure S1 *Bacteria* DGGE fingerprint of rhizosphere-samples were obtained from DS soil 2WAP of the lettuce (first experiment). Lanes a–d replicates of each of the treatments: untreated control (control), control inoculated with *Rhizoctonia solani* (control+*Rs*), inoculation with *Pseudomonas jessenii* RU47 (RU47), inoculation with *Serratia plymuthica* 3Re4-18 (3Re4-18), inoculation with RU47 and *R. solani* (RU47+*Rs*), inoculation with 3Re4-18 and *R. solani* (3Re4-18+*Rs*); M: marker [57].

Figure S2 *Bacteria* DGGE fingerprint of rhizosphere-samples were obtained from AL soil 2WAP of the lettuce (first experiment). Lanes a–d replicates of each of the treatments: untreated control (control), control inoculated with *Rhizoctonia solani* (control+*Rs*), inoculation with *Pseudomonas jessenii* RU47 (RU47), inoculation with *Serratia plymuthica* 3Re4-18 (3Re4-18), inoculation with RU47 and *R. solani* (RU47+*Rs*), inoculation with 3Re4-18 and *R. solani* (3Re4-18+*Rs*); P: inoculants (upper band 3Re4-18; lower band RU47); M: marker [57].

Figure S3 *Bacteria* DGGE fingerprint of rhizosphere-samples were obtained from LL soil 2WAP of the lettuce (first experiment). Lanes a–d replicates of each of the treatments: untreated control (control), control inoculated with *Rhizoctonia solani* (control+*Rs*), inoculation with *Pseudomonas jessenii* RU47 (RU47), inoculation with *Serratia plymuthica* 3Re4-18 (3Re4-18), inoculation with RU47 and *R. solani* (RU47+*Rs*), inoculation with 3Re4-18 and *R. solani* (3Re4-18+*Rs*); P: inoculants (upper band 3Re4-18; lower band RU47); M: marker [57].

Figure S4 *Bacteria* DGGE fingerprint of rhizosphere-samples were obtained from DS soil 2WAP of the lettuce (second experiment). Lanes a–d replicates of each of the treatments: untreated control (control), control inoculated with *Rhizoctonia solani* (control+*Rs*), inoculation with *Pseudomonas jessenii* RU47 (RU47), inoculation with *Serratia plymuthica* 3Re4-18 (3Re4-18), inoculation with RU47 and *R. solani* (RU47+*Rs*), inoculation with 3Re4-18 and *R. solani* (3Re4-18+*Rs*); P: inoculants (upper band 3Re4-18; lower band RU47); M: marker [57].

Figure S5 *Bacteria* DGGE fingerprint of rhizosphere-samples were obtained from LL soil 2WAP of the lettuce (second experiment). Lanes a–d replicates of each of the treatments: untreated control (control), control inoculated with *Rhizoctonia solani* (control+*Rs*), inoculation with *Pseudomonas jessenii* RU47 (RU47), inoculation with *Serratia plymuthica* 3Re4-18 (3Re4-18), inoculation with RU47 and *R. solani* (RU47+*Rs*), inoculation with 3Re4-18 and *R. solani* (3Re4-18+*Rs*); P: inoculants (upper band 3Re4-18; lower band RU47); M: marker [57].

Table S1 Concentration of nitrogen (N), phosphorus (P) and potassium (K) in soil measured before planting lettuce in three soil types (diluvial sand, DS; alluvial loam, AL; loess loam LL) in the season 2011 of the experimental plot system at the same field site.

Table S2 Soil temperature recorded on average of the upper 20 cm top soil in three soil types (diluvial sand, DS; alluvial loam, AL; loess loam, LL) in the season 2011 at the same field site.

Table S3 Comparison of the volumetric soil water content (VWC) of three soil types (diluvial sand, DS; alluvial loam, AL; loess loam, LL) in the season 2011 at the same field site.

Acknowledgments

We would also like to thank Petra Zocher, Ute Zimmerling, Sabine Breitkopf and Angelika Fandrey for their skilled technical assistance. We also thank Christin Zachow for providing the *Serratia plymuthica* 3Re4-18 strain and Ilse-Marie Jungkurth for helpful comments on the manuscript.

Author Contributions

Conceived and designed the experiments: KS RG. Performed the experiments: SS RG KS. Analyzed the data: SS RG MS. Contributed reagents/materials/analysis tools: KS RG SS MS. Contributed to the writing of the manuscript: SS RG KS. Designed software used in analysis of volumetric soil water content: MS.

References

1. Oerke EC (2006) Crop losses to pests. J Agr Sci 144: 31–43.
2. Alabouvette C, Olivain C, Steinberg C (2006) Biological control of plant diseases: the European situation. Eur J Plant Pathol 114: 329–341.
3. Leistra M, Matser AM (2004) Adsorption, transformation, and bioavailability of the fungicides carbendazim and iprodione in soil, alone and in combination. J Environ Sci Heal B 39: 1–17.
4. Wang YS, Wen CY, Chiu TC, Yen JH (2004) Effect of fungicide iprodione on soil bacterial community. Ecotox Environ Safe 59: 127–132.
5. Weller DM, Raaijmakers JM, Gardener BBM, Thomashow LS (2002) Microbial populations responsible for specific soil suppressiveness to plant pathogens. Annu Rev Phytopathol 40: 309–348.
6. Kazempour MN (2004) Biological control of *Rhizoctonia solani*, the causal agent of rice sheath blight by antagonistics bacteria in greenhouse and field conditions. Plant Pathol J 3: 88–96.
7. Scherwinski K, Grosch R, Berg G (2008) Effect of bacterial antagonists on lettuce: active biocontrol of *Rhizoctonia solani* and negligible, short-term effects on nontarget microorganisms. FEMS Microbiol Ecol 64: 106–116.
8. Tikhonovich IA, Provorov NA (2011) Microbiology is the basis of sustainable agriculture: an opinion. Ann Appl Biol 159: 155–168.
9. Quinlan RJ, Lisansky SG (2010) North America: Biopesticides Market. CPL Business Consultants, Oxfordshire, UK.
10. Mark GL, Morrissey JP, Higgins P, O'Gara F (2006) Molecular-based strategies to exploit *Pseudomonas* biocontrol strains for environmental biotechnology applications. FEMS Microbiol Ecol 56: 167–177.
11. Barret M, Morrissey JP, O'Gara F (2011) Functional genomics analysis of plant growth-promoting rhizobacterial traits involved in rhizosphere competence. Biol Fert Soils 47: 729–743.
12. Pierson III LS, Pierson EA (1996) Phenazine antibiotic production in *Pseudomonas aureofaciens*: role in rhizosphere ecology and pathogen suppression. FEMS Microbiol Lett 136: 101–108.
13. Steidle A, Allesen-Holm M, Riedel K, Berg G, Givskov M, et al. (2002) Identification and characterization of an N-acylhomoserine lactone-dependent quorum-sensing system in *Pseudomonas putida* strain IsoF. Appl Environ Microbiol 68: 6371–6382.
14. De Bellis P, Ercolani GL (2001) Growth interactions during bacterial colonization of seedling rootlets. Appl Environ Microbiol 67: 1945–1948.
15. Ghirardi S, Dessaint F, Mazurier S, Corberand T, Raaijmakers JM, et al. (2012) Identification of traits shared by rhizosphere-competent strains of fluorescent Pseudomonads. Microb Ecol 64: 725–737.
16. Bloemberg GV, Lugtenberg BJJ (2001) Molecular basis of plant growth promotion and biocontrol by rhizobacteria. Curr Opin Plant Biol 4: 343–350.
17. Lugtenberg BJJ, Dekkers L, Bloemberg GV (2001) Molecular determinants of rhizosphere colonization by *Pseudomonas*. Annu Rev Phytopathol 39: 461–493.
18. Persello-Cartieaux F, Nussaume L, Robaglia C (2003) Tales from the underground: molecular plant-rhizobacteria interactions. Plant Cell Enviro 26: 189–199.
19. Schnider U, Keel C, Blumer C, Troxler J, Defago G, et al. (1995) Amplification of the housekeeping sigma-factor in *Pseudomonas fluorescens* CHA0 enhances antibiotic production and improves biocontrol abilities. J Bacteriol 177: 5387–5392.
20. Dekkers LC, Mulders IHM, Phoelich CC, Chin-A-Woeng TFC, Wijfjes AHM, et al. (2000) The sss colonization gene of the tomato-*Fusarium oxysporum* f. sp *radicis-lycopersici* biocontrol strain *Pseudomonas fluorescens* WCS365 can improve root colonization of other wild-type *Pseudomonas* spp. bacteria. Mol Plant Microbe In 13: 1177–1183.
21. Patten CL, Glick BR (2002) Role of *Pseudomonas putida* indoleacetic acid in development of the host plant root system. Appl Environ Microbiol 68: 3795–3801.
22. Loper JE, Hassan KA, Mavrodi DV, Davis EW, II, Lim C, et al. (2012) Comparative genomics of plant-associated *Pseudomonas* spp.: insights into diversity and inheritance of traits involved in multitrophic interactions. PLoS Genet 8: e1002784–e1002784.
23. Raaijmakers JM, Mazzola M (2012) Diversity and natural functions of antibiotics produced by beneficial and plant pathogenic bacteria. Annu Rev Phytopathol 50: 403–424.
24. Haas D, Defago G (2005) Biological control of soil-borne pathogens by fluorescent Pseudomonads. Nat Rev Microbiol 3: 307–319.
25. Raaijmakers JM, Paulitz TC, Steinberg C, Alabouvette C, Moënne-Loccoz Y (2009) The rhizosphere: a playground and battlefield for soilborne pathogens and beneficial microorganisms. Plant Soil 321: 341–361.
26. Capdevila S, Martinez-Granero FM, Sanchez-Contreras M, Rivilla R, Martin M (2004) Analysis of *Pseudomonas fluorescens* F113 genes implicated in flagellar filament synthesis and their role in competitive root colonization. Microbiol-Uk 150: 3889–3897.
27. Rodriguez-Navarro DN, Dardanelli MS, Ruiz-Sainz JE (2007) Attachment of bacteria to the roots of higher plants. FEMS Microbiol Lett 272: 127–136.
28. Mavrodi DV, Joe A, Mavrodi OV, Hassan KA, Weller DM, et al. (2011) Structural and functional analysis of the type III secretion system from *Pseudomonas fluorescens* q8r1-96. J Bacteriol 193: 177–189.
29. Lottmann J, Heuer H, de Vries J, Mahn A, Düring K, et al. (2000) Establishment of introduced antagonistic bacteria in the rhizosphere of transgenic potatoes and their effect on the bacterial community. FEMS Microbiol Ecol 33: 41–49.
30. Chowdhury SP, Dietel K, Randler M, Schmid M, Junge H, et al. (2013) Effects of *Bacillus amyloliquefaciens* FZB42 on lettuce growth and health under pathogen pressure and its impact on the rhizosphere bacterial community. PLoS ONE 8: e68818–e68818. doi:10.1371/journal.pone.0068818.
31. Adesina MF, Lembke A, Costa R, Speksnijder A, Smalla K (2007) Screening of bacterial isolates from various European soils for in vitro antagonistic activity towards *Rhizoctonia solani* and *Fusarium oxysporum*: Site-dependent composition and diversity revealed. Soil Biol Biochem 39: 2818–2828.
32. Grosch R, Faltin F, Lottmann J, Kofoet A, Berg G (2005) Effectiveness of 3 antagonistic bacterial isolates to control *Rhizoctonia solani* Kühn on lettuce and potato. Can J Microbiol 51: 345–353.
33. Wibberg D, Jelonek L, Rupp O, Hennig M, Eikmeyer F, et al. (2013) Establishment and interpretation of the genome sequence of the phytopathogenic fungus *Rhizoctonia solani* AG1-IB isolate 7/3/14. J Biotechnol 167: 142–155.
34. Adesina MF, Grosch R, Lembke A, Vatchev TD, Smalla K (2009) In vitro antagonists of *Rhizoctonia solani* tested on lettuce: rhizosphere competence, biocontrol efficiency and rhizosphere microbial community response. FEMS Microbiol Ecol 69: 62–74.
35. Berg G, Krechel A, Ditz M, Sikora RA, Ulrich A, et al. (2005) Endophytic and ectophytic potato-associated bacterial communities differ in structure and antagonistic function against plant pathogenic fungi. FEMS Microbiol Ecol 51: 215–229.
36. Rühlmann J, Ruppel S (2005) Effects of organic amendments on soil carbon content and microbial biomass-results of the long-term box plot experiment in Grossbeeren. Arch Agron Soil Sci 51: 163–170.
37. Schreiter S, Ding G-C, Heuer H, Neumann G, Sandmann M, et al. (2014) Effect of the soil type on the microbiome in the rhizosphere of field-grown lettuce. Front Microbiol 5: 144.
38. Gutezeit B, Herzog FN, Wenkel KO (1993) Das Beregnungsbedarfssystem für Freilandgemüse. Gemüse 29: 106–108.
39. Grosch R, Schneider JHM, Kofoet A, Feller C (2011) Impact of continuous cropping of lettuce on the disease dynamics of bottom rot and genotypic diversity of *Rhizoctonia solani* AG 1-IB. J Phytopathol 159: 35–44.
40. Schneider JHM, Schilder MT, Dijst G (1997) Characterization of *Rhizoctonia solani* AG 2 isolates causing bare patch in field grown tulips in the Netherlands. Eur J Plant Pathol 103: 265–279.
41. Heuer H, Krsek M, Baker P, Smalla K, Wellington EMH (1997) Analysis of actinomycete communities by specific amplification of genes encoding 16S rRNA and gel-electrophoretic separation in denaturing gradients. Appl Environ Microbiol 63: 3233–3241.
42. Weinert N, Meincke R, Gottwald C, Heuer H, Gomes NCM, et al. (2009) Rhizosphere communities of genetically modified zeaxanthin-accumulating potato plants and their parent cultivar differ less than those of different potato cultivars. Appl Environ Microbiol 75: 3859–3865.
43. Kropf S, Heuer H, Grüning M, Smalla K (2004) Significance test for comparing complex microbial community fingerprints using pairwise similarity measures. J Microbiol Meth 57: 187–195.
44. Neumann G, Bott S, Ohler M, Mock H, Lippman R, et al. (2014) Root exudation and root development of lettuce (*Lactuca sativa* L. cv. Tizian) as affected by different soils. Front Microbiol 5: 2.
45. Paterson E, Gebbing T, Abel C, Sim A, Telfer G (2007) Rhizodeposition shapes rhizosphere microbial community structure in organic soil. New Phytol 173: 600–610.
46. Henry S, Texier S, Hallet S, Bru D, Dambreville C, et al. (2008) Disentangling the rhizosphere effect on nitrate reducers and denitrifiers: insight into the role of root exudates. Environ Microbiol 10: 3082–3092.
47. Benizri E, Nguyen C, Piutti S, Slezack-Deschaumes S, Philippot L (2007) Additions of maize root mucilage to soil changed the structure of the bacterial community. Soil Biol Biochem 39: 1230–1233.
48. Götz M, Gomes NCM, Dratwinski A, Costa R, Berg G, et al. (2006) Survival of *gfp*-tagged antagonistic bacteria in the rhizosphere of tomato plants and their effects on the indigenous bacterial community. FEMS Microbiol Ecol 56: 207–218.
49. Grosch R, Dealtry S, Schreiter S, Berg G, Mendonça-Hagler L, et al. (2012) Biocontrol of *Rhizoctonia solani*: complex interaction of biocontrol strains, pathogen and indigenous microbial community in the rhizosphere of lettuce shown by molecular methods. Plant Soil 361: 343–357.

50. Baudoin E, Benizri E, Guckert A (2002) Impact of growth stage on the bacterial community structure along maize roots, as determined by metabolic and genetic fingerprinting. Appl Soil Ecol 19: 135–145.
51. Bull CT, Weller DM, Thomashow LS (1991) Relationship between root colonization and suppression of *Gaeumannomyces graminis* var. tritici by *Pseudomonas fluorescens* strain 2–79. Phytopathology 81: 954–959.
52. Raaijmakers JM, Weller DM (2001) Exploiting genotypic diversity of 2,4-diacetylphloroglucinol-producing *Pseudomonas* spp.: Characterization of superior root-colonizing *P. fluorescens* strain Q8r1-96. Appl Environ Microbiol 67: 2545–2554.
53. Lemanceau P, Alabouvette C (1993) Suppression of fusarium-wilt by fluorescent pseudomonads: Mechanisms and applications. Biocontrol Sci Techn 3: 219–234.
54. Steinberg C, Edel-Hermann V, Alabouvette C, Lemanceau P (2007) Soil suppressiveness to plant diseases. Modern Soil Microbiology: 455–478.
55. Gomes NCM, Kosheleva IA, Abraham WR, Smalla K (2005) Effects of the inoculant strain *Pseudomonas putida* KT2442 (pNF142) and of naphthalene contamination on the soil bacterial community. FEMS Microbiol Ecol 54: 21–33.
56. Davis R, Subbarao K, Raid R, Kurtz E (1997) Compendium of Lettuce Diseases. APS Press: 15–16.
57. Heuer H, Wieland G, Schönfeld J, Schönwalder A, Gomes NCM, et al. (2001) Bacterial community profiling using DGGE or TGGE analysis; Rochelle PA, editor. 177–190 p.

20

Responses of Soil Microbial Communities to Experimental Warming in Alpine Grasslands on the Qinghai-Tibet Plateau

Bin Zhang[1]⁹, Shengyun Chen[2]⁹, Xingyuan He[1], Wenjie Liu[2], Qian Zhao[2], Lin Zhao[2,3], Chunjie Tian[1]*

1 Northeast Institute of Geography and Agroecology, Chinese Academy of Sciences, Changchun, China, **2** State Key Laboratory of Cryospheric Sciences, Cold and Arid Regions Environmental and Engineering Research Institute, Chinese Academy of Sciences, Lanzhou, China, **3** Observation and Research Station of Qinghai-Tibet Plateau, Cold and Arid Regions Environmental and Engineering Research Institute, Chinese Academy of Sciences, Golmud, China

Abstract

Global surface temperature is predicted to increase by at least 1.5°C by the end of this century. However, the response of soil microbial communities to global warming is still poorly understood, especially in high-elevation grasslands. We therefore conducted an experiment on three types of alpine grasslands on the Qinghai-Tibet Plateau to study the effect of experimental warming on abundance and composition of soil microbial communities at 0–10 and 10–20 cm depths. Plots were passively warmed for 3 years using open-top chambers and compared to adjacent control plots at ambient temperature. Soil microbial communities were assessed using phospholipid fatty acid (PLFA) analysis. We found that 3 years of experimental warming consistently and significantly increased microbial biomass at the 0–10 cm soil depth of alpine swamp meadow (ASM) and alpine steppe (AS) grasslands, and at both the 0–10 and 10–20 cm soil depths of alpine meadow (AM) grasslands, due primarily to the changes in soil temperature, moisture, and plant coverage. Soil microbial community composition was also significantly affected by warming at the 0–10 cm soil depth of ASM and AM and at the 10–20 cm soil depth of AM. Warming significantly decreased the ratio of fungi to bacteria and thus induced a community shift towards bacteria at the 0–10 cm soil depth of ASM and AM. While the ratio of arbuscular mycorrhizal fungi to saprotrophic fungi (AMF/SF) was significantly decreased by warming at the 0–10 cm soil depth of ASM, it was increased at the 0–10 cm soil depth of AM. These results indicate that warming had a strong influence on soil microbial communities in the studied high-elevation grasslands and that the effect was dependent on grassland type.

Editor: Liping Zhu, Institute of Tibetan Plateau Research, China

Funding: This work was financially supported by the National Basic Research Program of China (973 program, 2013CBA01807, http://www.973.gov.cn/Default_3.aspx), the Science Foundation of Chinese Academy of Sciences (KZZD-EW-TZ-16, http://www.cas.cn/), the National Natural Science Foundation of China (Y4E1041001,40901040, http://www.nsfc.gov.cn/) and the Natural Science Foundation of Jilin Province (20130101080JC, http://kjt.jl.gov.cn/kjt/4/tindex.shtml). The funders had no role in study design, data collection and analysis, decision to publish, or preparation of the manuscript.

Competing Interests: The authors have declared that no competing interests exist.

* Email: tiancj@iga.ac.cn

⁹ These authors contributed equally to this work.

Introduction

The global average surface temperature has increased by 0.85°C since 1880, due primarily to an anthropogenic increase in greenhouse gas concentrations, and is predicted to increase by at least 1.5°C by the end of the 21st century [1]. These temperature increases could have profound effects on biodiversity and ecosystem functioning [2,3], which subsequently influence feedback loops between ecosystem carbon (C) storage and climate warming [4]. While climate-modeling studies predict positive feedback from warming in the form of decreased C storage due to soil respiration [5–7], experimental warming studies on ecosystem C storage give controversial results [8,9]. Such controversy stems partially from uncertainties regarding the response of belowground microbial communities to climate warming as soil microorganisms play crucial roles in regulating ecosystem functioning and soil biogeochemistry [4,10,11]. Therefore, a better understanding of microbial responses to elevated temperature is critical to predicting the effects of future climate warming.

Experimental warming with a modest increase in soil temperature has been shown to influence the biomass and composition of soil microbial communities. While some warming studies showed a decrease in soil microbial biomass when exposed to elevated temperature [12–16], others reported an increase [4,9,17,18]. The picture was more complex with regards to the warming effect on the composition of soil microbial communities. For example, Zhang et al. [15] found that warming increased the presence of fungi in the soil of unclipped tallgrass prairie as evidenced by both microbial substrate utilization patterns and the profiles of phospholipid fatty acids (PLFAs). Deslippe et al. [12] noted a significant increase in the evenness of fungal communities in warmed Arctic tundra. However, it was also reported that warming decreased the relative abundance of fungi in a northeastern forest [13] and a subarctic heath ecosystem [14].

Furthermore, study by Frey et al. [13] indicated that the soil microbial community shifted towards gram-positive (G+) bacteria and actinomycetes after 12 years of warming. These changes suggest that the warming effects on biomass and composition of soil microbial communities are linked to a wider range of factors than temperature alone. In addition, warming studies are generally restricted to low-elevation areas, and less attention has been paid to the warming effect on soil microorganisms in high-elevation grasslands. Soils in high-elevation grasslands are exposed to harsher environmental conditions as a result of a colder climate and less favorable nutrient conditions [19,20], which may induce different responses by microbial communities to warming when compared with those in low-elevation areas.

In this study, we examined the impacts of experimental warming on abundance and composition of soil microbial communities in alpine grasslands on the Qinghai-Tibet Plateau. The Qinghai-Tibet Plateau is considered the "Third Pole" of the earth and has experienced a striking warming trend over the last half-century [21,22], providing a unique opportunity to explore the effects of warming in high-elevation grasslands. Three types of alpine grasslands (alpine swamp meadow, alpine meadow, and alpine steppe) were analyzed to determine whether the response of soil microbial communities to warming depended on grassland type. Soil microbial communities were characterized by PLFA analysis as it is a sensitive tool for measuring microbial biomass and fingerprinting microbial community composition [23,24].

Materials and Methods

Site Description and Experimental Setup

The study site is located near the Beiluhe Observation and Research Station (34°51′ N, 92°56′ E, Altitude: 4659 m), Cold and Arid Regions Environmental and Engineering Research Institute, Chinese Academy of Sciences. This region is typical of the continental climate of the plateau and is characterized by rarefied air, low air temperature, little but concentrated rainfall, low pressure, and strong ultraviolet radiation [25]. The mean annual air temperature is −3.8°C and the mean annual precipitation is 383 mm [26]. Precipitation falls mainly during the summer monsoon season and the freeze period lasts from September to April. The soils are classified as Calcic Kastanozem and Gelic Arenosols in the WRB soil classification system [27,28]. The three most typical alpine grasslands near this site are alpine swamp meadow (ASM), alpine meadow (AM), and alpine steppe (AS). ASM is populated by hardy perennial hygrophilous herbs, principally *Kobresia tibetica* and *Carex moorcroftii*, which thrive under the waterlogged or moist soil conditions. AM is characterized by cold meso-perennial herbs growing under soil conditions with moderate water availability and consists mainly of *K. pygmaea*, *K. robusta*, and *Androsace tapete*. AS is dominated by hardy perennial xeric herbs and dwarf shrubs such as *Stipa purperea*, *Carex moorcroftii*, and *Saussurea arenaria*.

An area of 30 m×30 m was fenced off in each type of alpine grassland to protect against grazing. Permission to conduct experiment here was issued by the Cold and Arid Regions Environmental and Engineering Research Institute, Chinese Academy of Sciences. In fall 2008, three replicated warming plots were set up within the fenced area using open-top chambers (OTCs) as a passive warming device, modeling the approach used in the International Tundra Experiment (ITEX) [29,30]. Parallel control plots were established 2 m apart from their adjacent warming plots. The 1 mm-thick fiberglass OTCs, each 40 cm tall, are shaped like truncated cones with inwardly inclined sides (60° with respect to the horizontal) and a 1.48 m bottom diameter (Sun-Lite HP, Solar Components Corporation, Manchester, USA). These OTCs remained on the warming plots throughout the experimental period.

Soil Sampling and Analysis

On October 11th, 2011, seven randomized soil cores (3.14 cm in diameter) were collected per plot from a depth of 0 to 20 cm using a hand auger. The soil cores were cut into two segments: 0–10 cm and 10–20 cm. Segments from the same plot were separated on the basis of depth, placed in plastic bags, and kept cool until processed in the laboratory. After removal of visible fresh roots and plant materials, the soils were homogenized, passed through a 2 mm sieve, and kept in a refrigerator at 4°C. One subsample was air-dried for analysis of soil properties and another was freeze-dried for PLFA extraction.

Organic carbon (C) and total nitrogen (N) in the soil were determined by dichromate oxidation using the Walkley-Black procedure [31] and the micro-Kjeldhal procedure [32], respectively. Soil pH was measured in a 1:5 soil/water suspension. Soil moisture was calculated from weight loss during oven drying at 105°C for 24 h. Soil temperature at the 0, 10, and 20 cm depths were recorded year-round on data-loggers (CR1000, Campbell Co., Ltd.) at half hour intervals by automatic thermal sensors.

Phospholipid Fatty Acid Analysis

PLFA extraction was conducted following the procedure described by Bossio and Scow [33]. Briefly, lipids were extracted in a single-phase chloroform-methanol- citrate buffer system. Phospholipids were separated from neutral lipids and glycolipids on solid phase extraction columns (Supelco, Inc., Bellefonte, PA). After methylation of the polar lipids, PLFA methyl esters were analyzed by an Agilent 6850N gas chromatograph (GC, Agilent Tech. Co., USA) equipped with an HP-5 capillary column (30 m×0.32 mm×0.25 μm) and a flame ionization detector (FID). The MIDI Sherlock Microbial Identification System (Microbial ID Inc., Newark, USA) was used to identify fatty acids. Nonadecanoic acid methyl ester (19:0, Sigma) was added as an internal standard and used to convert fatty acid peak areas to absolute abundance.

A total of 53 different PLFAs (C chain length of 12 to 20) were detected in this study. Thirty-six individual PLFAs, accounting for >98% of the total in mole percentages, were consistently present in all samples and therefore used for data analysis. The sum of all PLFAs indicates the total microbial biomass. We used the sum of i14:0, a15:0, i15:0, i16:0, a17:0, and i17:0 to determine Gm+ bacteria and the sum of 16:1 2OH, 16:1ω7c, 16:1ω9c, cy17:0, 17:1ω8c, 18:1ω7c, and cy19:0 to determine gram-negative bacteria (Gm-) [34,35]. The sum of 10Me16:0, 10Me17:0, and 10Me18:0 was used to ascertain total biomass of actinomycetes. We used 16:1ω5c to determine the amount of arbuscular mycorrhizal fungi (AMF), the sum of 18:1ω9c and 18:2ω6c to determine the amount of saprotrophic fungi (SF), 20:4ω6,9,12,15c to determine the amount of protozoa, and the sum of 16:1ω5c, 18:1ω9c, 18:2ω6c to determine the amount of fungi [36]. Ratios of (i17:0+i15:0)/(a17:0+a15:0) and cy17:0/16:1ω7c were used to ascertain nutritional or environmental stress [33,37].

Statistical Analysis

To explore variation in soil microbial community composition between treatments and among alpine grasslands, the mole percentages of individual PLFAs were subjected to principal component analysis (PCA) after standardizing to unit variance. PCA was carried out with the 'rda' function in the 'vegan' library using R statistical software (version 3.0.2) [38]. To assess the

Table 1. Effect of experimental warming on selected soil properties at the 0–10 and 10–20 cm depths of alpine swamp meadow, alpine meadow, and alpine steppe.

	Soil moisture (%)	Soil organic C (g kg^{-1})	Total nitrogen (g kg^{-1})	C/N	pH
Alpine swamp meadow					
0–10 cm					
Control	39.3 (0.51) b†	22.4 (4.27) a	1.61 (0.21) a	13.7 (0.81) a	8.77 (0.04) a
Warming	37.2 (0.60) a	29.1 (6.21) b	1.96 (0.28) b	14.6 (1.01) b	8.60 (0.15) a
10–20 cm					
Control	46.8 (0.94) a	22.6 (1.64) a	1.53 (0.07) a	14.7 (0.44) a	8.76 (0.02) a
Warming	44.4 (0.80) a	25.3 (2.46) a	1.61 (0.14) a	15.6 (0.24) b	8.80 (0.03) a
Alpine meadow					
0–10 cm					
Control	37.8 (0.59) b	16.5 (0.76) a	1.25 (0.06) a	13.2 (0.23) a	8.69 (0.08) a
Warming	35.8 (0.80) a	17.1 (1.88) a	1.23 (0.15) a	14.0 (0.20) a	8.73 (0.16) a
10–20 cm					
Control	43.7 (0.83) b	13.5 (1.06) a	1.16 (0.04) b	11.6 (0.49) a	8.80 (0.06) a
Warming	41.9 (0.65) a	11.1 (0.99) a	0.98 (0.04) a	11.3 (0.49) a	8.80 (0.08) a
Alpine steppe					
0–10 cm					
Control	7.08 (0.13) b	4.97 (0.20) a	0.39 (0.07) a	13.6 (2.00) a	8.58 (0.04) a
Warming	6.16 (0.14) a	4.77 (0.46) a	0.40 (0.05) a	12.3 (1.09) a	8.59 (0.10) a
10–20 cm					
Control	8.89 (0.36) b	3.95 (0.56) a	0.35 (0.04) a	11.3 (0.41) a	8.57 (0.05) a
Warming	7.91 (0.24) a	4.31 (0.31) a	0.27 (0.01) a	15.9 (1.80) b	8.59 (0.07) a

†Values in the parentheses are standard errors. Different small-case letters between control and warming indicate significant differences at the 0.05 probability level (two-tailed paired t-test).

compositional dissimilarity of microbial communities between control and experimental warming, the Euclidean distance was calculated. The significance of the difference in the Euclidean distances was determined by permutation tests (999 permutations). The loading scores for individual PLFAs were used to assess their relative importance. Two-tailed paired t-tests were performed by the R program to calculate the significant differences of various variables between each warming and control treatment. Figures were generated by Sigmaplot 10.0 (Systat Software Inc.).

To explore the PLFA data explained by a linear model of environmental variables, redundancy analysis (RDA) was carried out with the 'rda' function in the 'vegan' library of the R program. The ordination of the response variables (PLFA data) was constrained by multiple regression analysis of the explanatory variables (environmental data). The significance of the RDA results was assessed by permutation tests (999 permutations). Because $P<0.01$, we presented the ordination biplot, which shows site scores as points and environmental variables as vectors. The angles in the biplot between response and explanatory variables, and between explanatory variables themselves, reflect their correlations. The proportion of explained variation was calculated by using adjusted R-squared values as described by Peres-Neto et al. [39]. The biplot was generated by the R program.

Results

Environmental Variables

Plant coverage was significantly different ($P<0.05$) between the warming and control plots of AM (31.7% and 24.5%, respectively), but was only slightly higher in the warming than the control plots of ASM (46.2 and 44.0%, respectively) and AS (21.2 and 19.9%, respectively). Three years of experimental warming increased the annual mean soil temperature by 1.7, 0.5, and 0.4°C in ASM, by 2.3, 1.5, and 1.0°C in AM, and by 1.7, 1.4, and 0.3°C in AS at the 0, 10, and 20 cm soil depths, respectively. Soil moisture was significantly decreased by warming at the 0–10 cm depth of ASM and at the 0–10 and 10–20 cm depths of AM and AS ($P<0.05$, Table 1). Soil moisture was significantly higher at the 10–20 cm depth regardless of grassland type and significantly lower in AS than ASM and AM ($P<0.05$, Table 1).

Soil organic C and total N levels were highest in ASM and lowest in AS and were generally less in the lower soil depths (Table 1). Experimental warming increased the soil organic C and total N levels by 30% and 22%, respectively, at the 0–10 cm depth of ASM, but decreased total N by 16% at the 10–20 cm depth of AM ($P<0.05$, Table 1). Soil C to N (C/N) ratios were significantly increased by warming at the 0–10 and 10–20 cm depths of ASM and at the 10–20 cm depth of AS ($P<0.05$, Table 1). All soils were slightly alkaline with pH values ranging from 8.57 to 8.80.

Abundance of Soil Microbial Communities

In the ASM ecosystem, experimental warming significantly influenced the total microbial biomass (total PLFAs) at the 0–10 cm soil depth ($P<0.05$, Fig. 1). The total microbial biomass was significantly higher in the warming versus the control plots at the 0–10 cm depth, due primarily to a significant increase in the abundance of all investigated microbial groups except protozoa and AMF ($P<0.05$, Fig. 1). The ratios of fungi to bacteria (F/B),

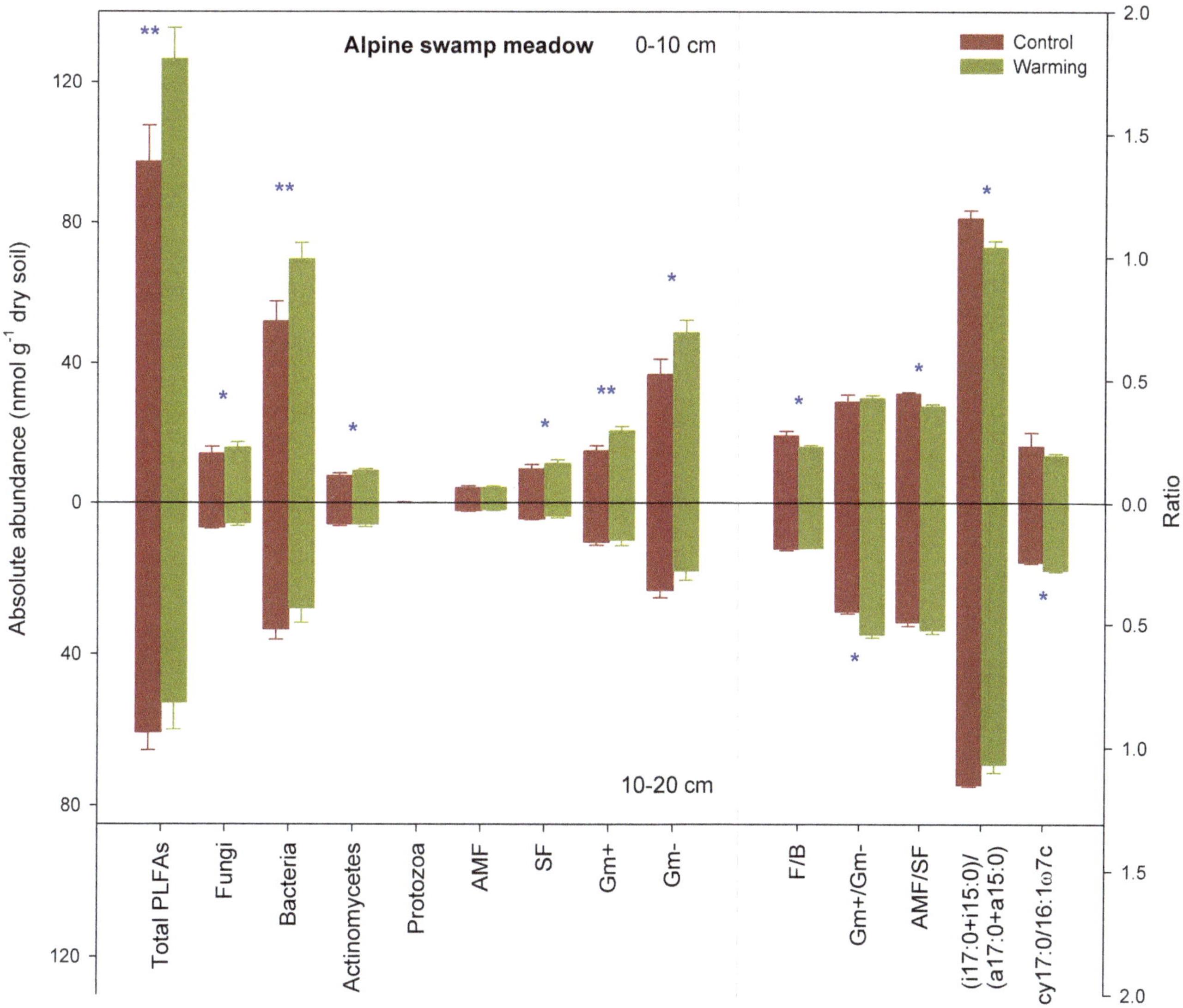

Figure 1. Sums and ratios of microbial groups in control and warming plots of alpine swamp meadow. Error bars indicate standard errors. Differences between the control and warming plots were analyzed by two-tailed paired t-tests, indicated by ** for P<0.01 and * for P<0.05. PLFA, phospholipid fatty acid; AMF, arbuscular mycorrhizal fungi; SF, saprotrophic fungi; Gm+, gram-positive bacteria; Gm-, gram-negative bacteria; F/B, the ratio of fungi to bacteria; Gm+/Gm-, the ratio of gram-positive to gram negative bacteria; AMF/SF, the ratio of arbuscular mycorrhizal fungi to saprotrophic fungi.

arbuscular mycorrhizal to saprotrophic fungi (AMF/SF), and (i17:0+i15:0)/(a17:0+a15:0) ratios were significantly decreased as a result of warming at the 0–10 cm depth (P<0.05, Fig. 1). Total microbial biomass and the abundance of specific microbial groups were significantly decreased at the lower soil depths (P<0.05, Fig. 1). Warming did not influence the total microbial biomass or abundance of specific microbial groups at the 10–20 cm depth, but significantly increased the Gm+/Gm- and cy17:0/16:1ω9c ratios at this layer (Fig. 1).

The total microbial biomass and abundance of specific microbial groups in AM were significantly lower than those in ASM (P<0.05, Figs. 1 and 2). Experimental warming significantly increased the total microbial biomass and abundance of specific microbial groups except protozoa at the 0–10 cm soil depth of AM (P<0.05, Fig. 2). In addition, although F/B and (i17:0+i15:0)/(a17:0+a15:0) ratios were significantly lower, AMF/SF ratios were significantly higher in the warming versus control plots at this layer (P<0.05, Fig. 2). At the 10–20 cm depth of AM, the increase in total microbial biomass as a result of warming mainly stemmed from an increased abundance of fungi and Gm+ and Gm- bacteria (P<0.05, Fig. 2). In addition, experimental warming significantly decreased the Gm+/Gm- and cy17:0/16:1ω9c ratios at this depth (P<0.05, Fig. 2).

Compared to the soils of ASM and AM, those of AS contained the lowest total microbial biomass and abundance of specific microbial groups (Figs. 1, 2, and 3). Experimental warming of AS significantly increased bacteria and protozoa at the 0–10 cm soil depth but decreased the Gm+ bacteria at the 10–20 cm depth (P< 0.05, Fig. 3). Furthermore, these warming plots had significantly higher cy17:0/16:1ω9c and (i17:0+i15:0)/(a17:0+a15:0) ratios but significantly lower Gm+/Gm- ratios at the 10–20 cm depth compared to the control plots (P<0.05, Fig. 3).

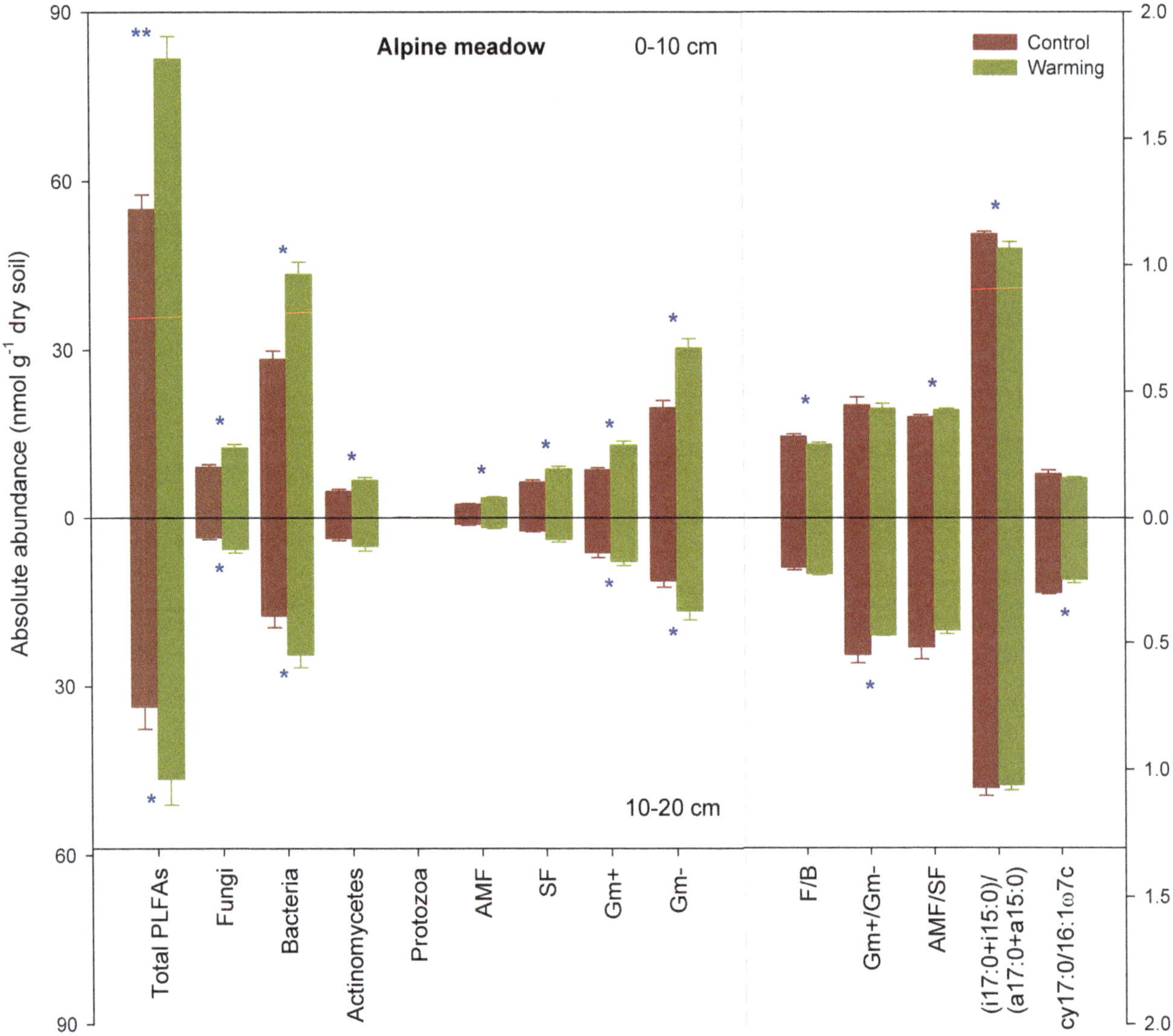

Figure 2. Sums and ratios of various microbial groups in control and warming plots of alpine meadow. Error bars indicate standard errors. Differences between the control and warming plots were analyzed by two-tailed paired t-tests, indicated by ** for P<0.01 and * for P<0.05. PLFA, phospholipid fatty acid; AMF, arbuscular mycorrhizal fungi; SF, saprotrophic fungi; Gm+, gram-positive bacteria; Gm-, gram-negative bacteria; F/B, the ratio of fungi to bacteria; Gm+/Gm-, the ratio of gram-positive to gram negative bacteria; AMF/SF, the ratio of arbuscular mycorrhizal to saprotrophic fungi.

Composition of Soil Microbial Communities

At the 0–10 cm soil depth, the first principal component (PC1) and the second principal component (PC2) explained 39.4% and 21.7% of the total variance in the PLFA data, respectively (Fig. 4A). The ASM plots with the highest soil organic C levels are shown on the left-hand side of Figure 4A, the AM plots with an intermediate organic C levels in the middle portion of Figure 4A, and the AS plots with the lowest organic C levels are on the right-hand side of Figure 4A. Along the PC2 axis, the warming plots of ASM and AM scored higher than their corresponding control plots (Fig. 4A). The PCA plot also showed that data points for AS were intermixed. The Euclidean distance of soil microbial community composition between control and warming was significant in the ASM and AM ecosystems at this layer ($P < 0.05$, Fig. 4B). The loading scores for individual PLFAs revealed 10Me17:0, 17:1ω8c, 18:1ω9c, and cy19:0 were most important for the separation of sites along the PC1 (Table S1).

At the 10–20 cm soil depth, the PC1 and PC2 explained, respectively, 47.6 and 11.6% of the total variance in the PLFA data (Fig. 4C). The data points for ASM and AM were on the left-hand side and those for AS on the right-hand side (Fig. 4C). Permutation tests suggested significant differences in soil microbial community composition between control and warming plots in the AM ecosystem at this layer ($P<0.05$, Fig. 4D). Lipid signatures 16:1ω5c, 17:1ω8c, and 18:1ω9c had higher positive loading scores while cy19:0 and 20:0 had lower negative loading scores along the PC1 axis (Table S1).

Relationship between PLFA and Environmental Data

Redundancy analysis showed that the first and second canonical axes explained 38.5 and 17.0%, respectively, of the total variance in the PLFA data at the 0–10 cm depth, and 44.8 and 8.3%, respectively, of the total variance at the 10–20 cm depth (Fig. 5). Because ecological data are generally quite noisy, we are confident

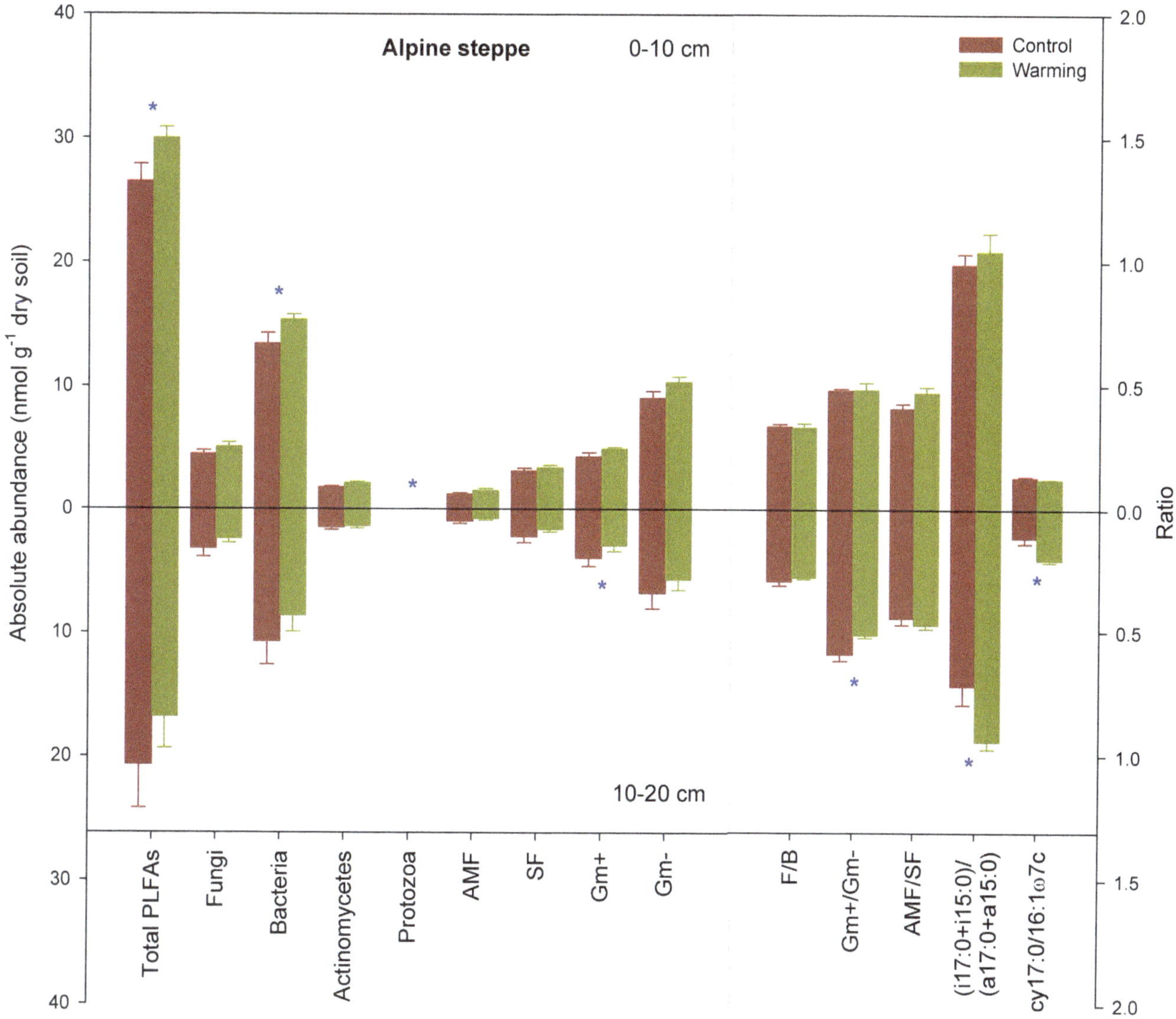

Figure 3. Sums and ratios of various microbial groups in control and warming plots of alpine steppe. Error bars indicate standard errors. Differences between the control and warming plots were analyzed by two-tailed paired t-tests, indicated by ** for P<0.01 and * for P<0.05. PLFA, phospholipid fatty acid; AMF, arbuscular mycorrhizal fungi; SF, saprotrophic fungi; Gm+, gram-positive bacteria; Gm-, gram-negative bacteria; F/B, the ratio of fungi to bacteria; Gm+/Gm-, the ratio of gram-positive to gram negative bacteria; AMF/SF, the ratio of arbuscular mycorrhizal fungi to saprotrophic fungi.

that the major trends have been modeled in this analysis. Furthermore, the first unconstrained eigenvalues are comparatively small, which means that they do not display any important residual structure of the PLFA data. Soil organic C, total N, soil moisture, and plant coverage had a significant effect on soil microbial communities at the 0–10 cm depth ($P<0.05$, Fig. 5A). For the 10–20 cm depth, organic C, total N, soil moisture, pH, and soil temperature were important environmental factors ($P<0.05$, Fig. 5B).

Discussion

Warming Effect on the Abundance of Soil Microbial Communities

Our results clearly show that 3 years of experimental warming consistently and significantly increased the abundance of microbial communities at the 0–10 cm soil depth of all studied alpine grasslands. These results align well with studies showing significant increases in microbial biomass in a range of Antarctic environments after 3 years of passive warming [9], but differ from the common finding of most warming studies that microbial biomass either remained at steady levels or decreased depending on the duration of warming [12–15,40,41]. These disparities most likely stem from the prevailing cold climate conditions in our study and that of Yergeau et al. [9]. Soil microorganisms in cold environments have optimal growing temperatures far above field conditions [19,42], which potentially enable them to increase their biomasses rapidly under elevated temperatures [43]. Therefore, temperature had a direct effect on microbial biomass. However, it seems that the direct effect of warming on microbial biomass accumulation was significant only when the soil temperature increase reaches a certain value. When we averaged the soil temperatures at 0 and 10 cm for the 0–10 cm soil depth and at 10 and 20 cm for the 10–20 cm soil depth, we obtained increases in

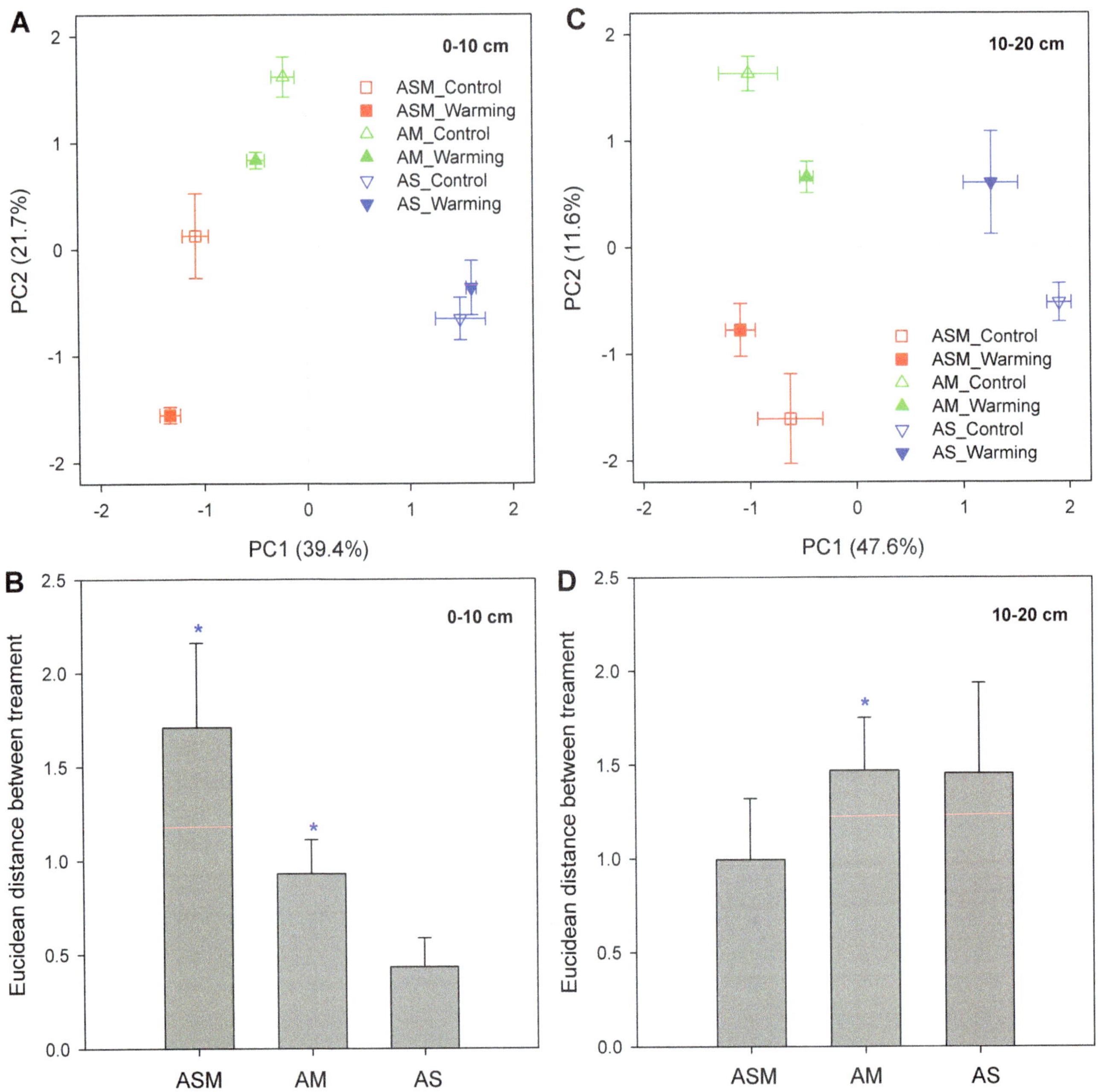

Figure 4. Principal component analysis (PCA) of the phospholipid fatty acid data in alpine grasslands. The composition of soil microbial communities (A and C) and the composition dissimilarity between the control and warming treatments (B and D) at the 0–10 and 10–20 cm soil depths of alpine swamp meadow (ASM), alpine meadow (AM), and alpine steppe (AS). Error bars indicate standard errors. Dissimilarity was assessed by calculating the Euclidean distances between the PC values. Star symbols indicate significant differences in the mean values of dissimilarity at the 0.05 probability level based on permutation tests (999 permutations).

soil temperature of 1.10, 1.90, and 1.55°C at the 0–10 cm soil depth and 0.45, 1.25, and 0.85°C at the 10–20 cm soil depth of ASM, AM, and AS, respectively. Given that there were generally no significant differences in microbial biomass between control and warming plots at the 10–20 cm soil depth of ASM and AS, these averaged temperature values suggest that microbial biomass did not accumulate when the temperature increased ≤0.85°C. Further research is needed to elucidate this point.

In addition to a direct temperature effect, warming may indirectly affect soil microbial abundance by influencing aboveground biomass [40]. In this study, warming increased total plant coverage by 4.9, 29.3, and 6.1% in ASM, AM, and AS, respectively. Increased aboveground biomass may introduce more organic C into soils (through plant litters and root exudates), which may have a strong influence on substrate availability and consequently on the growth and activity of soil microbial communities [44]. The warming effect on plant coverage was so profound in the AM grasslands that the increased organic C input may have provided an ample supply of substrates even at the 10–20 cm soil depth, and thus partially contributed to the significant increase in soil microbial biomass at this layer. Higher microbial biomass is usually accompanied by higher heterotrophic respira-

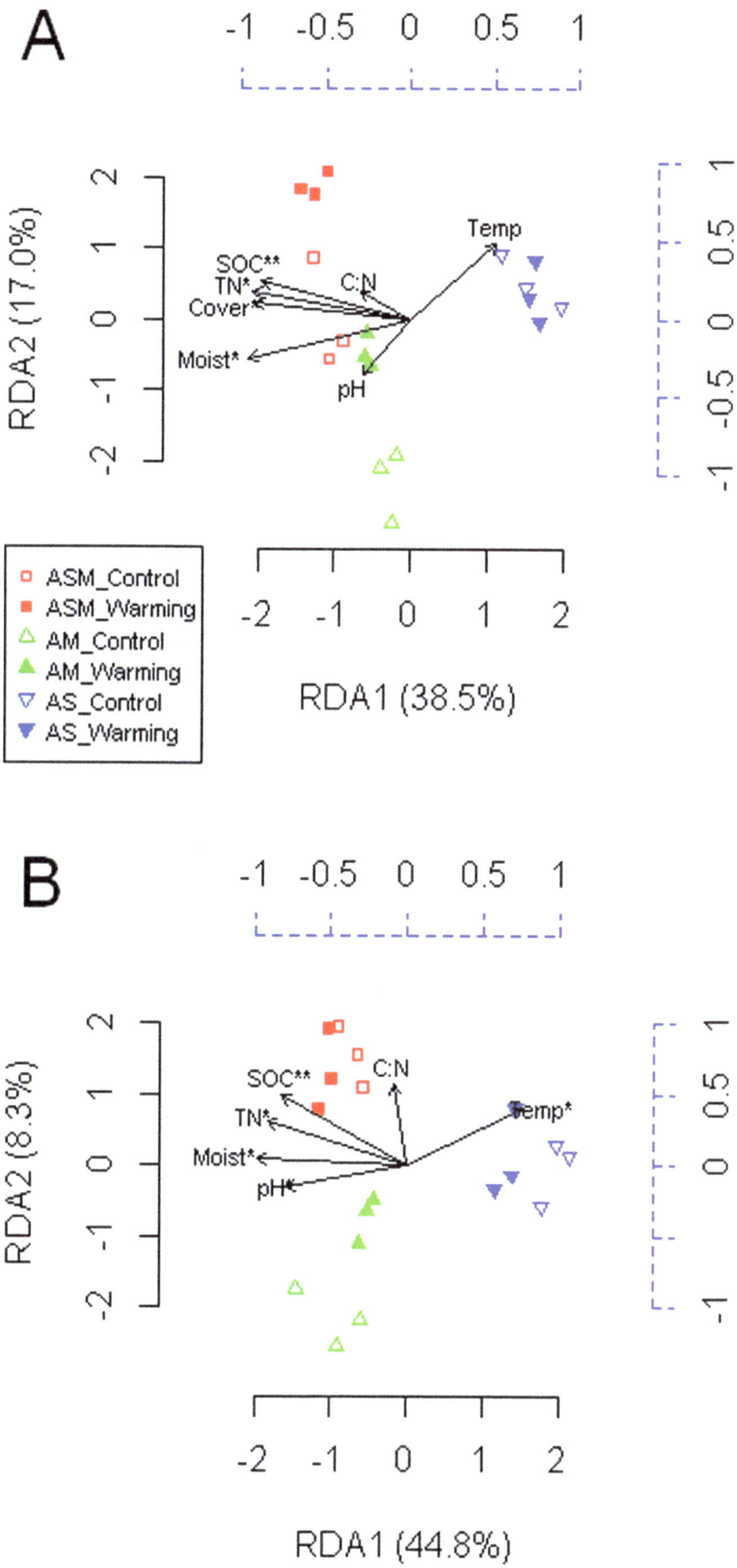
A
-1 -0.5 0 0.5 1
RDA2 (17.0%)
Temp
SOC**
TN*
C:N
Cover*
Moist*
pH
ASM_Control
ASM_Warming
AM_Control
AM_Warming
AS_Control
AS_Warming
-2 -1 0 1 2
RDA1 (38.5%)
B
-1 -0.5 0 0.5 1
RDA2 (8.3%)
C:N
SOC**
Temp*
TN*
Moist*
pH*
-2 -1 0 1 2
RDA1 (44.8%)

Figure 5. Redundancy analysis (RDA) of the phospholipid fatty acid (PLFA) data as explained by environmental variables. Seven and six environmental variables were used for the 0–10 (A) and 10–20 cm (B) soil depths, respectively. The PLFA data are scaled by the black solid axes (bottom and left), and the explanatory variables are scaled by the blue dashed axes (top and right). The proportion of explained variation was calculated by using adjusted R-squared values as described by Peres-Neto et al. (2006). The explanatory variables followed by an asterisk indicate significant influences on the PLFA data (*, $P<0.05$; **, $P<0.01$). ASM, alpine swamp meadow; AM, alpine meadow; AS, alpine steppe; SOC, soil organic C; TN, total nitrogen; Moist, soil moisture; Temp, soil temperature; Cover, plant coverage.

tion [10]. It seems that the increase in C loss through soil respiration was offset by the increase in C inputs from aboveground biomass at the 0–10 cm depth of AM and AS and at the 10–20 cm depth of AM. As a result, total organic C remained unchanged at these soil layers. However, warming significantly increased total organic C at the 0–10 cm depth of ASM, suggesting a negative feedback between soil microbial communities and C storage under experimental warming. These data highlight the importance of conserving the ASM ecosystem as a C sink to mitigate climate change in the studied region.

Soil moisture has considerable influence over soil microorganisms [45]. In this study, experimental warming significantly decreased soil moisture at the 0–10 cm depth of ASM and AM and at the 10–20 cm depth of AM. Because soil moisture is relatively high in ASM and AM ecosystems, a decrease in soil moisture most likely indicates reduced physiological stress, which might partially contribute to the observed increase in soil microbial biomass at these layers [18,46]. This is supported by significantly lower stress indicators at the 0–10 cm depth of warmed plots in ASM and AM ((i17:0+i15:0)/(a17:0+a15:0)) and at the 10–20 cm depth of warmed plots in AM (cy17:0/16:1ω9c).

Warming Effect on Composition of Soil Microbial Communities

Warming has been reported to markedly shift the composition of soil microbial communities across a wide range of ecosystems [4,9,15,16]. The results from our study also showed that the microbial composition had changed significantly following 3 years of experimental warming at the 0–10 cm depth of ASM and AM and at the 10–20 cm depth of AM, as indicated by PCA-based ordination of the PLFA data. These data suggest that the warming effects on soil microbial composition were dependent on grassland type and soil depth. RDA revealed that microbial composition at the 0–10 cm soil depth was primarily controlled by plant coverage, soil organic C, total N, and soil moisture, suggesting that the warming effect on soil microbial communities at this layer was indirect and mainly influenced by vegetation inputs and the microbial habitat of our study area. Alternatively, temperature effects may not be explained by a simple linear relationship. RDA also indicated that the composition of the soil microbial community at the 10–20 cm depth was heavily shaped by environmental factors.

Our results indicated a community shift towards bacteria at the surface layer of the warming plots, which is consistent with a significant reduction in the relative abundance of fungi after warming in a subarctic heath ecosystem [14] and within forest soil [13]. Competition for substrates between microorganisms likely explains this observation. Increased substrate availability and decreased environmental stress under warming likely give bacteria a competitive advantage over fungi with regards to available C and other nutrients [47]. Consequently, bacteria would grow faster than fungi, leading to the observed bacterial dominance in the soil microbial community at the 0–10 cm depth of ASM and AM ecosystems.

Another composition change in the warmed soil microbial communities was the relative abundance of AMF and SF at the 0–10 cm depth of ASM and AM. Saprotrophic fungi contribute to decomposition mostly through enzymatic action, promoting cell lysis and degrading cell constituents [48]. Higher microbial biomass is generally accompanied by higher production of microbial necromass [49], which most likely provides substrates for SF and thus contributed to greater SF abundance in warming plots of ASM and AM. However, AMF was not increased by warming in ASM due to their weak saprotrophic capacity [43], which led to significantly lower AMF/SF ratios in warming versus control plots. The significantly higher AMF/SF ratio in warming plots of AM suggests a greater increase in the abundance of AMF. Increased AMF most likely resulted from their association with plant roots, which increased due to significantly higher plant coverage [50,51]. Thus, plant coverage plays an important role in influencing the relative abundance of AMF and SF at the 0–10 cm soil depth of ASM and AM.

Conclusions

Results from our study indicate that 3 years of experimental warming leads to significant increases in microbial biomass of alpine grasslands, due primarily to changes in soil temperature, moisture, and plant coverage. However, soil organic C content did not decrease as a result of increased microbial biomass in alpine meadow and alpine steppe, suggesting that increased C loss through soil respiration was offset by an increase in C inputs from plants. The significantly higher soil organic C in warming plots of alpine swamp meadow highlights the importance of conserving this type of ecosystem on the Qinghai-Tibet Plateau. Warming also significantly decreased the fungal to bacterial ratio and induced changes in the relative abundance of arbuscular mycorrhizal and saprotrophic fungi at the 0–10 cm depth of ASM and AM. These changes mainly resulted from variations in the aboveground plant coverage and soil environmental conditions. Conversely, soil microbial community composition at the 10–20 cm depth was primarily influenced by environmental variables. Our results suggest that in response to climate warming, microorganisms within the soil may have a profound impact over long-term ecosystem feedback loops.

Supporting Information

Table S1 Top 5 phospholipid fatty acids (PLFAs) most responsible for the changes in lipid signatures along the first principal component (PC1).

Acknowledgments

We thank the Beiluhe Observation and Research Station on Frozen Soil Engineering and Environment in Qinghai-Tibet Plateau and the Cold and Arid Regions Environmental and Engineering Research Institute, Chinese Academy of Sciences for their help.

Author Contributions

Conceived and designed the experiments: SC. Performed the experiments: BZ WL QZ. Analyzed the data: BZ QZ. Contributed reagents/materials/analysis tools: SC XH LZ CT. Contributed to the writing of the manuscript: BZ SC.

References

1. C (2013) Climate Change 2013: The Physical Science Basis. Contribution of Working Group I to the Fifth Assessment Report of the Intergovernmental Panel on Climate Change. Cambridge, United Kingdom and New York: Cambridge University Press. 1535 p.
2. of JS, Alsterberg C, Havenhand JN, Sundback K, Wood HL, et al. (2012) Experimental climate change weakens the insurance effect of biodiversity. Ecol Lett 15: 864–872.
3. mm NB, Chapin III FS, Bierwagen B, Gonzalez P, Groffman PM, et al. (2013) The impacts of climate change on ecosystem structure and function. Front Ecol Environ 11: 474–482.
4. Zhou JZ, Xue K, Xie JP, Deng Y, Wu LY, et al. (2012) Microbial mediation of carbon-cycling feedbacks to climate warming. Nature Clim Change 2: 106–110.
5. Cox PM, Betts RA, Jones CD, Spall SA, Totterdell IJ (2000) Acceleration of global warming due to carbon-cycle feedbacks in a coupled climate model. Nature 408: 184–187.
6. Friedlingstein P, Cox P, Betts R, Bopp L, von Bloh W, et al. (2006) Climate-carbon cycle feedback analysis: Results from the C[4]MIP model intercomparison. J Climate 19: 3337–3353.
7. Heimann M, Reichstein M (2008) Terrestrial ecosystem carbon dynamics and climate feedbacks. Nature 451: 289–292.
8. Luo YQ (2007) Terrestrial carbon-cycle feedback to climate warming. Annu Rev Ecol Evol Syst 38: 683–712.
9. Yergeau E, Bokhorst S, Kang S, Zhou J, Greer CW, et al. (2012) Shifts in soil microorganisms in response to warming are consistent across a range of Antarctic environments. ISME J 6: 692–702.
10. Bardgett RD, Freeman C, Ostle NJ (2008) Microbial contributions to climate change through carbon cycle feedbacks. ISME J 2: 805–814.
11. Jackson RB, Fierer N, Schimel JP (2007) New directions in microbial ecology. Ecology 88: 1343–1344.
12. Deslippe JR, Hartmann M, Simard SW, Mohn WW (2012) Long-term warming alters the composition of Arctic soil microbial communities. FEMS Microbiol Ecol 82: 303–315.
13. Frey SD, Drijber R, Smith H, Melillo J (2008) Microbial biomass, functional capacity, and community structure after 12 years of soil warming. Soil Biol Biochem 40: 2904–2907.
14. Rinnan R, Michelsen A, Bååth E, Jonasson S (2007) Fifteen years of climate change manipulations alter soil microbial communities in a subarctic heath ecosystem. Glob Change Biol 13: 28–39.
15. Zhang W, Parker KM, Luo Y, Wan S, Wallace LL, et al. (2005) Soil microbial response to experimental warming and clipping in a tallgrass prairie. Glob Change Biol 11: 266–277.
16. Zogg GP, Zak DR, Ringelberg DB, MacDonald NW, Pregitzer KS, et al. (1997) Compositional and functional shifts in microbial communities due to soil warming. Soil Sci Soc Am J 61: 475–481.
17. Haugwitz MS, Bergmark L, Priemé A, Christensen S, Beier C, et al. (2014) Soil microorganisms respond to five years of climate change manipulations and elevated atmospheric CO_2 in a temperate heath ecosystem. Plant Soil 374: 211–222.
18. Sheik CS, Beasley WH, Elshahed MS, Zhou X, Luo Y, et al. (2011) Effect of warming and drought on grassland microbial communities. ISME J 5: 1692–1700.
19. Margesin R, Jud M, Tscherko D, Schinner F (2009) Microbial communities and activities in alpine and subalpine soils. FEMS Microbiol Ecol 67: 208–218.
20. Luo C, Xu G, Chao Z, Wang S, Lin X, et al. (2010) Effect of warming and grazing on litter mass loss and temperature sensitivity of litter and dung mass loss on the Tibetan plateau. Glob Change Biol 16: 1606–1617.
21. Guo D, Wang H (2011) The significant climate warming in the northern Tibetan Plateau and its possible causes. Int J Climatol 31: 1257–1413.
22. Liu X, Chen B (2000) Climatic warming in the Tibetan Plateau during recent decades. Int J Climatol 20: 1729–1742.
23. Frostegård A, Tunlid A, Bååth E (2011) Use and misuse of PLFA measurements in soils. Soil Biol Biochem 43: 1621–1625.
24. Wixon DL, Balser TC (2013) Toward conceptual clarity: PLFA in warmed soils. Soil Biol Biochem 57: 769–774.
25. Chen SY, Liu WJ, Sun ZZ, Zhao L, Liu YZ (2012) A quantitative study on the plant population phenology of alpine meadow in the permafrost regions of Qinghai-Tibetan plateau. J glaciol geocryol 34: 1142–1148.
26. Wen Z, Niu F, Yu Q, Wang D, Feng W, et al. (2014) The role of rainfall in the thermal-moisture dynamics of the active layer at Beiluhe of Qinghai-Tibetan plateau. Environl Earth Sci 71: 1195–1204.
27. Liu WJ, Chen SY, Zhao Q, Sun ZZ, Zhao L, et al. (2014) Vertical, seasonal variations and control of soil organic carbon and other nutrients in permafrost regions of the central Qinghai-Tibetan plateau. Environ Res Lett (under review).
28. IUSS Working Group WRB (2006) World reference base for soil resources 2006. World Soil Resources Reports No. 103. Rome: FAO.
29. Marion GM, Henry GHR, Freckman DW, Johnstone J, Jones G, et al. (1997) Open-top designs for manipulating field temperature in high-latitude ecosystems. Glob Change Biol 3: 20–32.
30. Klein J, Harte J, Zhao XQ (2005) Dynamic and complex microclimate responses to warming and grazing manipulations. Glob Change Biol 11: 1440–1451.
31. Nelson DW, Sommers LE (1982) Total carbon, organic carbon, and organic matter. In: Page AL, editor.Methods of soil analysis part II.Madison: American Society of Agronomy. pp. 539–579.
32. Institute of Soil Sciences, Chinese Academy of Sciences (1978) Physical and chemical analysis methods of soils. Shanghai: Shanghai Science Technology Press.pp. 7–59.
33. Bossio DA, Scow KM, Gunapala N, Graham KJ (1998) Determinants of soil microbial communities: effects of agricultural management, season, and soil type on phospholipid fatty acid profiles. Microb Ecol 36: 1–12.
34. Liang C, Jesus ED, Duncan DS, Jackson RD, Tiedje JM, et al. (2012) Soil microbial communities under model biofuel cropping systems in southern Wisconsin, USA: Impact of crop species and soil properties. Appl Soil Ecol 54: 24–31.
35. Zhang B, He H, Ding X, Zhang X, Zhang X, et al. (2012) Soil microbial community dynamics over a maize (*Zea mays* L.) growing season under conventional- and no-tillage practices in a rainfed agroecosystem. Soil Till Res 124: 153–160.
36. Vestal JR, White DC (1989) Lipid analysis in microbial ecology: quantitative approaches to the study of microbial community. BioScience 39: 535–541.
37. Kieft TL, Ringelberg DB, White DC (1994) Changes in ester-linked phospholipid fatty acid profiles of subsurface bacteria during starvation and desiccation in a porous medium. Appl Environ Microbiol 60: 3292–3299.
38. R Core Team (2013) R: A language and environment for statistical computing. Vienna: R Foundation for Statistical Computing. Available: http://www.R-project.org/.
39. Peres-Neto PR, Legendre P, Dray S, Borcard D (2006) Variation partitioning of species data matrices: estimation and comparison of fractions. Ecology 87: 2614–2625.
40. Zhang N, Liu W, Yang H, Yu X, Gutknecht JLM, et al. (2013) Soil microbial responses to warming and increased precipitation and their implications for ecosystem C cycling. Oecologia 173: 1125–1142.
41. Zhang X, Zhang G, Chen Q, Han X (2013) Soil bacterial communities respond to climate changes in a temperate steppe. PLoS ONE 8: e78616.
42. Yuan Y, Si G, Wang J, Luo T, Zhang G (2014) Bacterial community in alpine grasslands along an altitudinal gradient on the Tibetan Plateau. FEMS Microbiol Ecol 87: 121–132.
43. Rousk J, Bååth E (2011) Growth of saprotrophic fungi and bacteria in soil. FEMS Microbiol Ecol 78: 17–30.
44. Berg N, Steinberge Y (2008) Role of perennial plants in determining the activity of the microbial community in the Negev Desert ecosystem. Soil Biol Biochem 40: 2686–2695.
45. Brockett BFT, Prescott CE, Grayston SJ (2012) Soil moisture is the major factor influencing microbial community structure and enzyme activities across seven biogeoclimatic zones in western Canada. Soil Biol Biochem 44: 9–20.
46. Liu W, Zhang Z, Wan S (2009) Predominant role of water in regulating soil and microbial respiration and their responses to climate change in a semiarid grassland. Glob Change Biol 15: 184–195.
47. Mille-Lindblom C, Fischer H, Tranvik LJ (2006) Antagonism between bacteria and fungi: substrate competition and a possible tradeoff between fungal growth and tolerance towards bacteria. Oikos 113: 233–242.
48. Crowther TW, Boddy L, Jones TH (2012) Functional and ecological consequences of saprotrophic fungus-grazer interactions. ISME J 6: 1992–2001.
49. Zhang B, Liang C, He H, Zhang X (2013) Variations in soil microbial communities and residues along an altitude gradient on the northern slope of Changbai Mountain, China. PLoS ONE 8(6): e66184.
50. Jackson RB, Canadell J, Ehleringer JR, Mooney HA, Sala OE, et al. (1996) A global analysis of root distributions for terrestrial biomes. Oecologia 108: 389–411.
51. Posada RH, Madriñan S, Rivra EL (2012) Relationships between the litter colonization by saprotrophic and arbuscular mycorrhizal fungi with depth in a tropical forest. Fungal Biol 116: 747–755.

Spatial Variation in Carbon and Nitrogen in Cultivated Soils in Henan Province, China: Potential Effect on Crop Yield

Xuelin Zhang[1]*, Qun Wang[1], Frank S. Gilliam[2], Yilun Wang[1], Feina Cha[3], Chaohai Li[1]

1 The Incubation Base of the National Key Laboratory for Physiological Ecology and Genetic Improvement of Food Crops in Henan Province, Zhengzhou, China; Agronomy College of Henan Agricultural University, Zhengzhou, China, **2** Department of Biological Sciences, Marshall University, Huntington, West Virginia, United States of America, **3** Meteorological Bureau of Zhengzhou, Zhengzhou, China

Abstract

Improved management of soil carbon (C) and nitrogen (N) storage in agro-ecosystems represents an important strategy for ensuring food security and sustainable agricultural development in China. Accurate estimates of the distribution of soil C and N stores and their relationship to crop yield are crucial to developing appropriate cropland management policies. The current study examined the spatial variation of soil organic C (SOC), total soil N (TSN), and associated variables in the surface layer (0–40 cm) of soils from intensive agricultural systems in 19 counties within Henan Province, China, and compared these patterns with crop yield. Mean soil C and N concentrations were 14.9 g kg^{-1} and 1.37 g kg^{-1}, respectively, whereas soil C and N stores were 4.1 kg m^{-2} and 0.4 kg m^{-2}, respectively. Total crop production of each county was significantly, positively related to SOC, TSN, soil C and N store, and soil C and N stock. Soil C and N were positively correlated with soil bulk density but negatively correlated with soil porosity. These results indicate that variations in soil C could regulate crop yield in intensive agricultural systems, and that spatial patterns of C and N levels in soils may be regulated by both climatic factors and agro-ecosystem management. When developing suitable management programs, the importance of soil C and N stores and their effects on crop yield should be considered.

Editor: Dafeng Hui, Tennessee State University, United States of America

Funding: This study was supported by grants from Henan Science and Technology Department of China under the Key Research Project (30200051). The funder had no role in study design, data collection and analysis, decision to publish, or preparation of the manuscript.

Competing Interests: The authors have declared that no competing interests exist.

* Email: xuelinzhang1998@163.com

Introduction

Safeguarding food security and ensuring sustainable development are two fundamental goals of intensive agriculture in China [1,2]. Increasing soil C and N sequestration while reducing C and N emissions from agricultural fields are important aspects of sustainable farming and these goals can be achieved through improvement in soil quality [1,3]. This requires a better understanding of the functional relationship between crop yield and soil organic C and N stores.

Indeed, variations in soil C and N stores may closely regulate crop yield, although published data on the relationship between these parameters are inconsistent. Some studies have reported a positive correlation between soil C and N and crop yield [4,5], whereas other studies have found no significant relationship between these parameters [6,7]. Lal (2006) reported that the relationship between soil organic C and crop yield may vary between patterns that are sigmoidal, linear, or exponential [8]. Clearly, the existence of such variability warrants further investigation.

Soil C and N stores in crop lands, especially in the topsoil layer, are potentially greatly affected by human activity; thus, understanding the spatial pattern of soil C and N stores on a regional scale is crucial to developing a management strategy for improving soil fertility [1,2]. Spatial variation in soil C and N stores in agro-ecosystems has been widely reported [9,10,11], including from the northern [12,13], eastern [14], and southern [15,16] regions of China. Since these reports from China were based on two national surveys from 1960 and 1983, such data may have limited use in helping to develop management strategies based on current practices [17]. Therefore, in order to better understand the spatial patterns and their relationship to crop yield, it is necessary to update regional soil organic C and N information with contemporary measurements, especially for intensively-used crop land.

Henan Province is the second largest area of crop production in China (China National Bureau of Statistics). To produce an adequate supply of food for the domestic population, unsustainable production methods have often been used in this province. Historically, intensive production based on an annual wheat-maize system has been used to achieve high crop yield. This practice, however, has resulted in badly degraded agricultural soils, causing erosion and a loss of good soil structure. More than 600 kg N ha^{-1} annually has been applied in this production area, resulting in an increase in soil acidity [18]. Based on the determination that crop yields in China will need to increase from 50 billion in 2010 to 65 billion kg in 2020, the provincial crop lands in Henan Province will continue to play an important role in food production. Such goals create the challenge of improving soil quality, enhancing soil fertility, and mitigating C and N loss, while achieving food security and practicing sustainable agriculture. A better understanding of the spatial variability of soil organic C and N, and their relationship to crop yield, should help to develop management practices that are designed to meet this challenge [1,19].

The objective of the present study was to characterize the spatial distribution of C and N stores in intensively cultivated counties within the Henan Province of China and to determine the relationship between crop yield and soil organic C and N.

Materials and Methods

Statement: We have field permits for sampling soil in each of the field sites within each county of Henan Province, China. All of the sampling sites are privately owned, and there was no potential impact on any endangered or protected species among these sampling sites.

Study site

The study was carried out in 19 counties within Henan Province, located in central China (Figure 1). Map data were obtained from the National Geomatics Center of China (http://ngcc.sbsm.gov.cn/) using ArcGIS software. As of 2009, the human population of Henan was about 9.9×10^7 persons. The Province is

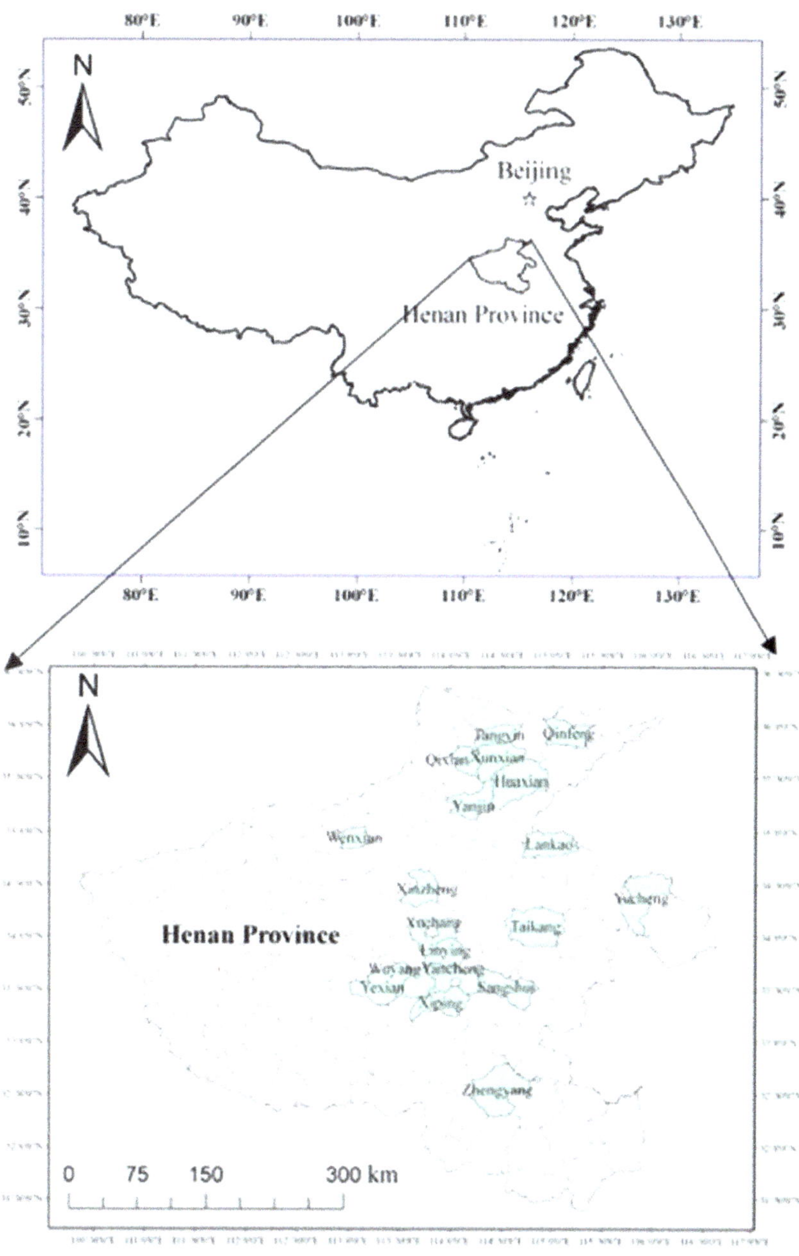

Figure 1. Map of China (top) showing location of Henan Province and counties (bottom) within Henan Province used in this study.

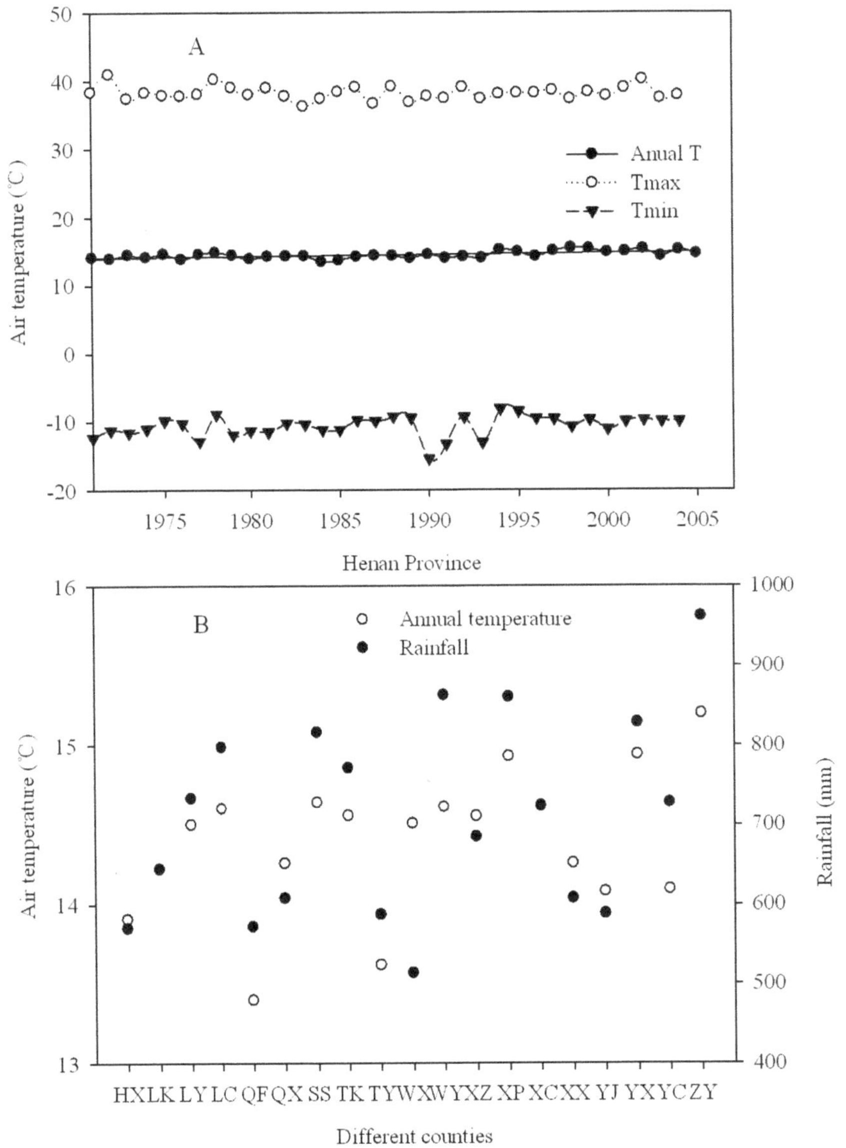

Figure 2. Average annual, maximum, and minimum temperature from 1971 to 2004 in Henan Province, China (A), and (B) average temperature and rainfall from 1975 to 2006 in different counties within Henan Province, China (B). See Table 1 for key to county name abbreviations. All these counties were arranged in English alphabetical order.

approximately 167,000 km^2 in land area, lying within the monsoonal temperate zone. It has a cultivated land area of 79, 260 km^2 for the production of wheat and maize. There are three dominant soil types in Henan Province: Yellow-cinnamon soil (Eutric Cambisols in FAO taxonomy), Sajiang black soil (Eutric Vertisols/Gleyic Cambisol), and Fluvo-aquic soil (Fluvisols in FAO taxonomy) [20]. Mean annual precipitation ranges from 400 to 1000 mm among the counties of the study, with ~70% of it occurring from July to September; mean annual temperature ranges from 13.6 to 15°C (Figure 2). Cultivated agricultural fields are the predominant land use, representing 60% of the total land area in Henan Province. A double cropping system of winter wheat (early October-early June) and maize (mid-June–later September) is the most common planting system used in this region.

Collection of crop yield and soil sampling and analysis

Data on total crop production (including wheat, maize and millet) and wheat yield from 1978–2009 (Figure. 3A) were obtained from the Henan Statistical Yearbook 2010 (13–17) (http://www.ha.stats.gov.cn/hntj/index.htm). Annual yield data for winter wheat and total crop production in 2009 were also obtained from Henan Statistical Yearbook 2010 (29-7) and the Agricultural Bureau of each of the 19 counties in which soil sampling took place (Figure. 3B). These counties, along with basic climatic information, are listed in Table 1. Climatic data of each county were obtained from Meteorological Bureau of Zhengzhou. All counties will be referred to by the two-letter codes presented in Table 1.

The 19 counties were selected as representative of the main agro-ecosystems of Henan Province. Soil samples were collected during June 1–15, 2009 following the wheat harvest but prior to the sowing of maize. Six representative, replicate field plots,

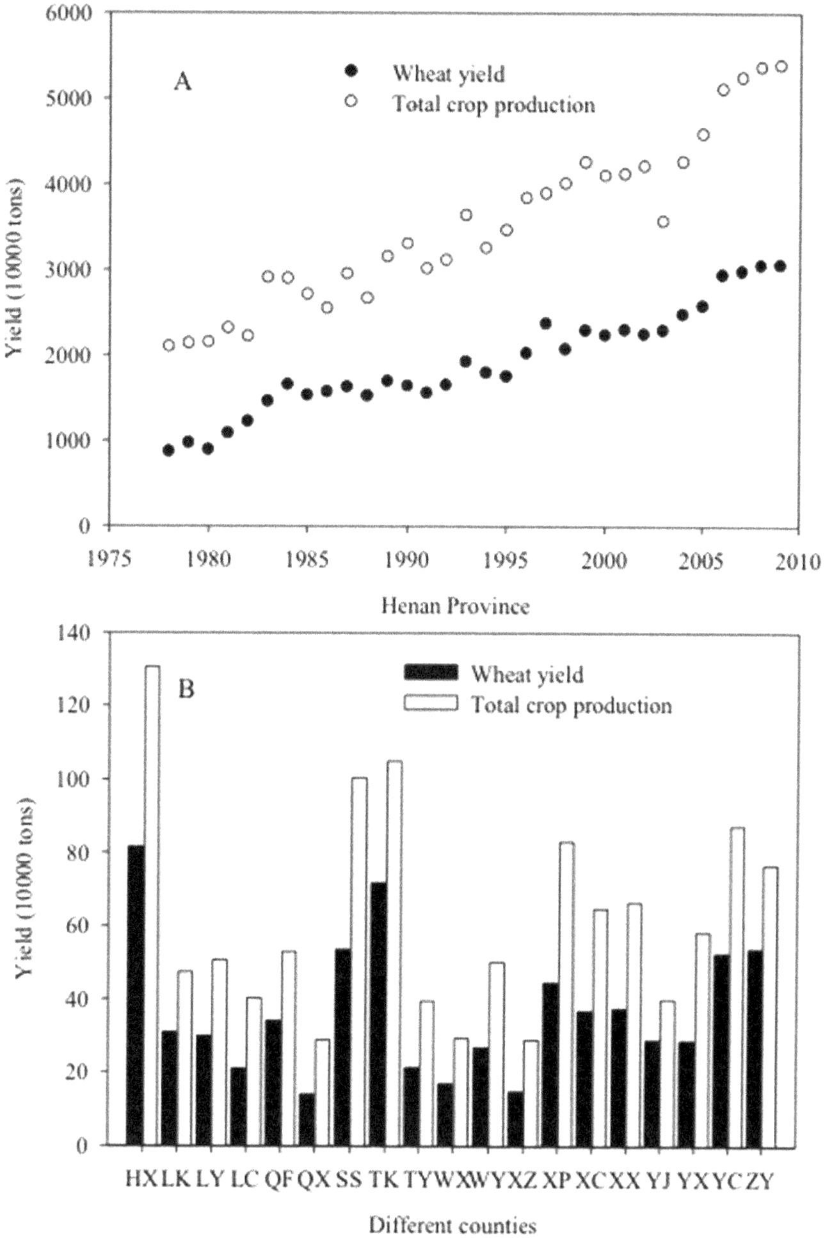

Figure 3. Wheat yield and total crop production (including wheat, maize, millet,) in Henan Province from 1978–2009 (A), and wheat yield and total crop production in different counties within Henan Province in 2009 (B). See Table 1 for key to county name abbreviations.

located at least 6 km apart, were selected within each county based on four criteria: (1) the field plots had been continuously cultivated for at least 30 yr with a native variety, (2) the cropland area was located within 5 km of native vegetation with a similar landscape, soil type and texture, and a relatively flat terrain, and (3) all of the sampling sites are privately owned, and (4) there was no potential impact on any endangered or protected species in the sampling site. Geographic coordinates of each sampling site was recorded by handed GPS of Magellan eXplorist 210(USA), and all of these data were attached in the supporting information.

Sample areas of ~1300 m^2 were established in each plot, with sixteen sampling points taken at random in each of two layers (0–20 cm and 20–40 cm) using a 70 mm - diameter auger. All of the soil samples taken at each layer within a sample plot were mixed together and treated as one sample to represent the value of the plot, yielding 114 soil samples at each layer.

Residual plant material was removed from the soil samples after the samples were air-dried at room temperature. The soil samples were then ground to pass a 2 mm sieve, and a portion of the ground sample was subsequently ground again in a porcelain mortar in order to pass through a 0.15–mm sieve. Organic C and total N measurements were obtained from the twice-ground soil samples. Soil organic C (SOC) was measured using a modified Mebius method. Briefly, 0.1 g soil samples were digested for 5 min

Table 1. Basic geographic coordinates for each county, along with climate data for 19 counties within Henan Province, China.

County	Latitude	Longitude	Sea level (m)	Average Temp (°C)	Rainfall (mm)	Sunshine (h)
Huaxian (HX)	35°44′	114°28′	68	13.9	570.0	2060.9
Lankao (LK)	34°55′	114°46′	70	14.2	644.5	2183.2
Linying (LY)	33°55′	113°55′	63	14.5	732.9	2141.3
Luoheyancheng (LC)	33°35′	114°02′	65	14.6	797.2	2273.0
Qingfeng (QF)	35°53′	115°06′	51	13.4	571.9	2209.1
Qixian (QX)	35°35′	114°12′	72	14.3	607.5	2133.8
Shangshui (SS)	33°39′	114°34′	52	14.6	815.8	1902.0
Taikang (TK)	34°05′	114°50′	53	14.6	770.9	1998.4
Tangyin (TY)	36°03′	114°19′	103	13.6	587.1	2159.3
Wenxian (WX)	35°01′	113°03′	109	14.5	513.2	2302.2
Wuyang (WY)	33°36′	113°32′	77	14.6	862.3	2060.4
Xinzheng (XZ)	34°30′	113°39′	159	14.6	684.6	2058.7
Xiping (XP)	33°29′	113°59′	65	14.9	859.8	2084.7
Xuchang (XC)	34°04′	113°52′	72	14.6	722.7	1959.8
Xunxian (XX)	35°40′	114°32′	59	14.3	607.5	2133.8
Yanjin (YJ)	35°13′	114°11′	69	14.1	588.0	2287.8
Yexiang (YX)	33°38′	113°21′	88	14.9	827.8	1972.4
Yucheng (YC)	34°25′	115°52′	46	14.1	727.3	2244.6
Zhengyang (ZY)	32°37′	114°24′	70	15.2	961.8	2004.4

Note: Counties are arranged in English alphabetical order.

with 5 mL of 1N $K_2Cr_2O_7$ and 10 mL of concentrated H_2SO_4 at 150°C, followed by titration of the digests with standardized $FeSO_4$. Total soil N (TSN) was measured using a modified Kjeldahl wet digestion procedure and a Tector Kjeltec System 1026 distilling unit. Soil available N was analyzed using a micro-diffusion technique after alkaline hydrolysis (1.8 mol L^{-1} NaOH). The Olsen method was used to determine available soil phosphorus (P), and available soil potassium (K) was measured in 1 mol L^{-1} NH_4OAc extracts by flame photometry (Table 2).

Three sampling points were used to determine soil bulk density in each plot. Samples were collected separately from four layers within a depth of 0–40 cm in each sampling point. Soil bulk density was measured using 100-cm^3 soil cores obtained from the four layers. Soil porosity was calculated from soil bulk density and specific gravity, with any stone material removed and not considered in bulk density calculations.

Calculation of soil organic C and N stores and SOC and TSN

Total soil organic C store (TSOCS) and total soil N stores (TSNS) at 0–40 cm depth were calculated as follows:

$$\mathrm{TSOC}(\mathrm{g.m^{-2}}) =$$
$$\text{Soil organic C}(\mathrm{g.kg^{-1}}) \times \text{soil bulk density}(\mathrm{g.cm^{-3}}) \times \text{sampling depth}(\mathrm{cm})$$
$$\mathrm{TSN}(\mathrm{g.m^{-2}}) =$$
$$\text{Soil total N}(\mathrm{g.kg^{-1}}) \times \text{soil bulk density}(\mathrm{g.cm^{-3}}) \times \text{sampling depth}(\mathrm{cm})$$

Given the cultivated area, the total cultivated topsoil (0–40 cm) C and N stocks of each county were estimated by the equation:

$$\mathrm{CS} = \sum area_i \times \mathrm{TSOC}$$
$$\mathrm{NS} = \sum area_i \times \mathrm{TSN}$$

where *area* is the given total cultivated area of each county, and CS and NS are C and N stocks, respectively. SOC and TSN were means of six sampling sites of each county.

Statistics

Analysis of variance was used to assess the significance of location (county) on soil C and N concentration and storage; means were compared using Duncan's multi-range test at $\alpha = 0.05$. Linear regression was used to determine the relationships between C and N stock versus wheat and total crop production. Principle components analysis was used to assess patterns of similarity/dissimilarity among counties with respect to several environmental variables [21]. All statistical analyses were performed using SPSS 10.0 (Chicago IL, USA).

Results

Wheat yields increased more than 250% from 1978 to 2009 while total annual crop production in Henan Province increased from 21 to 54 million tons over the same time period (Figure 3A). Wheat yield varied from 143 to 729 thousand tons among the different counties in 2009 (Figure 3B).

The absolute value of SOC concentration in the top 40 cm of soil varied from 8.13 to 27.89 g kg^{-1} among the 19 counties in 2009 (Table 2) while TSN concentration varied from 0.84 to 2.2 g kg^{-1}. Soil C/N varied from 6.4 to 20 (Table 2). Soil organic C stores (TSOCS) in the 0–40 cm soil layer varied from 2,322 g m^{-2} to 8,038 g m^{-2}, whereas total N stores (TSNS) varied from 221 to

Table 2. Spatial variation in soil (0–40 cm depth) properties, soil organic C (SOC), total soil N (TSN) concentration (g kg^{-1}), and C/N in the 0–40 cm soil layer in 19 counties within Henan province, China.

	Alkaline-extractable N (mg kg^{-1})	Olsen-extractable P (mg kg^{-1})	NH_4OAc-extractable K (mg kg^{-1})	Bulk density (g cm^{-3})	Soil porosity (%)	SOC (g kg^{-1})	TSN (g kg^{-1})	C/N
HX	48.9±3.2abc	1.8±0.7a	80.1±9.2abc	1.44±0.03de	38.3±1.4abcd	12.4±0.9abc	1.4±0.05abcd	8.8±0.7abc
LK	56.5±2.4abcd	7.6±1.9ab	71.9±11.9abc	1.42±0.02bcde	40.7±1.0bcdef	11.2±0.7ab	1.4±0.09abcd	7.9±0.5a
LY	49.9±3.1abc	4.2±0.7a	145.4±24.1ef	1.36±0.02abc	41.7±1.4cdefg	15.5±1.0bcd	1.1±0.07a	14.2±0.2efg
LC	49.0±1.6abc	11. 5±2.1abc	103.6±8.9abcde	1.39±0.02bcd	38.7±1.6abcd	14.6±1.4abcd	1.4±0.14bcd	10.5±1.1abcd
QF	47.5±2.8abc	10.9±4.9abc	71.7±6.3abc	1.39±0.01bcd	41.8±0.4cdefg	11.8±0.6ab	1.4±0.08abcd	8.7±0.8abc
QX	51.9±3.7abc	6.3±2.2ab	82.1±10.8abc	1.44±0.01de	38.7±0.3abcd	21.1±1.8f	1.5±0.21cd	16.2±3.1g
SS	45.1±1.9abc	11.7±3.7abc	169.3±33.9f	1.35±0.02abc	37.9±1.3abcd	14.5±0.9abcd	1.3±0.08abc	11.4±0.8bcde
TK	59.3±4.5cd	17.7±8.2bcd	140.6±23.9def	1.35±0.02ab	41.5±1.5cdef	13.4±1.0abcd	1.1±0.05ab	11.9±0.7cde
TY	59.2±2.4cd	6.4±2.6ab	110.9±12.8bcde	1.45±0.03de	38.6±1.3abcd	15.0±0.4abcd	1.7±0.09de	8.9±0.5abc
WX	56.8±3.5abcd	11.0±1.9abc	82.1±7.7abc	1.3±0.03a	43.6±1.5fg	17.1±1.9de	1.5±0.11bcd	11.5±0.6bcde
WY	47.9±1.1abc	10.3±2.9ab	84.2±10.5abc	1.38±0.02bcd	36.8±1.2ab	14.9±1.6abcd	1.6±0.09cd	9.5±0.6abc
XZ	72.9±7.4e	7.8±2.1ab	95.8±18.9abcd	1.47±0.02e	38.3±0.7abcd	16.1±1.6cd	1.1±0.12ab	14.5±1.1efg
XP	72.2±4.2e	27.3±5.8d	117.5±17.7cde	1.43±0.01cde	37.8±1.2abc	19.9±1.9ef	1.3±0.07abc	15.4±1.3fg
XC	43.2±3.4a	17.2±5.8bcd	66.2±9.6ab	1.42±0.03bcde	42.9±0.9efg	16.6±2.3cde	1.3±0.18abc	12.7±0.7def
XX	53.1±4.2abcd	5.1±0.5ab	89.9±3.0abc	1.41±0.02bcde	35.3±1.0a	14.8±0.6abcd	1.9±0.11e	7.9±0.3a
YJ	49.5±2.9abc	12.8±4abc	77.0±9.6abc	1.39±0.02bcd	38.2±1.0abcd	10.8±0.6a	1.3±0.07abc	8.2±0.4ab
YX	44.5±2.1ab	22.9±5.9cd	89.5±11.4abc	1.3±0.04a	45.5±1.9g	15.4±0.9bcd	1.2±0.07ab	13.6±1.1defg
YC	66.4±11.3de	3.5±1.1a	59.9±8.3a	1.44±0.02de	39±1.3abcde	14.8±0.9abcd	1.1±0.08ab	13.1±0.2defg
ZY	58.7±2.4bcd	6.3±1.4ab	73.8±5.7abc	1.35±0.02ab	41.9±0.9defg	12.7±0.3abc	1.5±0.06bcd	8.8±0.5abc

Different letters indicate significant differences ($p = 0.05$) among the 19 counties. Counties are arranged in English alphabetical order.

Table 3. Total C (TSOCS) and N (TSNS) stores in the surface soil layer (0–40 cm) of soils in 19 counties in Henan Province, China.

	C store (g m^{-2})	N store (g m^{-2})
HX	3541±261.8 abcd	410.1±18.9 cde
LK	3118.9±189.3 ab	399.7±23.6 bcde
LY	4106.7±294.1 abcd	290.4±22.2 a
LC	4023.8±372 abcd	398.6±35.6 bcde
QF	3229.7±140 abc	381.8±22.3 abcd
QX	5977.9±524.3 e	429.1±63.1 de
SS	3881.8±219 abcd	348.3±24.0 abcd
TK	3605.1±328.1 abcd	303.4±19.2 ab
TY	4300.4±98.5 bcd	494.3±24.9 ef
WX	4396.2±451.9 cd	379.4±23.4 abcd
WY	4081.3±455.2 abcd	429.3±25.3 de
XZ	4528.2±516.5 d	320.3±40.1 abc
XP	5709.2±582.9 e	369.8±21.9 abcd
XC	4614.5±609.9 d	366.3±47.7 abcd
XX	4323.8±224.7 bcd	558.2±35.1 f
YJ	3072.5±178.3 a	378.7±19.2 abcd
YX	4190.7±241.3 abcd	315.4±17.4 abc
YC	3926.2±304.8 abcd	299.3±26.3 ab
ZY	3413.4±104.3 abcd	398.1±21.7 bcde

Counties are arranged in English alphabetical order.

659 g m^{-2}. The highest value was in XX County and the lowest in LY County in N reserves (Table 3).

Linear regression analysis indicated that total crop production was significantly and positively correlated with SOC and TSN (Figure 4A), soil C and N store (Figure 4B), and soil C and N stocks (Figure 4C). Soil bulk density was significantly and positively correlated with soil N concentration ($r = 0.25$, $p = 0.008$, $n = 114$), soil C ($r = 0.21$, $p = 0.03$, $n = 114$) and N store ($r = 0.43$, $p = 0.001$, $n = 114$). While soil porosity was significantly and negatively correlated with soil N concentration ($r = -0.19$, $p = 0.05$, $n = 114$), soil C ($r = -0.25$, $p = 0.007$, $n = 114$) and N store ($r = -0.32$, $p = 0.001$, $n = 114$).

Principle components analysis revealed that Axis 1, which explained 98% of the variation in all data (eigenvalue = 0.98), was highly correlated with soil C, whereas Axis 2, explaining 1% of the variation (eigenvalue = 0.09), was highly correlated with soil N. Thus, counties such as QX and XP located highly positive on Axis 1 with high levels of soil C, but other counties, such as LK, YJ, and QF, occupied positions toward the negative end of Axis 1 with low soil C (Figure 5).

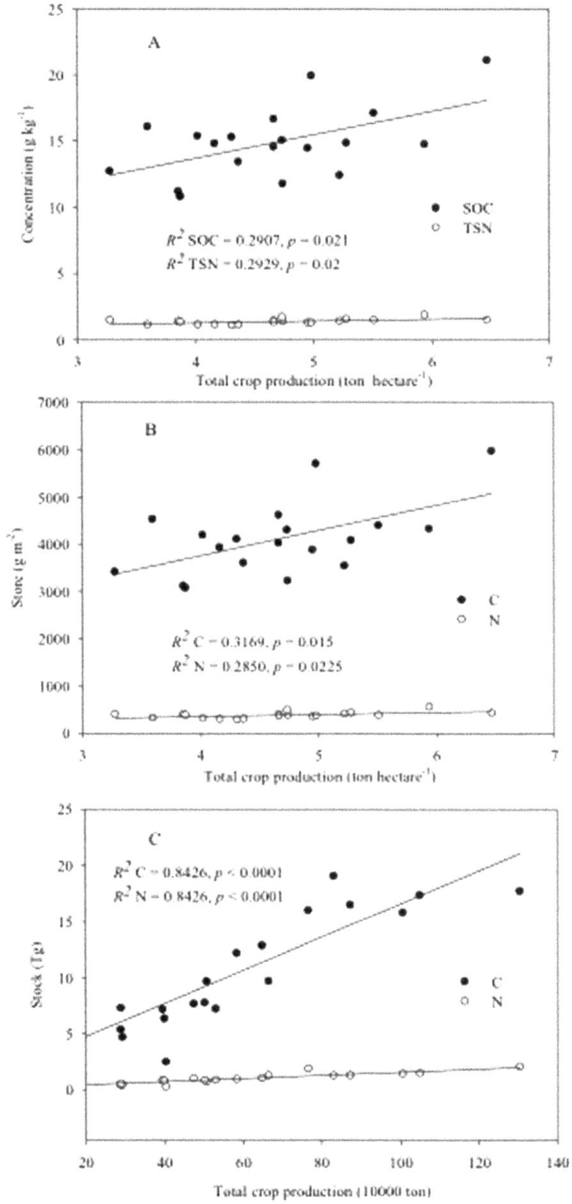

Figure 4. Linear regression analysis of total crop production in each county (ton ha^{-1}) with SOC and TSN (A) and with soil C and N store (0–40 cm) (B), and the total crop production of each county (10000 ton) with their soil C and N stock (C) (n = 19).

Discussion

Potential influences on crop yield

It is notable that 14 environmental (e.g., mean annual temperature and precipitation -Table 1) and soil variables (including extractable nutrients-Table 2) examined in our analysis of the data from the 19 counties in Henan Province were correlated with either wheat or total crop yield (data not shown), and total crop production were significantly, positively related to SOC and TSN, soil C and N store, and soil C and N stock (Figure 4). Part of this is likely related to the highly integrated nature of the measures of C and N stocks, i.e., their calculations combine soil concentrations of C and N, soil bulk density, sampling depth, and area of cultivation. However, all of these have been shown to directly influence crop performance. For example, increases in soil C have been shown to increase crop yield in other studies. Lal (2004, 2006) reported increases in yield from 20 to 70 kg ha^{-1} and 10 to 300 kg ha^{-1} for wheat and maize, respectively, following increases of 1 MT of C in agricultural soils in Africa [1,8]. Similarly, loss of soil C has been shown to decrease yield in agricultural soils of Canada and the U.S. [4,5].

Soil C-mediated increases of crop yields also may arise from improvements in soil structure and available water-holding

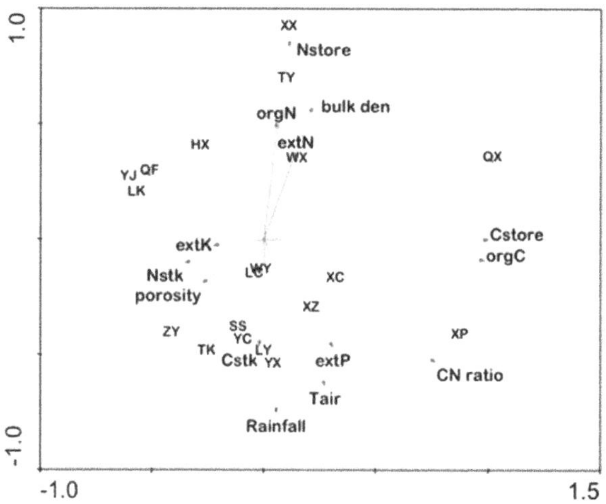

Figure 5. Principle components analysis of environmental and soil variables for agricultural soils in 19 counties within Henan Province. Length of arrows is directly proportional to their importance in explaining spatial patterns in the counties. Direction of the arrows indicates increasing values. Thus, the x-axis is primarily a gradient in soil C, whereas the y-axis is primarily a gradient in soil N and rainfall and secondarily a gradient in soil N. See Table 1 for key to county name abbreviations.

capacity. Enhanced soil structure, via increased soil C, generally arises from several processes, including increasing stability of soil aggregates [22,23,24]. As a result of the increased stability of the aggregates, soils become less prone to crusting, compaction, and erosion [25,26,28]. Emerson (1995) demonstrated that an increase of 1 g of soil organic matter (~50% of which is C) can increase available soil moisture by up to 10 g [27], which is enough to maintain crop growth between periods of rainfall of 5 to 10 days [8].

Spatial variation in cultivated soils

In this study, soil organic C concentration averaged 14.9 g kg^{-1} and total N averaged 1.4 g kg^{-1} in the 0–40 cm layer across all sites, while soil C and N stores averaged 4.1 kg C $m^{-2}$and 0.38 kg N m^{-2}, respectively. These values are comparable to published values from other regions of China, including 9–15 g C kg^{-1} and 1.2–1.8 g N kg^{-1} in northern China [12,29], and 16.1 g C $kg^{-1}$and 1.04 g N kg^{-1} in eastern and southern China [14,16,30]. Liu et al. (2011) reported soil C stores of 4.57 kg C m^{-2} in the Loess Plateau region in northwestern China [13].

Principal components analysis separated the 19 counties primarily along a gradient in soil C, with counties LK, YJ, QF, ZY, HX, and TK (mean soil C = 12.1 g C kg^{-1}) toward the lower end and XP and QX (mean soil C = 20.5 g C kg^{-1}) toward the upper end of Axis 1, which accounted for nearly 80% of the variation in soil and environmental data (Figure 5). Spatial variation in soil organic C in agricultural systems can be influenced by several factors, including microclimate, soil type, topography, and especially human activity [31].

Spatial variation in soil N was essentially orthogonal to that of soil C. This was surprising since typically, the two are highly correlated in terrestrial ecosystems [32]. As a result, the secondary gradient (i.e., Axis 2) was one of soil N, with counties TK, YC, SS, LY, YX, and XP (mean soil N = 1.15 g N kg^{-1}) located toward the lower end of Axis 2 (accounting for <10% of variation) and XX and TY (mean soil N = 1.81 g N kg^{-1}) located toward the upper end of Axis 2 (Figure 5). Although C and N are often correlated through their organic forms in plant detritus, spatial variation of N in soils of agro-ecosystems can also be greatly influenced by the extensive use of N fertilizers.

Management methods used in crop production systems, including tillage practices and fertilizer use, can affect soil C and N on broad spatial scales, including that of an entire Province [33]. Over the course of repeated seasons of crop growth in Henan Province, agricultural fields are repeatedly subjected to soil tillage, planting, fertilization, irrigation, and harvest, all of which potentially influence soil C and N stores [30,34]. In contrast, Zhang et al. (2012) reported that raised-bed planting, a viable alternative to conventional tillage, can significantly enhance the yield of summer maize while simultaneously improving soil structure, as well as the structure and function of microbial communities essential to the quality of agricultural soils [22].

Results presented in the current study underscore the complexity of factors that can impact agricultural soils and their ability to produce crops to meet the ever-increasing demand in China resulting from population growth. Some of the spatial pattern exhibited in ordination space (Figure 5) is clearly related to regional factors, such as microclimate. For example, WY and LC are adjacent to each other in Henan Province (Figure 1) and are also closely clustered in ordination space, indicating that they are very similar with respect to environmental and soil characteristics. XP and SS, however, are also adjacent counties; yet occur distant from each other in ordination space, indicating great dissimilarity in environmental and soil factors. Agronomists should take into account the large spatial variability in important components of the soils in Henan Province, especially in the variation of soil C and N, when considering appropriate agronomic management practices.

Author Contributions

Conceived and designed the experiments: XLZ QW CHL. Performed the experiments: XLZ QW YLW. Analyzed the data: XLZ QW FSG. Contributed reagents/materials/analysis tools: XLZ QW YLW FNC. Contributed to the writing of the manuscript: XLZ FSG CHL.

References

1. Lal R (2004) Soil carbon sequestration impacts on global climate change and food security. Science 304: 1623–1627.
2. Liu DW, Wang ZM, Zhang B, Song KS, Li XY, et al. (2006) Spatial distribution of soil organic carbon and analysis of related factors in croplands of the black soil region, Northeast China. Agriculture, Ecosystems and Environment 113: 73–81.
3. Smith WN, Desjardins RL, Pattey E (2000) The net flux of carbon from agricultural soils in Canada 1970–2010. Global Change Biology 6: 557–568.
4. Bauer A, Black AL (1994) Quantification of the effect of soil organic matter content on soil productivity. Soil Science Society of America Journal 58: 185–193.
5. Larney FJ, Janzen HH, Olson BM, Lindwall CW (2000) Soil quality and productivity response to simulated erosion and restorative amendments. Canadian Journal of Soil Science 80: 515–522.
6. Hairiah K, Van Noordwijk M, Cadisch G (2000) Crop yield, C and N balance of the three types of cropping systems on an Ultisol in northern Lampung. Netherland Journal of Agricultural Science 48: 3–17.
7. Duxbury JM (2001) Long-term yield trends in the rice-wheat cropping system: results from experiments in Northwest India. Journal of Crop Production 3: 27–52.
8. Lal R (2006) Enhancing crop yields in the developing countries through restoration of the soil organic carbon pool in agricultural lands. Land Degradation and Development 17: 197–209.
9. Batjes NH (2002) Carbon and nitrogen stocks in the soils of Central and Eastern Europe. Soil Use and Management 18: 324–329.

10. Maia SMF, Ogle SM, Cerri CC, Cerri CEP (2010) Changes in soil organic carbon storage under different agricultural management systems in the Southwest Amazon Region of Brazil. Soil and Tillage Research 106: 177–184.
11. Piao SL, Fang JY, Ciais P, Peylin P, Huang Y, et al. (2009) The carbon balance of terrestrial ecosystems in China. Nature 458, doi:10.1038/nature 07944.
12. Wang ZM, Zhang B, Song KS, Liu DW, Ren CY (2010) Spatial variability of soil organic carbon under maize monoculture in the Song-Nen plain, Northeast China. Pedosphere 20: 80–89.
13. Liu ZP, Shao MA, Wang YQ (2011) Effect of environmental factors on regional soil organic carbon stocks across the Loess Plateau region, China. Agriculture, Ecosystems and Environment 142: 184–194.
14. Liao QL, Zhang XH, Li ZP, Pan GX, Smith P, et al. (2009) Increase in soil organic carbon stock over the last two decades in China's Jiangsu Province. Global Change Biology 15: 861–875.
15. Zhang HB, Luo YM, Wong MH, Zhao QG, Zhang GL (2007) Soil organic carbon storage and changes with reduction in agricultural activities in Hong Kong. Geoderma 139: 412–419.
16. Feng S, Tan S, Zhang A, Zhang Q, Pan G, et al. (2011) Effect of household land management on cropland topsoil organic carbon storage at plot scale in a red earth soil area of South China. Journal of Agricultural Science 149: 557–566.
17. Harper RJ, Gilkes RJ (1995) Some factors affecting the distribution of carbon in soils of a dry land agricultural system in southwestern Australia. In: Lal R, Kimble JM, Follett RF, Stewart BA (editors). Assessment Methods for Soil Carbon. CRC Press. Boca Raton, FL, USA. PP.577–591.
18. Guo JH, Liu XJ, Zhang Y, Shen JL, Han WX, et al. (2010) Significant Acidification in Major Chinese Croplands. Science 327: 1008–1010.
19. Pan GX, Li LQ, Wu LS, Zhang XH (2003) Storage and sequestration potential of topsoil organic carbon in China's paddy soils. Global Change Biology 10: 79–92.
20. Wu HB, Guo ZT, Gao Q, Peng CH (2009) Distribution of soil inorganic carbon storage and its changes due to agricultural land use activity in China. Agriculture, Ecosystems and Environment 129: 413–421.
21. Gilliam FS, Saunders NE (2003) Making more sense of the order: A review of Canoco for Windows 4.5, PC-ORD version 4 and SYN-TAX 2000. Journal of Vegetation Science 14: 297–304.
22. Zhang XL, Ma L, Gilliam FS, Wang Q, Liu T, et al. (2012) Effects of raised-bed planting for enhanced summer maize yield on soil microbial functional groups and enzyme activity in Henan Province, China. Field Crops Research 130: 28–37.
23. Feller C, Beare MH (1997) Physical control of soil organic matter dynamics in tropics. Geoderma 79: 69–116.
24. Haynes RJ, Naidu R (1998) Influence of lime, fertilizer and manure applications on soil organic matter content and soil physical conditions: a review. Nutrient Cycling in Agroecosystems 51: 123–137.
25. Diaz-Zorita M, Grosso GA (2000) Effect of soil texture, organic carbon and water retention on the compatibility of soils from the Argentinean Pampas. Soil and Tillage Research 54: 121–126.
26. Schertz DL, Moldenhauer WC, Livingston SJ, Weeisies GA, Hintz AE (1989) Effect of past soil erosion on crop productivity in Indiana. Journal of Soil and Water Conservation 44: 604–608.
27. Emerson WW (1995) Water-retention, organic-carbon and soil texture. Australian Journal of Soil Research 33: 241–251.
28. Powlson DS, Hirsch PR, Brookes PC (2001) The role of soil micro-organisms in soil organic matter conservation in the tropics. Nutritional Cycling in Agroecosystems 61: 41–51.
29. Du ZL, Ren TS, Hu CS (2010) Tillage and residue removal effects on soil carbon and nitrogen storage in the North China Plain. Soil Science Society of American Journal 74: 196–202.
30. Pan GX, Li LQ, Zhang Q, Wang XK, Sun XB, et al. (2005) Organic carbon stock in topsoil of Jiangsu Province, China, and the recent trend of carbon sequestration. Journal of Environmental Sciences 17: 1–7.
31. Post WM, Pastor J, Zinke PJ, Stangenberger AG (1985) Global patterns of soil nitrogen storage. Nature 317: 613–616.
32. Gilliam FS, Dick DA, Kerr ML, Adams MB (2004) Effects of silvicultural practices on soil carbon and nitrogen in a nitrogen saturated Central Appalachian (USA) hardwood forest ecosystem. Environmental Management 33: S108–S119.
33. Pan GX, Zhao QG (2005) Study on evolution of organic carbon stock in agricultural soils of China: facing the challenge of global change and food security. Advances in Earth Science 20: 384–393 (in Chinese).
34. Dersch G, Böhm K (2001) Effects of agronomic practices on the soil carbon storage potential in arable farming in Austria. Nutrient Cycling in Agroecosystems 60: 49–55.

22

Effects of Nitrogen and Phosphorus Fertilization on Soil Carbon Fractions in Alpine Meadows on the Qinghai-Tibetan Plateau

Jin Hua Li[1]*, Yu Jie Yang[1], Bo Wen Li[1], Wen Jin Li[1], Gang Wang[1], Johannes M. H. Knops[2]

1 State Key Laboratory of Grassland Agro-Ecosystems, School of Life Sciences, Lanzhou University, Lanzhou, P.R. China, **2** School of Biological Sciences, University of Nebraska, Lincoln, Nebraska, United States of America

Abstract

In grassland ecosystems, N and P fertilization often increase plant productivity, but there is no concensus if fertilization affects soil C fractions. We tested effects of N, P and N+P fertilization at 5, 10, 15 g m^{-2} yr^{-1} (N_5, N_{10}, N_{15}, P_5, P_{10}, P_{15}, N_5P_5, $N_{10}P_{10}$, and $N_{15}P_{15}$) compared to unfertilized control on soil C, soil microbial biomass and functional diversity at the 0–20 cm and 20–40 cm depth in an alpine meadow after 5 years of continuous fertilization. Fertilization increased total aboveground biomass of community and grass but decreased legume and forb biomass compared to no fertilization. All fertilization treatments decreased the C:N ratios of legumes and roots compared to control, however fertilization at rates of 5 and 15 g m^{-2} yr^{-1} decreased the C:N ratios of the grasses. Compared to the control, soil microbial biomass C increased in N_5, N_{10}, P_5, and P_{10} in 0–20 cm, and increased in N_{10} and P_5 while decreased in other treatments in 20–40 cm. Most of the fertilization treatments decreased the respiratory quotient (qCO_2) in 0–20 cm but increased qCO_2 in 20–40 cm. Fertilization increased soil microbial functional diversity (except N_{15}) but decreased cumulative C mineralization (except in N_{15} in 0–20 cm and N_5 in 20–40 cm). Soil organic C (SOC) decreased in P_5 and P_{15} in 0–20 cm and for most of the fertilization treatments (except $N_{15}P_{15}$) in 20–40 cm. Overall, these results suggested that soils will not be a C sink (except $N_{15}P_{15}$). Nitrogen and phosphorus fertilization may lower the SOC pool by altering the plant biomass composition, especially the C:N ratios of different plant functional groups, and modifying C substrate utilization patterns of soil microbial communities. The N+P fertilization at 15 g m^{-2} yr^{-1} may be used in increasing plant aboveground biomass and soil C accumulation under these meadows.

Editor: Upendra M. Sainju, Agricultural Research Service, United States of America

Funding: This research was supported by Program for New Century Excellent Talents in University (NCET-11-0210), the Fundamental Research Funds for the Central Universities (LZUJBKY-2013-K12) and National Natural Science Foundation of China (No. 30871823). The funders had no role in study design, data collection and analysis, decision to publish, or preparation of the manuscript.

Competing Interests: The authors have declared that no competing interests exist.

* Email: lijinhuap@sohu.com

Introduction

In grassland ecosystems, soils represent the largest active pool of organic C which may have global implications [1–3]. In N-limited grasslands, N deposition through precipitation and/or dust accumulation or fertilization and sometimes P fertilization can increase the productivity of the plant community but also can decrease species richness [2,4–9]. However, N and P fertilization may affect soil C fractions and the ecological function of grassland soils in different ways [2,10]. Long-term N fertilization can decrease microbial biomass carbon (MBC) of grassland soil [11,12]. Fertilization effects on C mineralization are variable [13–15]. Mack et al. [16] showed that NP fertilization contributed to decreases in soil organic C (SOC) pools in tundra ecosystems. Phosphorus additions can lead to greater C sequestration in soils [17]. In addition, N additions can significantly enhanced soil C stocks, but combined NPKMg additions may not result in soil C sequestration [9]. Thus, it is clear that in contrast to the consistent increases of productivity due to N and P fertilization, the impacts of N and P fertilization on microbial composition, activity and soil C is not consistent and differs markedly between sites.

Alpine grasslands in the Qinghai-Tibetan Plateau are estimated to hold approximately 55% of China's total grassland C, out of which 93% is present in the soil [18]. Little is known about how fertilization types and rates affect soil C fractions in alpine meadows.

Here we specifically examined how 5 years of N and/or P additions had influenced soil organic C fractions. Our objective was to develop a better understanding of the effects of N and P fertilization on changes in soil organic C dynamics. Previous studies from the same experiment have shown that N or N+P fertilization increased plant aboveground productivity and there was a yearly variation in plant biomass [19–21]. Forbs and legumes richness decreased after N and P fertilization [21]. Nitrogen concentration and N/P ratios of grasses and forbs increased significantly but were relatively constant for legumes after N addition [22,23]. In this paper, we divided species into different functional groups to address the changes in soil organic C fractions in relation to the species changes. Specifically, we evaluated the effects of N and P fertilization on the aboveground plant biomass of different functional groups and soil C fractions at 0–20 cm and 20–40 cm.

Fertilization can affect the quantity of C input to soil by alteration of the plant productivity, diversity, and stoichiometry of plant tissues [16,24], as well as soil organic matter decomposition and respiration by altering soil microbial communities and activity [11,25]. We hypothesized that a combination of N and P fertilization will increase SOC but decrease MBC, microbial functional diversity and C mineralization rates compared to N and P fertilization alone or no fertilization.

Methods

Ethics Statement

Our study area was located at the Alpine Meadow Ecosystem Research Station in Hezuo, Gansu, eastern Qinghai-Tibetan Plateau, China (N34°55', E102°53' 3,000 m above sea level), which is managed by Lanzhou University. No specific permissions were required for conducting experiment in this location, and no endangered, protected species or vertebrate species were involved with this research.

Study site

This study was conducted from 2009 to 2013 in the Research Station of Alpine Meadow and Wetland Ecosystems of Lanzhou University, located in Hezuo, Gansu, eastern Qinhai-Tibetan Plateau of China (N34°55', E102°53', 3,000 m above sea level). Hezuo has a 30-year mean annual precipitation of 550 mm, with 85% of the precipitation occurring during the growing season from June through September (Institute of Hezuo Meteorology). Mean annual temperature is 2.4°C, ranging from −8.3°C during December-February to 11.9°C during June–August periods. The soils are classified as chestnut soils or Haplic Calcisols according to the FAO classification or sub-alpine meadow soil according to the Chinese soil classification system [26]. The vegetation is a typical alpine meadow, dominated by grasses such as *Festuca ovina* Linn., *Poa poophagorum* Bor and *Elymus nutans* Griseb; sedges such as *Scirpus pumilus* Vahl and *Kobresia capillifolia* (Decne.) C.B. *Kobresia pygmaea* C.B. Clarke in Hook; forbs such as *Anemone rivularis* Buch.-Ham, *Trollius farreri* Stapf and *Anemone obtusiloba* D. Don, *Taraxacum lugubre*, *Geranium pylzowianum*, *Polygonum viviparum*, and legumes such as *Astragalus polycladus* Bur., *Oxytropis ochrocephala* and *Gueldenstaedtia verna* Georgi.

Experimental design

Soil organic C fractions resulting from the increased plant production by N and P fertilization were evaluated using the following fertilization treatments: (i) N alone; (ii) P alone; and (iii) N+P together. There were three rates (5, 10, 15 g m^{-2} yr^{-1}) of N (urea with 46% N), and P (sodium dihydrogen phosphate anhydrous with 44.6% P). A control treatment (CK) without any N or P fertilization was also included. As a result, there were ten treatments (N_5, N_{10}, N_{15}, P_5, P_{10}, P_{15}, N_5P_5, $N_{10}P_{10}$, $N_{15}P_{15}$, and CK) with five replicates for a total of 50 treatment plots. Only four replicates were used in this study. All treatments were randomly assigned in 50 plots (5×5 m area each). In addition, there were four plots without fertilization but destroyed by zokors (*Myospalax fontanieri*) in 2009, but this didn't affect the neighbor plots. All plots were separated by 1 m unfertilized buffers. Fertilizer was broadcast evenly in the plot once per year on 20th August for 5 years since 2009. Fertilizer was applied in August for two reasons: (i) In our study site, the growing season was from June through September and 85% of the rainfall occurred during the growing season. (ii) Plant biomass peaked at this time, which was reasonable to be contrasted with that before fertilization application. However, applying fertilizers in the middle of the growing season might result in the inefficient use and wastage. Before applying fertilizers in August 2009, average SOC, soil total N and total P was 33.8±1.8 g kg^{-1}, 3.7±0.1 g kg^{-1}, 0.65±0.01 g kg^{-1} in 0–20 cm and 26.1±1.5 g kg^{-1}, 2.8±0.1 g kg^{-1}, 0.65±0.01 g kg^{-1} in 20–40 cm, respectively, across all plots. The soil had 200 g kg^{-1} sand, 600 g kg^{-1} silt, and 200 g kg^{-1} clay.

Plant sampling

In early August 2013 before applying fertilizers, four sampling quadrates of 50 cm×50 cm (selected at random from the central of each 5×5 m plot to avoid edge effects) were used to harvest plant samples at 1 cm above the ground level using scissors. Plant samples were separated into three functional groups: grasses, legumes, and forbs. At the same time, root samples were collected also within the quadrates after clipping the aboveground plant. Four soil cores (5 cm diameter by 20 cm depth) in each quadrate were collected randomly using a hand probe and composited by plot. Soil samples were washed gently with water over a 60 mesh screen until roots were separated from the soil. All above- and belowground plant samples were dried at 65°C to constant mass and ground to 1 mm prior to determination of total C and N. Aboveground plant samples were also collected from 2009 to 2012, oven dried, ground, and C and N concentrations determined as above.

Soil sampling

In addition to samples collected for root biomass, soil samples were also collected for determining C fractions. Five cores (5 cm in diameter) in 0–20 cm and 20–40 cm depths were taken randomly in each plot, and mixed into one composite sample by depth. After removing gravel, coarse fragments and roots, each soil sample was homogenized and divided into two portions. One portion was air dried and sieved to 2 mm to analyze abiotic parameters. The other portion was kept at 4°C for analysis of microbial parameters (soil MBC, functional diversity of microbial community, and C mineralization).

Plant and soil analysis

Plant C and SOC concentrations were determined using the dichromate oxidation method [27]. Plant and soil total N was determined following Kjeldehl digestion by a Nitrogen Analyzer System (KJELTEC 2300 AUTO SYSTEM II, Foss Tecator AB, Sweden).

Soil MBC was determined by using the chloroform fumigation extraction method [28]. Field-moist soils were adjusted to 50% water holding capacity, incubated at 25°C for 2 weeks for uniform rewetting and stabilizing the microbial activity after the initial disturbances. Three 25 g sub-samples were fumigated with alcohol-free $CHCl_3$ for 24 h at 25°C. Samples were then extracted for MBC by adding 100 ml of 0.5 M K_2SO_4, shaking for 60 min and filtered through Whatman No. 2 paper. Three 25 g non-fumigated soil sub-samples were processed in the same manner. Carbon concentration in the extract was determined using the same method as SOC and MBC was calculated as the difference in C concentrations between fumigated and non-fumigated samples divided by the efficiency factor 0.38 [28].

Soil incubation for C mineralization was carried out in the laboratory for 28 days. Dry weight equivalent of 30 g of the sieved field-moist soil samples were weighed into 250 ml Schott jars. Small beakers filled with 15 ml of 1 M NaOH were placed at the soil surface in the jars to trap the evolved CO_2. The jars were fastened airtight and incubated for 28 days at 25°C. Moisture content of the samples was periodically adjusted to a value of 30% of water holding capacity since this is generally considered to mimic the moisture content of the field conditions [29]. Constant soil water content was maintained by weighing each sample once a week and adding water to the soil. The CO_2 evolved from the soil was measured at 1, 7, 14, 21, and 28 day after incubation. After each day of incubation, the small beaker with NaOH solution was removed and replaced by a new one with fresh NaOH. The CO_2 absorbed in NaOH solution was titrated with 0.1 M HCl after the addition of $BaCl_2$. A NaOH solution without soil, incubated as above, was also titrated. Basal respiration was calculated as the amount of CO_2 evolution in the first 24-h incubation divided by the dry mass of soil. The metabolic quotient (qCO_2) was calculated as the ratio of basal respiration to MBC [30]. The cumulative C mineralization was calculated as the sum of carbon dioxide (CO_2) released during 28 days of incubation.

Bacterial functional diversity in Biolog EcoPlate

Field-moist soils were used to assess the functional diversity of the microbial community, measured by Biolog EcoPlate system [28,30–32] according to the procedure describes by Li et al [33]. Average well color development (AWCD) in each microplate was determined as described by Garland [34]. AWCD and Shannon-Wiener diversity index in Biolog EcoPlate were calculated using data of Biolog EcoPlate collected during 0–168 h and at 168 h respectively as described by Li et al [33].

AWCD = $[\sum(C\text{-}R)/31]$, C is sum of optical densities of 31 wells in a plate, R is optical density of control well in the same plate.

Shannon-Wiener diversity $H = -\Sigma P_i \ln P_i$, of which P_i is proportional color development of the ith well over total color development of all wells of a plate.

Data analysis

Data were analyzed using SPSS 16.0 and graphs were plotted using Sigma Plot 10.0. In this paper, biomass data for only in the fifth year (i.e. 2013) was analyzed because data for the first four years have already been published [20,21]. A one-way ANOVA was used to analyze data by considering all treatments (three fertilization types × three rates + unfertilized control = 10 treatments). When the analysis was significant, the control was taken out and a two-way MANOVA was used to test the overall effects of fertilization types and rates on measured parameters. The least significant difference (LSD) test was used to separate means at $P = 0.05$. Principal component analysis (PCA) was performed on AWCD data at 168 h of incubation in the BIOLOG Ecoplate as described by Zhang et al [35] to determine the carbon utilization patterns of soil microbial communities. Regression analysis was used to analyze the relationship between cumulative carbon mineralization rate at the end of incubation (28th day) and soil MBC and SOC in all treatments. Regression analysis was also used to analyze the relationship between SOC and C:N ratios of different plant functional groups, roots or soils.

Results

Fertilization effects on plant aboveground biomass

Both fertilization type and rate influenced plant community biomass, grass and legume biomass, with a significant type × rate interaction for grass and community biomass (Table 1).

Eight of nine fertilization treatments (except the 15 g m^{-2} yr^{-1} P fertilization) had higher community biomass than the control (Figure 1A). Combination of N+P at 5 g m^{-2} yr^{-1} and 10 g m^{-2} yr^{-1} had much higher community biomass than corresponding rates of N fertilization or any rates of P fertilization. Nitrogen or N+P at 15 g m^{-2} yr^{-1} had two times community biomass and

Table 1. Univariate tests: ANOVAs on plant biomass of community and three functional groups: grasses, legumes and forbs, and on soil microbial biomass C (MBC), qCO_2, cumulative C mineralization, soil organic C (SOC) in 0–20 cm and 20–40 cm for fertilization types and rates.

	Plant biomass				MBC		qCO_2		Cumulative C mineralization		SOC	
	Community	Grasses	Legumes	Forbs	0–20	20–40	0–20	20–40	0–20	20–40	0–20	20–40
Type	0.000	0.000	0.717	0.468	0.003	0.004	<0.001	0.073	<0.001	<0.001	0.049	0.016
Rate	0.018	0.001	0.663	0.009	<0.001	<0.001	0.016	0.003	0.573	0.693	0.807	0.009
Type* Rate	0.018	0.002	0.762	0.357	<0.001	<0.001	<0.001	0.118	0.031	0.503	0.095	0.052

four times grass biomass compared to the control (Figure 1A, B). Grasses comprised 70% of the total biomass, and its biomass trend with fertilization was similar to that of total biomass. All N+P rates had higher grasses biomass compared to corresponding N fertilization rates although no difference occurred between 15 g m^{-2} yr^{-1} N+P and 15 g m^{-2} yr^{-1} N. The N+P treatments at 10 and 15 g m^{-2} yr^{-1} had much higher grass biomass as compared to the corresponding P fertilization rates (Figure 1B). However, the biomass of legumes and forbs in all fertilization treatments (except forbs in 5 g m^{-2} yr^{-1} of N+P fertilization) were lower than the control (Figure 1C, D). The N+P fertilization at 15 g m^{-2} yr^{-1} had significantly lower legume and forb biomass than other treatments, except the forb biomass at 15 g m^{-2} yr^{-1} of N and P.

Fertilization effects on carbon/nitrogen ratio of plant and soil

Fertilization type influenced C:N ratios of the different plant functional groups. Fertilization rate and the fertilization type × rate interaction were significant for forb C:N ratio (Table 2). All fertilization at 5 and 10 g m^{-2} yr^{-1}decreased whereas fertilization at 15 g m^{-2} yr^{-1} increased grasses' C:N ratios compared to the control (Figure 2A). The N, P, and N+P rates decreased C: N ratios of legume and root (Figure 2B, 2D). The 5 and 10 g m^{-2} yr^{-1} N fertilization and 10 g m^{-2} yr^{-1} P decreased while N+P at all rates increased forbs' C:N ratios (Figure 2C). Fertilization type had significant effect on soil C concentrations in both layers and fertilization rate had significant effect on soil C concentrations in 20–40 cm. However, neither fertilization type nor fertilization rate had significant effects on soil N concentrations and C:N ratios in both layers (Table 2). Nevertheless, soil C:N was lower in fertilization treatments (except in 15 g m^{-2} yr^{-1} of N or N+P in 0–20 cm and 15 g m^{-2} yr^{-1} of N+P in 20–40 cm) than the control (Figure 2E, 2F).

Effect of fertilization on soil carbon fractions

Fertilization type and rate had significant effects on soil MBC, qCO_2, cumulative C mineralization, and SOC at 0–20 cm and 20–40 cm layers, except for the effect of rate on C mineralization at both layers and SOC at 0–20 cm (Table 1). The fertilization type × rate interaction was significant for all parameters in all layers, except for qCO_2 and C mineralization at 20–40 cm.

Nitrogen and P at rates of 5 and 10 g m^{-2} yr^{-1} increased MBC compared to other treatments in 0–20 cm (Figure 3A). In 20–40 cm, 10 g m^{-2} yr^{-1} of N fertilization and 5 g m^{-2} yr^{-1} of P

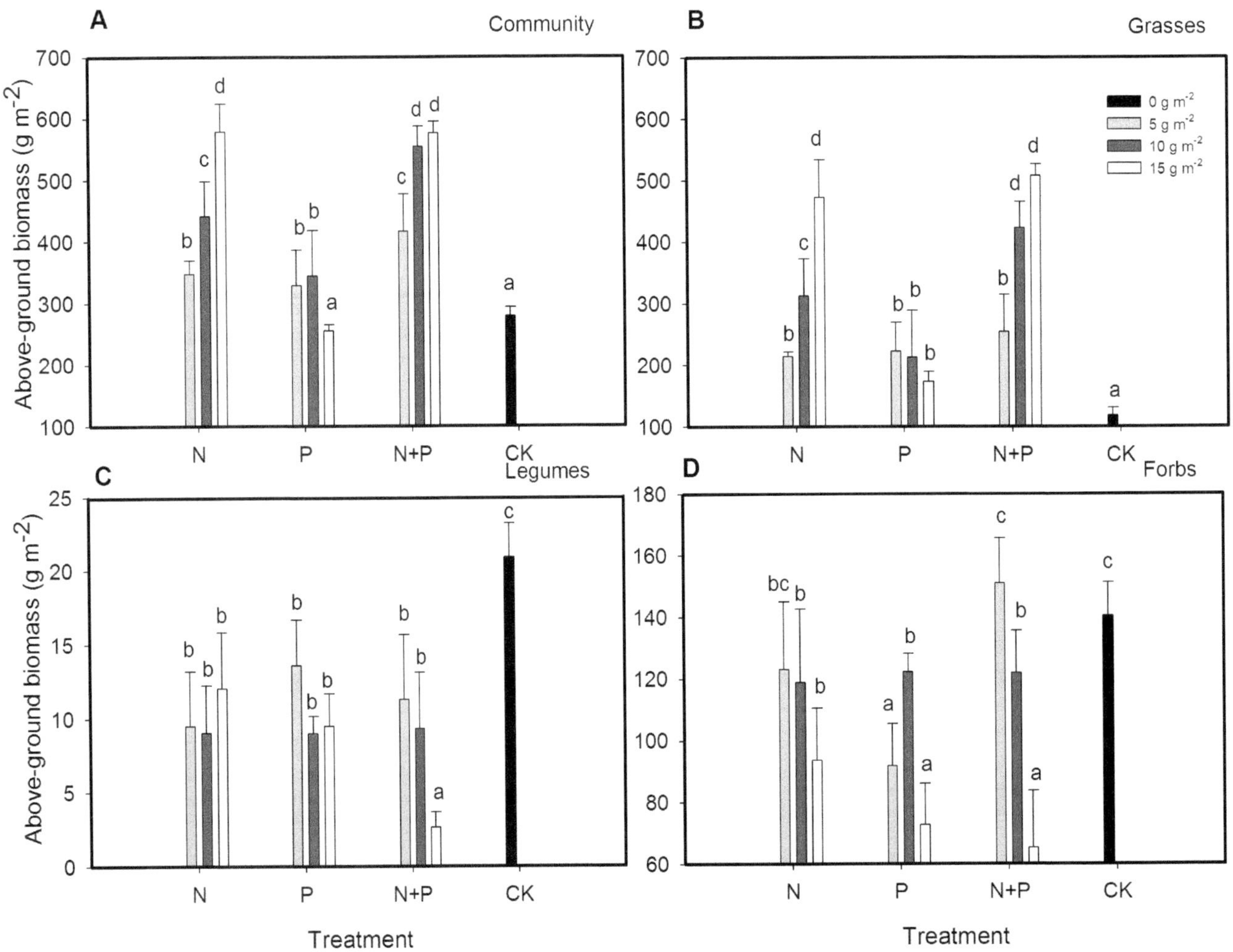

Figure 1. Mean ± SE of total aboveground plant biomass of (A) community, (B) grasses, (C) legumes, and (D) forbs as affected by N and P fertilization treatments. Different letters above bars indicate significantly different at $P = 0.05$.

Table 2. Univariate tests: ANOVAs on C:N ratios of plant (different functional groups and roots) and soil in 0–20 cm and 20–40 cm for fertilization types and rates

	Plant C:N ratio				Soil C:N ratio	
	Grasses	Legumes	Forbs	Roots	0–20 cm	20–40 cm
Type	<0.001	<0.001	<0.001	0.381	0.626	0.058
Rate	0.152	0.474	<0.001	0.695	0.807	0.169
Type* Rate	0.095	0.676	<0.001	0.390	0.890	0.346

fertilization significantly increased MBC compared to other treatments (Figure 3B). Soil MBC was significantly lower in 5 and 10 g m^{-2} yr^{-1} of N+P than those in corresponding rates of N or P fertilization in both layers (Figure 3A, B).

Nitrogen fertilization at 5 and 15 g m^{-2} yr^{-1} significantly increased qCO_2 while P fertilization at 5 and 15 g m^{-2} yr^{-1} and N+P at all rates significantly decreased qCO_2 in 0–20 cm (Figure 3C). The N+P at 5 and 10 g m^{-2} yr^{-1} decreased qCO_2 compared to other treatments in 20–40 cm (Figure 3D).

Most of the fertilization treatments (except the 15 g m^{-2} yr^{-1} N in 0–20 cm soil and 5 g m^{-2} yr^{-1} N in 20–40 cm soil) had lower cumulative C mineralization rates than the control. All N+P fertilization rates had lower cumulative C mineralization than other treatments (Figure 3E, F).

Soil organic C averaged across treatments was higher in 0–20 cm than 20–40 cm (Figure 4A, B). Compared to the control, 10 g m^{-2} yr^{-1} of N or P fertilization, and 15 g m^{-2} yr^{-1} of N+P fertilization significantly increased SOC in 0–20 cm (Figure 4A). The SOC at 20–40 cm was greater in 15 g m^{-2} yr^{-1} of N+P than other treatments (Figure 4B).

Effect of fertilization on functional diversity of soil microbial community

The AWCD was usually higher in 10 g m^{-2} yr^{-1} of N+P application throughout the incubation than other treatments (Figure 5). The AWCD was lower in N+P than N fertilization at 5 g m^{-2} yr^{-1}, but it was higher in N+P than N fertilization at 10 and 15 g m^{-2}. Eight of nine fertilization treatments (except 15 g m^{-2} yr^{-1} N) significantly increased the Shannon functional diversity of soil bacterial community compared to the control (Figure 6). The N+P fertilization at 10 g m^{-2} yr^{-1} had the highest diversity, followed by 10 and 5 g m^{-2} yr^{-1} N fertilization

In the PCA ordination diagram, samples with similar AWCD datasets were located close to one another, and those dissimilar were located far apart (Figure 7). The PCA results showed that the C substrate utilizing profiles were separated into four distinct groups: 1) 5 and 15 g m^{-2} yr^{-1} of P fertilization; 2) 10 g m^{-2} yr^{-1} of N or P fertilization, 15 g m^{-2} yr^{-1} of N+P fertilization; 3) 5 and 15 g m^{-2} yr^{-1} of N fertilization, 5 and 10 g m^{-2} yr^{-1} of N+P fertilization; and 4) unfertilized control. Metabolic profiles from the treatments within each group were similar to each other, but significantly different from other groups. The first two PCs (PC1 and PC2) explained 42.86% and 8.46% of the variance in AWCD data. The main loadings on the PC1 axis were carboxylic acids (0.87), polymers (0.83), amino acids (0.79), amine/amides (0.78), and carbohydrates (0.75), and on the PC2 axis were amino acids (0.77), and carboxylic acids (0.59).

Discussion

Response of plant biomass, functional group composition, plant and root C:N ratio to fertilization

Fertilization probably increased the growth and biomass of plants, especially grasses, by supplying essential nutrients, such as N and P [21]. The N and P-induced increase in plant biomass was consistent with previous study [21] and N addition experiments in other grasslands [4–9]. Previous study has shown that N and N+P fertilization significantly decreased species richness but N+P addition significantly increased species number of grasses, and species loss was mainly due to the loss of forbs and legumes after fertilization [21]. Studies from other grasslands also have shown that N addition significantly decreased species diversity [5–9]. Fertilization-induced increase in aboveground plant productivity and community coverage resulted in a decrease of light penetration in community canopy. This resulted in reduced growth of legumes and forbs under intense light competition, leading to local extinction of these species [21,36,37].

Fertilization-induced decrease in C:N ratios of all plant functional groups and roots was consistent with previous studies in alpine meadows that N concentrations of grass and forbs increased significantly thereby decreasing C/N ratios while these of legumes were relatively constant after N addition [22,23]. This might be due to the species specific ecological stoichiometry [38]. Grasses had higher N and P use efficiency than other functional groups, which resulted in increase in N concentration after fertilization [38]. The height and biomass of forbs didn't change or even decreased. Legume species were not sensitive to N additions for they could get additional N by N-fixing [22,38]. Studies from other grasslands also showed that N inputs consistently increased soil N availability, leading to tissue and litter N-enrichment in semiarid shrublands [15,39].

Response of soil microbial community to fertilization

Increase soil MBC in N or P fertilization treatments was contrary to our hypothesis and inconsistent with some studies that showed long-term N addition decreased MBC by 20–35% [25,40]. However, in our study site, N fertilization rate was much lower than those in other studies conducted in Europe and North America [25,40]. Increase in aboveground plant biomass and in turn C input to soil following N and P fertilization might be responsible for the observed increase in soil MBC. The increase in carbon substrate to microbes and alleviation of microbial N or P-limitation could increase the microbial growth [41], which was responsible for the increase in AWCD and functional diversity after fertilization in our study. Reduction in microbial biomass and activity with P and N+P fertilization might be due to the toxicity of nutrients that affected MBC and favored one species of microbe

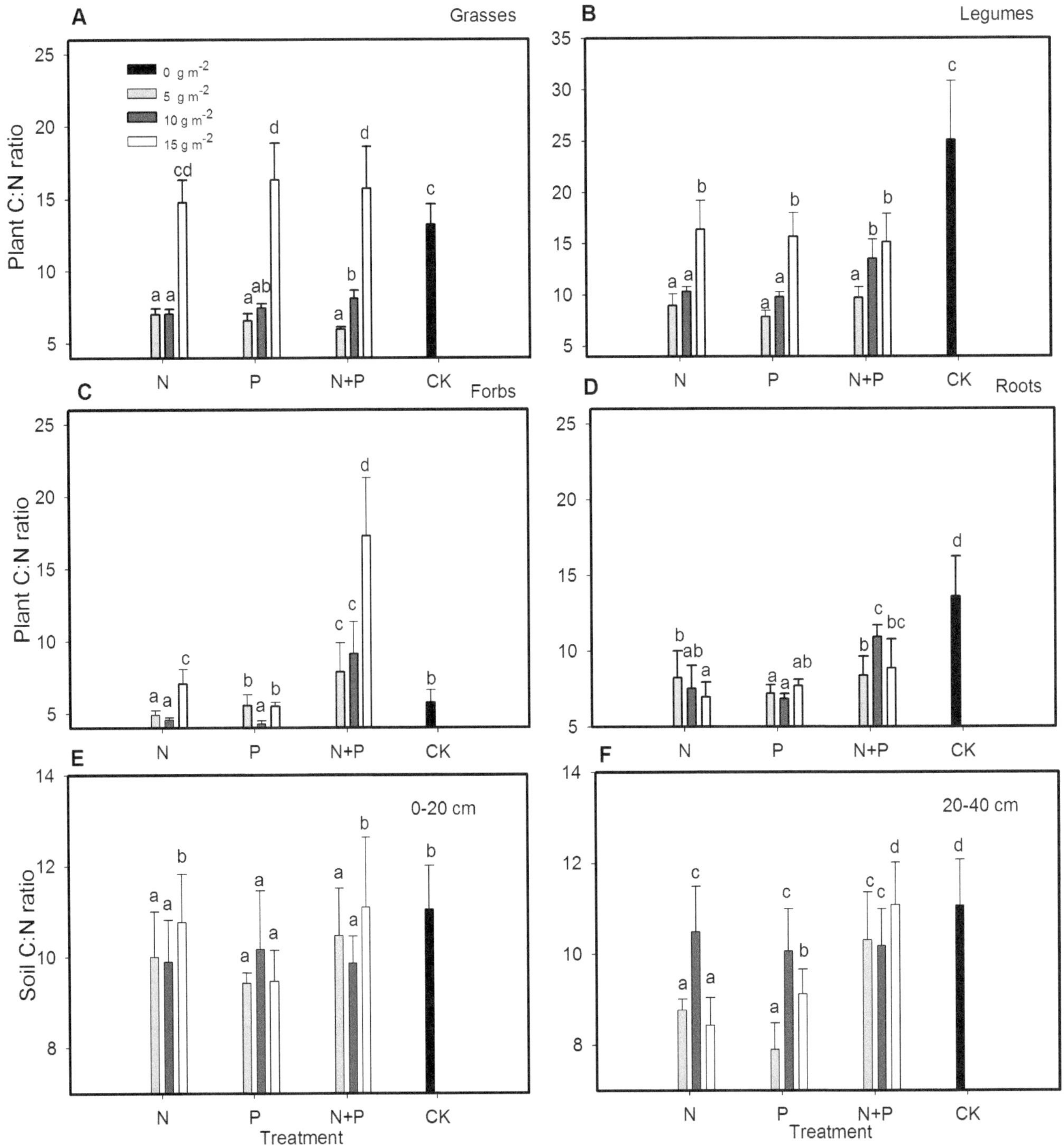

Figure 2. Mean ± SE of C:N ratios of (A) grasses, (B) legumes, (C) forbs, and (D) roots, (E) 0–20 cm soil, (F) 20–40 cm soil as affected by N and P fertilization treatments. Different letters above bars indicate significantly different at $P = 0.05$.

over another [42]. However, increased N availability could stimulate carbon-degrading enzyme activities [43,44] and microbial enzyme shifts could result in increased C mineralization and $q\mathrm{CO_2}$ [46].

Our study showed similar patterns of $q\mathrm{CO_2}$ and C mineralization with fertilization treatments, but inverse to the pattern of MBC. The decrease in C mineralization might be due to a decline in lignin degradation induced by reduction enzyme activity and/or fungal population densities, and modification of the soil microbe composition following fertilization [45–47]. Ramirez et al. [40] reported that there were consistent phylum-level changes in the bacterial communities across soils with the N-amended soil. The N-induced shifts in microbial community structure should yield corresponding shifts in the functional and metabolic potentials of the communities, resulting in a change in decomposition rates [40]. In our study, we did not measure soil microbial composition or enzyme activities; however the results of C substrate utilization data based on Biolog Ecoplate indicated that N, P and N+P

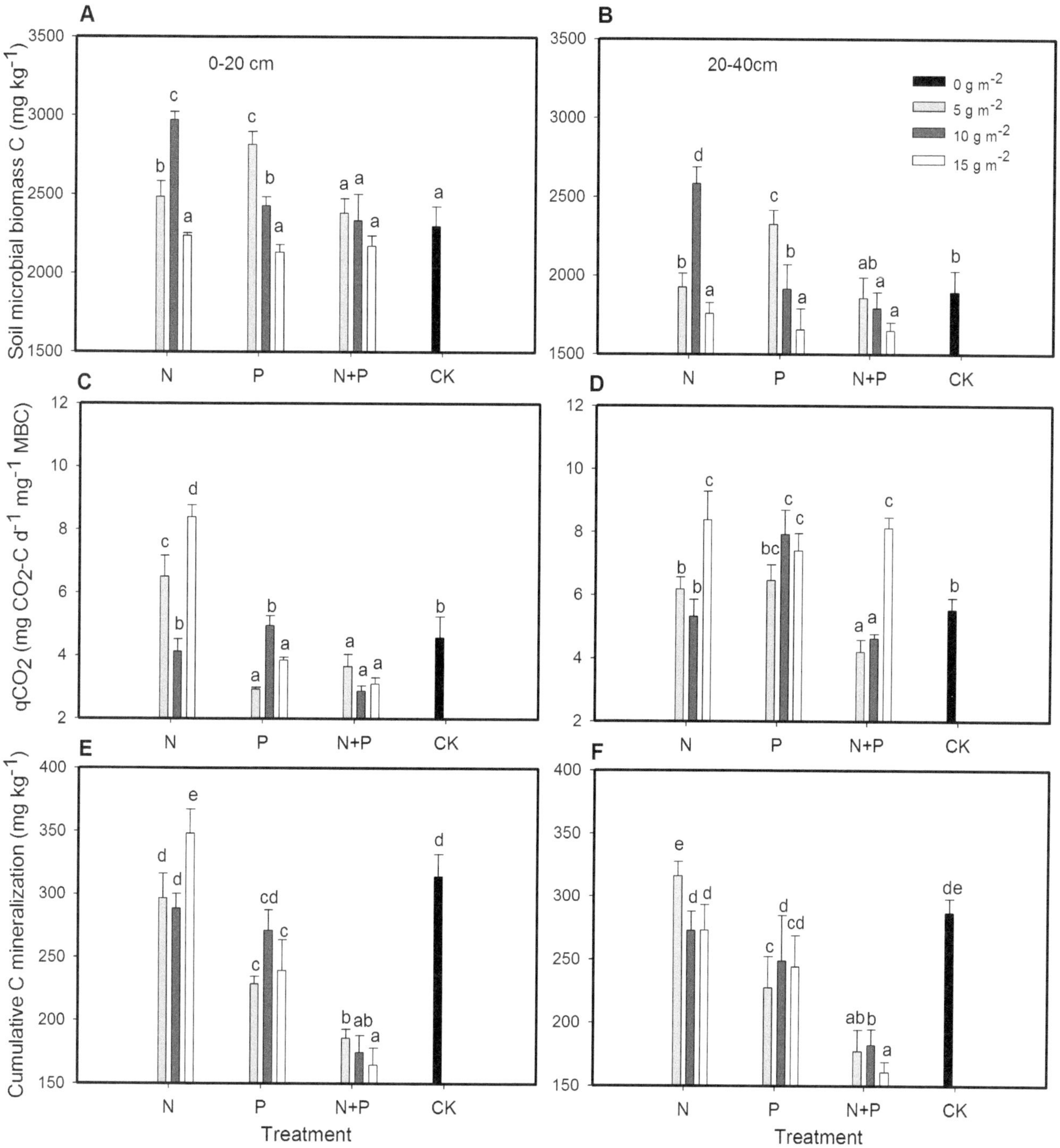

Figure 3. Mean ± SE of (A) and (B) soil microbial biomass C, (C) and (D) metabolic quotient (qCO_2), (E) and (F) cumulative C mineralization in 0–20 cm and 20–40 cm soil layers as affected by N and P fertilization treatments. Different letters above bars indicate significantly different at $P = 0.05$.

fertilization could increase microbial functional diversity and affect microbial carbon utilization and metabolism. Increased rates of N or P fertilization alone increased microbial biomass but reduced their activity, probably due to favorable growth of certain microbes over another due to fertilization. This was likely the negative impact on microbes by decreasing soil pH [42] and increasing toxicity [43] following N fertilization.

The combination of N+P fertilization however increased AWCD and Shannon diversity. This might be due to the following reasons: (i) Alleviation of microbial N and P-limitation and the alteration of community composition, resulting in increased growth in certain microbial population [48]. (ii) Changes in plant community structure and composition which affected soil microbial community. (iii) Grass roots provided unique attachment sites for certain microbial populations, e.g. mycorrhizal fungi, or grass growth stimulated the formation of certain microbial communities after fertilization [49]. This was supported by principal component analysis of carbon utilization data which revealed that microbial

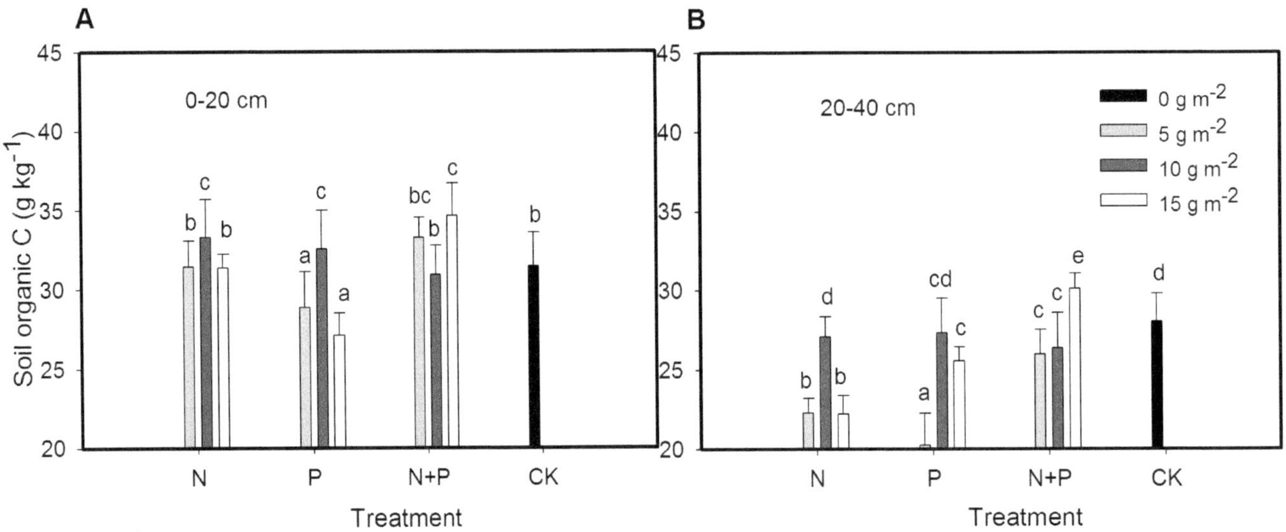

Figure 4. Mean ± SE of soil organic C in 0–20 cm (A) and 20–40 cm (B) as affected by N and P fertilization treatments. Different letters above bars indicate significantly different at $P = 0.05$.

functional diversity depended on the plant community components. These findings suggested that fertilization may alter the structure and composition of soil microbes and modify substrate utilization patterns by soil microbes in the alpine meadow, resulting in a change in C mineralization rates.

Soil organic C

Our data showed that plant community biomass was positively correlated with SOC in 0–20 cm ($R^2 = 0.115$, $P = 0.032$). The increased soil organic C only appeared in 10 g m^{-2} yr^{-1} of N, P and 15 g m^{-2} yr^{-1} of N+P compared to control. The increase in soil organic C agreed with previous studies, which showed that N

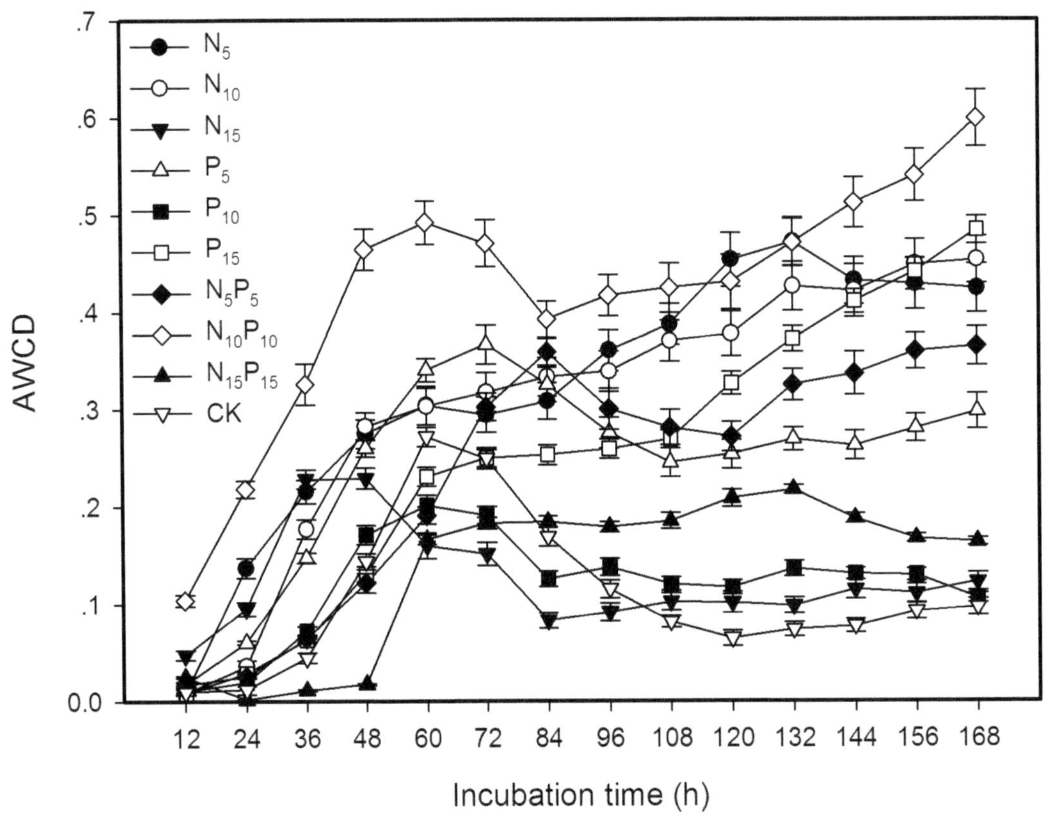

Figure 5. Average well color development (AWCD) (means ± SE) of soil microbial community in BIOLOG Ecoplate as affected by N and P fertilization treatments.

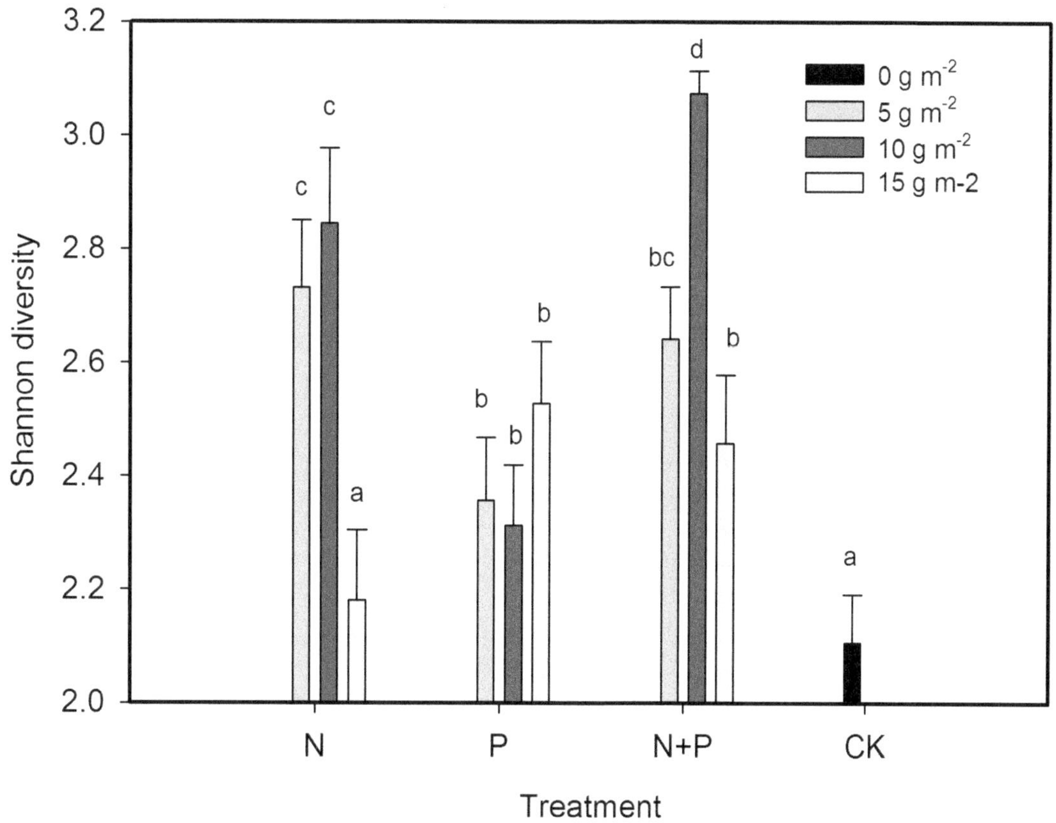

Figure 6. Functional diversity of soil microbial community in 0–20 cm as affected by N and P fertilization treatments Different letters above bars indicate significantly different at *P*=0.05.

and P additions significantly enhanced C stocks and C sequestration in grassland soils [34,49–51]. In our study, plant community biomass increased and grasses accounted for 70% of the total biomass in eight of nine fertilization treatments. So, it was the amount of C input returned mostly from grasses that increased SOC. Also C:N ratio was higher in grasses than forbs in two of the three fertilization treatments. Since 15 g m^{-2} yr^{-1} of N+P fertilization increased grass biomass, the increase in SOC with this treatment was most likely to be due to increased C input and slower decomposition of grasses due to higher C/N ratio. These findings suggested that 15 g m^{-2} yr^{-1} of N+P fertilization can be used to sequester C in such alpine meadow soils. The decrease in soil organic C with 5 and 15 g m^{-2} yr^{-1} of P in 0–20 cm and most of the fertilization treatments (except 15 g m^{-2} yr^{-1} of N+P) in 20–40 cm agreed with several studies which have found that N and P fertilization contributed to decreases in soil organic C pools in tundra ecosystems [16,45]. In our study, SOC concentration was positively correlated with C:N ratios of grasses ($R^2 = 0.112$, $P = 0.035$ in 0–20 cm soil and $R^2 = 0.261$, $P = 0.001$ in 20–40 cm soil respectively) and forbs ($R^2 = 0.124$, $P = 0.026$ in 0–20 cm soil and $R^2 = 0.185$, $P = 0.006$ in 20–40 cm soil respectively), suggesting that the decrease in soil organic C concentration was partially due to changes in plant diversity and productivity of functional groups and decrease in plant C:N ratios, in particular the dominant functional groups—grasses and forbs. Given that soil represented the largest active C pool in terrestrial ecosystems, and, in particular, alpine meadows with higher soil organic C [52], a decrease in SOC and C accumulation might lead to an increase in CO_2 emission to atmosphere.

After 5 years of fertilization, decreased SOC in 5 and 15 g m^{-2} yr^{-1} of P in the upper soil layer and most of the fertilization treatments (except 15 g m^{-2} yr^{-1} of N+P) in the deeper soil layer indicated that soils with these fertilization treatments in these meadows will not be a strong C sink, and that N or P fertilization may even lower the soil organic C pool. Combined with the increase in plant community productivity and changes in the C:N ratios of plant tissues, our results also suggested that a new balance between C sequestration by vegetation and soil organic C mineralization will determine whether alpine meadow ecosystems become a C sink, source or vary across space and time. A study from European grassland showed that multi-nutrient (N, P, K, Mg) additions increased aboveground plant productivity (APP), and N+P had higher APP than N alone, but lower APP than NPKMg fertilization [9]. However, soil C sequestration was increased by N-only additions, not multi-nutrient fertilization [9]. An experiment from our study site showed that APP increased with the increasing number of added limiting resources (N, P, K, and water) and APP was higher in N+P treatment than in N+P+K treatment or N, P alone treatment [19]. Therefore, in alpine meadows, both N and P fertilization need to be applied at adequate rates (15 g m^{-2} yr^{-1}) to increase C sequestration, not N or P alone. Potassium combination with N and P might not further increase plant biomass according to study of Ren et al. [21], but further studies were needed to examine their effects on SOC in alpine meadows.

Conclusions

Nitrogen alone and N+P fertilization at rates of 15 g m^{-2} yr^{-1} doubled aboveground biomass of plant communities, driven by a

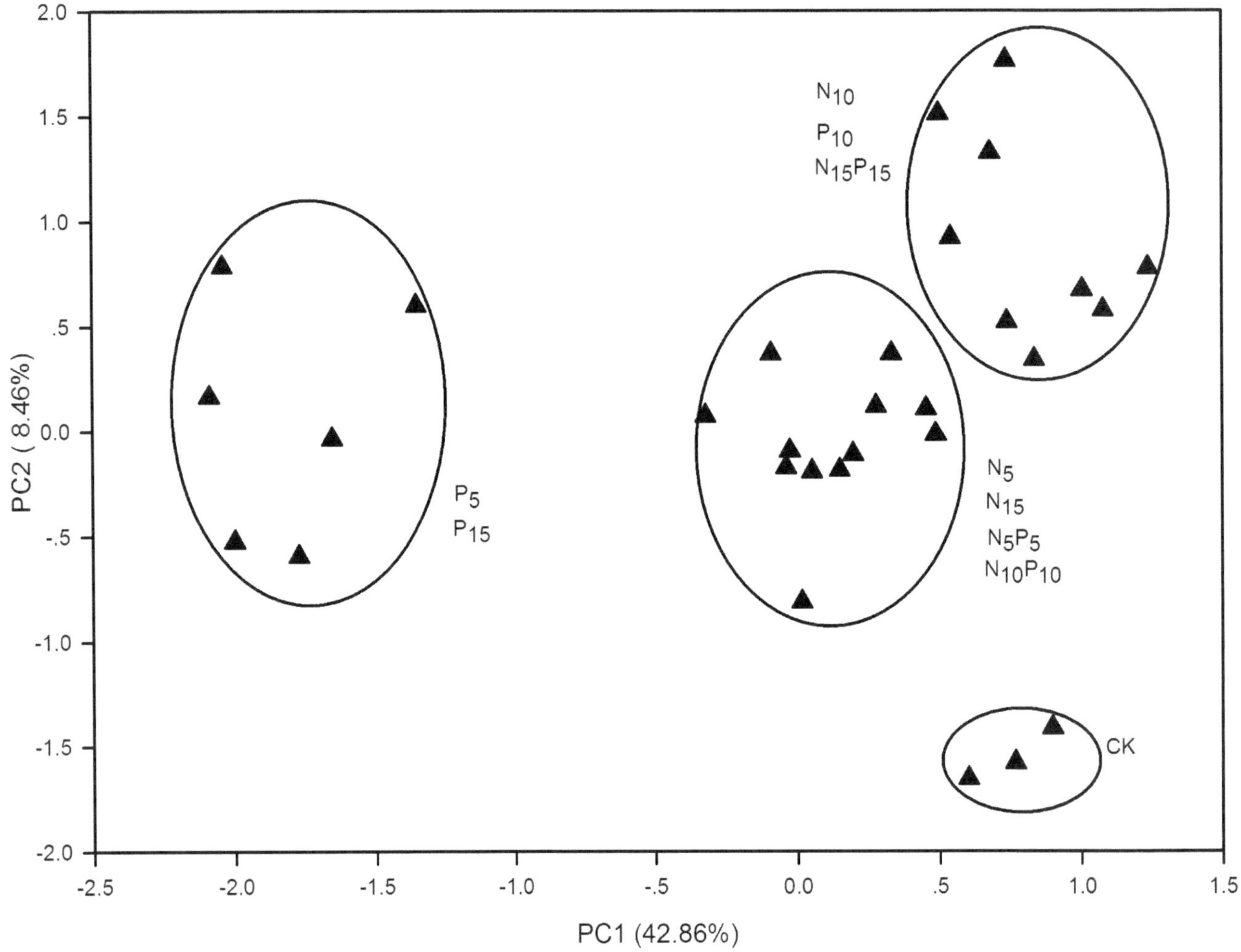

Figure 7. Principal component analysis results according to average well color development data of carbon substrate groups at 168 h in BIOLOG Ecoplates. N, nitrogen fertilization; P, phosphorus fertilization; CK, unfertilized control; subscript numbers 5, 10, 15 indicate corresponding fertilization rates (g m^{-2} yr^{-1}).

fourfold increase in grass biomass. The combination of N and P fertilizers increased SOC but reduced microbial biomass and activity and C mineralization compared to N or P alone or no fertilization. The N+P at rate of 15 g m^{-2} yr^{-1} can be used in increasing aboveground biomass of plant community and especially grass biomass and dominance, resulting in increased soil C accumulation at the surface layer in these alpine meadows.

Acknowledgments

We are grateful to Alpine Meadow Ecosystem Research Station of Lanzhou University for providing experimental site.

Author Contributions

Conceived and designed the experiments: JHL GW. Performed the experiments: YJY BWL. Analyzed the data: JHL JK. Contributed reagents/materials/analysis tools: WJL. Wrote the paper: JHL.

References

1. Post WM, Emanuel WR, Zinke PJ, Stangenberger AG (1982) Soil carbon pools and world life zones. Nature 298: 156–159.
2. Scurlock JMO, Hall DO (1998) The global carbon sink: a grassland perspective. Glob Chang Biol 4(2):229–233.
3. Yang H, Shaojie MU, Chengming SUN, Jianlong LI, Weimin JU (2011) Summary of research on estimation of organic carbon storage in grassland ecosystem. Chin J Grassland 33: 107–114.
4. Bai YF, Wu JG, Clark CM, Naeem S, Pan QM, et al. (2010) Tradeoffs and thresholds in the effects of nitrogen addition on biodiversity and ecosystem functioning: evidence from inner Mongolia Grasslands. Glob Chang Biol 16: 358–372.
5. Clark CM, Tilman D (2008) Loss of plant species after chronic low-level nitrogen deposition to prairie grasslands. Nature 45: 712–715.
6. Isbell F, Tilman D, Polasky S, Binder S, Hawthorne P (2013) Low biodiversity state persists two decades after cessation of nutrient enrichment. Ecol Lett 16: 454–460.
7. Stevens CJ, Dise NB, Mountford JD (2004) Impacts of nitrogen deposition on the species richness of grasslands. Science 303: 1676–1679.
8. Wedin DA, Tilman D (1996) Influence of nitrogen loading and species composition on the carbon balance of grasslands. Science 274: 1720–1723.
9. Fornara DA, Banin L, Crawley MJ (2013) Multi-nutrient vs. nitrogen-only effects on carbon sequestration in grassland soils. Glob Chang Biol 19: 3848–3857.
10. Wang GX, Qian J, Cheng GD, Lai YM (2002) Soil organic carbon pool of grassland soils on the Qinghai-Tibetan Plateau and its global implication. Sci Total Environ 291: 207–217.

11. Allison SD, Lu Y, Weihe C, Goulden ML, Martiny AC, et al. (2013) Microbial abundance and composition influence litter decomposition response to environmental change. Ecology 94(3):714–725.
12. Lovell RD, Jarvis SC, Bardgett RD (1995) Soil microbial biomass and activity in long term grassland: Effects of management changes. Soil Biol Biochem 27(7):969–975.
13. Li YC, Song CC, Hou CC, Song YY (2010) Effects of nitrogen input on meadow marsh soil N_2O emission and organic carbon mineralization. Chin J Ecol 29(11):2091–2096.
14. Bai JB, Xu XL, Song MH, He YT, Jiang J, et al. (2011) Effects of temperature and added nitrogen on carbon mineralization in alpine soils on the Tibetan Plateau. Ecol Environ Sci 20(5):855–859.
15. Vourlitis GL, Zorba G (2007) Nitrogen and carbon mineralization of semi-arid shrubland soil exposed to long-term atmospheric nitrogen deposition. Biol Fert Soils 43: 611–615.
16. Mack MC, Schuur EAG, Bret Harte MS, Shaver GR, Chapin III FS (2004) Ecosystem carbon storage in arctic tundra reduced by long term nutrient fertilization. Nature 431: 440–443.
17. Bradford MA, Fierer N, Jackson RB, Maddox TR, Reynolds JF (2008) Nonlinear root-derived carbon sequestration across a gradient of nitrogen and phosphorous deposition in experimental mesocosms. Glob Chang Biol 14: 1113–1124.
18. Ni J (2002) Carbon storage in grasslands of China. J Arid Environ 50: 205–218.
19. Ren ZW, Li Q, Chu CJ, Zhao LQ, Zhang JQ, et al. (2010) Effects of resource additions on species richness and *ANPP* in an alpine meadow community. J Plant Ecol 3: 25–31.
20. Xin XJ (2011) Effects of N, P additions on above/below-ground biomass allocation and plant functional types' composition in a sub-alpine meadow. Thesis, Lanzhou University.
21. Qi R (2013) Response of plant community to nitrogen and phosphorous additions in sub-alpine meadows of the Qinghai Tibetan Plateau. Thesis, Lanzhou University.
22. Chen LY (2010) Effect of N, P addition on N:P stioehiometry of different functional groups in *Potentilla fruticosa* community in a sub-alpine meadow. Thesis, Lanzhou University.
23. Zhang RY, Gou X, Bai Y, Zhao J, Chen LY, et al. (2011) Biomass fraction of graminoids and forbs in N-limited alpine grassland N:P stoichiometry. Polish J Ecol 59(1):105–114.
24. Fornara DA. Tilman D (2012) Soil carbon sequestration in prairie grasslands increased by chronic nitrogen addition. Ecology 93: 2030–2036.
25. Liu L, Greaver TL (2010) A global perspective on belowground carbon dynamics under nitrogen enrichment. Ecol Lett 13: 819–828.
26. Gong ZT (1999) Chinese Soil Taxonomy: Theories Methods and Applications. Science Press, Beijing.
27. Kalembasa SJ, Jenkinson DS (1973) A comparative study of titrimetric and gravimetric methods for the determination of organic carbon in soil. J Sci Food Agric 24: 1085–1090.
28. Vance ED, Brookes PC, Jenkinson DS (1987) An extraction method for measuring soil microbial biomass C. Soil Biol Biochem 19: 703–707.
29. Song MH, Jiang J, Cao GM, Xu XL (2010) Effects of temperature, glucose and inorganic nitrogen inputs on carbon mineralization in a Tibetan alpine meadow soil. Eur J of Soil Biol 46: 375–380.
30. Anderson TH, Domsch KH (1990) Application of ecophysiological quotients (qCO_2 and qD) on microbial biomass from soils of different cropping histories. Soil Biol Biochem 22: 251–255.
31. Garland JL, Mills AL (1991) Classification and characterization of heterotrophic microbial communities on the basis of patterns of community-level sole-carbon-source utilization. Appl Enviro Microbio 7: 2351–2359.
32. Insam H (1997) A new set of substrates proposed for community characterization in environmental samples. In: Insam H, Rangger A, editors. Microbial Communities. Berlin Heidelberg: Springer pp. 259–260.
33. Li JH, Jiao SM, Gao RQ, Bargett RD (2012) Differential effects of legume species on the recovery of soil microbial communities, and carbon and nitrogen contents in abandoned fields of the Loess Plateau, China. Environ Manage 50: 1193–1203.
34. Garland JL (1996) Analytical approaches to the characterization of samples of microbial communities using patterns of potential C source utilization. Soil Biol Biochem 28: 213–221.
35. Zhang CB, Wang J, Liu WL, Zhu SX, Ge HL, et al. (2010) Effects of plant diversity on microbial biomass and community metabolic profiles in a full-scale constructed wetland. Ecol Eng 36 (1):62–68.
36. Suding KN, Collins SL, Gough L, Clark C, Cleland EE, et al. (2005) Functional- and abundance- based mechanisms explain diversity loss due to N fertilization. PNAS 102: 4387–4392.
37. Hautier Y, Niklaus PA, Hector A (2009) Competition for light causes plant diversity loss after eutrophication. Science 324: 636–638.
38. Güsewell S (2004) N:P ratios in terrestrial plants: variation and functional significance. New Phytol 164: 243–266.
39. Vourlitis GL, Pasquini SC, Mustard R (2009) Effects of dry-season N input on the productivity and N storage of Mediterranean-type shrublands. Ecosystems 12: 473–488.
40. Ramirez KS, Craine JM, Fierer N (2012) Consistent effects of nitrogen amendments on soil microbial communities and processes across biomes. Glob Chang Biol 18: 1918–1927.
41. Stapleton LM, Crout NMJ, Sawstrom C, Marshall WA, Poulton PR, et al. (2005). Microbial carbon dynamics in nitrogen amended Arctic tundra soil: measurement and model testing. Soil Biol Biochem 37: 2088–2098.
42. Vitousek PM, Aber JD, Howarth RW, Likens GE, Matson PA, et al. (1997) Human alteration of the global nitrogen cycle: sources and consequences. Ecol Appl 7: 737–750.
43. Keeler B, Hobbie S, Kellogg L (2009) Effects of long-term nitrogen addition on microbial enzyme activity in eight forested and grassland sites: Implications for litter and soil organic matter decomposition. Ecosystems 12: 1–15.
44. Koyama A, Wallenstein MD, Simpson RT, Moore JC (2013) Carbon-Degrading Enzyme Activities Stimulated by Increased Nutrient Availability in Arctic Tundra Soils. PLoS ONE 8(10): e77212. doi:10.1371/journal.pone.0077212
45. Carreiro MM, Sinsabaugh RL, Repert DA, Parkhurst DE (2000) Microbial enzyme shifts explain litter decay responses to simulated N deposition. Ecology 81: 2359–2365.
46. Liu K, Crowley D (2009) Nitrogen deposition effects on carbon storage and fungal:bacterial ratios in coastal sage scrub soils of southern California. J Environ Qual 38: 2267–2272.
47. Bradley K, Drijber RA, Knops J (2006) Increased N availability in grassland soils modifies their microbial communities and decreases the abundance of arbuscular mycorrhizal fungi. Soil Biol Biochem 38: 1583–1595.
48. Wardle DA, Bardgett RD, Klironomos JN, Setala H, van der Putten WH, et al. (2004) Ecological linkages between aboveground and belowground biota. Science 304: 1629–1633.
49. Malhi SS, Harapiak JT, Nyborg M, Gill KS, Monreal CM, et al. (2003) Total and light fraction organic C in a thin Black Chernozemic grassland soil as affected by 27 annual applications of six rates of fertilizer N. Nutr Cycl Agroecosys 66: 33–41.
50. Biudes MS, Vourlitis GL (2012) Carbon and Nitrogen Mineralization of a Semiarid Shrubland Exposed to Experimental Nitrogen Deposition. Soil Sci. Soc. Am. J. 76: 2068–2073.
51. Li LJ, Zeng DH, Yu ZY, Fan ZP, Mao R (2010) Soil microbial properties under N and P additions in a semi-arid, sandy grassland. Biol Fert Soils 46(6):653–658.
52. Zhang JX, Cao GM, Zhou DW, Hu QW, Zhao XQ (2003) The carbon storage and carbon cycle among the atmosphere, soil, vegetation and animal in the *Kobresia humilis* alpine meadow ecosystem. Acta Ecol Sin 23(4):627–634.

Characterization of Arbuscular Mycorrhizal Fungus Communities of *Aquilaria crassna* and *Tectona grandis* Roots and Soils in Thailand Plantations

Amornrat Chaiyasen[1], J. Peter W. Young[2], Neung Teaumroong[3], Paiboolya Gavinlertvatana[4], Saisamorn Lumyong[1]*

1 Department of Biology, Faculty of Science, Chiang Mai University, Chiang Mai, Thailand, **2** Department of Biology, University of York, York, United Kingdom, **3** Schoool of Biotechnology, Institute of Agricultural Technology, Suranaree University of Technology, Nakhon Ratchasima, Thailand, **4** Thai Orchid Labs Co. Ltd., Khannayao, Bangkok, Thailand

Abstract

Aquilaria crassna Pierre ex Lec. and *Tectona grandis* Linn.f. are sources of resin-suffused agarwood and teak timber, respectively. This study investigated arbuscular mycorrhizal (AM) fungus community structure in roots and rhizosphere soils of *A. crassna* and *T. grandis* from plantations in Thailand to understand whether AM fungal communities present in roots and rhizosphere soils vary with host plant species and study sites. Terminal restriction fragment length polymorphism complemented with clone libraries revealed that AM fungal community composition in *A. crassna* and *T. grandis* were similar. A total of 38 distinct terminal restriction fragments (TRFs) were found, 31 of which were shared between *A. crassna* and *T. grandis*. AM fungal communities in *T. grandis* samples from different sites were similar, as were those in *A. crassna*. The estimated average minimum numbers of AM fungal taxa per sample in roots and soils of *T. grandis* were at least 1.89 vs. 2.55, respectively, and those of *A. crassna* were 2.85 vs. 2.33 respectively. The TRFs were attributed to Claroideoglomeraceae, Diversisporaceae, Gigasporaceae and Glomeraceae. The Glomeraceae were found to be common in all study sites. Specific AM taxa in roots and soils of *T. grandis* and *A. crassna* were not affected by host plant species and sample source (root vs. soil) but affected by collecting site. Future inoculum production and utilization efforts can be directed toward the identified symbiotic associates of these valuable tree species to enhance reforestation efforts.

Editor: Zhengguang Zhang, Nanjing Agricultural University, China

Funding: This research was supported by the Thailand Research Fund for Research-Team Promotion Grant (RTA5580007) and the Commission of Higher Education for National Research University (A1) (http://www.trf.or.th/). AC was funded by the Thailand Research Fund; The Royal Golden Jubilee PhD Program (PHD/0150/2550: http://rgj.trf.or.th/indexth.asp). The funders had no role in study design, data collection and analysis, decision to publish, or preparation of the manuscript.

Competing Interests: Paiboolya Gavinlertvatana is employed by a commercial company (Thai Orchid Labs Co. Ltd).

* Email: saisamorn.l@cmu.ac.th

Introduction

Tropical forests are disappearing at the rate of 13.5 million hectares each year owing to logging, burning and clearing for agriculture and shifting cultivation [1]. At present, managed woodlands are required for timber and non-timber products in many countries. *Aquilaria crassna* Pierre ex Lec. (agarwood) and *Tectona grandis* Linn.f. (teak) are perennial plants that are used extensively to provide aromatic resin-infused wood products [2] and good quality teak wood products [3], respectively. The depletion of wild trees from indiscriminate cutting of *Aquilaria* species has resulted in the trees being listed and protected as endangered species. All *Aquilaria* species were listed in Appendix II of the Convention on International Trade in Endangered Species of Wild Fauna and Flora in 2005 [4]; however, a number of countries have outstanding reservations regarding that listing. Plantlets of *A. crassna* and *T. grandis* are produced in Thailand for domestic and foreign markets such as Jamaica, Guatemala, Mozambique, Sri Lanka, Indonesia, Laos, Malaysia, and Australia. Most *T. grandis* plantations in Thailand are planted in the northern provinces such as Chiang Mai, Chaing Rai and Phetchabun, while *A. crassna* plantations are mostly in eastern (Rayong, Trat and Chanthaburi provinces) and central (Nakhon-Nayok) Thailand.

Arbuscular mycorrhizal fungi (AMF) are soil fungi in the phylum Glomeromycota [5] that are mutualistically associated with roots of a wide spectrum of tropical and temperate tree species [6]. AM fungi have major effects on plant growth such as enhance the nutrient uptake by plant roots (especially phosphorus), particularly in low fertility soils [7,8], protected plant against drought stress [9,10], protect plant from soil-borne plant pathogenic infection [11], and improve soil aggregate stability through the action of mycelia and glomalin [12,13,14]. AMF inocula applied to plantlets and plant seedlings increased growth during early tree establishment in the field [2,15,16]. AM fungi have been used to inoculate and enhance growth of *T. grandis* [3,17] and *Aquilaria* spp. [2,18] prior to planting out. Therefore, studying the AM fungal communities of these plants in the field should aid plantation establishment and reforestation efforts. Information about the diversity of AM fungi associated with both plants has been reported mostly from natural forests in India [19,20,21,22,23] and only in *T. grandis* from Thailand [24]. These studies characterized communities based upon spore morphology. However, there are no reports of AM fungal communities of either tree using molecular tools. Identification of AM fungi based on spore morphology inevitably has some limitations, e.g. omission of AM fungi that did not produce spores during the sampling period and inability to identify the AM fungi within the roots.

PCR-based methods have been widely used in AM fungal community studies. Various studies have designed sets of specific primers for AM fungi [25,26,27] to facilitate rapid detection and identification directly from field-grown plant roots. Previously, Terminal restriction fragment length polymorphism (T-RFLP) has been used to study the AM fungi community in roots of arable crops [28], perennial herbs [29], herbaceous flowering plants [30], grass species [31,32], grass species with herbaceous flowering plants [33,34], and temperate deciduous trees [35]. Populations of AM fungi have been well studied in a number of ecosystems around the world, but there is scant information available for tropical forests and plantations of tropical and sub-tropical species.

This study provides the first molecular community analysis of AM fungi associated with field-collected roots and rhizosphere soils of the tropical trees *A. crassna* and *T. grandis*, and is part of a long term goal of optimizing AM fungus inoculation strategies to enhance reforestation efforts with these trees. It also provides an early insight into the biodiversity of AM fungi in Thailand to test the hypothesis that differences in AM fungal communities present in the roots and rhizosphere soils are determined by collecting sites, host plant species, and local environmental factors.

Materials and Methods

Ethics Statement

No specific permits were required to carry out research in the plantations: Chiang Mai (99°15' E, 18°58' N), Chiang Rai (99°29′ E/19°14′ N), Nakhon-Nayok (101°16′ E, 14°9′ N), Phetchabun (100°47' E, 16°2' N) and Thai Orchids Lab Ltd. (101°7′ E, 14°16′ N). The field studies did not involve endangered or protected species in Thailand. *Aquilaria crassna* is defined to be the forbidden forest item in only the forest area as the Forest Act. Therefore, the *A. crassna* planting and deforestation in the land of ownership is legal. All *A. crassna* samples were obtained from privately-owned plantations and are therefore not subject to the restrictions of the Forest Act of Thailand. Permission to sample the *T. grandis* and *A. crassna* were granted by the landowner.

Study sites and sampling

Rhizosphere soils and roots were sampled from plantations of *T. grandis* and *A. crassna* in four provinces of Thailand (Table 1). Two sampling sites were located in Chiang Mai and Chiang Rai provinces in the northern region. These sites are monocultures of *T. grandis* planted at 2 m spacings and left to grow naturally with accumulated leaf litter and negligible understory perennial plants. Only roots attached to the main roots of *T. grandis* were sampled. At the sites in the central region; Nakhon-Nayok and Thai Orchids Lab Ltd., Nakhon-Nayok province, and in the northern region; Phetchabun province, *T. grandis* and *A. crassna* were planted alternately 2 m apart at Thai Orchids Lab Ltd. and Phetchabun. At both sites, weeds were controlled by ploughing and herbicide treatment. Thus, both species were planted without any above-ground vegetation, while in Nakhon-Nayok site, *A. crassna* was left to grow naturally. Paired soil and root samples from each plant species were randomly collected from 3 locations per site at 0–15 cm depth within 50 m^2 and taken to the laboratory. All collections were carried out in July 2010. Root fragments were washed free of soil and air dried on tissue paper. Root fragments and soil samples were stored frozen at −20°C until further analysis.

Soil analyses

Soil pH and electrical conductivity (EC) were determined in a 1:1 soil: water slurry by direct measurement with pH-meter (Waterproof EC Testr, EUTECH instruments). Available phosphorus was measured employing the Bray II method [36]. Total inorganic nitrogen, exchangeable potassium and soil organic carbon were quantified following the methods of soil analysis outlined in Sparks et al. [37].

Molecular analysis

Three replicate rhizosphere soil and root samples from each plant species were used to represent each site of collection. DNA was extracted from rhizophere soils and roots using the PowerSoil DNA isolation kit (MoBio Laboratories, CA) and Nucleospin Plant II (Macherey-Nagel GmbH & Co. KG, Düren), respectively according to the manufacturers' protocols. DNAs were amplified separately by nested PCR and then 20 µl of each of the three replicates from each sampling site were pooled and purified before restriction digestion [38]. The first round of AMF-specific PCR amplification was performed using the unlabelled primers AML1 and AML2 [26] with 30 cycles. In this first PCR, 40 µl reactions were carried out and each mixture contained 10 pmol of each primer, 1 unit of Taq polymerase (Promega) and 25 mM of each dNTP (Invitrogen) in manufacturer's reaction buffer (Promega). PCR was performed on a PTC100 thermocycler (MJ Research) with an initial denaturation at 94°C for 15 min, followed by 30 cycles of denaturation at 94°C for 30 s, annealing at 57°C for 45 s, extension at 72°C for 45 s, followed by a final extension of 72°C for 5 min. PCR products were visualized on a 1% agarose gel containing 0.1 × SybrSafe (Invitrogen). The second round primers, 0.5 unit of Taq polymerase (Promega) and 20 pmol of HEX-labeled NS31 and FAM-labeled AML3 were added directly into 24 µl of each resulting product. Second-round PCR was conducted with 5 additional cycles using the same PCR conditions as the first PCR. The PCR products were purified using the QIAquick PCR purification kit (Qiagen). The purified PCR products were digested separately with the selected restriction enzymes, HinfI, Hsp92II and MboI (Promega) [31,39] for 3 h at 37°C. Digested products were purified as mentioned above. Terminal restriction fragment (TRF) sizes from each sample were determined using the ABI PRISM 3130 Genetic Analyzer System (Applied Biosystems) with GeneScan LIZ-600 (Applied Biosystems) as internal size standards. The GeneMapper software (Applied Biosystems) was used for the analysis of fragment data. To reduce data noise, only fragments containing intensity above a baseline threshold (50 fluorescence units) were recorded. Relative peak heights were calculated and fragments with an average relative abundance <5% were excluded from further analysis.

Screening and DNA sequence analysis

The remainders of the first PCR products were combined and purified using the QIAquick PCR purification kit (Qiagen).

Table 1. Chemical characteristic of soils (mean value ± SEM) in wet season (July 2010) which soils and roots were sampled.

Study plot	Soil pH[a]	Electrical conductivity [a]	Soil organic carbon (%)[a]	Total inorganic N (g kg^{-1} soil)[a]	Available P (mg kg^{-1} soil)[a]	Exchangeable K (mg kg^{-1} soil)[a]	Agricultural management
Phetchabun (PB)	5.77±0.25ab	0.18±0.02a	4.26±0.45ab	222±15a	156±72ab	449±163a	plowing, organic fertilizer, herbicide
Thai Orchid Lab (TO)	6.68± 0.28a	0.14±0.01a	4.15±1.48ab	188±57a	370±158a	296±81a	plowing, organic fertilizer, herbicide
Chiang Mai (CM)	5.70± 0.26bc	0.11±0.05a	6.10 ±0.87a	210±68a	171±133ab	284±10a	No management
Chiang Rai (CR)	5.23±0.11c	0.10±0.04a	2.83±0.25b	140±17a	24±3b	243±28a	No management
Nakhon Nayok (NN)	6.19±0.14ab	0.21±0.07a	3.08±0.66b	200±8a	149±63ab	347±78a	No management

[a]Means of three observations. Means in the same column followed by the same letter are not significantly different ($\alpha = 0.05$).

Purified DNA was cloned into pGEM-T Easy Vector (Promega) and transformed into *Escherichia coli* JM109. One hundred transformants were selected randomly and their insertion checked by PCR using the same primers, AML1 and AML2. The amplified DNAs were digested by the restriction enzymes HinfI and Hsp92II separately. One clone of each RFLP type was screened and sequenced using sequencing primers SP6 and T7 on an ABI PRISM 3130 Genetic Analyzer System (Applied Biosystems). Sequences were trimmed to the NS31-AML3 region and virtually digested with the restriction enzymes HinfI, Hsp92II, and MboI using an online restriction mapping website (RestrictionMapper).

Phylogenetic analysis

Phylogenetic analysis was carried out on the sequences obtained in this study and those corresponding to the closest matches from Genbank, as well as sequences from cultured AMF taxa including representatives of the major groups of Glomeromycota from GenBank. All sequences obtained from this study were aligned by ClustalX using the BioEdit sequence alignment editor [40] along with 28 AMF sequences from GenBank. The aligned SSU rRNA dataset was trimmed to 450 bp by excluding the terminal primer sequences. A neighbour-joining (NJ) phylogeny was constructed using PAUP*4b10 [41] with the Kimura 2-parameter model and 1000 bootstraps. The nucleotide sequences of the clones retrieved in this study have been deposited in GenBank (accession numbers JQ8643324-JQ864355).

Statistical analysis

The total number of TRFs was used as an AM fungal community diversity measurement [31]. The main and interaction effects of collecting sites, host plant species and sample source (root vs. soil) on number of TRFs using three restriction enzymes were tested with two-way factorial ANOVA using SPSS 11.5 for Windows (SPSS Inc., Chicago, IL, USA). Jaccard similarity coefficients were calculated for the T-RFLP patterns of root and soil samples of both plants, which were clustered by the unweighted pair-group average (UPGMA) algorithm with 1000 bootstrap replicates to obtain confidence estimates. These calculations were performed using FreeTree [42] and the results displayed using TreeView [43].

Results

Soil analyses and correlation with TRFs

Chemical characteristics of soil varied among sites (Table 1). Soil pH values ranged from 5.23 to 6.68. No significant different was observed in electrical conductivity, exchangeable potassium, and total inorganic nitrogen. Available phosphorus in soils tended to be highest at the Thai Orchid Lab site (370 mg kg^{-1}soil) and differed significantly from the Chiang Rai site (24 mg kg^{-1}soil). Soil organic carbon was highest at the Chiang Mai site (6.10%) and differed significantly from the Chiang Rai and Nakhon Nayok sites. Pearson correlation analysis between the soil properties measured and TRFs showed that TRFs were positively correlated with available phosphorus, organic matter, and pH (Table S1).

AM fungal community of root and soil samples from *T. grandis* and *A. crassna*

The total number of different TRFs was used as a measure of AM fungal community diversity. Thirty eight TRFs were found in total for the AML3 (FAM-labelled) primer, while the NS31 (HEX-labelled) primer identified 30 TRFs. Since the AML3 primer revealed many more TRFs than the NS31 primer, only the AML3 fragments were used. Overall, in the roots and soils of *T. grandis* and *A. crassna*, we found 13 different AML3 TRFs after digestion with HinfI, 14 after digestion with Hsp92II and 11 after digestion with MboI. The mean number of TRFs in *T. grandis* root and soil samples was 5.67 and 7.67, respectively when the TRF data of the three enzymes were pooled (Figure 1). It is possible to estimate the minimum average number of AM fungi colonizing the *T. grandis* root samples by dividing the average number of TRFs by 3 (three enzymes and one labeled end) [31]. Thus, there were on average at least 1.89 fungal taxa colonizing each *T. grandis* root sample and 2.55 fungal taxa in surrounding soils, respectively. The values for *A. crassna* were at least 2.85 fungal taxa per root sample and 2.33 fungal taxa in surrounding soils. The mean number of TRFs per sample was not significantly affected by source of samples (root and soil) ($F = 0.159$, $P = 0.693$) and host plant ($F = 3.452$, $P = 0.074$) (Table S2), but there was a statistically significant effect of collecting sites ($F = 42.77$, $P < 0.01$), and a significant interaction among those three factors (Table S2). The cluster analysis of TRF patterns in roots (R-) and rhizosphere soils (S-) of *A. crassna* and *T. grandis*, based on Jaccard similarities, showed no significant grouping of samples by sites and source of samples (root and soil) (Figure 2a). This suggested that the AM fungal community in roots and rhizosphere soils was almost independent in *A. crassna* (A) and *T. grandis* (T) plots. Some TRF patterns in roots and rhizosphere soils that were collected from the same site were similar, e.g. R-CRT versus S-CRT and R-TOA versus R-TOT. Combining roots and rhizosphere soils of each plant by sampling site (CM: Chiang Mai, CR: Chiang Rai, NN: Nakhon-Nayok, PB: Phetchabun and TO: Thai Orchids Lab) indicated a tendency for *T. grandis* plots to be grouped together (PBT, CMT and TOT) as well as some *A. crassna* plot samples (PBA and TOA) (Figure 2b). This suggests that the AM fungal community associated with each tree species was more similar across plots than were communities for different trees species at the same location. The response for CRT and NNA, however, does not support this.

Occurrence of AM fungi in soils and roots of both plants

Nearly all of the distinct TRFs (31 out of 38) were found in both host plant species (Figure 3). There were some differences in AM fungal communities between *T. grandis* and *A. crassna* because the TRF 329c (TRFs are identified by their relative mobility and a code indicating the restriction enzyme that generated them: a: MboI, b: HinfI and c: Hsp92II) was not found in *T. grandis*, while 5 TRFs (135c, 141b, 158c, 176b, and 435b) were not found in *A. crassna*. Comparison of the population in roots and soils of *T. grandis* (Fig 3a) showed that 6 TRFs (135c, 158c, 176b, 181c, 435b and 438b) were found only in roots, while 141b and 281a were only found in soils. In *A. crassna* (Figure 3b), TRFs 176c, 181c and 438b were only found in root samples.

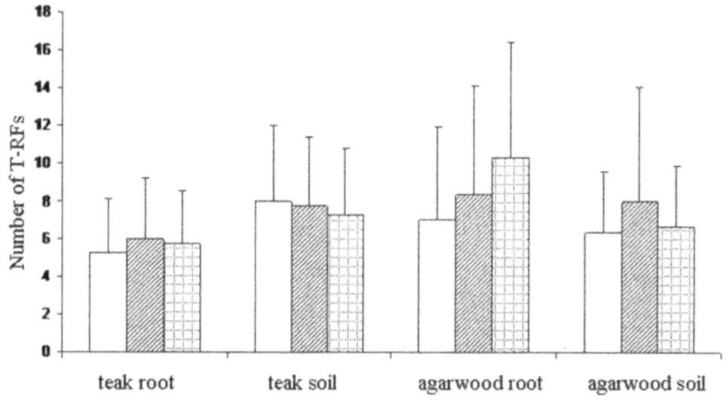

Figure 1. Effects of host plant, *Aquilaria crassna* (agarwood) and *Tectona grandis* (teak), and source of samples (root and soil) on mean number of terminal restriction fragments (TRFs) per sample using three restriction enzymes *MboI* (open bars), *HinfI* (hatched bars) and *Hsp92II* (cross-hatched bars). Values are mean ± SEM (n = 4 for teak and n = 3 for agarwood).

Sequence and phylogenetic analysis

Clones were selected for sequencing on the basis of HinfI and Hsp92II RFLP typing. DNA sequences of 32 selected clones were determined, 7 clones from *A. crassna* and 25 clones from *T. grandis*. Predicted TRFs from the 32 virtually digested clone sequences were compared to observed TRFs from all three restriction enzymes (Table S3). A difference in size of up to 7 nucleotides was accepted as a match, because migration in capillary electrophoresis is sequence-specific, so that mobility (in rmu) is only approximately equivalent to sequence length (in bp). All predicted TRFs were observed, and the great majority of the observed TRFs were represented in the cloned sequences.

Our phylogenetic analysis was based on the new classification of Krüger et al. [44]. The 32 clone sequences were aligned with 23 sequences identified as closely related reference sequences in GenBank and a phylogenetic tree was constructed using the 18S rRNA gene sequences of *Paraglomus occultum* (GenBank accessions AJ276081 and JN687477) as outgroup. This indicated the presence of five AM fungal clades belonging to the families Claroideoglomeraceae, Diversisporaceae, Gigasporaceae, and Glomeraceae (Figure 4), the most frequent sequences corresponding to Glomeraceae. The subclusters contained close matches to taxa previously identified by Singh et al. [22] based on spore morphology of AM fungi in rhizosphere soils of *T. grandis*: TR1-16, TR1-43, TS4-4, AR5-7 and TS6-1 are close to *Rhizophagus intraradices* or *R. irregularis*, while TR1-27 is close to *Redeckera fulvum*. Clone sequences TS4-9 and TS4-32 are similar to *Diversispora aurantia*, while TR3-R10 is probably *Gigaspora margarita*. When sequence data are compared with individual TRFs (Table S3 and Figure 4), it is clear that individual TRFs cannot be used to identify sequence type, because many different species may generate a TRF of the same size. For example, the FAM fragment at 164b could equally well be from *G. indicum*, *Re. fulvum* or *Claroideoglomus etunicatum*.

Discussion

This study examined the AM fungal communities of *A. crassna* and *T. grandis* plantations in Thailand. The estimated numbers of AM fungal taxa in roots and soils of *T. grandis* seedlings were 1.89 and 2.55 respectively, while in roots and soils of *A. crassna* there were 2.85 and 2.33 respectively. The AM fungal diversity was low compared with other plants. Using similar methodologies and definitions, Vandenkoornhuyse et al. [31] reported an average of 6.1 AM fungal taxa colonizing grass roots in a temperate seminatural grassland system, and 5.5 AM fungal taxa were found colonizing each *Solidago virgaurea* L. seedling root sample in low-Arctic meadow habitat [29].

Previous studies quantified the AM fungal diversity in rhizospheres of *T. grandis* and *A. crassna* mainly based on spore morphology and aimed to select efficient AM fungal isolates for growth enhancement. For example, Singh et al. [22] found an average of nine species per 100 g dry soil in a Jhum fallow site at which *T. grandis* was the dominant tree species, and most species belonging to the genus Glomus. Tamuli and Boruah [21] studied the AM fungi association of agarwood (*Aquilaria malaccensis*) plantations in Jorhat District of the Brahmaputra Valley, India. They found that the genus *Glomus* was dominant; among these

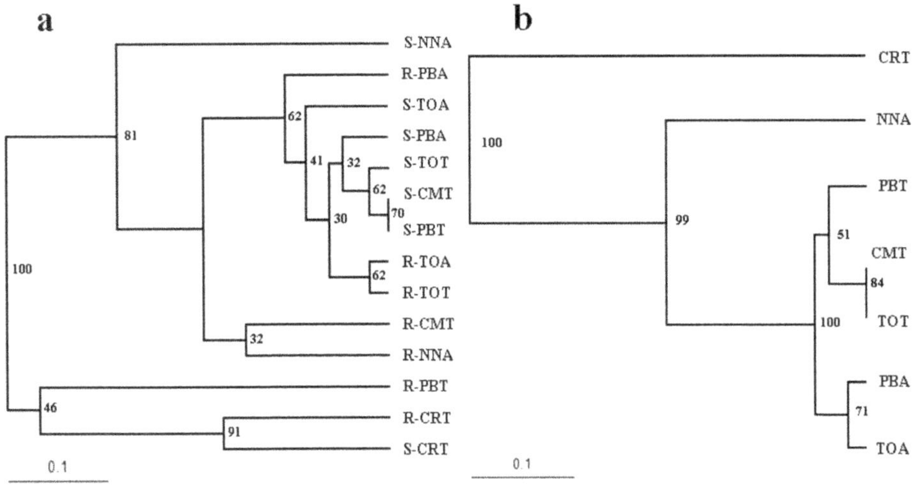

Figure 2. Cluster analysis of terminal restriction fragment length polymorphism patterns from AM fungal communities associated with *Aquilaria crassna* (A) and *Tectona grandis* (T); a) TRFs patterns in roots (R-) and rhizosphere soils (S-) and b) TRFs patterns in five sites (CM: Chiang Mai, CR: Chiang Rai, NN: Nakhon-Nayok, PB: Phetchabun and TO: Thai Orchid Labs). The unweighted pair-group average (UPGMA) algorithm was used to cluster patterns based on Jaccard similarities. Percentage values based on 1000 bootstrap replicates are given at each node.

G. fasciculatum (now known as *Rhizophagus fasciculatus*; [45]) was the most dominant followed by *G. aggregatum*. We are not aware of any information on the diversity of AM fungi on *A. crassna*. According to previous studies, we also found that most sequences belonged to the family Glomeraceae that includes *Glomus* and *Rhizophagus*. This result is consistent with previously

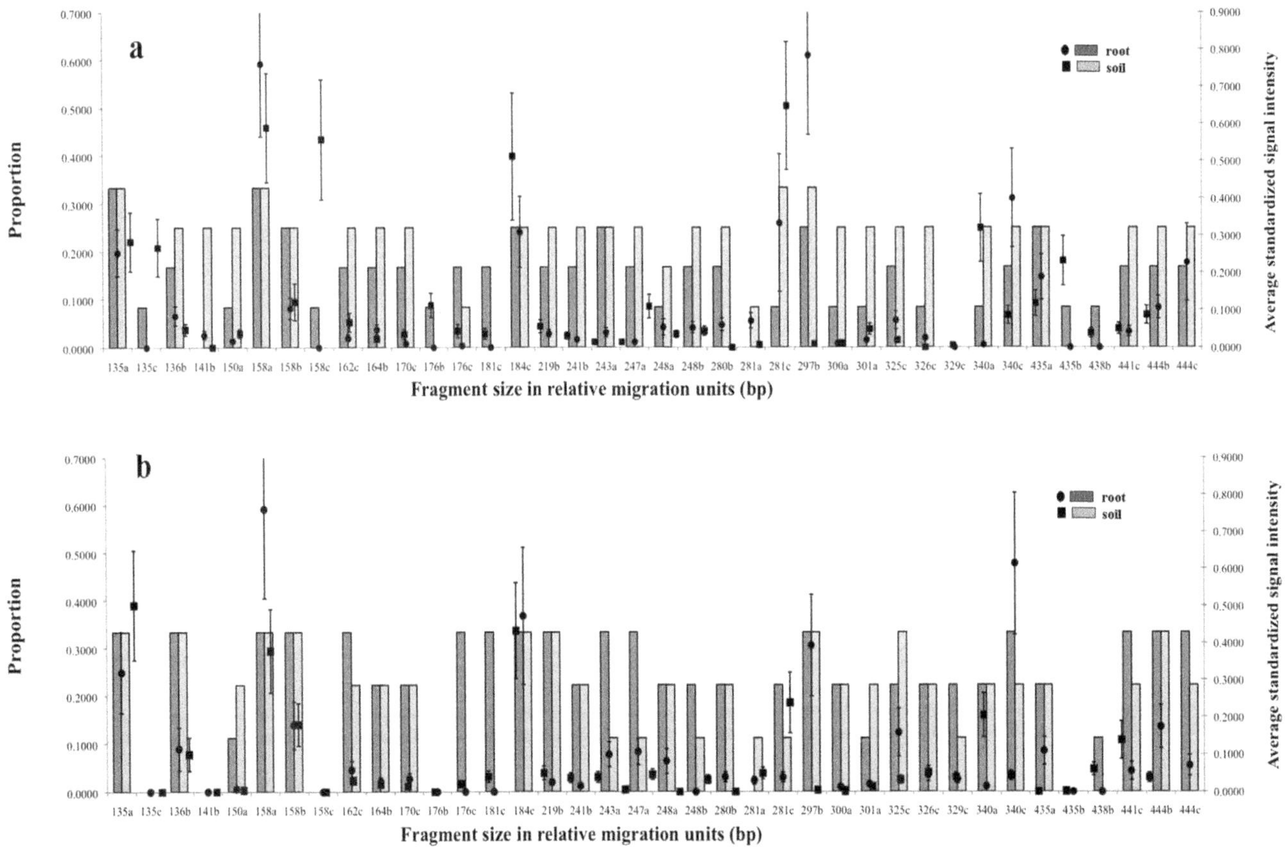

Figure 3. Occurrence of TRFs from roots and soils in (a) *Tectona grandis* and (b) *Aquilaria crassna*. Bars indicate the proportion of samples that yielded each TRF; dots indicate the average intensity of that fragment (± SEM) in those samples. The letters indicate the restriction enzyme involved in each fragment size, a: *MboI*, b: *HinfI* and c: *Hsp92II*.

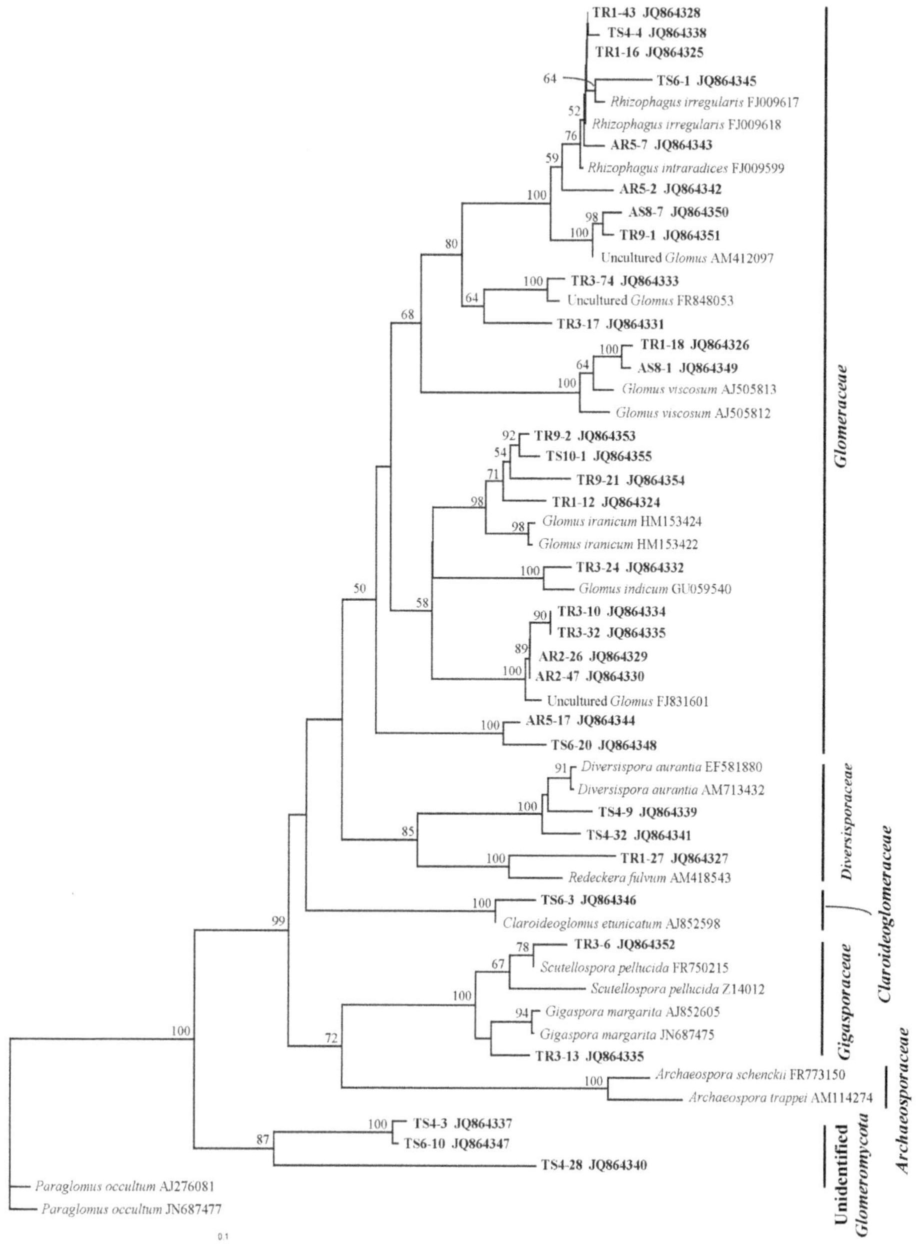

Figure 4. Neighbour-joining (NJ) phylogenetic tree of partial small subunit rRNA gene. Phylogeny was constructed using the region from NS31 to AML3. The percentage support values are based on 1000 bootstraps.

published phylogenies [29,39,46]. The dominance of this family suggests that they able to survive under various agricultural conditions such as soil disturbance from plowing and cultivation and pesticide usage like that used here in the Phetchabun and Nakhon-Nayok sites. Those conditions may be unfavorable for other AM fungi. One possible reason why *Glomus* species have the ability to survive in a disturbed system is related to differences in propagation strategies [29]. Glomeraceae are capable of colonizing via fragments of mycelium, mycorrhizal root pieces, and spores, while Gigasporaceae are only capable of propagation via spores because they do not produce intra-radical vesicles: lipid-rich storage structures which allow for re-growth of hyphae from previously colonized root pieces [46,47,48,49]. This difference can explain the dominance of the Glomeraceae over Gigasporaceae members in an environment with repetitive agricultural disturbance. Oehl et al. [50] revealed a clear seasonal and successional AMF sporulation dynamics and implied that different life strategies of different ecological AMF groups could be defined on the basis of diverging temporal sporulation dynamics.

This study shows that the choice of restriction enzymes (HinfI, Hsp92II, MboI) did not significantly affect AM fungal diversity found per sample. Using a combination of those three restriction enzymes could detect possible species of AM fungi in the samples, even if they resulted in similar-sized fragments. HinfI and Hsp92II were chosen in this study because they showed the highest polymorphism of cleavage sites at the extremities of the amplified DNA fragment [31]. Mummey and Rillig [39] and Wolfe et al. [51] also found that HinfI and MboI can separate different closely-related species of AM fungi identified from phylogenetic analyses. For example, *R. irregularis* and *R. intraradices* are closely related species that group in the same clade (Figure 4). Six clone sequences (TR1-16, TR1-43, TS4-4, AR5-2, AR5-7and TS6-1) that were related to both species were not completely separated using phylogenetic analysis, but virtual digesting with those three enzymes did separate them by using the combination of restriction pattern of each enzyme (Table S3). Clone sequences TR1-16, 1-43, and 6-1 grouped with *R. irregularis* and TS4-4, 5-2, 5-7 grouped with *R. intraradices*.

Some TRFs were only found in roots or only in soils, suggesting that some AM fungi may be rare in soil but produce fungal structures in roots that are rich enough for T-RFLP detection, while some were found only as spores in soils and did not colonize roots. While the majority of TRFs were associated with both *T. grandis* and *A. crassna*, some TRFs were associated with just one plant (i.e. 135c, 141b, 158c, 176b, 329c and 435b). In clustering analysis, samples from each plant species were grouped together even if they were collected from different sites. *A. crassna* samples seemed to group together, but since many AMF taxa were shared by both trees, *A. crassna* shared some AM fungal community patterns with *T. grandis* (Figure 2). Statistical analysis revealed significant effects of collecting sites and the interaction between collecting sites, host plant species and source of samples on TRFs (Table S2). Thus, specific AM taxa in roots and soils of *T. grandis* and *A. crassna* were affected by site but not affected by host plant species and source of samples (root and soil). This is in accordance with the observation of Bever et al. [52] that the host-dependence of the relative growth rates of fungal populations may play an important role in the maintenance of fungal species diversity. Previously, it has been reported that neighboring plants may have a significant impact on the AM fungal colonization and community composition of AM fungi in plant roots [34]. Although *T. grandis* at the Chiang Mai site had other *T. grandis* as closest neighbors with some negligible understory perennial plants, and at the other two sites the closest neighbors were *A. crassna*, the cluster analysis did not reveal any effect of this difference in neighbors. AM fungal community patterns in CMT were grouped with PBT and TOT sites in which weeds were controlled by agricultural management.

In conclusion, we demonstrated here that AM fungal community patterns in rhizosphere soils and roots of *T. grandis* and *A. crassna* were similar even if they were collected from different sites. AM fungal communities of *T. grandis* samples from different sites were similar, as were those in *A. crassna* samples. We also found that most sequences represented Glomeraceae, including *Glomus* spp. and *Rhizophagus* spp. Virtual digestion of sequences using the target sequences of the restriction enzymes HinfI, Hsp92II and MboI yielded expected fragments that mostly matched observed TRFs, linking possible AM fungal species to each TRF. Specific AM taxa in roots and soils of *A. crassna* and *T. grandis* were affected by site but were not affected by host plant species and source of samples (root and soil). Although the T-RFLP technique can provide important information about the AM fungal diversity associated with plant species of interest, trap cultures and cultured spores from the field site are still important in order to assess the ability of the AM fungi to enhance the growth of the plants, and to provide effective candidates for inoculum production targeted for these economically important tree species.

Supporting Information

Table S1 Correlation matrix of soil factors and terminal restriction fragments (TRFs) of study areas in wet season (July 2010) which soils were sampled.

Table S2 Summary of two-way analysis of variance for main and interaction effects of host plants (*Aquilaria crassna* and *Tectona grandis*), sites, and source of samples (root and soil) on AM fungal community diversity measured as the number of different TRFs per sample. Significant P-values are shown in bold.

Table S3 Clone sequences and TRFs derived from roots and rhizosphere soils of *T. grandis* and *A. crassna*. Values in bold indicate TRFs that match the sizes of virtual digest fragments (with differences ranging from 0 to 7 bp).

Author Contributions

Conceived and designed the experiments: SL JY. Performed the experiments: AC JY. Analyzed the data: AC JY. Contributed reagents/materials/analysis tools: JY NT PG SL. Wrote the paper: AC JY SL. Collected samples: PG.

References

1. Kobayashi S (2004) Landscape rehabilitation of degraded tropical forest ecosystems, Case study of the CIFOR/Japan project in Indonesia and Peru. Forest Ecol Manag 201: 13–22.
2. Turjaman M, Tamai Y, Santoso E, Osaki M, Tawaraya K (2006) Arbuscular mycorrhizal fungi increased early growth of two nontimber forest product species *Dyera polyphylla* and *Aquilaria filaria* under greenhouse conditions. Mycorrhiza 16: 459–464.
3. Rajan SK, Reddy BJD, Bagyaraj DJ (2000) Screening of arbuscular mycorrhizal fungi for their symbiotic efficiency with *Tectona grandis*. Forest Ecol Manag 126: 91–95.
4. CITES (2004) Amendments to Appendices I and II of the Convention on International Trade in Endangered Species of Wild Flora and Fauna (CITES). Thirteenth meeting of the conference of the parties, Bangkok, Thailand, 3–14 October, pp. 1–9.
5. Schüßler A, Schwarzott D, Walker C (2001) A new fungal phylum, the *Glomeromycota*: phylogeny and evolution. Mycol Res 105 (12): 1413–1421.
6. Habte M (2000) Mycorrhizal fungi and plant nutrition. In: Silva JA, Uchida R, editors. Plant Nutrient Management in Hawaii's Soils, Approaches for Tropical and Subtropical Agriculture. College of Tropical Agriculture and Human Resources, Manoa: University of Hawaii. pp. 127–131.
7. Brundrett MC, Bougher N, Dell B, Grove T, Malajczuk N (1996) Working with mycorrhizas in forestry and agriculture. Australian Centre for International Agricultural Research Monograph 32, Canberra.
8. Smith S, Read D (2008) Mycorrhizal Symbiosis. New York: Academic Press.
9. Porcel R, Ruiz-Lozano JM (2004) Arbuscular mycorrhizal influence on leaf water potential, solute accumulation, and oxidative stress in soybean plants subjected to drought stress. J Exp Bot 55: 1743–1750.
10. Doubková P, Vlasáková E, Sudová R (2013) Arbuscular mycorrhizal symbiosis alleviates drought stress imposed on *Knautia arvensis* plants in serpentine soil. Plant Soil 370: 149–161.
11. Azcón-Aguilar C, Barea JM (1996) Arbuscular mycorrhizas and biological control of soil-borne plant pathogens – an overview of the mechanisms involved. Mycorrhiza 6: 457–464.
12. Wright SF, Upadhyaya A (1998) A survey of soils for aggregate stability and glomalin, a glycoprotein produced by hyphae of arbuscular mycorrhizal fungi. Plant Soil 198: 97–107.
13. Rillig MC (2004) Arbuscular mycorrhizae, glomalin, and soil aggregation. Can J Soil Sci 84: 355–363.
14. Rillig MC, Ramsey PW, Morris S, Paul EA (2003) Glomalin, an arbuscular-mycorrhizal fungal soil protein, responds to land-use change. Plant Soil 253: 293–299.
15. Habte M, Miyasaka SC, Matsuyama DT (2001) Arbuscular mycorrhizal fungi improve early forest-tree establishment. In: Horst WJ et al., editors. Plant nutrition–Food Security and Sustainability of Agro-ecosystems. Dordrecht: Kluwer Academic Publishers. pp. 644–645.
16. Urgiles N, Loján P, Aguirre N, Blaschke H, Günter S, et al. (2009) Application of mycorrhizal roots improves growth of tropical tree seedlings in the nursery: a

step towards reforestation with native species in the Andes of Ecuador. New Forest 38: 229–239.
17. Swaminathan C, Srinivasan VM (2006) Influence of microbial inoculants on seedling production in teak (*Tectona grandis* L.f.). J Sustain Forest 22 (3): 63–76.
18. Tabin T, Arunachalam A, Shrivastava K, Arunachalam K (2009) Effect of arbuscular mycorrhizal fungi on damping-off disease in *Aquilaria agallocha* Roxb. seedlings. Trop Ecol 50 (2): 243–248.
19. Thapar HS, Khan SN (1988) Seasonal frequency of *Endogone* spores in new-forest soils. In: Khosla PK, Sehgal RN, editors. Trends in Tree Sciences. Solan: Indian Society of Tree Scientists. pp.161–162.
20. Kanakadurga VV, Manoharachary D, Rama RP (1990) Occurrence of endomycorrhizal fungi in teak. In: Bagyaraj DJ, Manjunath A, editors. Mycorrhizal symbiosis and plant growth.Bangalore: University of Agricultural Sciences. pp.17.
21. Tamuli P, Boruah P (2002) Vesicular-abuscular mycorrhizal (VAM) association of agarwood tree in Jorhat District of the Brahmatputra Valley. Indian For 128 (9): 991–994.
22. Singh SS, Tiwari SC, Dkhar MS (2003) Species diversity of vesicular-arbuscular mycorrhizal (VAM) fungi in Jhum fallow and natural forest soils of Arunachal Pradesh, north eastern India. Trop Ecol 44 (2): 207–215.
23. Dhar PP, Mridha MAU (2012) Arbuscular mycorrhizal associations in different forest tree species of Hazarikhil forest of Chittagong, Bangladesh. J For Res 23 (1): 115–122.
24. Ramanwong K (1998) Species diversity of vesicular-arbuscular mycorrhizal fungi of teak (*Tectona grandis* Linn.f.) and their effects on growth of teak seedlings. M.S. Thesis, Kasetsart University, Chatuchak, Bangkok.
25. Helgason T, Daniell TJ, Husband R, Fitter AH, Young JPW (1998) Ploughing up the wood-wide web? Nature 394: 431.
26. Lee J, Lee S, Young JPW (2008) Improved PCR primers for the detection and identification of arbuscular mycorrhizal fungi. FEMS Microbiol Ecol 65: 339–349.
27. Krüger M, Stockinger H, Krüger C, Schüßler A (2009) DNA-based species level detection of Glomeromycota: one PCR primer set for all arbuscular mycorrhizal fungi. New Phytol 183: 212–223.
28. Daniell TJ, Husband R, Fitter AH, Young JPW (2001) Molecular diversity of arbuscular mycorrhizal fungi colonising arable crops. FEMS Microbiol Ecol 36: 203–209.
29. Pietikäinen A, Kytöviita MM, Husband R, Young JPW (2007) Diversity and persistence of arbuscular mycorrhizas in a low-Arctic meadow habitat. New Phytol 176: 691–698.
30. Mummey DL, Rillig MC (2006) The invasive plant species Centaurea maculosa alters arbuscular mycorrhizal fungal communities in the field. Plant Soil 288: 81–90.
31. Vandenkoornhuyse P, Ridgway KP, Watson IJ, Duck M, Fitter AH, et al. (2003) Co-existing grass species have distinctive arbuscular mycorrhizal communities. Mol Ecol 12: 3085–3095.
32. Johnson D, Vandenkoornhuyse PJ, Leake JR, Gilbert L, Booth RE, et al. (2004) Plant communities affect arbuscular mycorrhizal fungal diversity and community composition in grassland microcosms. New Phytol 161: 503–515.
33. van der Heijden MGA, Wiemken A, Sanders IR (2003) Different arbuscular mycorrhizal fungi alter coexistence and resource distribution between co-occurring plant. New Phytol 157: 569–578.
34. Mummey DL, Rillig MC, Holben WE (2005) Neighbouring plant influences on arbuscular mycorrhizal fungal community composition as assessed by T-RFLP analysis. Plant Soil 271: 83–90.
35. Barto EK, Antunes PM, Stinson K, Koch AM, Klironomos JN, et al. (2011) Differences in arbuscular mycorrhizal fungal communities associated with sugar maple seedlings in and outside of invaded garlic mustard forest patches. Biol Invasions 13: 2755–2762.
36. Houba VJG, Van Der Lee JJ, Novozamsky I, Wallinga J (1988) Soil and Plant analysis. Part 5: Soil Analysis Procedure. Agricultural University, Wageningen.
37. Sparks DL, Page AL, Helmke PA, Loeppert RH, Soltanpour PN, et al. (1996) Methods of Soil Analysis. Part 3. Chemical Methods. Madison, Wisconsin: Soil Science Society of America
38. Render C, Weißhuhn K, Kellner H, Buscot F (2006) Rationalizing molecular analysis of field-collected roots for assessing diversity of arbuscular mycorrhizal fungi: to pool, or not to pool, that is the question. Mycorrhiza 16: 525–531.
39. Mummey DL, Rillig MC (2007) Evaluation of LSU rRNA-gene PCR primers for analysis of arbuscular mycorrizal fungal communities via terminal restriction fragment length polymorphism analysis. J Microbiol Meth 70: 200–204.
40. Hall T (1999) BioEdit: a user-friendly biological sequence alignment editor and analysis program for Windows95/98/NT. Nucl Acid S 41: 95–98.
41. Swofford DL (2002) PAUP* Phylogenetic analysis using parsimony (*and other methods). version4b10. Sunderland. MA: Sinauer Associates
42. Hampl V, Pavlícek A, Flegr J (2001) Construction and bootstrap analysis of DNA fingerprinting-based phylogenetic trees with a freeware program Free-Tree: Application to trichomonad parasites. Int J Syst Evol Micr 51: 731–735.
43. Page RDM (1996) TreeView: An application to display phylogenetic trees on personal computers. Comput Appl Biosci 12: 357–358.
44. Krüger M, Krüger C, Walker C, Stockinger H, Schüßler A (2012) Phylogenetic reference data for systematics and phylotaxonomy of arbuscular mycorrhizal fungi from phylum to species-level. New Phytol 193: 970–984.
45. Schüßler A, Walker C (2010) The Glomeromycota: A species list with new families and new genera (Libraries at the Royal Botanic Garden Edinburgh, Edinburgh, UK; The Royal Botanic Garden Kew, Kew, UK; Botanische Staatssammlung Munich, Munich, Germany; and Oregon State University, Corvallis, Oregon, pp.1–56.
46. Helgason T, Fitter AH, Young JPW (1999) Molecular diversity of arbuscular mycorrhizal fungi colonising *Hyacinthoides non-scripta* (bluebell) in a seminatural woodland. Mol Ecol 8: 659–666.
47. Gazey C, Abbott KK, Robson AD (1993) VA mycorrhizal spores from 3 species of *Acaulospora*—germination, longevity and hyphal growth. Mycol. Res. 97: 785–790.
48. INVAM Newsletter 3 (1993) Properties of infective propagules at the suborder level (*Glomineae* versus *Gigasporineae*). West Virginia University Web Services. Available: http//invam.caf.wvu.edu/articles/propagules.htm. Accessed 9 July 2012.
49. Brundrett M, Abbott LK, Jasper DA (1999) Glomalean mycorrhizal fungi from tropical Australia. I. Comparison of the effectiveness and specificity of different isolation procedures. Mycorrhiza 8: 305–314.
50. Oehl F, Sieverding E, Ineichen K, Mäder P, Wiemken A, et al. (2009) Distinct sporulation dynamics of arbuscular mycorrhizal fungal communities from different agroecosystems in long-term microcosms. Agr Ecosyst Environ 134: 257–268.
51. Wolfe BE, Mummey DL, Rillig MC, Klironomos JN (2007) Small-scale spatial heterogeneity of arbuscular mycorrhizal fungal abundance and community composition in a wetland plant community. Mycorrhiza 17: 175–183.
52. Bever JD, Morton JB, Antonovics J, Schultz PA (1996) Host-dependent sporulation and species diversity of arbuscular mycorrhizal fungi in a mown grassland. J Ecol 84: 71–82.

Effects of Grazing Regimes on Plant Traits and Soil Nutrients in an Alpine Steppe, Northern Tibetan Plateau

Jian Sun[1], Xiaodan Wang[2]*, Genwei Cheng[2], Jianbo Wu[2], Jiangtao Hong[2], Shuli Niu[1]

1 Synthesis Research Centre of Chinese Ecosystem Research Network, Key Laboratory of Ecosystem Network Observation and Modelling, Institute of Geographic Sciences and Natural Resources Research, Chinese Academy of Sciences, Beijing, China, **2** The key laboratory of mountain surface processes and eco-regulation, Institute of Mountain Hazard and Environment, Chinese Academy of Sciences, Chengdu, China

Abstract

Understanding the impact of grazing intensity on grassland production and soil fertility is of fundamental importance for grassland conservation and management. We thus compared three types of alpine steppe management by studying vegetation traits and soil properties in response to three levels of grazing pressure: permanent grazing (M1), seasonal grazing (M2), and grazing exclusion (M3) in the alpine steppe in Xainza County, Tibetan Plateau. The results showed that community biomass allocation did not support the isometric hypothesis under different grassland management types. Plants in M1 had less aboveground biomass but more belowground biomass in the top soil layer than those in M2 and M3, which was largely due to that root/shoot ratios of dominant plants in M1 were far greater than those in M2 and M3. The interramet distance and the tiller size of the dominant clonal plants were greater in M3 than in M1 and M2, while the resprouting from rhizome buds did not differ significantly among the three greezing regimes. Both soil bulk density and soil available nitrogen in M3 were greater than in M1 at the 15–30 cm soil depth ($P = 0.05$). Soil organic carbon and soil total nitrogen were greater in M3 than in M1 and M2 ($P = 0.05$). We conclude that the isometric hypothesis is not supported in this study and fencing is a helpful grassland management in terms of plant growth and soil nutrient retention in alpine steppe. The extreme cold, scarce precipitation and short growing period may be the causation of the unique plant and soil responses to different management regimes.

Editor: Cheng-Sen Li, Institute of Botany, China

Funding: This research was jointly funded by the Knowledge Innovation Program of the Chinese Academy of Sciences (KZCX2-XB3-08), CAS Strategic Priority Research Program (XDA05050602), and the Open Fund of Key Laboratory of Ecosystem Network Observation and Modelling (110301A1PA). The funders had no role in study design, data collection and analysis, decision to publish, or preparation of the manuscript.

Competing Interests: The authors have declared that no competing interests exist.

* Email: wxd@imde.ac.cn

Introduction

Alpine grasslands make up the dominant ecosystem occupying approximately 94% of Northern Tibet [1]. The natural environment of the region is extremely harsh, and the alpine steppe, a fragile ecosystem, is extremely susceptible to the impacts of human activities [2]. It suffers from overgrazing, deforestation, and the harvesting of numerous herbs commonly used in traditional medicines [3–5]. Studies examining the response of above- and belowground biomass, the root/shoot ratio, the morphological characteristics of dominant plants, and the soil properties to human disturbance offer important insights that can contribute to adopting the most effective approach to grassland management in an alpine steppe, in which it is particularly difficult to recover from ecosystem degradation due to the region's long period of frost and relatively short growing season [3,6].

Although grazing and fencing, both of which have a substantial affect on vegetation traits and soil properties [7–9], are the most prevalent management regimes for grasslands worldwide, and although the effects of herbivores on soil properties in (sub)alpine ecosystems have recently been reported [10,11], knowledge about plants and soils in response to grassland management regimes (i.e., fencing and grazing) in the Tibetan Plateau remains limited due to an extremely difficult geographic situation [9,12]. With regard to soil properties, it has been documented that grazing depresses soil carbon storage by changing the plant biomass and composition of a Tibetan alpine meadow [13]. In contrast, Shi et al. [14] found grazing exclusion to decrease soil organic carbon storage in an alpine grassland of the Tibetan Plateau, while another report suggested that seasonal grazing might enrich soil nutrients [15]. Such conflicting results indicate that different grazing intensities may have varying impacts on soil properties.

Understanding the influence of different management types on grassland production is essential for improving grassland conservation and management [16]. Previous studies have yielded varying results for aboveground biomass changes. Grazing thus increased [17] or decreased [18] aboveground production in different cases. The fencing optimization hypothesis posits that fencing significantly enhances above- and belowground biomass by means of carbon reallocation to plant growth and promotes increases in soil nutrient concentrations [19,20]. Belowground biomass has also been shown to be affected by grazing, with some studies indicating that plants reduce the proportion of aboveground parts and allocate more biomass to belowground parts, so as to geminate and resist environmental stress (i.e., grazing pressure) [21]. Conversely, more belowground biomass was

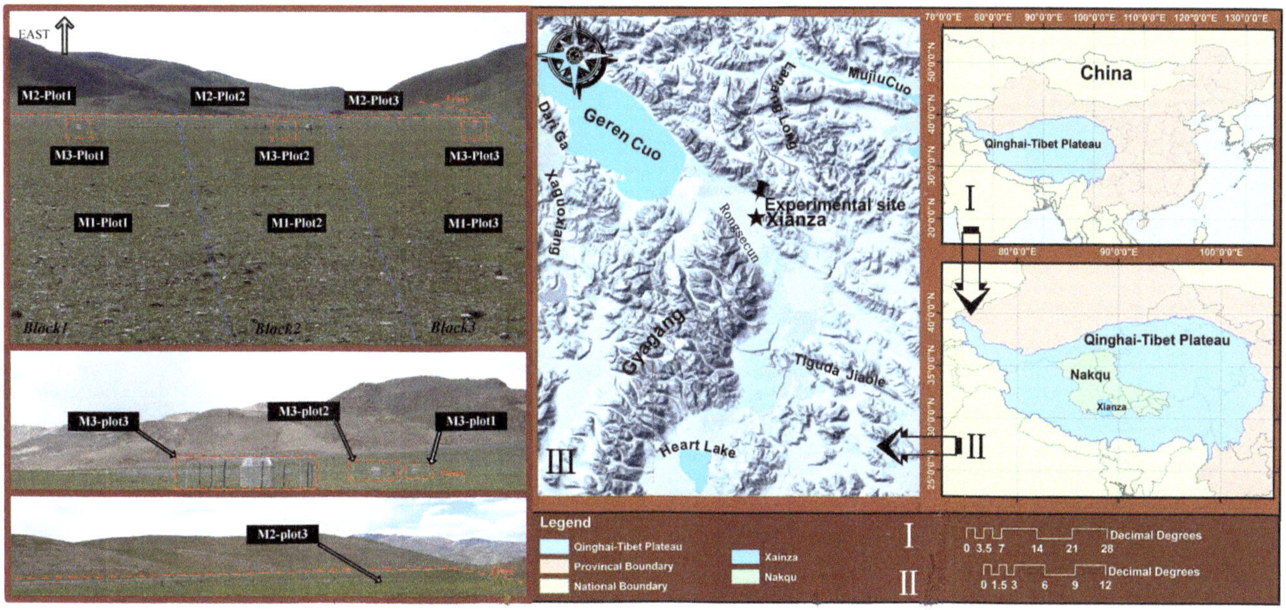

Figure 1. The experiment sites in alpine steppe, northern Tibetan Plateau.

allocated to the surface layer than the subsurface of the soil profile with increasing grazing pressure [22]. For instance, the root to shoot ratio in the grazing pattern was significantly higher than for the mowing and fencing patterns in temperate grassland [23].

The root:shoot ratio provides the basis for understanding the response or adaptive strategies of plants to environmental stress [24,25], and reflects how plants respond to different selection pressures [26,27]. Theory predicts the ratio of roots to shoots to respond isometrically across different individuals and community types under varying environmental conditions [28–30]. While biomass allocation has been widely examined from individual to community to ecosystem levels [31–35], few studies compare the biomass partitioning of dominant alpine species under different grassland management types.

Aside from biomass, the morphological characteristics of clonal plants are also regarded as a sensitive index for assessing the possible effect of grazing pressure [36,37]. The traditional theory is that a clone with a guerrilla-like foraging strategy should be able to locate and exploit favorable patches in which competitors are absent. In a habitat in which potential competitors are more homogeneously distributed, a phalanx clone exhibits a "consolidation" strategy and may be able to persist and monopolize locally available resources [38]. However, we know little about the responses of the morphological characteristics of clonal plants to grazing pressure in alpine steppes.

In this study, we examined plant traits and soil properties in response to different grassland management strategies in the Northern Tibetan Plateau. The specific aims are: (1) to reveal the impact of different grassland managements on plant traits and soil properties and to test the grazing optimization hypothesis; (2) to test the isometric partitioning theory on different alpine species under varying grassland management patterns; and (3) to explore the effects of grazing stress on clonal growth.

Materials and Methods

Study area

Our study was conducted in the Alpine Steppe Experiment Station (N 30°57′, E 88°42′, 4675 m a.s.l.) in Xainza County, Northern Tibet (Fig. 1). The annual mean air temperature of this region is 0°C, while the mean air temperature ranges from −10.1°C in January and 9.6°C in July. The average annual precipitation is 300 mm. There is no absolute frost-free season, and the frosty period can last up to 279.1 days [39]. The region's soil is classified as steppe soil with sandy loam [40]. Vegetation coverage of the alpine steppe is approximately 20% and is dominated by *Stipa purpurea* (Bunch grass) and *Carex moorcrofii* (sedge family, creeping rhizomes) (Fig. 2A). Other common species include *Stellera chamaejasme Linn.* (Bunch weeds), *Oxytropis glacialis* (Bunch weeds), and *Leontopodium alpinum* (Fig. 2B). All of these plants are perennial herbs.

Grazing treatments and experimental plots

There were three grazing treatments in this study. i.e, M1 (grazing): The pasture is always used for grazing. M2 (seasonal grazing): The pasture has been used for grazing in the non-growing seasons but not in the growing season (from May to September) since 2007. M3 (grazing exclusion): Herbivores have been excluded from the pasture since 2010.

The study area was divided into three blocks. Each block contained three plots, the eastern plot having a grazing intensity of M2, the middle plot a grazing intensity of M3, and the western plot a grazing intensity of M1 (Fig. 1). The dimensions of each M3 fenced plot were 6×6 m; the dimensions for each M2 plot were 200×500 m. Three M1 plots were located outside of the fenced area in each block, with each plot having dimensions of 200×500 m. In each plot, a quadrat (30 cm×30 cm) was used to measure biomass and take samples. The estimated average livestock density in both M1 and M2 was approximately 120 sheep units per km^2. (The "sheep unit" is the most frequently used unit of measurement for evaluating the carrying capacity of pasture areas in China and includes figures for cattle, which are converted

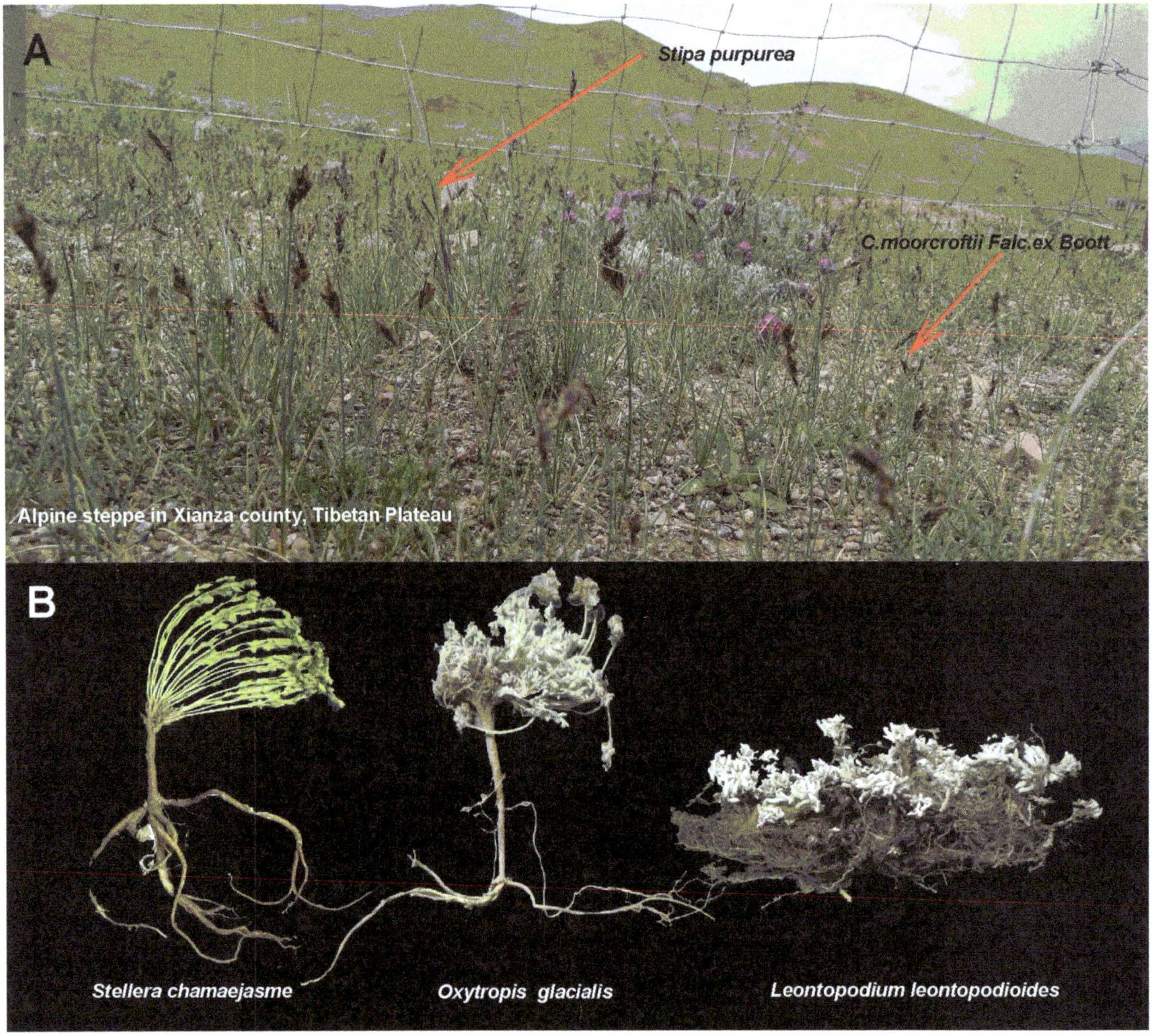

Figure 2. The vegetation composition diagram is in the study area. The Vegetation coverage of the alpine steppe was about 20%, which are dominated by *Stipa purpurea* and *Carex moorcrofii* (panel A) and the accompanying species are *Stellera chamaejasme Linn.*, *Leontopodium alpinum*, *Oxytropis microphylla* (panel B) and so on.

to sheep; thus, e.g., one cow is equivalent to five sheep based on the average weight of a cow) [41].

Ethics statement

No specific permits were required for the samples collected from any of the sites, and the field studies did not involve endangered or protected species.

Biomass measure and sample taking

We established a 30 cm×30 cm quadrat in each of the sampling plots to harvest aboveground biomass (AGB) in late August of 2012. After harvesting the aboveground biomass, we took soil samples by auger with a 5-cm core diameter at two depths (0–15 cm and 15–30 cm) to estimate belowground biomass (BGB). According to previous research, most belowground biomass is located in the top 30 cm layer [42]. Soil samples were also taken at depths of 0–15 cm and 15–30 cm in order to measure soil elements. After being air-dried and sieved (2 mm mesh), the soil samples were carefully handpicked to extract the surface organic materials and fine roots for an analysis of soil chemical properties. Soil nutrients (including soil organic carbon, soil total nitrogen content, available nitrogen, total phosphorus and available phosphorus) were determined using standard protocol [43].

In order to estimate the root:shoot ratio of the dominant species (*S. purpurea*, *C. moorcrofii*, *O. glacialis*, and *L. alpinum*), we took three soil blocks in each sampling plot of different management regimes. The patch was excavated with a spade with a diameter of 20 cm to a depth of 30 cm, which was selected based on the root morphological traits of the species [42]. The root samples obtained from the sites were immediately placed in a cloth bag and then soaked in water to remove the residual soil by means of a 0.5 mm sieve. The roots of each sample were carefully separated from soil and other belowground materials first. Then individuals of the all species were carefully separated from other plant roots. Finally, we

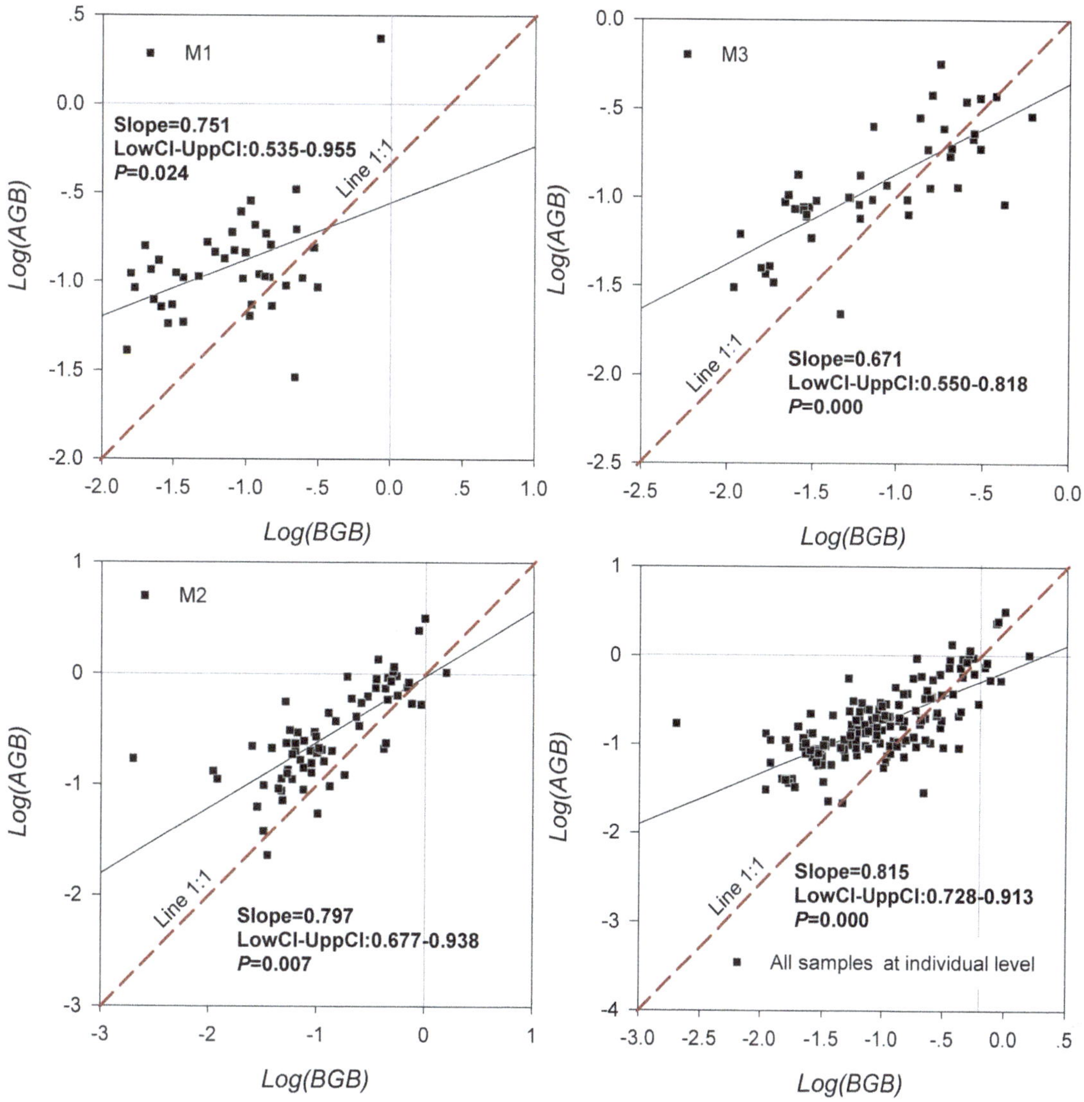

Figure 3. Relationships between AGB and BGB under grazing (M1), seasonal grazing (M2), and grazing exclusion (M3) regimes.

separated the roots and shoots of each individual. Individual biomass (shoots and roots) and community biomass (AGB and BGB) was oven-dried at 65°C until a constant weight was achieved. Individual biomass data were used to analyze allometric functions under different management regimes (with a total of 40, 45, and 68 individuals in M1, M2, and M3, respectively).

The plant morphology of *C. moorcrofii* is regarded as a fairly good indicator that reflects the different grassland management regimes [44]. We thus sampled *C. moorcrofii*. from other 30 cm×30 cm quadrats at each of the 9 plots. The samples were then taken back to the laboratory, where the morphological characteristics (plant height, interramet distance, number of rhizome buds, and tillering number) were immediately measured. Special attention was paid to avoid destroying the structural integrity of the plant throughout this process.

Statistical analysis

We performed a standardized major axis (SRMA) regression to examine whether AGB and BGB scale isometrically at an individual level across all samples, and then explored their relationships under different management types. The regression relationship of the form $LogAGB = \alpha + \beta LogBGB$ was used to describe the allometric relationship between AGB and BGB, where x is BGB, y is AGB, α is the intercept, and β is the scaling slope [25,28,45,46]. The scaling slope and y-intercept of the allometric function were determined using the software package titled "Standardized Major Axis Tests & Routines Version 2.0" [47]. If a 95% confidence interval of the scaling slope covered 1.0, the relationship between AGB and BGB was considered to be isometric.

One-way ANOVA was used to test differences in soil properties and biomes (AGB, BGB, and R/S) among the different treatments, and the Tukey test was used to distinguish differences at a $P = 0.05$ level.

Results

Biomass partitioning

The relationship between BGB and AGB under different grazing pressures was characterized by the linear function $LogAGB = \alpha + \beta LogBGB$ (Fig. 3). The slopes of the allometric

Table 1. Aboveground biomass (AGB), belowground biomass (BGB), and their ratios (R/S) under different grassland managements.

Management Types	Category	M1		M2		M3	
		Mean	STDEV	Mean	STDEV	Mean	STDEV
AGB ($g \cdot cm^{-2}$)		48.09	12.39	51.09	35.48	62.25	3.21
	0–15 cm	312.92	65.6	238.07	53.95	247.16	38.96
BGB ($g \cdot cm^{-2}$)	15–30 cm	11.95	7.64	6.77	5.64	14.72	10.27
	Ratio(0–15 cm/15–30 cm)	26.19	8.59	35.17	9.57	16.79	3.79
	0–30 cm	324.87	58.18	244.84	59.59	261.88	48.67
R/S		7.37	3.58	5.79	2.23	4.2	0.67

Note: The differences of biomes under the different grassland management type were insignificant at 0.05 level.

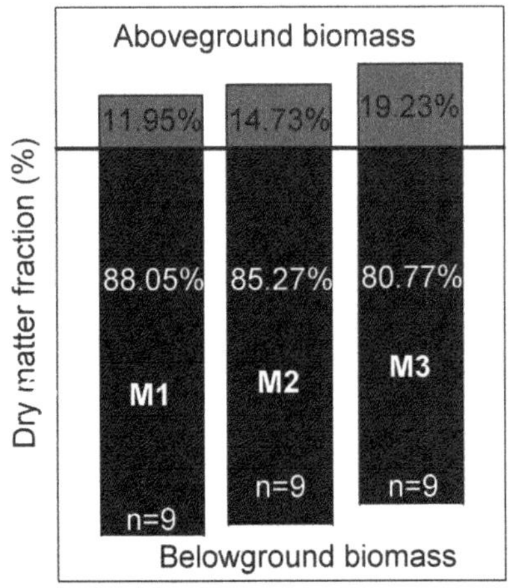

Figure 4. Biomass fractions (% of total biomass) in herbaceous plants in M1, M2 and M3 of Tibetan Plateau.

relationship were 0.751 (for M1), 0.797 (for M2), 0.671 (for M3), and 0.815 (for all samples at an individual level), respectively. All of the slopes differed significantly from 1 at $P = 0.05$.

Responses of AGB, BGB, and R/S to different management types

AGB in M1 was slightly but not significantly lower than in M2 and M3 (Table 1). In contrast, BGB of M1, M2, and M3 amounted to 324.87, 244.84, and 261.88 $g \cdot m^{-2}$, respectively. The ratio of root to shoot in M1 was also higher than in M2 and M3. The root biomass was mainly distributed in a soil depth of 0–15 cm, while the root biomass in a soil depth of 15–30 cm only accounted for 2.77–5.62% of total BGB. The ratios of BGB in the 0–15 cm-layer to BGB in the 15–30 cm-layer of M1, M2, and M3 were 26.19, 35.17, and 16.79, respectively.

To investigate the characteristics of biomass allocation further, we analyzed the dry matter fraction of M1, M2, and M3 in the Tibetan Plateau (Fig. 4). The results indicated that the roots of total plants in the Tibetan Plateau amounted to 88.05% (in M1), 85.27% (in M2), and 80.77% (in M3).

Traits of clonal plants under different management types

C. moocroftii was tallest in M3 and shortest in M1 ($P<0.05$, Fig. 5A). Both interramet distance (Fig. 5C) and the tiller number (Fig. 5D) in M3 were significantly larger than in M1 and M2. No significant differences were found in the number of rhizome buds among M1, M2, and M3 treatments ($P<0.05$, Fig. 5B).

Soil properties in response to management types

At a soil depth of 0–15 cm, soil bulk density, soil organic carbon, total nitrogen, and available nitrogen were non-significant among M1, M2, and M3 (Table 2). At a soil depth of 15–30 cm, both soil bulk densiy and soil available nitrogen in M3 were greater than in M1 at a $P = 0.05$ level. Soil organic carbon and soil total nitrogen in M3 were greater than in M1, and M2 ($P = 0.05$).

Discussion

Biomass partitioning

Based on the results of our SMA analysis, biomass allocation at the community level did not agree with the isometric hypothesis under different grassland management types. Plants consistently sense changes in their environment and often allocate a greater

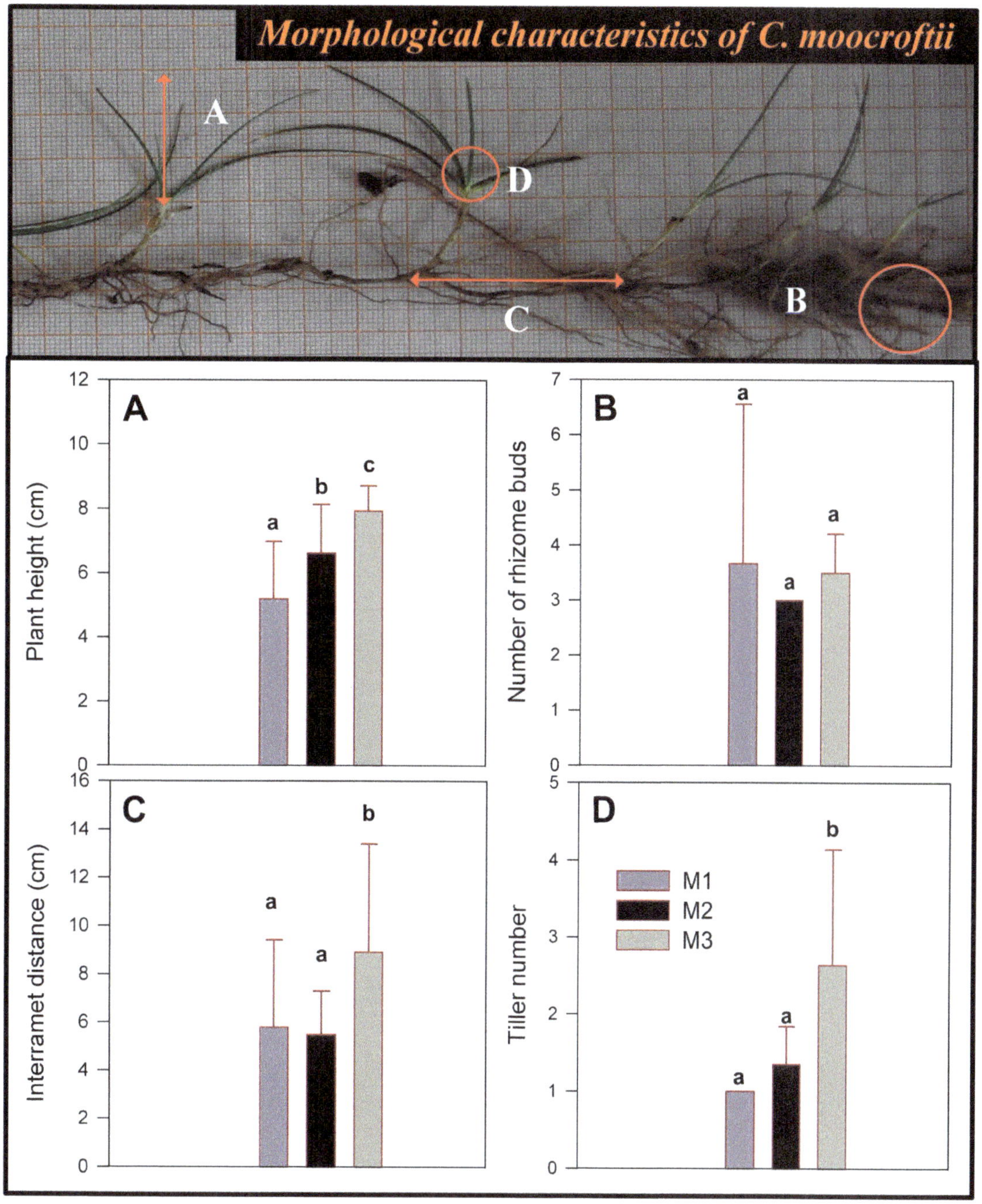

Figure 5. The morphological characteristics of *C. moocroftii* in the different grassland management types. A, represent plant height, B, number of rhizome buds, C, interramet distance, and D tiller number.

proportion of their biomass to the root system when water or mineral elements are scarce [48]. Moreover, roots have also been found to store carbohydrates in alpine grasslands [49]. In the alpine steppe, greater plant biomass allocation to belowground reflects the plants' response to the harsh alpine environment (limited precipitation and low temperatures) for survival. Another reason may be the relatively narrow variation in plant size in that region. Moreover, many species in the Tibetan alpine grasslands do not have typical hierarchical branching structures [42], which did not meet the assumption that stem length scales isometrically with respect to root length. This characteristic is clearly reflected in the root morphology structure shown in Fig. 2B and Fig. 5. Biomass allocation in these areas may thus support the allometric biomass partitioning hypothesis, rather than the isometric allocation hypothesis.

Responses of AGB, BGB, and R/S to different management types

AGB in the grazing plots was found to be lower than in the non-grazing plots, which is in agreement with the findings of Medina-Roldan et al. [50]. We suggested that alpine grassland production might be decreased to some degree by grazing as a result of the

Table 2. Soil properties under the different grassland managements.

Management Types	Depth	M1		M2		M3	
		Mean	STDEV	Mean	STDEV	Mean	STDEV
Soil bulk density (g·cm^{-3})	**0–15 cm**	1.71	0.09	1.48	0.08	1.60	0.11
Soil organic carbon (g·kg^{-1})		10.09	3.38	12.72	2.45	11.02	0.85
Soil total nitrogen(g·kg^{-1})		1.12	0.26	1.54	0.40	1.19	0.11
Soil available Nitrogen(mg·kg^{-1})		62.96	24.14	84.65	17.24	81.61	7.14
Soil total phosphorous(g·kg^{-1})		0.34	0.03	0.33	0.04	0.28	0.02
Soil available phosphorous(mg·kg^{-1})		3.91	0.72	3.83	0.16	3.39	0.29
Soil bulk density(g·cm^{-3})	**15–30 cm**	1.76 **a**	0.04	1.58 **ab**	0.09	1.48 **b**	0.14
Soil organic carbon (g·kg^{-1})		5.21 **a**	0.87	5.81 **a**	0.25	10.16 **b**	2.58
Soil total nitrogen(g·kg^{-1})		0.64 **a**	0.13	0.57 **a**	0.04	1.07 **b**	0.27
Soil available Nitrogen(mg·kg^{-1})		32.36 **a**	4.87	40.57 **ab**	8.51	68.01 **b**	17.37
Soil total phosphorous(g·kg^{-1})		0.27	0.01	0.24	0.04	0.37	0.09
Soil available phosphorous(mg·kg^{-1})		2.95	0.54	3.25	0.61	3.33	0.11

Note: The bold lower case represents the result of variance analysis at 0.05 level.

consumption effect of the livestock. The higher BGB in M1 than M2 and M3 in our alpine steppe conflict with the results of Cheng et al. [51], who found grazing to significantly decrease belowground biomass in the Loess Plateau. Gao et al. [52] also reported significantly less belowground biomass under conditions of heavy grazing in comparison with a non-grazing site in Inner Mongolia. The conflicting results between our study and previous studies in other areas probably stem from differences in environmental conditions. In the extreme cold region of alpine grassland, plants reduce the proportion of aboveground parts and allocate larger amounts of biomass to belowground parts in order to germinate and resist grazing pressures [21]. More BGB in the top soil layer of M1 than M2 and M3 suggests that grazing results in more belowground biomass being allocated to the topsoil [53], possibly due to the alpine environment inhibiting root growth in ungrazed plots. We did not find an ameliorating effect of reduced grazing (M2) or grazer exclusion (M3), suggesting that the effects of grazing cessation may only be detectable in the long term, as proposed by Steffens et al. [54].

The root:shoot ratio (R/S) in M1 was much larger than in M2 and M3, suggesting that plants allocate more to BGB than AGB in order to maximize resources for optimal growth in the grazed environment [21]. We compared the dry matter fraction of M1, M2, and M3 in the Tibetan Plateau (Fig. 4) with dry matter fractions in Argentina, Bolivia, Ecuador, the Arctic, and the Alps [49], and we found that the proportion of BGB was much higher in our study sites than for plants in a semi-arid grassland ecosystem (Bolivia), mountain grassland ecosystem (Argentina), and a humid mountain grassland ecosystem (Ecuadorian Andes). The main reason may be associated with the comparatively slow depletion of carbohydrates in roots, resulting from low respiration rates in the extremely cold winter, and the slower root turnover in colder environments [55]. Meanwhile, long-term grazing results in a higher BGB fraction in the Tibetan Plateau than for plants in other cold alpine environments and the Arctic.

The clonal plant in response to different management types

Clonal plants defend against large herbivores by regulating their morphological characteristics [56]. Clonal plant *C. moocroftii* in M3 was notably taller than in M1 and M2 because it was not bitten by livestock, and they can plastically respond to environmental heterogeneity by placing ramets and changing tiller size in favorable sites [57,58]. We found the interramet distance and tiller number in M3 to be greater than for plants in M1 and M2. Tillers may produce more daughter tillers as a result of increased nutrient availability in grazing exclusion. *C. moocroftii* thus responds to grazing by decreasing the tiller number [37]. The results also indicate that *C. moocroftii* may have both phalanx and guerrilla strategies. The phalanx strategy involves the production of a compact structure of closely spaced ramets (i.e., M1), and the guerrilla strategy involves the production of loosely arranged and more widely spaced ramets in order to seek out more soil nutrients in a relatively suitable habitat (i.e., M3).

Clonal growth by means of lateral roots is regarded as a typical trait of opportunistic species growing in disturbed habitats, as such growth potentially produces a large number of buds on lateral roots [59]. However, our results implied that different grassland management systems did not affect resprouting from rhizome buds, which suggests that reprouting from rhizome buds is not a major adjustment stragegy in response to grazing intensity. This finding is probably due to the short time period since the implementation of fencing and seasonal grazing.

Soil properties in response to different management types

Grazing intensity is one of the most important factors influencing soil properties [50,60]. It has been documented that the trampling action of grazing animals impacts the soil by increasing bulk density [61] and mechanical resistance, and reducing porosity, water infiltration, and aggregate stability [62], and native perennial cover and litter cover, with the consequence of changing the soil nutrient concentrations [63]. In our study, bulk density was found to increase with increasing grazing intensity, and all soil organic carbon, total nitrogen, and available nitrogen contents in M3 were slightly higher than in M1 and M2. We thus report that grazing exclusion may improve soil N availability. Arevalo et al. [64] found an increase in soil phosphorus content in response to increased goat grazing pressure in pastures. In contrast, our results indicated that soil total phosphorous and available phosphorous contents in M1 did not differ significantly from those of M2 and M3. The extreme cold, limited precipitation and short growing period may lead to no obvious change in phosphorus content.

Conclusions

By studying plant traits and soil nutrients in response to grazing intensity in an alpine steppe, we found that the isometric hypothesis is not supported in this particular region. The extreme cold, limited amount of precipitation and short growing period may lead to the plants' and soils' unique to different management types. Overall, the study suggests that the implementation of grazing exclusion played a positive role in the sustainable development of the alpine steppe region.

Acknowledgments

The authors are grateful to two anonymous reviewers all gave very helpful editorial comments.

Author Contributions

Conceived and designed the experiments: XDW GWC. Performed the experiments: JS JBW JTH. Analyzed the data: JS. Contributed reagents/materials/analysis tools: JS GWC JTH. Wrote the paper: JS SLN.

References

1. Lu XY, Fan JH, Yan Y, Wang XD (2013) Responses of Soil CO_2 Fluxes to Short-Term Experimental Warming in Alpine Steppe Ecosystem, Northern Tibet. Plos One 8.
2. Gao QZ, Li Y, Wan YF, Qin XB, Jiangcun WZ, et al. (2009) Dynamics of alpine grassland NPP and its response to climate change in Northern Tibet. Climatic Change 97: 515–528.
3. Sun J, Cheng GW, Li WP, Sha YK, Yang YC (2013) On the Variation of NDVI with the Principal Climatic Elements in the Tibetan Plateau. Remote Sensing 5: 1894–1911.
4. Wang GX, Cheng GD, Shen YP, Qian J (2003) Influence of land cover changes on the physical and chemical properties of alpine meadow soil. Chinese Science Bulletin 48: 118–124.
5. Sun J, Cheng GW, Li WP (2013) Meta-analysis of relationships between environmental factors and aboveground biomass in the alpine grassland on the Tibetan Plateau. Biogeosciences 10: 1707–1715.
6. Yu HY, Luedeling E, Xu JC (2010) Winter and spring warming result in delayed spring phenology on the Tibetan Plateau. Proceedings of the National Academy of Sciences of the United States of America 107: 22151–22156.
7. Zheng SX, Ren HY, Li WH, Lan ZC (2012) Scale-Dependent Effects of Grazing on Plant C: N: P Stoichiometry and Linkages to Ecosystem Functioning in the Inner Mongolia Grassland. Plos One 7.
8. Wu GL, Liu ZH, Zhang L, Chen JM, Hu TM (2010) Long-term fencing improved soil properties and soil organic carbon storage in an alpine swamp meadow of western China. Plant and Soil 332: 331–337.
9. Wang XD, Yan Y, Cao YZ (2012) Impact of historic grazing on steppe soils on the northern Tibetan Plateau. Plant and Soil 354: 173–183.
10. Haynes AG, Schutz M, Buchmann N, Page-Dumroese DS, Busse MD, et al. (2014) Linkages between grazing history and herbivore exclusion on decomposition rates in mineral soils of subalpine grasslands. Plant and Soil 374: 579–591.
11. Risch AC, Haynes AG, Busse MD, Filli F, Schutz M (2013) The Response of Soil CO2 Fluxes to Progressively Excluding Vertebrate and Invertebrate Herbivores Depends on Ecosystem Type. Ecosystems 16: 1192–1202.
12. Fan YJ, Hou XY, Shi HX, Shi SL (2013) Effects of grazing and fencing on carbon and nitrogen reserves in plants and soils of alpine meadow in the three headwater resource regions. Russian Journal of Ecology 44: 80–88.
13. Sun DS, Wesche K, Chen DD, Zhang SH, Wu GL, et al. (2011) Grazing depresses soil carbon storage through changing plant biomass and composition in a Tibetan alpine meadow. Plant Soil and Environment 57: 271–278.
14. Shi XM, Li XG, Li CT, Zhao Y, Shang ZH, et al. (2013) Grazing exclusion decreases soil organic C storage at an alpine grassland of the Qinghai-Tibetan Plateau. Ecological Engineering 57: 183–187.
15. Whitford WG, Steinberger Y (2012) Effects of seasonal grazing, drought, fire, and carbon enrichment on soil microarthropods in a desert grassland. Journal of Arid Environments 83: 10–14.
16. Yang XH, Guo XL, Fitzsimmons M (2012) Assessing light to moderate grazing effects on grassland production using satellite imagery. International Journal of Remote Sensing 33: 5087–5104.
17. Hilbert DW, Swift DM, Detling JK, Dyer MI (1981) Relative Growth-Rates and the Grazing Optimization Hypothesis. Oecologia 51: 14–18.
18. Lkhagva A, Boldgiv B, Goulden CE, Yadamsuren O, Lauenroth WK (2013) Effects of grazing on plant community structure and aboveground net primary production of semiarid boreal steppe of northern Mongolia. Grassland Science 59: 135–145.
19. Wu GL, Du GZ, Liu ZH, Thirgood S (2009) Effect of fencing and grazing on a Kobresia-dominated meadow in the Qinghai-Tibetan Plateau. Plant and Soil 319: 115–126.
20. Pandey CB, Singh JS (1992) Influence of Rainfall and Grazing on Belowground Biomass Dynamics in a Dry Tropical Savanna. Canadian Journal of Botany-Revue Canadienne De Botanique 70: 1885–1890.
21. Ma WL, Shi PL, Li WH, He YT, Zhang XZ, et al. (2010) Changes in individual plant traits and biomass allocation in alpine meadow with elevation variation on the Qinghai-Tibetan Plateau. Science China-Life Sciences 53: 1142–1151.
22. Burke IC, Lauenroth WK, Riggle R, Brannen P, Madigan B, et al. (1999) Spatial variability of soil properties in the shortgrass steppe: The relative importance of topography, grazing, microsite, and plant species in controlling spatial patterns. Ecosystems 2: 422–438.
23. Liu M, Liu GH, Wu X, Wang H, Chen L (2014). Vegetation traits and soil properties in respones to utilization patterns of grassland in Hulun Buir City, Inner Mongolia, China. Chinese Geographical Science, 24(4):471–478.
24. Xie JB, Tang LS, Wang ZY, Xu GQ, Li Y (2012) Distinguishing the Biomass Allocation Variance Resulting from Ontogenetic Drift or Acclimation to Soil Texture. Plos One 7.
25. Yang YH, Luo YQ (2011) Isometric biomass partitioning pattern in forest ecosystems: evidence from temporal observations during stand development. Journal of Ecology 99: 431–437.
26. Baruch Z (1994) Responses to Drought and Flooding in Tropical Forage Grasses.1. Biomass Allocation, Leaf Growth and Mineral Nutrients. Plant and Soil 164: 87–96.
27. Sanchez-Vilas J, Bermudez R, Retuerto R (2012) Soil water content and patterns of allocation to below- and above-ground biomass in the sexes of the subdioecious plant Honckenya peploides. Annals of Botany 110: 839–848.
28. Enquist BJ, Niklas KJ (2002) Global allocation rules for patterns of biomass partitioning in seed plants. Science 295: 1517–1520.
29. Niklas KJ (2005) Morphogenesis and biomechanics: The roles of mechanical perturbation and other environmental stresses in plant development and evolution. Evolving Form and Function: Fossils and Development: 25–41.
30. Cheng DL, Niklas KJ (2007) Above- and below-ground biomass relationships across 1534 forested communities. Annals of Botany 99: 95–102.
31. McConnaughay KDM, Coleman JS (1999) Biomass allocation in plants: Ontogeny or optimality? A test along three resource gradients. Ecology 80: 2581–2593.
32. Shipley B, Meziane D (2002) The balanced-growth hypothesis and the allometry of leaf and root biomass allocation. Functional Ecology 16: 326–331.
33. Reich PB, Oleksyn J, Wright IJ, Niklas KJ, Hedin L, et al. (2010) Evidence of a general 2/3-power law of scaling leaf nitrogen to phosphorus among major plant groups and biomes. Proceedings of the Royal Society B-Biological Sciences 277: 877–883.
34. McCarthy MC, Enquist BJ (2007) Consistency between an allometric approach and optimal partitioning theory in global patterns of plant biomass allocation. Functional Ecology 21: 713–720.
35. Enquist BJ (2002) Universal scaling in tree and vascular plant allometry: Toward a general quantitative theory linking plant form and function from cells to ecosystems. Tree Physiology 22: 1045–1064.
36. Tolvanen A, Schroderus J, Henry GHR (2001) Age- and stage-based bud demography of Salix arctica under contrasting muskox grazing pressure in the High Arctic. Evolutionary Ecology 15: 443–462.

37. Jonsdottir IS (1991) Effects of Grazing on Tiller Size and Population-Dynamics in a Clonal Sedge (Carex-Bigelowii). Oikos 62: 177–188.
38. Cheplick GP (1997) Responses to severe competitive stress in a clonal plant: Differences between genotypes. Oikos 79: 581–591.
39. Lu XY, Fan JH, Yan Y, Wang XD (2011) Soil water soluble organic carbon under three alpine grassland types in Northern Tibet, China. African Journal of Agricultural Research 6: 2066–2071.
40. Wang XD, Li MH, Liu SZ, Liu GC (2006) Fractal characteristics of soils under different land-use patterns in the and and semiarid regions of the Tibetan Plateau, China. Geoderma 134: 56–61.
41. Xie YZ, Wittig R (2004) The impact of grazing intensity on soil characteristics of Stipa grandis and Stipa bungeana steppe in northern China (autonomous region of Ningxia). Acta Oecologica-International Journal of Ecology 25: 197–204.
42. Wang LA, Niu KC, Yang YH, Zhou P (2010) Patterns of above- and belowground biomass allocation in China's grasslands: Evidence from individual-level observations. Science China-Life Sciences 53: 851–857.
43. Bao SD (2000) Soil chemical anlysis of agriculture. Beijing: Chinese Agriculture Press.
44. Hu JY (2007) Morphological and structural traits and their ecological adaptation of two dominant species in alpine grassland in Tibetan Plateau. Master's degree thesis. Institute of botany, Chinese academy of sciences.
45. Yang YH, Fang JY, Ji CJ, Han WX (2009) Above- and belowground biomass allocation in Tibetan grasslands. Journal of Vegetation Science 20: 177–184.
46. Li HT, Han XG, Wu JG (2006) Variant scaling relationship for mass-density across tree-dominated communities. Journal of Integrative Plant Biology 48: 268–277.
47. Warton DI, Wright IJ, Falster DS, Westoby M (2006) Bivariate line-fitting methods for allometry. Biological Reviews 81: 259–291.
48. Hermans C, Hammond JP, White PJ, Verbruggen N (2006) How do plants respond to nutrient shortage by biomass allocation? Trends in Plant Science 11: 610–617.
49. Patty L, Halloy SRP, Hiltbrunner E, Korner C (2010) Biomass allocation in herbaceous plants under grazing impact in the high semi-arid Andes. Flora 205: 695–703.
50. Medina-Roldan E, Paz-Ferreiro J, Bardgett RD (2012) Grazing exclusion affects soil and plant communities, but has no impact on soil carbon storage in an upland grassland. Agriculture Ecosystems & Environment 149: 118–123.
51. Cheng J, Wu GL, Zhao LP, Li Y, Li W, et al. (2011) Cumulative effects of 20-year exclusion of livestock grazing on above- and belowground biomass of typical steppe communities in arid areas of the Loess Plateau, China. Plant Soil and Environment 57: 40–44.
52. Gao YZ, Giese M, Lin S, Sattelmacher B, Zhao Y, et al. (2008) Belowground net primary productivity and biomass allocation of a grassland in Inner Mongolia is affected by grazing intensity. Plant and Soil 307: 41–50.
53. Rodriguez MA, Brown VK, Gomezsal A (1995) The Vertical-Distribution of Belowground Biomass in Grassland Communities in Relation to Grazing Regime and Habitat Characteristics. Journal of Vegetation Science 6: 63–72.
54. Steffens M, Kolbl A, Totsche KU, Kogel-Knabner I (2008) Grazing effects on soil chemical and physical properties in a semiarid steppe of Inner Mongolia (PR China). Geoderma 143: 63–72.
55. Gill RA, Jackson RB (2000) Global patterns of root turnover for terrestrial ecosystems. New Phytologist 147: 13–31.
56. Smit C, Bakker ES, Apol MEF, Olff H (2010) Effects of cattle and rabbit grazing on clonal expansion of spiny shrubs in wood-pastures. Basic and Applied Ecology 11: 685–692.
57. Yu FH, Dong M, Krusi B (2004) Clonal integration helps Psammochloa villosa survive sand burial in an inland dune. New Phytologist 162: 697–704.
58. Gao Y, Xing F, Jin YJ, Nie DD, Wang Y (2012) Foraging responses of clonal plants to multi-patch environmental heterogeneity: spatial preference and temporal reversibility. Plant and Soil 359: 137–147.
59. Sosnova M, van Diggelen R, Klimesova J (2010) Distribution of clonal growth forms in wetlands. Aquatic Botany 92: 33–39.
60. Gao YH, Schumann M, Chen H, Wu N, Luo P (2009) Impacts of grazing intensity on soil carbon and nitrogen in an alpine meadow on the eastern Tibetan Plateau. Journal of Food Agriculture & Environment 7: 749–754.
61. Evans CRW, Krzic M, Broersma K, Thompson DJ (2012) Long-term grazing effects on grassland soil properties in southern British Columbia. Canadian Journal of Soil Science 92: 685–693.
62. Zhou ZC, Gan ZT, Shangguan ZP, Dong ZB (2010) Effects of grazing on soil physical properties and soil erodibility in semiarid grassland of the Northern Loess Plateau (China). Catena 82: 87–91.
63. Yates CJ, Norton DA, Hobbs RJ (2000) Grazing effects on plant cover, soil and microclimate in fragmented woodlands in south-western Australia: implications for restoration. Austral Ecology 25: 36–47.
64. Arevalo JR, Chinea E, Barquin E (2007) Pasture management under goat grazing on Canary Islands. Agriculture Ecosystems & Environment 118: 291–296.

25

Bayesian Inference of Baseline Fertility and Treatment Effects via a Crop Yield-Fertility Model

Hungyen Chen[1], **Junko Yamagishi**[2], **Hirohisa Kishino**[1]*

1 Graduate School of Agricultural and Life Sciences, The University of Tokyo, Tokyo, Japan, **2** Institute for Sustainable Agro-ecosystem Services, The University of Tokyo, Tokyo, Japan

Abstract

To effectively manage soil fertility, knowledge is needed of how a crop uses nutrients from fertilizer applied to the soil. Soil quality is a combination of biological, chemical and physical properties and is hard to assess directly because of collective and multiple functional effects. In this paper, we focus on the application of these concepts to agriculture. We define the baseline fertility of soil as the level of fertility that a crop can acquire for growth from the soil. With this strict definition, we propose a new crop yield-fertility model that enables quantification of the process of improving baseline fertility and the effects of treatments solely from the time series of crop yields. The model was modified from Michaelis-Menten kinetics and measured the additional effects of the treatments given the baseline fertility. Using more than 30 years of experimental data, we used the Bayesian framework to estimate the improvements in baseline fertility and the effects of fertilizer and farmyard manure (FYM) on maize (*Zea mays*), barley (*Hordeum vulgare*), and soybean (*Glycine max*) yields. Fertilizer contributed the most to the barley yield and FYM contributed the most to the soybean yield among the three crops. The baseline fertility of the subsurface soil was very low for maize and barley prior to fertilization. In contrast, the baseline fertility in this soil approximated half-saturated fertility for the soybean crop. The long-term soil fertility was increased by adding FYM, but the effect of FYM addition was reduced by the addition of fertilizer. Our results provide evidence that long-term soil fertility under continuous farming was maintained, or increased, by the application of natural nutrients compared with the application of synthetic fertilizer.

Editor: Wenju Liang, Chinese Academy of Sciences, China

Funding: The Grant-in-Aid for Scientific Research (grant number 59440009, PI: Wataru Sunohara) by the Ministry of Education, Science and Culture (MEXT, http://www.mext.go.jp) supported the initial phase (1984–1987) of the long-term fertilizer experiment. The University Farm continued the experiment until 2010 by the regular budget supplied by the University of Tokyo. This work was supported by Grants-in-Aid for Scientific Research (grant number 25280006) by the Japan Society for the Promotion of Science (JSPS, http://www.jsps.go.jp/) to H. K. The funders had no role in study design, data collection and analysis, decision to publish, or preparation of the manuscript.

Competing Interests: The authors have declared that no competing interests exist.

* Email: kishino@lbm.ab.a.u-tokyo.ac.jp

Introduction

Agricultural crop production is directly related to food supply, so agricultural soil productivity must be maintained. Balanced fertilization provides all the essential nutrients for crops to remain healthy and grow productively [1–3]. In a world facing increasing population pressure [4], our highest priority must be to increase crop productivity to ensure food security [2,5–7]. In this context, there has been increasing concern about the long-term productivity of soils on a global scale [8–11].

The relationship between crop yield and soil fertility under different fertilization regimes has been studied for decades. Aref and Wander [12] investigated the long-term trends of corn yield and soil organic matter in different crop sequences and soil fertility treatments. Merick and Németh [13] used results from 60-year field experiments to provide information on the relationship between fertilization and yields of rye and potato. Hallin et al. [14] investigated the relationship between microbial communities and total crop yield and nitrogen content in the crop in a 50-year-old fertilization experiment. Fan et al. [15] studied the trends in grain yields and soil organic carbon (SOC) in a 26-year dryland fertilization trial. By carrying out a 27-year experiment with various fertilization treatments in a rotation cropping system with wheat and maize in a red soil, Zhang et al. [16] investigated trends in SOC, soil nitrogen, and grain yield.

Many researchers have suggested that soil fertility under continuous farming is maintained, or increased, by the application of farmyard manure (FYM) compared with the application of synthetic fertilizer [2,11,17–19]. To ensure gains in crop productivity, it is necessary to maintain and even improve the soil fertility of a field under continuous farming practice [20–23]. In some cases, soils do not contain sufficient amounts of the essential nutrients required for rapid crop growth and high productivity [2,7,24]. As a result, supplemental nutrients, applied as fertilizers, manure or compost, are needed [25–27]. Many studies have analyzed the effects of manure [28–30], chemical fertilizer [31,32], and both [33,34] on soil fertility and crop yield.

Previous models have predicted the crop-yield response to fertilizer application [19] and climate change [35] and have greatly contributed to the software used in agricultural systems research [36,37]. Myers [38] used a static model to estimate the nitrogen fertilizer requirements of cereal crops. Deng et al. [39] proposed a theoretical framework that predicts the optimum planting density and maximal yield for an annual crop plant. Cong et al. [40] indicated that the CENTURY model [41] can simulate fertilization effects on SOC dynamics under different climate and soil conditions.

Soil quality is a complex combination of biological, chemical, and physical properties. In this paper, we focus on the application of these concepts to agriculture and define the baseline fertility as the level of soil fertility that crops utilize for their growth. Therefore, our soil quality is an interaction of the above properties and the crop activities. With this strict definition of baseline fertility, we propose a new crop yield-fertility model that enables quantification of the process of improving baseline fertility and the effects of treatments solely from the time series of crop yields. The model was modified from Michaelis-Menten kinetics and measured the additional effects of the treatments given the baseline fertility. Using more than 30 years of experimental data of maize (*Zea mays*), barley (*Hordeum vulgare*), and soybean (*Glycine max*) yield, we estimated in the Bayesian framework the improvements in baseline fertility and the effects of fertilizer and farmyard manure (FYM). We compared the efficiency of separate applications of fertilizer or FYM to the three crops (i.e., maize, barley and soybean) using estimated model parameters. The temporal variations in the baseline fertility of each crop were estimated and compared for six treatments and two soil types.

Materials and Methods

Field experiments

The long-term fertilizer experiment was conducted between 1980 and 2010 at the University Farm at the Institute for Sustainable Agro-ecosystem Services at the University of Tokyo, Nishitokyo, Tokyo, Japan (35°43′ N latitude and 139°32′ E longitude and an altitude of 53 m above mean sea level). The field site was located in the Kantō Plain, where the soil is covered with pyroclastic material from volcanoes that surrounded the western Kantō region 126,000 years ago. The soil parent material is Tachikawa loam. The surface soil is a black-colored fertile andosol containing a high percentage of humus and the subsurface soil (25 cm below the surface) is red-colored barren clay. Andosols are soils found in volcanic areas formed in volcanic tephra. Andosols have a different composition from chernozems and are not commonly found outside the Pacific "ring of fire". At the beginning of the experiment, the percent nitrogen was 2.23 g/kg in the surface soil and 0.97 g/kg in the subsurface soil. A 2-year crop rotation of maize-barley-soybean-barley was maintained throughout the experimental period and the crop yields were measured. The maize crop was sown at the beginning of July and harvested at the end of September every second year from 1980 onwards. The barley crop was sown in the first half of November and harvested either at the end of May or at the beginning of June of the next year each year from 1980 onwards. The soybean crop was sown at the end of June and harvested in the second half of October every second year from 1981 onwards. The seeds of the three crops were ridge sown using a seeding machine. For maize and soybean, one seed was sown in each hole, and the widths between the ridges and stocks were 71 cm and 23 cm, and 71 cm and 17.5 cm, respectively. For barley, row seeding was applied using a width of 17.8 cm between the ridges.

Six treatment plots containing the NPK fertilizer applications and two levels of farmyard manure (FYM) combined with compost were established in both of the fields with the surface soil (64 m^2 per plot, 8×8 m) and the subsurface soil (56 m^2 per plot, 7×8 m) and were replicated four times. The percent nitrogen, phosphorus, and potassium in the fertilizer treatments were N:P:K = 12:7.9:13.3 for maize and barley and 3:4.4:8.3 for soybean. Fertilizer was added to the soil at a rate of 1 t ha^{-1} for all crops. The FYM comprised wheat straw and cow dung. The percent nitrogen in the FYM ranged from 0.2–0.4. FYM was added to the plots at two levels, i.e., 20 and 60 t ha^{-1}. The FYM treatments were added to the soil twice a year from 1980 to 1990 before the crops were sown and once a year (before the barley was sown) from 1991 onwards. After the barley harvest in 2008, fertilizer and FYM applications were discontinued for all treatments. Phosphate fertilizer was added simultaneously with the NPK fertilizer and FYM at a rate of 2 t ha^{-1} after the soybean harvest from 1980 until 2001. All aboveground components were removed from the field during harvesting. The dead foliage, below-ground components, and remnant stem sections (i.e., 15 cm for maize and 10 cm for barley and soybean) remained in the field. The harvested material from the six treatments for both the surface and subsurface soils were weighed after drying at 80°C. The dry weight (g m^{-2}) of the aboveground maize components, the barley spike, and the soybean seed were measured.

Testing significance of yield differences in different soils

The crop yields of the surface and subsurface soils without the addition of fertilizer or farmyard manure (FYM) reflect the yield differences that occurred purely as a result of the soil. The crop yields observed for the subsurface soil were 39±25%, 49±20%, and 19±31% (±SD) lower than the yields observed for the surface soil for maize (Fig. S1A), barley (Fig. S1B), and soybean (Fig. S1C), respectively. There was a difference in yield between the two types of soils but it was only significant for barley ($t = -3.69$, $P<0.001$; for maize, $t = -1.58$, $P = 0.13$ and for soybean, $t = -0.94$, $P = 0.35$). Although the crops were planted in the same fields with the same soil, they obtained and used nutrients from the soil in a manner unique to each crop.

The maize and soybean yields did not differ significantly ($P>0.2$) between the surface and subsurface soils when fertilizer or FYM or a combination of fertilizer and FYM were applied (Fig. S1A and C for maize and soybean, respectively). The barley yields differed significantly between the surface and subsurface soils when both fertilizer and FYM ($t = 2.27$, $P<0.05$) or FYM alone ($t = 2.18$, $P<0.05$) were added, but did not differ significantly when only fertilizer was added ($t = -0.63$, $P = 0.52$) (Fig. S1B). The difference in yields between the surface and subsurface soils was reduced when either fertilizer or FYM was added to the soil. The application of fertilizer and FYM improved the yields for both the surface and subsurface soils and reduced the yield differences between the soil types. This finding indicates that productivity can be increased to an average level by fertilization even in a field with barren soil. While this may have been prior knowledge, the manner in which fertilizer and FYM contribute nutrients to the soil and how the crops make use of the nutrients obtained from these sources remains unclear.

Testing significance of yearly variance

The maize and soybean yields increased significantly ($P<0.05$) for both soil types over the 30-year period. The barley yield increased significantly in the subsurface soil ($P<0.05$) but not in the surface soil ($P = 0.15$). The soils were fertilized by crop residues (i.e., roots, fallen leaves, and stem sections (15 cm above ground for maize and 10 cm for barley and soybean)) that remained in and on the soil after harvesting. Organic matter accumulated in the soil over time, providing a long-term, slow-release source of nitrogen, phosphorus, sulfur, and other important nutrients for crop growth, which resulted in increased soil productivity.

Testing significance of the treatment effects

We used a nested analysis of variance to compare the yields of the three crops in the surface and subsurface soils in response to the different treatments. The surface soil produced significantly greater barley crop yields compared with the subsurface soil (Fig. S1B, $F = 14.68$, $P < 0.001$). Maize and soybean yields (Fig. S1A, $F = 2.91$, $P = 0.09$ and Fig. S1C, $F = 2.19$, $P = 0.14$, respectively) did not differ significantly between the two soil types. The interaction effects of fertilizer and FYM were significant for maize ($F = 3.32$, $P < 0.05$) and barley ($F = 10.71$, $P < 0.001$), but not for soybean ($F = 0.69$, $P = 0.60$). The soybean yield did not differ significantly among the treatments ($P > 0.1$ for both fertilizer and FYM).

The derivation of the crop yield-fertility model

Our model is based on the general pattern of the observed crop yield. The temporal variation in maize yield in response to four treatments applied to the surface soil over 30 years is shown in Fig. 1A. There were clear differences in the yield between the four treatments prior to 1982 (the first two data points). The response of the crop yield descended in the following order: the addition of NPK fertilizer and FYM (F:1, M:1), the addition of NPK fertilizer only (F:1, M:0), the addition of FYM only (F:0, M:1), and no nutrient application (F:0, M:0). The yield differences between the treatments receiving NPK fertilizer and FYM (F:1, M:1) and NPK fertilizer only (F:1, M:0) could not be detected from 1984 onwards. The effect of FYM could no longer be detected once the fertilizer application was continued. From 1994, similar crop yields were observed in the treatments with fertilizer, FYM or a combined application of fertilizer and FYM. In the last year of the study, there were negligible differences between treatments because fertilizer and FYM applications were discontinued in 2008. The soybean data are presented in Fig. 1C. The pattern of the soybean yield differed from maize. There were almost no differences in crop yield among the three treatments that received fertilizer, FYM or a combined application of fertilizer and FYM in the first and later years. The differences in crop yield between the treatments with and without fertilizer were negligible from 1995 onwards. The inherent soil fertility may be sufficient to grow soybean without the addition of fertilizer. The pattern of barley yield (Fig. 1B) was similar to maize but was not as well defined.

Our model can predict crop yields given the total fertility level of the soil. The model can be used to describe the variations in crop yield and the effect of different fertilizer treatments over different years for each crop using the estimated soil fertility. The crop yield-fertility curves for maize, barley, and soybean are presented in Fig. 2A, B, and C, respectively. Soil fertility was increased by the application of fertilizer over a 20-year period. Furthermore, the effect of fertilizer decreased over time for all three crops from 1980–2000.

The crop yield-fertility model

We developed a simple mathematical model to estimate the soil fertility level and to quantify the contributions of fertilizer and FYM to the improvements in crop yields. The model was modified from Michaelis-Menten kinetics [42]:

$$Y_{i,t} \cong \frac{V}{1+\dfrac{K}{BF_{i,t}+a\times \mathrm{F}+b\times \mathrm{M}}}. \tag{1}$$

We did not apply the model to describe the kinetics. Rather, we used a form of the model to interpret the effects of baseline fertility, fertilizer, and FYM on the crop yield and the extent of saturation of additional inputs. The model related crop yields ($\mathrm{Y}_{i,t}$) to the total fertility ($BF_{i,t}+a\times \mathrm{F}+b\times \mathrm{M}$) of the soil for treatment i at time t (Fig. 2). The value of the baseline fertility (BF) for the subsurface soil without the addition of fertilizer or FYM in the first year was normalized by setting this value to 1. V is the maximum yield in response to the maximum fertilizer input; K is the fertility level before fertilization that is required to produce half of the maximum yield, V. A large K indicates that more nutrients need to be added to the soil for the crop to grow, so K can infer the sterility of the soil prior to fertilization. $BF_{i,t}$ is the baseline fertility of the crop for treatment i at time t and is assumed to vary gradually over time; F is the level of fertilizer application (0 and 1), M is the level of FYM application (0, 1/3, and 1), and a and b represent the contributions of F and M relative to the baseline fertility, respectively. Hereafter, we refer to V, K, a, and b as the maximum yield, half-saturated fertility, fertilizer contribution, and FYM, respectively. A Bayesian framework was adopted for parameter estimation assuming a gradual change in the baseline fertility.

The likelihood and priors

The likelihood of the yield for treatment i at time t followed a normal distribution with the mean $\dfrac{V}{1+\dfrac{K}{BF_{i,t}+a\times \mathrm{F}+b\times \mathrm{M}}}$ and the variance δ:

$$\mathrm{Yield}_{i,t} \sim N\left(\frac{V}{1+\dfrac{K}{BF_{i,t}+a\times \mathrm{F}+b\times \mathrm{M}}},\delta\right) \tag{2}$$

This value was normalized by setting the BF of the subsurface soil without the addition of fertilizer or FYM in the first year to 1. The priors of the BF for the other treatments in the first year followed a gamma distribution with a shape parameter of 1 and a scale parameter of 1. The smoothness priors of the BF from the second year followed a normal distribution with the mean equal to the value of the BF in the preceding year and the variance τ:

$$BF_{i,t} \sim N(BF_{i,t-1},\tau), \quad t>1. \tag{3}$$

The inverse of δ followed a gamma distribution with a shape parameter of 0.1 and a scale parameter of 10. The inverse of τ followed a gamma distribution with a shape parameter of 0.1 and a scale parameter of 10. The prior of V followed a normal distribution with a mean of 0 and a standard deviation of 1000. The prior of K followed a gamma distribution with a shape parameter of 0.1 and a scale parameter of 10. The priors of a and b followed a gamma distribution with a shape parameter of 1 and a scale parameter of 1. The priors of the estimates were designed to be as non-informative as possible within a realistic range of the

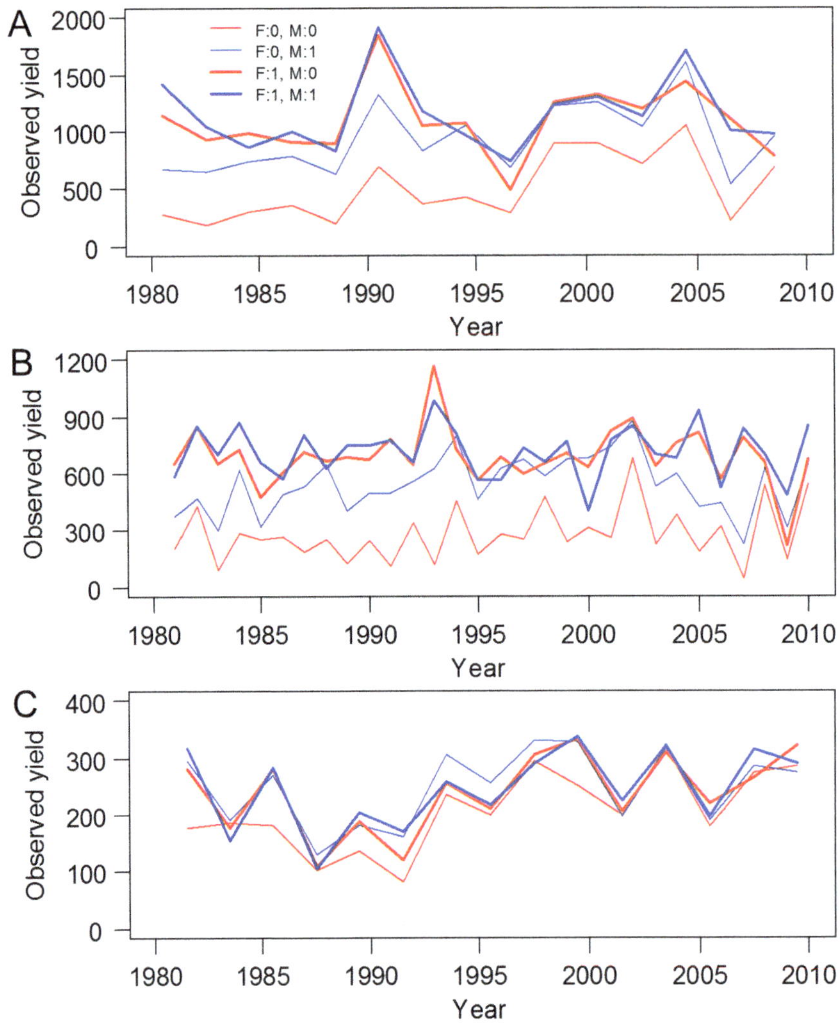

Figure 1. Temporal variations in observed crop yield (g m^{-2}) for four treatments in the surface soil. (A) Maize; (B) Barley; (C) Soybean; F, level of fertilizer; M, level of farmyard manure. See Fig. S1 for all treatments in surface and subsurface soils.

parameter values. All calculations and data analyses were performed using R v2.13.2 [43]. The raw data is available as (Data S1).

Results

Increasing trend of soil fertility

The baseline fertility of the surface and subsurface soils in all treatments increased over the 30-year period for all three crops (Figs. 3, S5–6). The baseline fertility of the subsurface soil was very low for maize and barley but was close to half-saturated for soybean when fertilization was initiated.

The treatment differences in the baseline fertility (i.e., maximum minus minimum) were increased by continuous cropping over the experimental period. The difference in the baseline fertility between the treatments was 0.89 in 1980 and 3.24 in 2008 for maize; for barley, it was 1.20 in 1980 and 2.54 in 2010, and for soybean, it was 0.96 in 1980 and 3.19 in 2009.

We used a 2×2 factorial analysis of variance to compare the baseline fertility of the three crops in response to the different treatments. The interaction effects of fertilizer and FYM were significant for maize ($F = 15.09$, $P<0.001$) and barley ($F = 50.67$, $P<0.001$), but not for soybean ($F = 1.77$, $P = 0.19$). The baseline fertility of soybean differed significantly for fertilizer ($F = 10.26$, $P<0.01$) but not for FYM ($F = 4.00$, $P = 0.05$).

The effects of fertilizer and FYM

The Bayesian estimates of the model for the three crops are shown in Table 1. The traces of the MCMC (Markov Chain Monte Carlo) samples show the well-mixing and convergence to the posterior distributions (Fig. S7). We compared the mean values of the parameters for the three crops. The maize crop had the highest half-saturated fertility level and maximum yield. The soybean crop had the lowest half-saturated fertility level and maximum yield. For maize, the size of the fertilizer contribution was almost equal to the size of the half-saturated fertility, and FYM contribution was 81% smaller than the half-saturated fertility. For barley, the fertilizer contribution was 40% higher than the half-saturated fertility and the FYM contribution was 65% lower than the half-saturated fertility. This indicates that the amount of fertilizer applied in the experiment was 40% more than the fertility level at which the barley yield was half of the maximum yield, but the amount of FYM applied was 65% less than required. For soybean, the values of the fertilizer and FYM

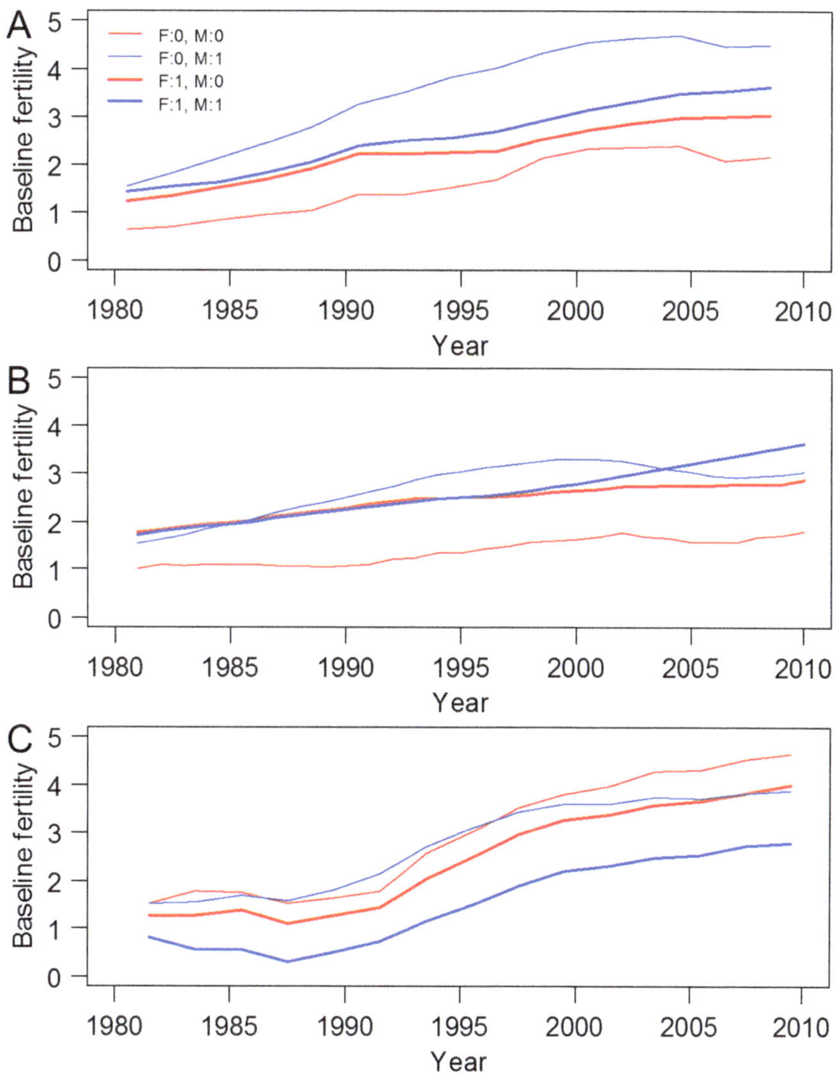

Figure 2. Crop yield-fertility curves for (A) Maize, (B) Barley, and (C) Soybean. The value of 1 for fertility is the fertility without the addition of fertilizer or FYM in the initial year for the subsurface soil. F, level of fertilizer. See Figs. S2A, S3A, and S4A for the band that corresponds to the standard deviation of the curves.

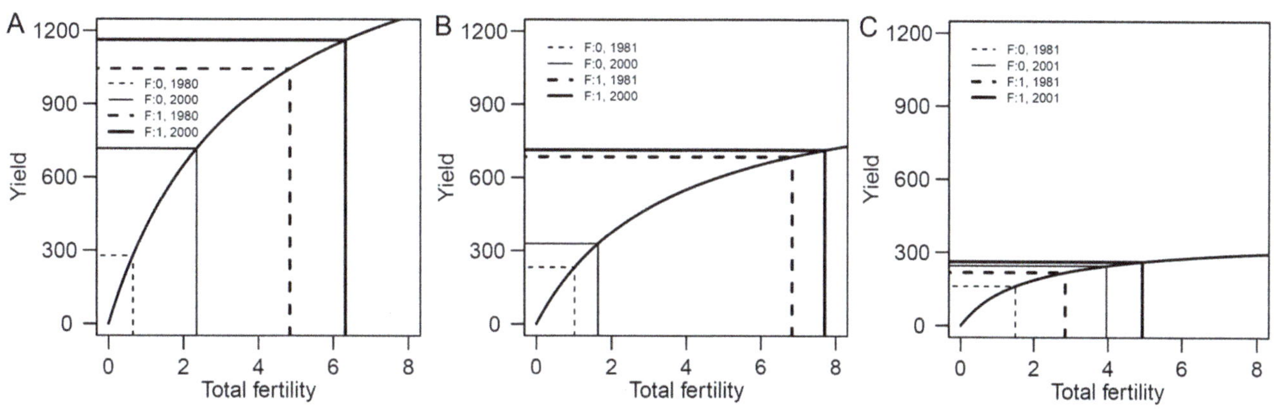

Figure 3. Temporal variations in the baseline fertility for four treatments in the surface soil. (A) Maize; (B) Barley; (C) Soybean; F, level of fertilizer; M, level of farmyard manure. See Fig. S5 for all treatments in surface and subsurface soils.

Table 1. The posterior mean and standard deviation (SD) of the Bayesian estimates.

Crop	V (g m^{-2})		K		A		B	
	Mean	SD	Mean	SD	Mean	SD	Mean	SD
Maize	1833.01	316.56	3.65	1.68	3.61	1.27	0.68	0.53
Barley	1042.87	134.97	3.56	1.18	5.07	1.44	1.25	0.63
Soybean	360.17	65.54	1.87	1.01	1.58	0.58	1.62	0.94

V, maximum yield; *K*, half-saturated fertility; *a*, fertilizer contribution; *b*, FYM contribution.

contributions were similar, i.e., 16 and 13%, respectively, and were lower than the half-saturated fertility.

In contrast with maize and barley, the value of the FYM contribution was higher than the fertilizer contribution for the soybean crop. The application rate of fertilizer and FYM approximated the fertility level at which the soybean yield was half of the maximum yield. The half-saturated fertility level for maize was 3.65, and the half-saturated fertility levels for barley and soybean were 2 and 49% lower, respectively. The fertilizer contribution for barley was 5.07, and the fertilizer contributions for maize and soybean were 29 and 69% lower, respectively. The FYM contribution for soybean was 1.62, and the FYM contributions for maize and barley were 58 and 23% lower, respectively. The model estimates revealed how the different crops used the nutrients from fertilizer and FYM applications to the soil. Maize needed the highest level of fertility to reach half of the maximum yield (i.e., the largest half-saturated fertility) compared with barley and soybean. Fertilizer contributed the maximum amount to the barley yield, and the FYM contributed the maximum amount to the soybean yield.

The total fertility levels among the treatments for both surface and subsurface soils during the 30-year period ranged from 0.54 to 7.22 for maize (Fig. S2A), 0.41 to 9.78 for barley (Fig. S3A), and 0.90 to 5.56 for soybean (Fig. S4A). Barley may be the most efficient crop in terms of fertilizer use among the three crops. The maximum total fertility obtained for soybean was 2.97 × the half-saturated fertility, which is larger than the values obtained for maize (1.98) and barley (2.75).

Using the total fertility range, we estimated the predicted range in crop yield using the crop yield-fertility model for each crop. The predicted crop yields ranged from 236.24 to 1217.51 g m^{-2} for maize, 107.70 to 764.56 g m^{-2} for barley, and 117.02 to 269.52 g m^{-2} for soybean. Because of fluctuations in crop yield that could not be described by the model, the predicted yields covered 49, 57, and 42% of the observed yield ranges for maize (35 and 2023 g m^{-2}), barley (14 and 1168 g m^{-2}), and soybean (16 and 381 g m^{-2}), respectively.

The decreasing effect of fertilization over time

In the sterile subsurface soil in the initial year, the application of fertilizer resulted in 153, 189, and 62% increases (the FYM resulted in 42, 65, and 61% increases) in crop yield for maize, barley, and soybean, respectively. In the fertile surface soil in the initial year, the application of fertilizer resulted in 275, 196, and 35% increases (FYM resulted in 149, 98, and 40% increases) in crop yield for maize, barley, and soybean, respectively. After 20 years of farming, the application of fertilizer resulted in 89, 219, and 11% increases (FYM resulted in 74, 150, and 13% increases) in the subsurface soil, and 62, 113, and 7% increases (FYM resulted in 50, 75, and 8% increases) in the surface soil for maize, barley, and soybean, respectively, because of the increased baseline fertility. The greatest yield increases were observed in the initial year in the fertile surface soil with the application of both fertilizer and FYM (compared with the treatment without fertilizer or FYM), and then the yield increment decreased each year (Fig. 4). In the sterile subsurface soil, it took 5–10 years for the yield increase to reach the maximum value.

Predicted crop yields reflected the trend of the observed yields

The estimated fertility levels were used to predict the crop yield using the yield-fertility curves for each crop (Figs. S2A, S3A, and S4A). The trends of the predicted yield were similar to the trends of the observed yield for all three crops. The correlation

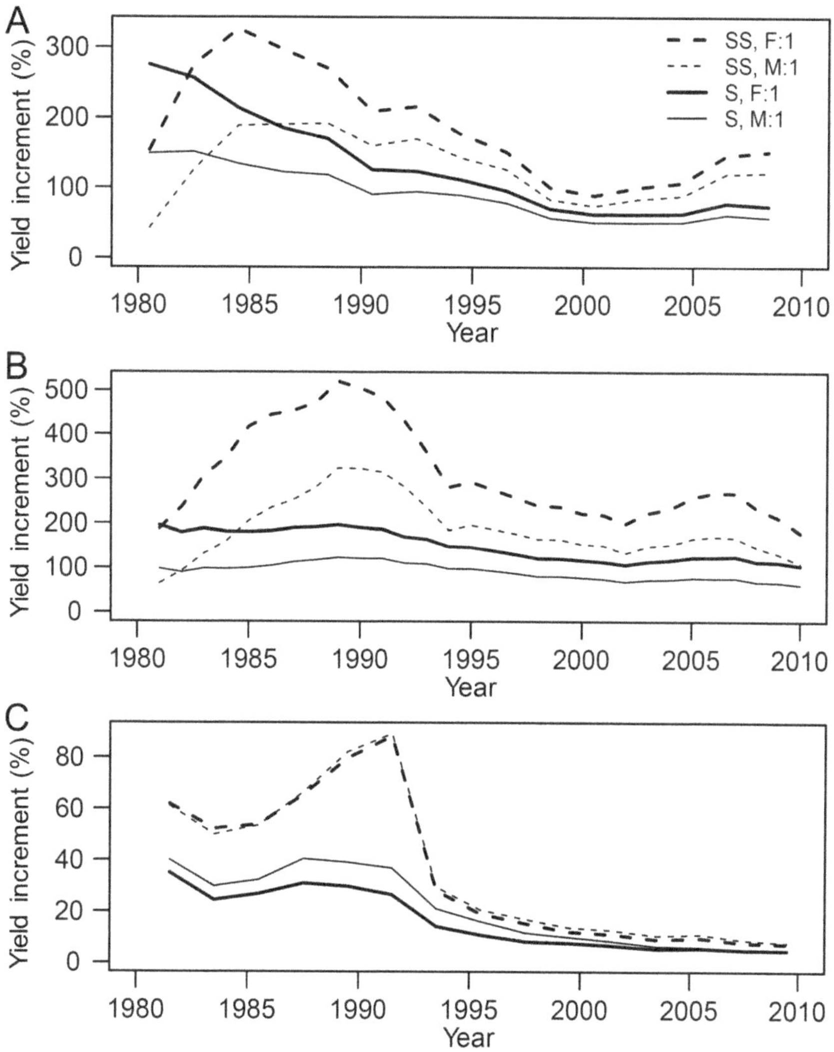

Figure 4. Temporal variations in the yield increment resulting from the application of fertilizer and farmyard manure. (A) Maize; (B) Barley; (C) Soybean; SS, subsurface soil; S, surface soil; F, level of fertilizer; M, level of farmyard manure.

coefficients were 0.758 (Fig. S2B, $P<0.001$) for maize, 0.801 (Fig. S3B, $P<0.001$) for barley, and 0.622 (Fig. S4B, $P<0.001$) for soybean.

Yield fluctuations were observed among the treatments but these fluctuations were not reflected by the predicted crop yields. For maize, the peak yields occurred in 1990 and 2004, and a reduced yield occurred in 1996. For barley, the peak yield for the fertilizer-treated plots occurred in 1993, but the peak yield occurred 1 year later in the plots without fertilizer treatment. The FYM treatments were discontinued in 1991 for the summer harvested crops (maize and soybean). Clear reductions in maize yield in 1991 and soybean yield in 1992 are shown in Fig. 1A and C, respectively, after which the yields for both crops increased until 2008. The estimated fertility levels and predicted yields represented the temporal variations that coincided with treatment changes during the experimental period (Fig. 1). Termination of the phosphate fertilizer application in 2001 was well described by the temporal variations in the baseline fertility of maize and barley. The baseline fertility levels declined, especially in the soil treated with FYM only (Fig. 4). Similar patterns were not observed for the fertilizer-treated soils. A decline in fertility was not observed for soybean. As previously mentioned, soybean yields were not significantly different among treatments.

The effect of climatic factors

Crop yield can be affected by climatic factors [35] such as precipitation [44,45], temperature [46,47] and recent warming [48]. The impact of climate change may damage crop yields on a global scale [48,49] and lead to decreases in crop production [50]. However, weather variation may play a small role in crop yield on a regional scale [45] and the effect may depend on fertilization and soil type [45,50]. Monthly precipitation and average monthly temperature recorded at the experimental site are shown in Fig. S8A and B. We performed a correlation analysis between the residual of the crop yield (i.e., predicted yield minus observed yield) and the climatic variables for the three crops (Table S1 in File S1). The simple residual analysis indicated that temperature had some effect on maize and soybean yields. We then conducted additional analyses that included the effect of temperature in the crop yield-fertility model (see File S1 and Table S2 in File S1). Based on the results of the minor contribution of temperature, we concluded that the annual fluctuations in crop yields were mainly caused by unknown factors other than the climatic factors.

Discussion

The baseline fertility

The baseline fertility of the subsurface soil was very low for maize and barley. This suggests that the fertility in the subsurface soil prior to the application of fertilizer or FYM was not sufficient for maize or barley.

The increasing baseline fertility trend was also observed in soil without additional fertilizer or FYM. The increased fertility may have resulted from the addition of crop residues to the soil and nitrogen fixation by the soybean crop. These fertility increases may indicate soil maturation (i.e., the accumulation of nutrients) in the experimental fields.

The baseline fertility of maize indicated that when FYM was applied alone, soil fertility increased more than when both FYM and fertilizer were applied, for both soil types (Fig. 3A). The application of FYM increased the organic matter content of the soil, which resulted in increased soil fertility. Similar trends were observed for barley and soybean (Fig. 3B and C), supporting the results observed for maize. The results for soybean suggested that the application of fertilizer could decrease the soil fertility when soybean was planted in a soil with a saturated level of fertilization (Fig. 3C).

The effects of fertilizer and FYM

Our result indicates that the amount of fertilizer applied in the experiment reached the fertility level at which the maize yield was half of the maximum yield, but the amount of FYM required to reach this value was much lower. Barley obtained the highest fertility from the soil and fertilization despite the fact that maize produced the greatest crop yield and had the largest half-saturated fertility level. By using nutrients from the soil and fertilizer, soybean was the crop that most often approached the maximum yield.

Our findings indicate that the sterile subsurface soil needed more time to accumulate nutrients than the surface soil. The percentage increments in crop yield were larger in the subsurface soil than in the surface soil, which indicates that fertilizer application effectively improved the crop yield in the barren field. The fertilizer-induced increment in maize and barley crop yield increased after 2001. This is because the phosphate fertilizer application was discontinued in 2001 and the baseline soil fertility subsequently declined. The soybean crop did not show a similar reduction in yield after 2001.

Long-term field crop fertilization experiment and crop models

Long-term experiments examining fertilizer treatments and crop yield have been widely conducted to assess the effects of applying fertilizer [51], manure [52], or a combination of fertilizer and manure [14,53,54]. In addition, experiments have been conducted to determine the impacts of nutrient addition on the fertility status of different soils [55–57]. However, it is important to note that our long-term record of crop yields with various types of treatments has made this the first study to simultaneously examine the temporal variation in baseline fertility and the contribution of nutrient applications.

Conclusions

By allowing the gradual temporal variation in baseline fertility, it was possible to estimate the process of soil fertilization and the short-term effects of different fertilizer treatments. Our results support the proposal that naturally derived nutrients should be used to maintain soil fertility and synthetic fertilizer should be used to maintain productivity. Our results provide a clear description of the relationship between crop yield and soil, which can be easily understood.

Supporting Information

Figure S1 Temporal variations in observed crop yield (g m^{-2}) for six treatments in a field with a fertile surface soil and a field with a barren subsurface soil from 1980 to 2010. (A) Maize; (B) Barley; (C) Soybean; SS, subsurface soil; S, surface soil; F, level of fertilizer; M, level of farmyard manure.

Figure S2 Temporal variations in the observed and predicted yields (g m^{-2}) of maize for six treatments in a field with a fertile surface soil and a field with a barren subsurface soil presented every second year from 1980 to 2008. (A) A crop yield-fertility map. The crop yield-fertility model translates total fertility into predicted yield. The gray band represents the band that corresponds to the standard deviation (±SD) of the crop yield-fertility curve. (B) The relationship between the observed and predicted yields. SS, subsurface soil; S, surface soil; F, level of fertilizer; M, level of farmyard manure; BF, baseline fertility.

Figure S3 As for Figure S2 but for barley every year from 1980 to 2010.

Figure S4 As for Figure S2 but for soybean every 2 years from 1981 to 2009.

Figure S5 Temporal variations in the baseline fertility estimated using the crop yield-fertility model for six treatments in a field with a fertile surface soil and a field with a barren subsurface soil from 1980 to 2010. (A) Maize; (B) Barley; (C) Soybean; SS, subsurface soil; S, surface soil; F, level of fertilizer; M, level of farmyard manure.

Figure S6 Temporal variations in the posterior means and the bands that correspond to the standard deviation (±SD) of the Bayesian estimates of baseline fertility for six treatments in a field with a fertile surface soil and a field with a barren subsurface soil. (A) Maize; (B) Barley; (C) Soybean; SS, subsurface soil; S, surface soil; F, level of fertilizer; M, level of farmyard manure.

Figure S7 Traces of the MCMC samples of V (g m^{-2}), K, a, and b. (A) Maize; (B) Barley; (C) Soybean. The chain length was set to 1,000,000 steps logging every 100th step.

Figure S8 Temporal variation of the climatic variables recorded at the experimental site from January 1980 to December 2011. (A) Monthly precipitation. (B) Average monthly temperature.

File S1 Contribution of the climatic factor. The analyses by including a temperature effect in the crop yield-fertility model.

Data S1 The raw data of the crop yields of maize, barley, and soybean from the long-term fertilizer experiment from 1980 to 2010. As for the detail of the experiment, see text.

Author Contributions

Conceived and designed the experiments: HC HK. Performed the experiments: JY. Analyzed the data: HC. Wrote the paper: HC HK. Confirmed the manuscript: HC JY HK.

References

1. Mäder P, Fliessbach A, Dubois D, Gunst L, Fried P (2002) Soil fertility and biodiversity in organic farming. Science 296: 1964–1967.
2. Tilman D, Cassman KG, Matson PA, Naylor R, Polasky S (2002) Agricultural sustainability and intensive production practices. Nature 418: 671–677.
3. Baligar VC, Fageria NK, He ZL (2001) Nutrient use efficiency in plants. Commun Soil Sci Plant 32: 921–950.
4. Holdren JP, Ehrlich PR (1974) Human population and the global environment: population growth, rising per capita material consumption, and disruptive technologies have made civilization a global ecological force. Am Sci 62: 282–292.
5. Glover JD, Reganold RP, Cox CM (2012) Plant perennials to save Africa's soils. Nature 489: 359–361.
6. Alexandratos N (1999) World food and agriculture: outlook for the medium and longer term. PNAS 96: 5908–5914.
7. Cassman KG (1999) Ecological intensification of cereal production systems: yield potential, soil quality, and precision agriculture. PNAS 96: 5952–5959.
8. Isbella F, Reichb PB, Tilmana D, Hobbiea SE, Polasky S, et al. (2013) Nutrient enrichment, biodiversity loss, and consequent declines in ecosystem productivity. PNAS 110: 11911–11916.
9. Cordell D, Drangert J-O, White S (2009) The story of phosphorus: global food security and food for thought. Glob Environ Change 19: 292–305.
10. Rasmussen PE, Goulding KWT, Brown JR, Grace PR, Janzen H, et al. (1998) Long-term agroecosystem experiments: assessing agricultural sustainability and global change. Science 282: 893–896.
11. Matson PA, Parton WJ, Power AG, Swift MJ (1997) Agricultural intensification and ecosystem properties. Science 277: 504–509.
12. Aref S, Wander MM (1997) Long-term trends of corn yield and soil organic matter in different crop sequences and soil fertility treatments on the morrow plots. Adv Agron 62: 153–197.
13. Mercik S, Németh K (1985) Effects of 60-year N, P, K and Ca fertilization on EUF-nutrient fractions in the soil and on yields of rye and potato crops. Plant Soil 83: 151–159.
14. Hallin S, Jones CM, Schloter M, Philippot L (2009) Relationship between N-cycling communities and ecosystem functioning in a 50-year-old fertilization experiment. ISME J 3: 597–605.
15. Fan T, Xu M, Song S, Zhou G, Ding L (2008) Trends in grain yields and soil organic C in a long-term fertilization experiment in the China Loess Plateau. J Plant Nutr Soil Sci 171: 448–457.
16. Zhang H, Xu M, Zhang F (2009) Long-term effects of manure application on grain yield under different cropping systems and ecological conditions in China. J Agr Sci 147: 31–42.
17. Power AG (2010) Ecosystem services and agriculture: tradeoffs and synergies. Philos Soc Trans R Soc Lond B 365: 2959–2971.
18. Altieri MA, Nicholls CI (2003) Soil fertility management and insect pests: harmonizing soil and plant health in agroecosystems. Soil Till Res 72: 203–211.
19. Clark MS, Horwath WR, Shennan C, Scow KM (1998) Changes in soil chemical properties resulting from organic and low-input farming practices. Agron J 90: 662–671.
20. Fan R, Zhang X, Liang A, Shi X, Chen X, et al. (2012) Tillage and rotation effects on crop yield and profitability on a Black soil in northeast China. Can J Soil Sci 92: 463–470.
21. Yanni YG, Dazzo FB (2010) Enhancement of rice production using endophytic strains of *Rhizobium leguminosarum* bv. trifolii in extensive field inoculation trials within the Egypt Nile delta. Plant Soil 336: 129–142.
22. Neupane RP, Thapa GB (2001) Impact of agroforestry intervention on soil fertility and farm income under the subsistence farming system of the middle hills, Nepal. Agric Ecosyst Environ 84: 157–167.
23. Peoples MB, Herridge DF, Ladha JK (1995) Biological nitrogen fixation: an efficient source of nitrogen for sustainable agricultural production? Plant Soil 174: 3–28.
24. Chapin HS III (1980) The mineral nutrition of wild plants. Ann Rev Ecol Syst 11: 233–60.
25. Altieri MA (2002) Agroecology: the science of natural resource management for poor farmers in marginal environments. Agric Ecosyst Environ 93: 1–24.
26. Watson CA, Atkinson D, Gosling P, Jackson LR, Rayns FW (2002) Managing soil fertility in organic farming systems. Soil Use Manage 18: 239–247.
27. Watson CA, Bengtsson H, Ebbesvik M, Lùes A-K, Myrbeck A (2002) A review of farm-scale nutrient budgets for organic farms as a tool for management of soil fertility. Soil Use Manage 18: 264–273.
28. Khan AUH, Iqbal M, Islam KR (2007) Dairy manure and tillage effects on soil fertility and corn yields. Bioresource Technol 98: 1972–1979.
29. Zhang H, Xu M, Zhang F (2009) Long-term effects of manure application on grain yield under different cropping systems and ecological conditions in China. J Agr Sci 147: 31–42.
30. Yadvinder-Singh Bijay-Singh, Ladha JK, Khind CS, Gupta RK, et al. (2004) Long-term effects of organic inputs on yield and soil fertility in the rice-wheat rotation. Soil Sci Soc Am J 68: 845–853.
31. Li J, Xu M, Qin D, Li D, Yasukazu H, et al. (2005) Effects of chemical fertilizers application combined with manure on ammonia volatilization and rice yield in red paddy soil. Plant Nutrit Fertilizer Sci 11: 51–56.
32. Malhi SS, Lemke R (2007) Tillage, crop residue and N fertilizer effects on crop yield, nutrient uptake, soil quality and nitrous oxide gas emissions in a second 4-yr rotation cycle. Soil Till Res 96: 269–283.
33. Min DH, Islam KR, Vough LR, Weil RR (2003) Dairy manure effects on soil quality properties and carbon sequestration in alfalfa–orchardgrass systems. Commun Soil Sci Plant 34: 781–799.
34. Körschensa M, Albert E, Armbruster M, Barkusky D, Baumecker M, et al. (2013) Effect of mineral and organic fertilization on crop yield, nitrogen uptake, carbon and nitrogen balances, as well as soil organic carbon content and dynamics: results from 20 European long-term field experiments of the twenty-first century. Arch Agron Soil Sci 59: 1017–1040.
35. Lobell DB, Burke MB (2010) On the use of statistical models to predict crop yield responses to climate change. Agric For Meteorol 150: 1443–1452.
36. McCown RL, Hammer GL, Hargreaves JNG, Holzworth DP, Freebairn DM (1996) APSIM: a novel software system for model development, model testing and simulation in agricultural systems research. Agric Syst 50: 255–271.
37. International Benchmark Sites Network for Agrotechnology Transfer (1993) The IBSNAT decade. Honolulu, Hawaii: Department of Agronomy and Soil Science, College of Tropical Agriculture and Human Resources, University of Hawaii.
38. Myers RJK (1984) A simple model for estimating the nitrogen fertilizer requirement of a cereal crop. Fertil Res 5: 95–108.
39. Deng J, Ran J, Wang Z, Fan Z, Wang G (2012) Models and tests of optimal density and maximal yield for crop plants. PNAS 109: 15823–15828.
40. Cong R, Wang X, Xu M, Ogle SM, Parton WJ (2014) Evaluation of the CENTURY model using long-term fertilization trials under corn-wheat cropping systems in the typical croplands of China. PLOS ONE 9: e95142.
41. Denef K, Six J, Merckx R, Paustian K (2004) Carbon sequestration in microaggregates of no-tillage soils with different clay mineralogy. Soil Sci Soc Am J 68: 1935–1944.
42. Michaelis L, Menten ML (1913) Die Kinetik der Invertinwirkung. Biochem Z 49: 333–369.
43. R Development Core Team (2011) R: a language and environment for statistical computing. Vienna, Austria: R Foundation for Statistical Computing.
44. Spiecker H (1995) Growth dynamics in a changing environment – long-term observations. Plant Soil 168–169: 555–561.
45. Drury CF, Tan CS (1994) Long-term (35 years) effects of fertilization, rotation and weather on corn yields. Can J Soil Sci 75: 355–362.
46. Lobell DB, Hammer GL, McLean G, Messina C, Roberts MJ (2013) The critical role of extreme heat for maize production in the United States. Nat Clim Chang 3: 497–501.
47. Welch JR, Vincent JR, Auffhammerc M, Moyae PF, Dobermann A (2010) Rice yields in tropical/subtropical Asia exhibit large but opposing sensitivities to minimum and maximum temperatures. PNAS 107: 14562–14567.
48. Lobell DB, Field CB (2007) Global scale climate–crop yield relationships and the impacts of recent warming. Environ Res Lett 2: 014002.
49. Schlenker W, Roberts MJ (2009) Nonlinear temperature effects indicate severe damages to U.S. crop yields under climate change. PNAS 106: 15594–15598.
50. Rosenzweig C, Parry ML (1994) Potential impact of climate change on world food supply. Nature 367: 133–138.
51. Shen J, Li R, Zhang F, Fan J, Tang C (2004) Crop yields, soil fertility and phosphorus fractions in response to long-term fertilization under the rice monoculture system on a calcareous soil. Field Crops Res 86: 225–238.
52. Canali S, Trinchera A, Intrigliolo F, Pompili L, Nisini L, et al. (2004) Effect of long term addition of composts and poultry manure on soil quality of citrus orchards in Southern Italy. Biol Fertil Soils 40: 206–210.
53. Poulton PR (1996) The Rothamsted long-term experiments: are they still relevant? Can J Soil Sci 76: 559–571.

54. Jenkinson DS (1991) The Rothamsted long-term experiments: are they still of use? Agron J 83: 2–10.
55. Huang S, Peng X, Huang Q, Zhang W (2010) Soil aggregation and organic carbon fractions affected by long-term fertilization in a red soil of subtropical China. Geoderma 154: 364–369.
56. Zhu P, Ren J, Wang L, Zhang X, Yang X (2007) Long-term fertilization impacts on corn yields and soil organic matter on a clay-loam soil in Northeast China. J Plant Nutr Soil Sci 170: 219–223.
57. Vanlauwe B, Diels J, Sanginga N, Merckx R (2005) Long-term integrated soil fertility management in South-western Nigeria: crop performance and impact on the soil fertility status. Plant Soil 273: 337–354.

PERMISSIONS

All chapters in this book were first published in PLOS ONE, by The Public Library of Science; hereby published with permission under the Creative Commons Attribution License or equivalent. Every chapter published in this book has been scrutinized by our experts. Their significance has been extensively debated. The topics covered herein carry significant findings which will fuel the growth of the discipline. They may even be implemented as practical applications or may be referred to as a beginning point for another development.

The contributors of this book come from diverse backgrounds, making this book a truly international effort. This book will bring forth new frontiers with its revolutionizing research information and detailed analysis of the nascent developments around the world.

We would like to thank all the contributing authors for lending their expertise to make the book truly unique. They have played a crucial role in the development of this book. Without their invaluable contributions this book wouldn't have been possible. They have made vital efforts to compile up to date information on the varied aspects of this subject to make this book a valuable addition to the collection of many professionals and students.

This book was conceptualized with the vision of imparting up-to-date information and advanced data in this field. To ensure the same, a matchless editorial board was set up. Every individual on the board went through rigorous rounds of assessment to prove their worth. After which they invested a large part of their time researching and compiling the most relevant data for our readers.

The editorial board has been involved in producing this book since its inception. They have spent rigorous hours researching and exploring the diverse topics which have resulted in the successful publishing of this book. They have passed on their knowledge of decades through this book. To expedite this challenging task, the publisher supported the team at every step. A small team of assistant editors was also appointed to further simplify the editing procedure and attain best results for the readers.

Apart from the editorial board, the designing team has also invested a significant amount of their time in understanding the subject and creating the most relevant covers. They scrutinized every image to scout for the most suitable representation of the subject and create an appropriate cover for the book.

The publishing team has been an ardent support to the editorial, designing and production team. Their endless efforts to recruit the best for this project, has resulted in the accomplishment of this book. They are a veteran in the field of academics and their pool of knowledge is as vast as their experience in printing. Their expertise and guidance has proved useful at every step. Their uncompromising quality standards have made this book an exceptional effort. Their encouragement from time to time has been an inspiration for everyone.

The publisher and the editorial board hope that this book will prove to be a valuable piece of knowledge for researchers, students, practitioners and scholars across the globe.

LIST OF CONTRIBUTORS

Liming Lai, Xuechun Zhao, Linhai Zhu and Yuanrun Zheng
Key Laboratory of Resource Plants, Beijing Botanical Garden, West China Subalpine Botanical Garden, Institute of Botany, Chinese Academy of Sciences, Xiangshan, Beijing, China

Yufei Li
Institute of RS and GIS, Peking University, Beijing, China

Yuan Tian and Xi Chen
Xinjiang Institute of Ecology and Geography, Chinese Academy of Sciences, Urumqi, Xinjiang, China

Lianhe Jiang
State Key Laboratory of Vegetation and Environmental Change, Institute of Botany, Chinese Academy of Sciences, Beijing, China

Yong Gao
Inner Mongolia Agricultural University, Hohhot, China

Shaoming Wang
Key Laboratory of Oasis Ecological Agriculture of Xinjiang Bingtuan, Shihezi, Xinjiang, China

Glyn M. Rimmington
Global Learning Office, College of Liberal Arts and Sciences, Wichita State University, Wichita, Kansas, United States of America

Atanu Mukherjee and Rattan Lal
Carbon Management and Sequestration Center, School of Environment and Natural Resources, The Ohio State University, Columbus, Ohio, United States of America

Rose Gageler, Mark Bonner, Gunnar Kirchhof, Nicole Robinson and Susanne Schmidt
School of Agriculture and Food Sciences, The University of Queensland, St Lucia, Australia

Mark Amos
Lake Baroon Catchment Care Group, Maleny, Australia

Luke P. Shoo
School of Biological Sciences, The University of Queensland, St Lucia, Australia

Yuki Fujita
Team Ecohydrology, KWR Watercycle Research Institute, Nieuwegein, The Netherlands

Peter M. van Bodegom
Department of Systems Ecology, VU University Amsterdam, Amsterdam, The Netherlands

Jan-Philip M. Witte
Team Ecohydrology, KWR Watercycle Research Institute, Nieuwegein, The Netherlands
Department of Systems Ecology, VU University Amsterdam, Amsterdam, The Netherlands

Markus Lange, Maike Habekost, Ernst-Detlef Schulze and Gerd Gleixner
Max Planck Institute for Biogeochemistry, Jena, Germany

Nico Eisenhauer
Institute of Ecology, Friedrich- Schiller-University Jena, Jena, Germany

Christiane Roscher
UFZ, Helmholtz Centre for Environmental Research, Department of Community Ecology, Halle, Germany

Christof Engels and Holger Bessler
Department of Plant Nutrition, Faculty of Agriculture and Horticulture, Humboldt-Universität zu Berlin, Berlin, Germany

Yvonne Oelmann
Geoecology/Geography, Eberhard Karls University of Tübingen, Tübingen, Germany

Stefan Scheu
J.F. Blumenbach Institute of Zoology and Anthropology, Georg August University Göttingen, Göttingen, Germany

Wolfgang Wilcke
Soil Science Group, Geographic Institute, University of Berne, Bern, Switzerland

Sikander Khan Tanveer, Xiaoxia Wen, Xing Li Lu, Junli Zhang and Yuncheng Liao
College of Agronomy, Northwest A&F University Yangling, Shaanxi, P.R. China

Stephanie G. Yelenik, Benjamin P. Colman, Jonathan M. Levine and Janneke HilleRisLambers
Department of Ecology, Evolution and Marine Biology, University of California Santa Barbara, Santa Barbara, California, United States of America

Richard V. Scholtz III and Allen R. Overman
Agricultural & Biological Engineering Department, University of Florida, Gainesville, Florida, United States of America

Huhe, Shinchilelt Borjigin, Nobukiko Nomura, Toshiaki Nakajima, Toru Nakamura and Hiroo Uchiyama
Graduate School of Life and Environmental Sciences, University of Tsukuba, Tsukuba, Ibaraki, Japan

Yunxiang Cheng
State Key Laboratory of Grassland Agro-Ecosystems, College of Pastoral Agriculture Science and Technology, Lanzhou University, Lanzhou, China

Jun Wang
College of Urban and Environmental Sciences, Northwest University, Xian, Shaanxi Province, China

Upendra M. Sainju
U.S. Department of Agriculture, Agricultural Research Service, Northern Plains Agricultural Research Laboratory, Sidney, Montana, United States of America

Lubo Gao, Huasen Xu, Biao Bao, Xiaoyan Wang, Chao Bi and Yifang Chang
College of Water and Soil Conservation, Beijing Forestry University, Beijing, P.R. China

Huaxing Bi
College of Water and Soil Conservation, Beijing Forestry University, Beijing, P.R. China
Key Laboratory of Soil and Water Conservation, Ministry of Education, Beijing, P.R. China

Weimin Xi
Department of Biological and Health Sciences, Texas A&M University-Kingsville, Kingsville, Texas, United States of America

Ana Novoa and Luís González
Departamento de Bioloxía Vexetal e Ciencias do Solo, Facultade de Bioloxía, Universidade de Vigo, Vigo, Spain

Olivier Dangles
Institut de Recherche pour le Développement (IRD), UR 072, Laboratoire Evolution, Génomes et Spéciation, UPR 9034, Centre National de la Recherche Scientifique (CNRS), Gif sur Yvette, France et Université Paris-Sud 11, Orsay, France
Facultad de Ciencias Exactas y Naturales, Pontificia Universidad Católica del Ecuador, Quito, Ecuador
Instituto de Ecología, Universidad Mayor San Andrés, Cotacota, La Paz, Bolivia

Jean-François Silvain
Institut de Recherche pour le Développement (IRD), UR 072, Laboratoire Evolution, Génomes et Spéciation, UPR 9034, Centre National de la Recherche Scientifique (CNRS), Gif sur Yvette, France et Université Paris-Sud 11, Orsay, France

Emile Faye
Institut de Recherche pour le Développement (IRD), UR 072, Laboratoire Evolution, Génomes et Spéciation, UPR 9034, Centre National de la Recherche Scientifique (CNRS), Gif sur Yvette, France et Université Paris-Sud 11, Orsay, France
UPMC Univ Paris06, Sorbonne Universités, Paris, France
Facultad de Ciencias Exactas y Naturales, Pontificia Universidad Católica del Ecuador, Quito, Ecuador

Mario Herrera
Facultad de Ciencias Exactas y Naturales, Pontificia Universidad Católica del Ecuador, Quito, Ecuador

Lucio Bellomo
Mediterranean Institute of Oceanography (MIO) CNRS/INSU, IRD, UM 110, Université de Toulon, La Garde, France

Xiaoqin Dai, Zhu Ouyang, Yunsheng Li and Huimin Wang
Key Laboratory of Ecosystem Network Observation and Modeling, Institute of Geographic Sciences and Natural Resources Research, Chinese Academy of Sciences, Beijing, China

Min-yun Xu
Department of Ecology, College of Urban and Environmental Sciences, Peking University, Beijing, P.R. China
College of Animal Science & Technology, Hebei Agricultural University, Baoding, Hebei province, P.R. China

Fan Xie
Department of Party Affairs, China National Light Industry Council, Beijing, P.R. China

Kun Wang
College of AnimalScience & Technology, China Agricultural University, Beijing, P.R. China

Rajan Ghimire, Jay B. Norton and Peter D. Stahl
Department of Ecosystem Science and Management, University of Wyoming, Laramie, Wyoming, United States of America

Urszula Norton
Department of Plant Sciences, University of Wyoming, Laramie, Wyoming, United States of America

Yongge Yuan, Jianjun Tang, Dong Leng and Xin Chen
College of Life Sciences, Zhejiang University, Hangzhou, China

Shuijin Hu
Department of Plant Pathology, North Carolina State University, Raleigh, North Carolina, United States of America

Jean W. H. Yong
Singapore University of Technology and Design, Singapore, Singapore

Mengyi Wang, Cuinan Wu, Zhihui Cheng, Huanwen Meng, Mengru Zhang and Hongjing Zhang
College of Horticulture, Northwest A&F University, Yangling, Shaanxi, China

Kornelia Smalla
Julius Kühn-Institut - Federal Research Centre for Cultivated Plants (JKI), Institute for Epidemiology and Pathogen Diagnostics, Braunschweig, Germany

Rita Grosch and Martin Sandmann
Leibniz Institute of Vegetable and Ornamental Crops Groβbeeren/Erfurt e.V., Department Plant Health, Groβbeeren, Germany

Susanne Schreiter
Julius Kühn-Institut - Federal Research Centre for Cultivated Plants (JKI), Institute for Epidemiology and Pathogen Diagnostics, Braunschweig, Germany
Leibniz Institute of Vegetable and Ornamental Crops Groβbeeren/Erfurt e.V., Department Plant Health, Groβbeeren, Germany

Bin Zhang, Xingyuan He and Chunjie Tian
Northeast Institute of Geography and Agroecology, Chinese Academy of Sciences, Changchun, China

Shengyun Chen, Wenjie Liu and Qian Zhao
State Key Laboratory of Cryospheric Sciences, Cold and Arid Regions Environmental and Engineering Research Institute, Chinese Academy of Sciences, Lanzhou, China

Lin Zhao
State Key Laboratory of Cryospheric Sciences, Cold and Arid Regions Environmental and Engineering Research Institute, Chinese Academy of Sciences, Lanzhou, China
Observation and Research Station of Qinghai-Tibet Plateau, Cold and Arid Regions Environmental and Engineering Research Institute, Chinese Academy of Sciences, Golmud, China

Xuelin Zhang, Qun Wang, Yilun Wang and Chaohai Li
The Incubation Base of the National Key Laboratory for Physiological Ecology and Genetic Improvement of Food Crops in Henan Province, Zhengzhou, China; Agronomy College of Henan Agricultural University, Zhengzhou, China

Frank S. Gilliam
Department of Biological Sciences, Marshall University, Huntington, West Virginia, United States of America

Feina Cha
Meteorological Bureau of Zhengzhou, Zhengzhou, China

Jin Hua Li, Yu Jie Yang, Bo Wen Li, Wen Jin Li and Gang Wang
State Key Laboratory of Grassland Agro-Ecosystems, School of Life Sciences, Lanzhou University, Lanzhou, P.R. China

Johannes M. H. Knops
School of Biological Sciences, University of Nebraska, Lincoln, Nebraska, United States of America

Amornrat Chaiyasen and Saisamorn Lumyong
Department of Biology, Faculty of Science, Chiang Mai University, Chiang Mai, Thailand

J. Peter W. Young
Department of Biology, University of York, York, United Kingdom

Neung Teaumroong
Schoool of Biotechnology, Institute of Agricultural Technology, Suranaree University of Technology, Nakhon Ratchasima, Thailand

Paiboolya Gavinlertvatana
Thai Orchid Labs Co. Ltd., Khannayao, Bangkok, Thailand

Jian Sun and Shuli Niu
Synthesis Research Centre of Chinese Ecosystem Research Network, Key Laboratory of Ecosystem Network Observation and Modelling, Institute of Geographic Sciences and Natural Resources Research, Chinese Academy of Sciences, Beijing, China

Xiaodan Wang, Genwei Cheng, Jianbo Wu and Jiangtao Hong
The key laboratory of mountain surface processes and eco-regulation, Institute of Mountain Hazard and Environment, Chinese Academy of Sciences, Chengdu, China

Hungyen Chen and Hirohisa Kishino
Graduate School of Agricultural and Life Sciences, The University of Tokyo, Tokyo, Japan

Junko Yamagishi
Institute for Sustainable Agro-ecosystem Services, The University of Tokyo, Tokyo, Japan

Index

O

P

Q

R

S

T

V

W

Y